NETWORK ANALYSIS WITH APPLICATIONS

Second Edition

William D. Stanley
Old Dominion University

Prentice Hall
Upper Saddle River, New Jersey | *Columbus, Ohio*

Library of Congress Cataloging-in-Publication Data

Stanley, William D.
    Network analysis with applications / William D. Stanley. — 2nd ed.
      p.   cm.
    Includes index.
    ISBN 0-13-260910-X (alk. paper)
    1. Electric network analysis. I. Title.
TK454.2.S72 1997
621.319′2—dc20   96-18803
                                            CIP

Cover photo: © 1993 Mike & Carol Werner/Comstock
Editor: Charles E. Stewart, Jr.
Production Editor: Alexandrina Benedicto Wolf
Cover Designer: Brian Deep
Production Manager: Patricia A. Tonneman
Marketing Manager: Debbie Yarnell

This book was set in Times Roman by Bi-Comp, Inc. and was printed and bound by
R.R. Donnelley & Sons. The cover was printed by Phoenix Color Corp.

 © 1997, 1985 by Prentice-Hall, Inc.
Simon & Schuster/A Viacom Company
Upper Saddle River, New Jersey 07458

Printed in the United States of America

10 9 8 7 6 5 4 3 2

ISBN: 0-13-260910-X

Prentice-Hall International (UK) Limited, *London*
Prentice-Hall of Australia Pty. Limited, *Sydney*
Prentice-Hall of Canada, Inc., *Toronto*
Prentice-Hall Hispanoamericana, S. A., *Mexico*
Prentice-Hall of India Private Limited, *New Delhi*
Prentice-Hall of Japan, Inc., *Tokyo*
Simon & Schuster Asia Pte. Ltd., *Singapore*
Editora Prentice-Hall do Brasil, Ltda., *Rio de Janeiro*

In memory of my father,
John D. Stanley

CONTENTS

3 CAPACITIVE AND INDUCTIVE TRANSIENTS AND EQUIVALENT CIRCUITS   137

4 INITIAL, FINAL, AND FIRST-ORDER CIRCUITS   219

PREFACE

The primary objective of this book is to present the general methods of circuit and network analysis by employing differential and integral calculus and transform methods with a strong *applied* emphasis. The level is particularly suited for an upper-division (junior-level) engineering technology course in circuit or network analysis. At this level, it should serve to unify the reader's lower-division circuit analysis and extend the capability to solve transient problems as well as utilize Laplace transforms and computer-aided analysis in more complex circuit problems. However, the book should also be useful in applied engineering and science and for practicing technical personnel who need to strengthen and extend their basic circuit analysis capabilities.

A unique feature of the book is that the first two chapters provide a mini-course in basic resistive circuit analysis for the purpose of strengthening and solidifying the reader's background. This is not a sketchy or superficial treatment. It is an in-depth study of the basic circuit theorems and network analysis methods, with the treatment limited to those concepts essential for advanced study. A reader without a formal electrical background could conceivably acquire a sufficient background from these chapters to deal with the remainder of the book.

The use of differential and integral calculus is introduced gradually as the level of network analysis increases. The physical significance of each mathematical operation is carefully explained in practical terms as it is introduced so that the reader acquires a feel for the operation. Transform methods are introduced for dealing with second-order and some higher-order transient solutions. Computer-aided analysis is emphasized for very complex circuits.

A special feature of the book is that problems at the end of most of the chapters are divided into three categories: (1) *drill problems,* consisting of exercises chosen to reinforce and solidify the basic principles given in the text, (2) *derivation problems,* involving proofs and extensions of the text principles, and (3) *application problems,* consisting of situations in which the theory is applied to solve real-life problems.

Chapter 1 provides a treatment of the basic circuit laws, including Ohm's law, Kirchhoff's laws, voltage and current dividers, units, and definitions. Equivalent resistance and complete circuit solutions for certain types of circuits are performed using step-by-step intuitive methods.

Chapter 2 deals with the more formal general circuit analysis methods, including mesh current analysis and node voltage analysis. Both Thevenin's and Norton's theorems are covered in detail. Techniques for dealing with both independent and dependent (controlled) sources are included. The superposition principle is also discussed.

Transient phenomena are first introduced in Chapter 3. The voltage–current relationships for capacitance and inductance are given. The major emphasis is on the graphical solution for piecewise linear voltage and current waveforms, to strengthen the reader's background in calculus and convey a feel for the physical significance of the phenomena. Analysis methods for mutual inductance and transformers are also developed.

Chapter 4 is devoted to establishing the various initial conditions and final conditions (steady-state behavior) for electrical circuits. The complete solutions of first-order circuits with dc excitations are also considered.

The Laplace transform is introduced in Chapter 5. Transform pairs and operations are given, and the various manipulations for determining transforms and inverse transforms are covered.

The use of transforms in determining complete circuit responses is given in Chapter 6. Major emphasis is on first-order circuits with arbitrary excitations and second-order circuits.

The transfer function concept is developed in Chapter 7. The techniques of s-plane analysis using poles and zeros are developed, and the concept of stability is introduced. Some system-level considerations using block diagram algebra are given.

The important special case of steady-state sinusoidal circuits is the topic of Chapter 8. This chapter constitutes a mini-course in ac circuits, but the extensive background established earlier in the book will provide better understanding of the nature of solutions.

Steady-state frequency response considerations are covered in Chapter 9. The steady-state or phasor transfer function is developed, and amplitude and phase response functions are defined. The powerful method of Bode plot analysis is developed in detail.

Chapter 10 is devoted to a special set of topics from various parts of the book that were delayed because of their more elaborate mathematical nature. This includes impulse circuit conditions, delayed functions, and the synthesis of complex circuit waveforms.

Chapter 11 deals with the subject of Fourier analysis, which involves determining the frequency components (or spectrum) of a given signal. Both the Fourier series for a periodic signal and the Fourier transform for a nonperiodic signal are covered.

PSPICE and MATLAB

An important addition to the second edition is the inclusion of numerous examples of circuits analyzed with either PSPICE or MATLAB, grouped at the end of each chapter. This coverage is optional, and is not a prerequisite to other topics within the text.

PSPICE is probably the most widely employed circuit analysis program used in education, and there are whole books available on the subject. Appendix D provides an overview of the elements of the program, and this treatment, along with the procedures developed within examples, should be sufficient to develop the reader's facility with the program for the types of examples within the text.

Both DOS-based and Windows-based versions of PSPICE are available. Although the field is certainly moving in the direction of Windows-based software, the decision was made to emphasize the DOS version in this text for several reasons.

1. The Windows version requires that the schematic be created on the screen, which is a procedure that, in the opinion of this author, takes a great deal of time and effort to learn properly. In contrast, the DOS version uses a fairly straightforward *code* or *netlist* to describe the circuit.
2. The Windows version is memory intensive and is slower in execution. If the primary focus of this text were PSPICE, the Windows version would definitely have been covered, but since the treatment here is designed to be supplementary, it was felt that the DOS version would be more suitable.

Said differently, it is this author's opinion that for the casual user, the DOS version is easier to learn and use. Obviously, those with an intense interest in computer-aided circuit analysis and design should investigate the Windows version, and all the examples in the text can be readily adapted to that format.

MATLAB is one of the most popular software packages for performing mathematical operations. It has a wide array of capabilities, including operations in algebra, calculus, differential equations, and matrix manipulations. Some of the most useful operations that support circuit analysis will be illustrated.

As powerful circuit analysis programs and mathematical software packages have become readily available, the question of which topics to stress in circuit analysis books is a real, relevant issue. Certainly, there is no need to stress long, unwieldy problems that require an answer but provide little insight, since such operations can best be achieved with a computer. On the other hand, the computer has reinforced more than ever the need to understand the basic principles of circuit analysis. Without this basis, a user has little insight into what assumptions must be made in formulating the circuit model and how to properly interpret the results of the computer analysis. This book aims to convey this understanding through carefully selected major topics and appropriate use of the computer.

Acknowledgment

I would like to thank the following reviewers for their invaluable feedback: Barry D. Bullard, Purdue University; Jane Jackson, Rochester Institute of Technology; and Francis M. Turner, Oregon Institute of Technology.

William D. Stanley

AUTHOR'S NOTE ON END-OF-CHAPTER PROBLEMS

At the end of most chapters are three sets of problems. These problems are designated as *drill problems, derivation problems,* and *application problems.* The nature of these categories will be discussed here.

Drill Problems

This set usually has more problems than the other categories, and represents the major focus of the chapter material. Drill problems have been designed to solidify and reinforce the major principles presented in the text. Many of the drill problems have been developed with simple element values in order to delineate principles rather than numerical manipulations. Although many drill problems represent realistic situations, some have been designed to illuminate a particular principle rather than to show a necessarily workable application.

Derivation Problems

The derivation problems are those in which the reader is asked to prove or develop some result in general terms. In some cases, these are simply verifications of rules or theorems given in the text, whereas in other cases, they are extensions or generalizations of the text results. The outcomes of many derivation problems are formulas or equations, rather than numerical values.

Application Problems

Application problems are selected exercises in which the reader is expected to apply principles of the chapter to accomplish some meaningful, practical objectives. The end result of a problem might be the design of a circuit or the analysis of some practical outcome. Element values are chosen to be representative of real-life cases in these problems (though they are still often rounded to simplify the computations). Some of the design application problems have more than one correct answer.

1

BASIC CIRCUIT LAWS

OBJECTIVES

After completing this chapter, the reader should be able to:

- Define the basic circuit quantities and state the symbols and units used to represent them.
- Define the basic active and passive circuit models and show their schematic forms.
- Explain power conventions and determine when power is delivered and absorbed.
- State Ohm's law and apply it in various ways.
- State Kirchhoff's voltage law and apply it in various ways.
- State Kirchhoff's current law and apply it in various ways.
- Determine the equivalent resistance of a passive circuit containing only resistors.
- State and apply the voltage divider rule.
- State and apply the current divider rule.
- Analyze certain complete resistive circuits by applying voltage and current divider rules along with the equivalent resistance concept to achieve a step-by-step solution.
- Define the form types of controlled (or dependent) sources and discuss their significance in circuit modeling.
- Use PSPICE to solve problems in the chapter.

1–1 GENERAL PLAN OF THE BOOK

The purpose of this section is to orient the reader toward the overall philosophy and organization of the book. The major objective of the book is to develop general methods of circuit analysis employing differential and integral calculus, transform methods, and computer-aided methods. The level is particularly suited for an upper-division (primarily junior-level) engineering technology course in circuit or network

1

analysis. However, the book should also be useful for applied engineering students, physics and science students, and practicing scientific and technical personnel who need to strengthen and extend their circuit analysis methods.

The first two chapters are devoted to developing the basic circuit laws and general network analysis methods. Most readers will probably have covered much, if not all, of this material in basic courses in circuit analysis and/or electricity. Consequently, not all the "fine points," particularly those relating to the physical basis of electricity, are covered here. However, the treatment is not sketchy or superficial. Rather, the emphasis is on establishing an in-depth understanding of the important laws and theorems that are used at all levels in circuit analysis. The author's experience is that very few students establish a full working understanding of these concepts in basic courses, so this material has been designed to circumvent this problem. Thus, although many readers will find the topics in the first two chapters familiar, a sincere effort to develop a stronger understanding of these important concepts will pay dividends.

Although these first two chapters do not provide all the details of basic circuits and electricity, the treatment is sufficiently complete to allow some mature readers without a formal electrical background to reach a level appropriate for pursuing more advanced circuit analysis later in the book. The details that are missing from this treatment could be filled in with a parallel effort while reading the remainder of the book, rather than as a necessary first step. In short, the first two chapters constitute a mini-course in basic circuit analysis methods for readers with sufficient motivation.

Although a number of the definitions and laws in the first two chapters use general time-varying quantities, virtually all the circuit analysis examples in these chapters are developed with direct circuit resistive circuits. *Direct current* (abbreviated *dc*) circuits are those in which all energy sources have constant values of voltages and currents. By restricting the circuits to have only resistive parameters, all circuit equations are algebraic in form. Thus, circuit equations involving derivative and integral relationships do not appear in this book before a sufficient mathematical basis is established.

Readers with a reasonably good background in circuits may be familiar with *alternating current* (abbreviated *ac*) circuit analysis employing complex phasor techniques (e.g., the use of *j* operator, and the like). This is an important area of circuit analysis, and one that will definitely be treated in the book. However, since they constitute a special case, ac circuit analysis methods will be delayed until after the general methods of complete circuit analysis are established. Thus, the reader may better understand the full meaning and limitations of ac circuit theory.

The basic circuit laws and analysis methods will be developed in Chapters 1 and 2, and the detailed circuit analysis methods and examples will primarily employ dc resistive circuit forms. The development of the more general circuit analysis forms involving time-varying voltages and currents will start in Chapter 3. After a comprehensive treatment of the general cases, the steady-state ac circuit forms will be developed as a special case later in the book.

The emphasis in Chapter 1 will be on establishing the basic circuit laws and models. However, a few complete circuits will be analyzed later in the chapter using intuitive methods. Although these methods are not general, they can be used in a sufficient number of practical cases to justify their coverage. In particular, they convey a "feeling" to the reader of what is happening in the circuit, which is often lacking in the more sophisticated general methods.

The emphasis in Chapter 2 will be on the more formal network analysis methods. Except for certain special cases, these methods tend to be more general in usage than the techniques in Chapter 1. However, as previously discussed, the methods will be developed with dc resistive circuit examples.

1–2 BASIC ELECTRICAL QUANTITIES

An electric circuit is an interconnection of electrical components in which energy may be transferred among the various elements within the circuit. Components may be classified as either active or passive, according to their energy status. An *active* device is one that is a primary source of energy in its operational form, whereas a *passive* device is not a basic source of energy. However, some active devices may accept energy from other active devices, and some passive devices may temporarily store energy, so the lines between these categories are sometimes blurred. Examples of active devices are batteries, rotating electrical generators, and transistor amplifiers (when connected to a power supply). Examples of passive devices are light bulbs and electric heaters.

Branches and Nodes

A complete electrical circuit consists of a combination of branches and nodes. A *branch* usually consists of one particular component with two terminals, but the definition is sometimes extended to include more than one component when the combination is to be viewed as equivalent to a single component. A *node* is a junction point between two or more branches.

The concepts of branches and nodes are illustrated in Figure 1–1 for a relatively simple circuit. This circuit consists of three branches, each of which contains one component, and two nodes. In (a), the branch terminals are brought together both at the top and bottom to illustrate the two nodes. For complex circuits, it may be awkward to show the schematic diagram with all connections to a given node at a single point. An alternate arrangement is shown in (b), and this form will be used in most schematic diagrams throughout the book. However, the circuit should still be considered as one having three branches and two nodes because the additional "wires" along the top and bottom do not contain any components. From a different point of view, the electrical potential all along the top of the circuit is the same, and the electrical potential all along the bottom of the circuit is the same (but different from the top). Thus, there are effectively only two points with different electrical potentials in the circuit. In searching for nodes, the criterion should be that all points that *could* be brought together without disturbing the circuit constitute a single node, irrespective of how the circuit is drawn.

FIGURE 1–1
Illustration of branches and nodes drawn two ways.

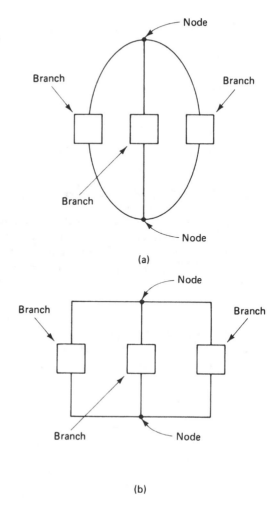

(a)

(b)

Circuit Quantities

The various quantities used in electrical circuit analysis will now be discussed. Each quantity is represented by one or more symbols, a unit of measurement, and an abbreviation for the unit. The symbols are used in various mathematical equations to refer to the quantity, and the units (along with associated abbreviations) are used to specify the actual value of the quantity in a given application.

The most common quantities used in circuit analysis, their symbols, units, and abbreviations for units are listed in Table 1–1. The most common prefixes used in circuit analysis appear in Table 1–2. The units and their abbreviations are used in accordance with the International System of Units (abbreviated as SI). Not all the quantities given in these tables will be discussed or used at this point in the book. Rather, the reader may refer back to these tables as the need arises later.

TABLE 1–1

Most common quantities used in electrical circuit analysis

Quantity	Symbols	Unit	Abbreviation of Unit
Time	t	second	s
Energy	w, W	joule	J
Power	p, P	watt	W
Charge	q, Q	coulomb	C
Current	i, I	ampere	A
Voltage	v, V, e, E	volt	V
Resistance	R	ohm	Ω
Conductance	G	siemens	S
Inductance	L	henry	H
Capacitance	C	farad	F
Impedance	$\overline{Z}$	ohm	Ω
Reactance	X	ohm	Ω
Admittance	$\overline{Y}$	siemens	S
Susceptance	B	siemens	S
Frequency (cyclic)	f	hertz	Hz
Frequency (radian)	ω	radians/second	rad/s

In the second column of Table 1–1, note that both lowercase and uppercase symbols are listed for many of the quantities. As a general rule, *lowercase* symbols are used whenever a quantity *is* or *could be varying with time;* i.e., it is considered a time-varying or instantaneous function. *Uppercase* symbols are generally used for either a dc quantity or a *constant value* associated with a time-varying quantity (e.g., the peak value of a time-varying quantity). However, there are many exceptions to

TABLE 1–2

Most common prefixes used in electrical systems

Value	Prefix	Abbreviation
10^{-18}	atto	a
10^{-15}	femto	f
10^{-12}	pico	p
10^{-9}	nano	n
10^{-6}	micro	μ
10^{-3}	milli	m
10^{3}	kilo	k
10^{6}	mega	M
10^{9}	giga	G
10^{12}	tera	T

this rule, so the reader should not interpret these guidelines as rigid. Subscripts may be added to these symbols as required when a number of quantities appear. For example, appropriate symbols for a dc circuit with three voltages might be V_1, V_2, and V_3 or, alternately, V_a, V_b, and V_c.

Functional Notation

When it is desirable or necessary to emphasize that a quantity is time-varying, functional notation of the form $v(t)$ or $i(t)$ can be used. The quantity t in parentheses represents time, and the notation indicates that the quantity is a function of time. To simplify expressions, the functional notation will be omitted in some routine equations, but if the quantity has a lowercase symbol, it should be interpreted as possibly being a time-varying quantity unless otherwise indicated. The functional notation will be maintained in many situations when the possibility of a time-varying quantity is to be strongly emphasized.

Symbols

Note that dual use is made of certain letters as both a symbol for one quantity and the abbreviation of the unit for the same or a different quantity. For example, W is a possible symbol for energy, and W is the abbreviation for watt. As an additional example, C is the symbol for capacitance, and C is also the abbreviation for coulomb. Finally, V is a possible symbol for voltage, and V is the abbreviation for volt. These dual uses need not cause any difficulty, since the use as a symbol for a quantity usually occurs within an equation or as a principal part of a sentence, whereas the use as an abbreviation for a unit always follows a specific value or number. Observe also that the styles of the symbols are different for the two uses.

It must be emphasized that strict use of uppercase and lowercase symbols as abbreviations for certain prefixes and units is necessary. For example, the prefix m represents 10^{-3}, whereas M represents 10^6. Confusion between these prefixes in certain applications would lead to serious errors! Finally, note that s is the abbreviation for second, whereas S is the abbreviation for siemens.

The most common quantities used in basic electrical circuit analysis will now be defined and discussed. The reader should refer directly to Table 1–1 (and possibly to Table 1–2) as a complement to the discussion that follows.

Starting at the top of Table 1–1, the first three quantities listed are time, energy, and power. These three quantities are obviously not just electrical variables, since they are among the most fundamental of all physical quantities. *Time* is the most common of all independent variables in nature, and many physical variables change with and are functions of the independent variable time.

Energy

Energy represents the capacity for, or the actual performance of, work. In mechanics, a force of 1 newton moving through a distance of 1 meter performs work equal to 1 joule.

Power

Power represents the rate of performing work or the rate of change of energy. If work equal to 1 joule is performed in 1 second, the power is 1 watt. Letting $p(t)$ represent the instantaneous power and $w(t)$ represent the instantaneous energy, the quantities are related by

$$p(t) = \frac{dw(t)}{dt} \tag{1-1}$$

Conversely, the total work W_{12} performed over the interval from t_1 to t_2 can be determined by the definite integral

$$W_{12} = \int_{t_1}^{t_2} p(t)\, dt \tag{1-2}$$

Many electrical devices are rated in terms of power capability, so the power levels in electrical circuits are often of more immediate interest than the energy.

A power level of 1 watt represents a change of energy of 1 joule per second, or equivalently, 1 W = 1 J/s. Although the joule is the most basic unit of energy and the watt is the most basic unit of power, many commercial power system energy levels are stated in terms of the *kilowatt-hour* (*kWh*). The number of kilowatt-hours is related to the number of joules by the relationship

$$\frac{\text{number of}}{\text{kilowatt-hours}} = \frac{\text{number of joules}}{3600 \times 1000} \tag{1-3}$$

The denominator of (1–3) is specified as the product of two constants to assist the reader in recognizing that 3600 is the number of seconds in one hour, and 1000 is the number of watts in one kilowatt.

Charge

Charge represents an actual quantity of electricity in its most fundamental form. Although charge is probably the most basic of all electrical quantities, it is not very convenient either to measure or to use directly in practical circuit analysis. Instead, most circuit responses are referred to as voltage and current variables.

Current

Current is a measure of the rate of flow of charge through a circuit. A flow of 1 coulomb/second past a certain point in a circuit constitutes a current of 1 ampere, or equivalently, 1 A = 1 C/s. Mathematically speaking, the instantaneous current $i(t)$ is related to the instantaneous charge $q(t)$ by the relationship

$$i(t) = \frac{dq(t)}{dt} \tag{1-4}$$

Conversely, the total charge Q_{12} passing a point over the interval from t_1 to t_2 can be determined by the the definite integral

$$Q_{12} = \int_{t_1}^{t_2} i(t)\, dt \tag{1-5}$$

This book will assume conventional current flow within a circuit. This is the most widely used convention, and it is opposite to the electron flow of current.

Voltage

Voltage is the electrical pressure or, as it has been traditionally called, the *electromotive force* (emf) between two points in an electrical circuit. Voltage is always measured between two points, and the term *potential difference* is often used to refer to the voltage between two points within a circuit. If 1 joule of work is performed on 1 coulomb of charge in moving between two points in a circuit, a potential difference of 1 volt exists between the points.

In many practical circuits, most voltages are measured with respect to and are referred to a single potential level, which is denoted as the *common* or *ground* potential. In such situations, expressions such as "the voltage at point A" are frequently made. However, the ground reference is understood as the other point in the circuit to which all voltages are referred.

By tradition, the symbols v, V, e, and E have all been employed to represent voltage. The major symbols used in this book will be v and V.

Current and voltage are the two most common variables that are measured and calculated in electrical circuit analysis. Note that voltage is always measured *across* a portion of a circuit, whereas current is measured *through* a portion of a circuit.

Resistance and Conductance

Resistance is the opposition to current flow present in all conducting materials. A lumped package of resistance is called a *resistor,* and resistors are used throughout electrical circuits for many purposes. Further, many actual electrical loads intended for various applications act like resistors at their terminals, and their models employed in circuit analysis appear as resistances.

The symbol for resistance is R, and the unit is the *ohm.* An alternate way to characterize resistance is through the concept of *conductance.* The symbol for conductance is G, and the unit is the *siemens.* The relationship between conductance and resistance is

$$G = \frac{1}{R} \tag{1-6}$$

Thus, the greater the resistance, the lower the conductance and vice versa.

EXAMPLE 1-1

A certain electric heater has a power rating of 1500 W. For 24 hours of continuous operation, determine **(a)** the total consumed energy in joules, **(b)** the corresponding energy in kilowatt-hours, and **(c)** the cost of operating the heater if the energy rate is 8¢ per kilowatt-hour.

Solution

(a) In general, the energy is the integral of the power over the time interval as indicated by Equation (1–2). However, when the power is constant, the integral is simply the product of the constant power and the total time. To determine the energy in joules, the time must be expressed in seconds. A time of 24 hours is $24 \times 3600 = 86,400$ s. The energy W_{12} in joules is

$$W_{12} = 1500 \text{ W} \times 86,400 \text{ s} = 1.296 \times 10^8 \text{ J} \qquad (1\text{–}7)$$

(b) The energy can be converted to kilowatt-hours by use of Equation (1–3). The result is

$$\text{Energy in kWh} = \frac{1.296 \times 10^8}{3600 \times 1000} = 36 \text{ kWh} \qquad (1\text{–}8)$$

Alternately, the product of 1.5 kW for the heater and 24 hours yields the result directly from the given information.

(c) The total cost is

$$\text{Cost} = 36 \text{ kWh} \times \$0.08/\text{kWh} = \$2.88 \qquad (1\text{–}9)$$

1-3 CIRCUIT MODELS

In general, electrical circuits may be classified as either *linear* or *nonlinear*. Complete definitions of these terms will be deferred until a more rigorous mathematical basis is established. However, the terms will be used somewhat intuitively early in the text so that the reader may gain an appreciation of the concepts involved. The vast majority of all circuit analysis theorems and laws is based on the linear models of circuit theory, and most of this book is based on that assumption. In fact, analytical solutions of nonlinear circuits are possible only for a limited number of cases. The analysis of many nonlinear circuits is achieved by using linear techniques modified to meet the restrictions of the nonlinear conditions. For example, a well-designed transistor amplifier may be assumed to be a linear device over a small signal operating range, and linear circuit models are used to predict its performance. However, if the signal levels exceed certain limits, the linear model is no longer valid.

Complex linear circuits may be represented as a combination of two active models and three passive models. The two active models are (1) the ideal voltage source and (2) the ideal current source. The three passive models are (1) resistance, (2) capacitance, and (3) inductance.

(a) Voltage source   (b) Current source

FIGURE 1–2
Active circuit parameter models.

Source Models

The models for the two ideal active devices are illustrated in Figure 1–2. The ideal voltage source model is shown in (a), and it is recognized by the presence of the (+) and (−) signs. The notation $v_s(t)$ emphasizes that this voltage is, in general, a time-varying quantity. The (+) terminal is the assumed positive reference terminal, and the (−) terminal is the assumed negative reference terminal. However, this does not mean that the (+) terminal always has a higher potential than the (−) terminal since this will depend on the $v_s(t)$ function. For example, at some value of time, suppose that $v_s(t) = -5$ V. The negative sign means that the positive reference terminal is 5 V lower in potential than the negative reference terminal at that particular time.

The ideal current source model is shown in Figure 1–2(b), and it is recognized by the presence of the arrow. The arrow denotes the direction of positive conventional current flow. The notation $i_s(t)$ emphasizes that this current is, in general, a time-varying quantity. This does not mean that the current is always flowing in the given direction, since this will depend on the $i_s(t)$ function. Thus, if $i_s(t)$ is negative at some instant of time, the current is actually flowing in the direction opposite to the arrow.

Independent vs. Dependent Sources

Either an ideal voltage source or an ideal current source may be further classified as an independent source or a dependent source. An *independent* ideal source is one whose specified value (voltage or current) is independent of any other variables in the circuit. A *dependent* source (also called a *controlled* source) is one whose specified value is a function of some other variable in the circuit. Models of transistors, for example, use dependent sources to represent the amplification process. In the absence of any adjective modifier, a source will be interpreted as an independent source. Dependent sources will be introduced in Section 1–10.

Independent Sources

The ideal independent *voltage* source model is characterized by the fact that *the voltage across its terminals is independent of any load connected to it or current flowing through it.* Likewise, the ideal independent *current* source model is characterized by the fact that *the current is independent of any load connected to it or voltage across it.* An ideal independent *voltage* source model maintains its specified voltage across an open circuit, and an ideal independent *current* source maintains its specified current through a short circuit. However, a contradiction appears when either a short circuit is considered for an ideal independent voltage source or an open circuit is considered for an ideal independent current source. These contradictions are results of the presumed ideal models, and more realistic models have to be used to explain these phenomena. These situations will be avoided at this point in the text.

Circuit Parameter Models

The first of the passive circuit parameters is *resistance,* which was introduced in Section 1–2. The schematic symbol is shown in Figure 1–3(a). The voltage–current relationship for a resistor is determined by Ohm's law, which will be discussed in Section 1–5.

The second and third passive circuit parameters are *capacitance* and *inductance.* Two acceptable symbols for capacitance are shown in Figure 1–3(b). The one on the right, using parallel lines, will be employed in this text. The schematic symbol for inductance is shown in Figure 1–3(c). The voltage–current relationship for these two parameters involves calculus, and a full development will be given in Chapter 3.

Linear circuits are characterized by the fact that the three passive circuit parameters are independent of the levels of the voltage and current variables in the circuit. For example, a resistor whose resistance changes with the current through it is nonlinear. A varactor diode, whose capacitance is a function of the voltage across it, is a nonlinear element. Finally, an inductance that changes with the current through it is nonlinear.

Virtually all of the parameter values considered in this book will be assumed to be constant values, and this constitutes a linear circuit. It is also possible to have special time-varying parameter values in a linear circuit, but the emphasis in this book will be on constant parameter values, the most common situation in the majority of linear circuits.

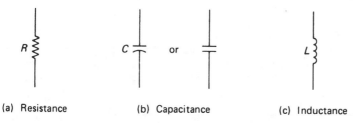

(a) Resistance   (b) Capacitance   (c) Inductance

FIGURE 1–3
Passive circuit parameter models.

1–4   POWER CONVENTIONS

When one or more electrical devices having the capacity to provide energy are connected in an electrical circuit, there will generally be a transfer of energy within the circuit. One or more active devices will deliver or supply energy to other devices that absorb or accept the electrical energy and possibly convert it to some other form of energy or work. A fundamental concept in electrical circuit analysis is to determine which elements are supplying power and which are absorbing power.

First, we will define the terms *delivered* and *absorbed* in relation to power or energy flow as follows: *Power delivered* by a device represents actual power being released by that device to the electrical circuit in which it is connected. *Power absorbed* by a device represents power accepted by that device from some other part of the circuit. Conservation of energy requires that at a given time and for a complete electrical circuit,

$$\text{power delivered} = \text{power absorbed} \qquad (1\text{–}10)$$

An elementary case to initiate this discussion is shown in Figure 1–4. A dc voltage of V_s volts is connected across a resistance. The battery delivers a certain amount of power, which in turn is absorbed by the resistance. The power delivered by the battery is exactly equal to the power absorbed by the resistance (assuming negligible resistance in the connecting wires). If the resistance is a simple lumped circuit resistor, the power that it absorbs is converted to heat. On the other hand, if this resistance represents some other electrical load, the power absorbed may be converted to a different form of energy. For example, if the resistance represents a light bulb, a portion of the power is converted to light and the remainder is converted to heat.

From this simple example, one might make the erroneous conclusion that an electrical source, such as a battery, always delivers power, and a passive device always absorbs the power. However, this is not the case at all, because the general situation can be more complex.

First, among passive devices, a true positive resistance always absorbs power. However, the other passive parameters (capacitance and inductance) may either deliver or absorb power, as will be seen later in the book. (Any power delivered

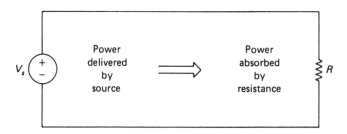

FIGURE 1–4
Simple circuit used to illustrate power transfer.

by these passive devices would have been absorbed at an earlier time and stored.) Further, certain electronic devices exhibit negative resistance effects, and this phenomenon may be thought of as a resistance that produces power. Since this negative resistance concept is somewhat specialized, we will restrict our consideration of resistance to passive positive resistance. Thus, we will assume in further discussions that resistance is the only passive parameter that always absorbs power.

In general, electrical sources may either deliver or absorb power. In the special case of a single source in a circuit with no energy previously stored, the source can only deliver power because there is no other "supplier" available. However, a circuit with several sources of power may present various combinations, with all sources delivering power or certain sources delivering power and others absorbing power.

The practical implication of a source absorbing power depends on the type of device. For example, a rechargeable battery is simply being recharged during the time it is absorbing power. Certain rotating generators act as motors when absorbing power. On the other hand, many electronic power supplies are not capable of absorbing power and will either fail to operate or become damaged if connected in a circuit where conditions indicate that the unit should absorb power. The linear circuit model will no longer be valid in such a case.

The treatment in this text is directed toward circuit modeling rather than the physical principles of different devices. We will assume throughout the book that all sources in which power is absorbed are those for which the linear model is valid.

Absorbed or Delivered?

To determine the direction of power flow, refer to the branch models shown in Figure 1–5. The blocks may represent either passive components or sources. If the conventional current direction is *into* the more *positive* terminal of the device as

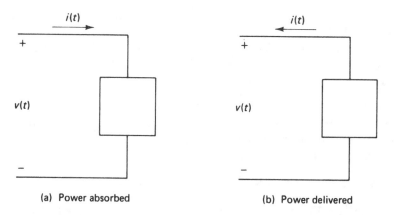

(a) Power absorbed   (b) Power delivered

FIGURE 1–5
Directions of voltage and current for power absorbed and power delivered.

shown in (a), the branch is *absorbing* power. If, however, the current direction is *out of* the more *positive* terminal as shown in (b), the branch is *delivering* power. The value of the instantaneous power $p(t)$ either absorbed or delivered is the product of the instantaneous voltage $v(t)$ and the instantaneous current $i(t)$, that is,

$$p(t) = v(t)i(t) \qquad (1\text{--}11)$$

For the dc case with constant values of voltage and current, the power P is a constant value and is

$$P = VI \qquad (1\text{--}12)$$

where V and I are the constant dc values of the voltage and current, respectively.

The preceding statements refer to the labeled directions of the branch quantities, and they are correct as far as these directions are concerned. However, if either the voltage or the current is negative, the actual direction of that variable is opposite to that shown. If the other variable is positive, the sense of power flow is opposite to that initially implied by the labels. This can be readily inferred by reversing the direction of the negative variable or by simply using a negative sign in the original power statement, which produces the change by negation.

For example, suppose that the labels were as given in Figure 1–5(a), for which power absorbed is being assumed. Suppose it is determined after analysis that at some value of time t_1, $v(t_1) = 6$ V and $i(t_1) = -2$ A, meaning that the current is actually flowing out of the positive terminal of the device. We could reverse the direction of the current and define $i'(t_1) = -i(t_1) = 2$ A, in which case reference to Figure 1–5(b) could then be made for power delivered. However, we could also retain the original sense and say that the power is $p(t_1) = 6$ V $\times$ $(-2$ A$) = -12$ W absorbed, which is equivalent to $+12$ W delivered. From the preceding discussion, we infer that

$$+\text{power delivered} = -\text{power absorbed} \qquad (1\text{--}13)$$

EXAMPLE 1–2

For each of the branches shown in Figure 1–6, determine the power and whether it is being delivered or absorbed. Each branch is a portion of some other circuit not shown.

Solution

(a) This branch contains an ideal 12-V voltage source, and a current of 2 A is flowing. Since the current is flowing out of the more positive terminal, power is delivered by this source. The power P is

$$P = 12 \text{ V} \times 2 \text{ A} = 24 \text{ W delivered} \qquad (1\text{--}14)$$

(b) This branch contains an ideal 4-A current source, and a voltage of 5 V appears across it. The portion of the circuit not shown must supply 4 A into the lower terminal of the source. Thus, current is flowing into the more positive terminal

FIGURE 1–6

Circuits for Example 1–2.

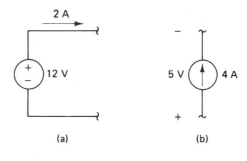

(a)   (b)

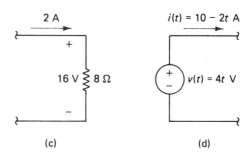

(c)   (d)

of the source, and the source is absorbing power. This power is

$$P = 5 \text{ V} \times 4 \text{ A} = 20 \text{ W absorbed} \qquad \textbf{(1–15)}$$

(c) This branch contains a positive resistance, which always absorbs power. The given directions of the voltage and current substantiate that assumption, and the power is

$$P = 16 \text{ V} \times 2 \text{ A} = 32 \text{ W absorbed} \qquad \textbf{(1–16)}$$

(d) This branch contains an ideal voltage source, for which the voltage $v(t)$ varies with time. The current $i(t)$ flowing from the source also varies with time. Consequently, the power varies with time, and $p(t)$ will be used to denote it. Since the direction of the current shown is leaving the more positive terminal of the branch, it is logical to define $p(t)$ as power delivered, and it is

$$p(t) = v(t)i(t) = 4t(10 - 2t) \text{ W} \qquad \textbf{(1–17)}$$

As long as $p(t)$ is positive, power is being delivered as assumed from the directions. For positive time, the voltage is always positive. However, by studying the function for $i(t)$, which is a straight line with negative slope, it can be deduced that the current is positive for $t < 5$ s, but it is negative for $t > 5$ s. Thus, for $t > 5$ s, the power delivered is negative, meaning that power is actually absorbed in that region of time.

1–5 OHM'S LAW AND RESISTIVE POWER

Ohm's law is among the most basic of all the laws of electricity. Refer to Figure 1–7, and assume a resistance R. Assume an instantaneous voltage $v(t)$ across the resistance and a current $i(t)$ flowing through it. Ohm's law states that the relationship between these quantities is

$$v(t) = Ri(t) \qquad \textbf{(1–18)}$$

or alternately,

$$i(t) = \frac{v(t)}{R} = Gv(t) \qquad \textbf{(1–19)}$$

where G is the conductance as defined in Equation (1–6).

From Figure 1–7, observe that the *most positive terminal of the voltage across a resistance is the terminal at which the current is assumed to enter.* This convention should be followed in all circuit analysis techniques. Note that this convention is compatible with the assumption that power is always absorbed by a resistance.

In many problems, the actual direction of the current or the polarity of the voltage will not be known, and an arbitrary assumption must be made. However, when either the current direction or the voltage polarity is assumed, the other quantity must be chosen to be compatible with it according to the preceding convention. If the wrong direction or polarity has been assumed, the corresponding quantity will turn out to be negative, and the literal use of Ohm's law for the other variable will also result in a negative quantity for that variable. Thus, the "bookkeeping" will work out correctly, provided care is taken with the sign convention and direction.

In defining Ohm's law, no mention was made of whether $v(t)$ or $i(t)$ was a source. It does not matter whether a voltage source is causing a current to flow or whether a current source is causing a voltage to exist. Rather, Ohm's law always expresses the relationship between the variables. In most cases, neither $v(t)$ nor $i(t)$ is a source, because a typical resistance will be one small portion of a complex electrical circuit containing a number of sources and components.

Variations of Resistive Power

The power $p(t)$ absorbed by a resistance can be expressed as the product of the voltage across the resistor and the current through it in accordance with Equation

FIGURE 1–7

Schematic representation for resistance and associated voltage and current variables.

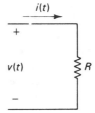

(1–11), that is,

$$p(t) = v(t)i(t) \qquad \textbf{(1–20)}$$

If $v(t) = Ri(t)$ is substituted in Equation (1–20), the power expressed in terms of the current and the resistance is

$$p(t) = Ri^2(t) \qquad \textbf{(1–21)}$$

Finally, if $i(t) = v(t)/R$ is substituted in Equation (1–20), the power expressed in terms of the voltage and the resistance is

$$p(t) = \frac{v^2(t)}{R} \qquad \textbf{(1–22)}$$

Waveforms for Resistance

A practical application concerning the form of Ohm's law for instantaneous voltage and current will be noted. From Equation (1–18), the voltage function is always a constant times the current function, so the waveforms of voltage and current for a resistor have the same forms. Frequently, it is desired to monitor the waveform of a certain time-varying current on an oscilloscope. However, an oscilloscope is basically a voltage-measuring device. If the circuit contains a resistor (normally with one terminal grounded) through which the current of interest flows, the voltage across it will have the same waveform as the current through it, and the oscilloscope can be made to respond to current indirectly. In fact, it may be possible to insert a small sampling resistor in the circuit for that purpose. Obviously, the resistance must be sufficiently small that it does not disturb the circuit. The oscilloscope voltage scale can be calibrated so the current level can be determined by a simple application of Ohm's law. Other passive circuit parameters are characterized by the fact that time-varying voltage and current waveforms are different, but resistance has the unique property that these two waveforms always have identical forms.

EXAMPLE 1–3

It is desired to monitor the waveform of a current in a certain electronic circuit by means of a voltage-sensitive oscilloscope. It is determined that the resistance level of the circuit is so large that an added 100-Ω precision sampling resistor presents negligible disturbance to the circuit. The resistor is inserted at a point where one side is connected to the common ground, so that the grounded side of the oscilloscope may be maintained at the circuit common ground. A satisfactory display is obtained when the oscilloscope vertical calibration factor is set at 50 mV/cm. Determine the equivalent calibration factor for the measured current.

Solution
This problem takes much more effort to state than to solve. By Ohm's law, with $R = 100 \ \Omega$ and $V = 50$ mV, the current is $I = 0.05$ V/100 Ω = 5×10^{-4} A = 0.5 mA. The calibration factor on this scale for current is thus 0.5 mA/cm.

Incidentally, the current in mA could have been determined directly by dividing the voltage in mV by the resistance in ohms. However, the author wishes to promote a philosophy to readers of returning to basic units for calculations unless one is absolutely confident about working with mixed units.

1–6 KIRCHHOFF'S LAWS

The two circuit laws to be discussed in this section are Kirchhoff's voltage law and Kirchhoff's current law. For brevity, these laws will be designated in many places in the text as KVL and KCL, respectively.

Kirchhoff's Voltage Law

Kirchhoff's voltage law states that the algebraic sum of the voltages around a closed loop is zero. This can be stated mathematically as

$$\sum_n v_n = 0 \qquad\qquad \textbf{(1–23)}$$

where $\sum_n$ represents a summation of all voltages v_n in the loop. To apply KVL, an algebraic convention must be applied to all voltages. Voltage drops can be assigned as positive, and voltage rises can be assigned as negative, which is a common convention, but the opposite convention will work just as well, provided it is consistent. In fact, it is also possible to restate KVL as "the sum of all voltage drops is equal to the sum of all voltage rises," with all terms being positive, but with rises and drops on opposite sides of the equation. Algebraically, all the forms are equivalent.

To illustrate KVL, consider the loop shown in Figure 1–8. The convention that drops are treated as positive quantities and rises as negative quantities will be

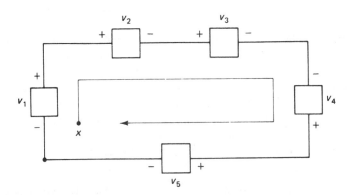

FIGURE 1–8
Loop used to illustrate Kirchhoff's voltage law (KVL).

followed. Starting at point x and moving clockwise around the loop, we have

$$-v_1 + v_2 + v_3 - v_4 + v_5 = 0 \qquad \text{(1-24)}$$

Kirchhoff's Current Law

Kirchhoff's current law states that the algebraic sum of the currents at a node is zero. This can be stated mathematically as

$$\sum_n i_n = 0 \qquad \text{(1-25)}$$

where $\sum_n$ represents a summation of all currents i_n at a node. To apply KCL, an algebraic convention must be applied to all currents. Currents leaving can be assigned as positive, and currents entering can be assigned as negative, which is a common convention, but the opposite convention will work just as well. Alternately, it is possible to restate KCL as "the sum of all currents entering a node is equal to the sum of all currents leaving the node," with all terms being positive, but with currents leaving on one side of the equation and currents entering on the other side. Algebraically, all the forms are equivalent.

To illustrate KCL, consider the node shown in Figure 1–9. The convention that currents leaving are treated as positive quantities and currents entering are treated as negative quantities will be followed. For this node, we have

$$-i_1 + i_2 - i_3 - i_4 + i_5 = 0 \qquad \text{(1-26)}$$

EXAMPLE 1–4

The dc resistive circuit of Figure 1–10 is assumed to be part of a certain electrical system. The resistance values of two of the resistors are known, but the resistance R is initially unknown. None of the branch currents is initially known. A voltmeter with one lead is grounded is available, so only voltages with respect to the ground connection shown can be measured. The two voltages measured are 21 V and 9 V as shown. From this information, determine I_0, I_1, I_2, and R. (*Note:* Since this is a dc circuit, uppercase symbols are used for the unknown voltage and currents.)

FIGURE 1–9
Node used to illustrate Kirchhoff's current law (KCL).

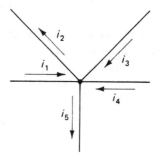

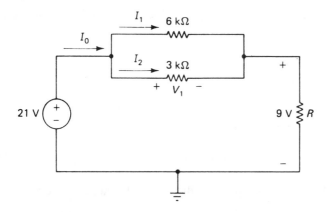

FIGURE 1–10
Circuit for Example 1–4.

Solution

First, V_1 can be determined by an application of KVL around the lower loop, because two of the three voltages are known. Taking a clockwise path with drops considered positive and rises negative, we have

$$-21 + V_1 + 9 = 0 \qquad (1\text{–}27)$$

which leads to

$$V_1 = 12 \text{ V} \qquad (1\text{–}28)$$

This voltage appears across both the 3-kΩ and the 6-kV resistors. Thus, I_1 and I_2 can be determined by Ohm's law as

$$I_1 = \frac{12 \text{ V}}{6000 \ \Omega} = 2 \times 10^{-3} \text{ A} = 2 \text{ mA} \qquad (1\text{–}29)$$

and

$$I_2 = \frac{12 \text{ V}}{3000 \ \Omega} = 4 \times 10^{-3} \text{ A} = 4 \text{ mA} \qquad (1\text{–}30)$$

The current I_0 can be determined by an application of KCL to the node just to the left of the two known resistances. Assuming currents leaving as positive and currents entering as negative, we have

$$I_1 + I_2 - I_0 = 0 \qquad (1\text{–}31)$$

or

$$I_0 = I_1 + I_2 = 2 + 4 = 6 \text{ mA} \qquad (1\text{–}32)$$

The current flowing through the unknown resistance R is also I_0, so R can be determined as

$$R = \frac{9\,\text{V}}{6\,\text{mA}} = 1.5\,\text{k}\Omega \qquad (1\text{--}33)$$

This example illustrates how various unknown quantities in certain circuits may be determined by appropriate applications of KVL, KCL, and Ohm's law. Not all problems are completely solvable in this manner, of course, but many practical measurement problems lend themselves to this mixed process. There is no particular pattern to this type of problem, so each case must be carefully reviewed in terms of the quantities known, the quantities desired, and the pertinent circuit laws that apply.

1–7 EQUIVALENT RESISTANCE

If a number of resistors are connected together in some arbitrary fashion, it is possible to represent their combined effect at a single set of terminals by a single equivalent resistance. This equivalent resistance will absorb the same power as the net combination of resistors in the circuit it replaces. This reduction scheme is very useful in many circuit analysis problems.

The equivalent resistance of a large number of practical circuits can be determined by the process of *successive series and parallel reduction*. Two rules are required to use this procedure, and they will be discussed next. Derivations of these rules are straightforward and will be left as guided exercises for the reader (Problems 1–29 and 1–30).

Resistors in Series

The first rule is used to determine the equivalent resistance of two or more resistors connected in *series*. Referring to Figure 1–11, the equivalent resistance R_{eq} as viewed from the external terminals is

$$R_{eq} = R_1 + R_2 + \cdots + R_n \qquad (1\text{--}34)$$

Resistors in Parallel

The second rule is used to determine the equivalent resistance of two or more resistors connected in *parallel*. Referring to Figure 1–12, the equivalent resistance R_{eq} is determined from the relationship

$$\frac{1}{R_{eq}} = \frac{1}{R_1} + \frac{1}{R_2} + \cdots + \frac{1}{R_n} \qquad (1\text{--}35)$$

Alternately, by working with conductances and letting $G_{eq} = 1/R_{eq}$, $G_1 = 1/R_1$, $G_2 = 1/R_2$, and so on, we have

$$G_{eq} = G_1 + G_2 + \cdots + G_n \qquad (1\text{--}36)$$

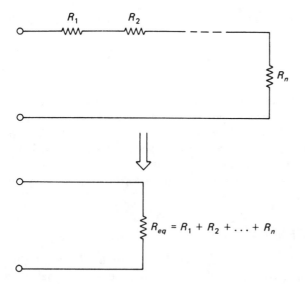

FIGURE 1–11
Equivalent resistance of *n* resistors connected in series.

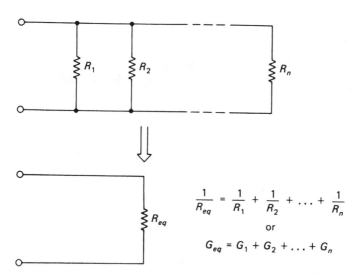

FIGURE 1–12
Equivalent resistance of *n* resistors connected in parallel.

As a special case for two resistors in parallel, the following useful formula can be readily deduced:

$$R_{eq} = \frac{R_1 R_2}{R_1 + R_2} \qquad (1\text{--}37)$$

Summarizing the preceding two paragraphs, *resistances* in *series add,* and *conductances* in *parallel add.*

The process of series and parallel reduction consists of working with individual segments and determining equivalent resistances on a step-by-step basis using the preceding rules. Not all circuits can be reduced this way, but the process works in a large number of useful cases. An example of a circuit whose equivalent resistance cannot be determined directly with this method is the bridge circuit shown in Figure P2–34 at the end of Chapter 2. The problem associated with that circuit (Problem 2–34) deals with an approach that can be used when the methods of this section fail.

The techniques of this section will be illustrated by the example that follows.

EXAMPLE 1–5
Using successive series and parallel reduction, determine the equivalent resistance at the input terminals for the circuit of Figure 1–13.

Solution
The process consists of starting on the right-hand side of the circuit and working back toward the input, making appropriate series and parallel combinations at each step. The various steps are illustrated by the different parts of Figure 1–14. The calculations will not be given here because they consist of simple applications of the formulas of this section, but the reader should verify the results.

First, the parallel combination of the 6-Ω and 3-Ω resistors is determined as shown in (a) of Figure 1–14, and this result is 2 Ω. Next, the series combination of

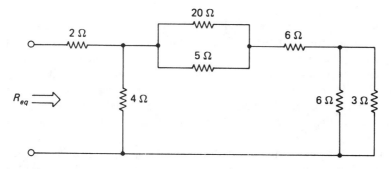

FIGURE 1–13
Circuit for Example 1–5.

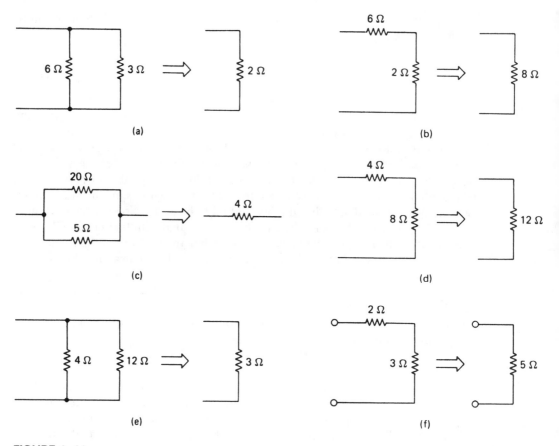

FIGURE 1–14
Successive steps in determining the equivalent resistance of the circuit for Example 1–5.

the 6-Ω and 2-Ω resistances is determined, and the result is 8 Ω as shown in (b). Before combining this value with other resistances, it is first necessary to determine the parallel combination of the 20-Ω and 5-Ω resistors, and the result is 4 Ω as shown in (c). The series combination of the 4-Ω and 8-Ω resistances is then determined, and this value is 12 Ω, as shown in (d). This resistance is in parallel with the 4-Ω resistor, and the resulting combination is 3 Ω, as shown in (e). Finally, the series combination of this value and the 2-Ω resistor is determined to be 5 Ω, as shown in (f), and thus $R_{eq} = 5\ \Omega$.

This problem, like many in the book, was "rigged" to provide simple integer values for all series and parallel combinations. Seldom do real-life problems have such simple values.

1-8 VOLTAGE AND CURRENT DIVIDERS

In this section, two very useful simplified circuit analysis rules for resistive circuits will be discussed, the voltage divider rule and the current divider rule. Derivations of these rules are straightforward and will be left as guided exercises for the reader (Problems 1–31 and 1–32). The emphasis here will be on the interpretation and application of the rules.

Voltage Divider Rule

Consider the simple circuit of Figure 1–15(a). In this figure, an input source voltage $v_s(t)$ is connected to the left of the circuit, and it is desired to determine the voltage $v_0(t)$ across the resistance R_0. The voltage $v_s(t)$ is shown as a source to emphasize that this voltage causes the voltage $v_0(t)$ across the output. However, the voltage as the input need not be a source; rather, it could be a result of one or more other sources whose net effects have been represented at the input. The point is that $v_s(t)$ is assumed to be the voltage that causes the output voltage $v_0(t)$. The voltage $v_0(t)$ is the open-circuit voltage measured or sensed across R_0, and any external load must be combined with R_0 to apply the voltage divider rule.

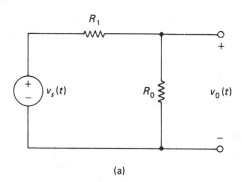

(a)

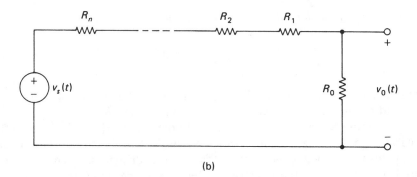

(b)

FIGURE 1–15
Circuits used to illustrate the voltage divider rule.

The *voltage divider rule* states that the voltage $v_0(t)$ across R_0 is

$$v_0(t) = \frac{R_0}{R_0 + R_1} v_s(t) \tag{1-38}$$

Note that $v_0(t)$ is always smaller than $v_s(t)$, a result of the fact that a portion of $v_s(t)$ is dropped across R_1, and the remainder appears across R_0. As R_0 increases relative to R_1, the quantity $R_0/(R_0 + R_1)$ approaches unity, and a larger fraction of $v_s(t)$ appears across R_0. Conversely, if R_0 decreases relative to R_1, the quantity $R_0/(R_0 + R_1)$ decreases, and a smaller fraction of $v_s(t)$ appears across R_0.

The voltage divider rule is readily extended to a series connection of several resistors as shown in Figure 1–15(b). In effect, all resistances except R_0 are treated as a single equivalent resistance, and the voltage divider rule becomes

$$v_0(t) = \frac{R_0}{R_0 + R_1 + R_2 + \cdots + R_n} v_s(t) \tag{1-39}$$

From Equations (1–38) or (1–39), it is noted that the voltages $v_s(t)$ and $v_0(t)$ have the same waveform shapes because the divider ratio simply changes the level. This property is true of all resistive voltage divider circuits; that is, the level of the signal, but not the shape, is changed. However, when more than one type of circuit parameter is used in a voltage divider, the shape of the output may be quite different from that of the input. This concept will appear in later chapters.

Current Divider Rule

Consider the simple circuit of Figure 1–16(a). In this figure, an input current source $i_s(t)$ is connected to the left of the circuit, and it is desired to determine the current $i_0(t)$ through the resistance R_0. The current $i_s(t)$ is shown as a source to emphasize that this current causes the current $i_0(t)$. However, the current at the input need not be a source; rather, it could be a result of one or more sources whose net effects have been represented at the input. The point is that $i_s(t)$ is assumed to be the current that causes the current $i_0(t)$.

The *current divider rule* states that the current $i_0(t)$ through R_0 is

$$i_0(t) = \frac{R_1}{R_1 + R_0} i_s(t) \tag{1-40}$$

Note that $i_0(t)$ is always smaller than $i_s(t)$, a result of the fact that a portion of $i_s(t)$ is shunted through R_1, and the remainder flows through R_0.

Comparing the current divider rule with the voltage divider rule, note that in the current divider rule, it is the "other resistor" that appears in the numerator, whereas for the voltage divider rule, it is the resistor across which the voltage is desired that appears in the numerator. Because of the opposite effects, a larger fraction of the current $i_s(t)$ flows through R_0 when R_0 is smaller than R_1 and vice versa.

The current divider rule may be extended to the case of several resistances

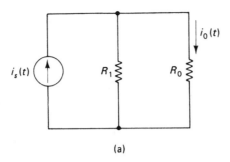

(a)

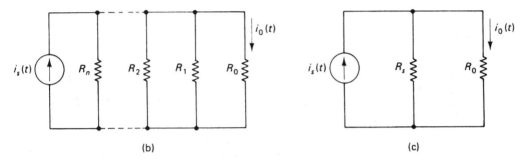

(b)   (c)

FIGURE 1–16
Circuits used to illustrate the current divider rule.

in parallel, but some caution must be exercised in the process. A form analogous to Equation (1–39) with *resistances* is *not correct.* There are two methods in which this process can be achieved. The first method is to combine all resistances except R_0 into a single resistance, as illustrated in (b) and (c) of Figure 1–16. The resistance R_s is determined as

$$\frac{1}{R_s} = \frac{1}{R_1} + \frac{1}{R_2} + \cdots + \frac{1}{R_n} \tag{1–41}$$

The current divider rule can then be applied as follows:

$$i_0(t) = \frac{R_s}{R_s + R_0} i_s(t) \tag{1–42}$$

Current Divider with Conductances

The second method for working with more than two resistances in parallel is to use conductances. Let $G_0 = 1/R_0$, $G_1 = 1/R_1$, and so on. The current divider rule can then be expressed as

$$i_0(t) = \frac{G_0}{G_0 + G_1 + G_2 + \cdots + G_n} i_s(t) \tag{1–43}$$

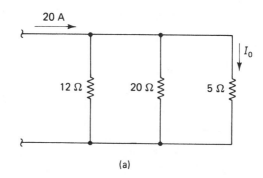

(a)

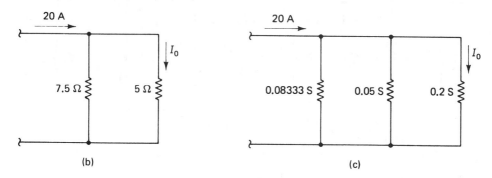

(b)   (c)

FIGURE 1–17
Circuit for Example 1–6.

This conductance form is analogous to Equation (1–39) for the voltage divider rule in that the numerator conductance is the conductance through which the current $i_0(t)$ is to be measured, and the denominator is the sum of all conductances.

As in the case of the voltage divider rule, the shapes of the two current waveforms for a resistive current divider have the same form.

EXAMPLE 1–6

In the circuit of Figure 1–17(a), a dc current of 20 A is known to be flowing into the parallel combination of the three resistors shown. Determine the dc current I_0 using the current divider rule in two different ways: **(a)** using resistance values and **(b)** using conductance values.

Solution

(a) Working directly with resistance values, it is necessary first to reduce the circuit to one with two equivalent resistances. The parallel combination of the two left-hand resistances is calculated, and the model shown in Figure 1–17(b) is determined. The reader is invited to verify that the parallel combination of 12 Ω and 20 Ω is 7.5 Ω. Application of the current divider rule to this circuit yields

$$I_0 = \frac{7.5}{7.5 + 5} \times 20 = 0.6 \times 20 = 12 \text{ A} \qquad (1\text{-}44)$$

(b) To work with conductances, the circuit is redrawn in Figure 1–17(c) with element values labeled in siemens. Application of the current divider rule in conductance form yields

$$I_0 = \frac{0.2}{0.2 + 0.05 + 0.08333} \times 20 \text{ A} = 12 \text{ A} \qquad (1\text{-}45)$$

This result obviously agrees with Equation (1–44).

The first approach may be slightly preferable if only one current is to be determined, but that method tends to be rather cumbersome if several currents must be determined. This is especially true if there are more than three resistors, because several different parallel combinations must be calculated. However, we will leave it up to the reader to make the choice as to which of these different approaches is easiest to apply.

1–9 SIMPLIFIED CIRCUIT ANALYSIS

By combining the equivalent resistance simplification with the voltage and current divider rules, it is possible to analyze a number of complete circuits using a step-by-step approach. At each step in the process, the circuit is represented in both directions by a simplified form, allowing one of the simple rules stated earlier to be applied.

The method under consideration can be classified as an intuitive method, as opposed to a more formal mathematical approach. Consequently, it is difficult to state any precise set of rules that always works. Most of the methods to be developed in the next chapter are somewhat general, but they do not always convey a "feeling" as to what is taking place in the analysis process. For that reason, the more intuitive technique to be discussed here is useful. However, it should be stressed at the outset that the approach of the section works only for a limited class of problems, but it works in enough practical cases to justify its coverage.

Circuits under consideration are generally those in which the circuit as viewed by the source can be easily represented by a single equivalent resistance. The current flowing from a voltage source or the voltage across a current source can then be determined. By moving away from the source and successively determining resistive equivalents at various points, voltage and current divider rules may lead to a complete analysis. The method is best illustrated by a representative example.

EXAMPLE 1–7

Consider the dc resistive circuit of Figure 1–18. The passive part of the circuit is the same as that of Example 1–5, but a 30-V dc source has been connected to the

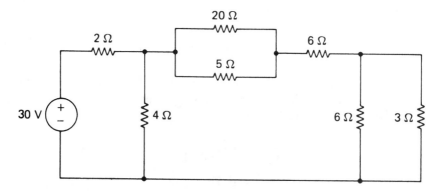

FIGURE 1–18
Circuit for Example 1–7.

input. Using a step-by-step approach as discussed in this section, determine all branch voltages and currents in the circuit.

Solution

The equivalent resistances at various points in the circuit were calculated in Example 1–5, so the reader may need to refer to that example (including Figure 1–14). In addition, some equivalent circuits with certain voltages and currents identified are shown in Figure 1–19 to aid in the steps that follow.

First, the equivalent resistance "seen" by the voltage is 5 Ω, so the current leaving the 30-V source is readily determined by Ohm's law as 30 V/5 Ω = 6 A. This is illustrated in Figure 1–19(a) with the resistance represented as the series combination of the 2-Ω resistor and the additional 3-Ω equivalent resistance. With this form, the voltage across the 2-Ω resistor is readily determined to be 2 Ω × 6 A = 12 V, and the voltage across the circuit to the right of the 2-Ω resistor is determined to be 3 Ω × 6 A = 18 V. Alternately, the voltage divider rule could have been used twice to determine these voltages.

With the 18-V voltage across 3 Ω determined, this resistance is next represented as the parallel combination of the 4-Ω resistor and the additional 12-Ω equivalent resistance, as shown in Figure 1–19(b).The current through the 4-Ω resistor is determined to be 18 V/4 Ω = 4.5 A, and the current flowing into the circuit to the right of the 4-Ω resistor is determined to be 18 V/12 Ω = 1.5 A. Alternately, the current divider rule could have been used twice to determine these currents, since the input current of 6 A is known.

With the 1.5-A current known, the equivalent circuit shown in Figure 1–19(c) is next used. In this case, the current divider rule will be applied twice to determine how this current divides between the two parallel resistors. The current through the 20-Ω resistor is [5/(5 + 20)] × 1.5 = 0.3 A, and the current through the 5-Ω resistor is [20/(20 + 5)] × 1.5 = 1.2 A. Either of these currents times the resistance through which it flows results in a 6-V drop, or, alternately, the net current of 1.5 A times the equivalent parallel resistance of 4 Ω also results in 6 V.

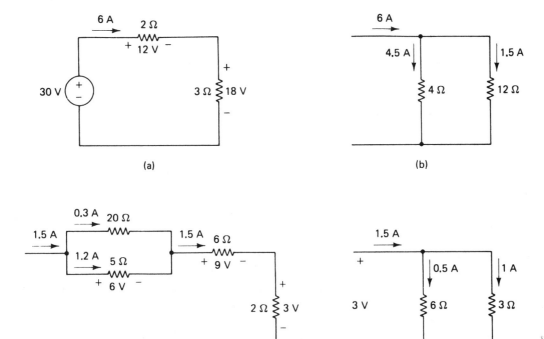

FIGURE 1–19
Steps involved in analyzing circuit of Example 1–7.

The voltage across the 6-Ω resistor is determined to be 6 Ω × 1.5 A = 9 V, and the voltage across the equivalent resistance on the right is 2 Ω × 1.5 A = 3 V. This 3-V voltage appears across the two right-hand parallel resistors, as shown in Figure 1–19(d). Ohm's law applied to these final branches results in currents of 3 V/6 Ω = 0.5 A and 3 V/3 Ω = 1 A.

All of the currents and voltages calculated in the preceding steps are identified on the original circuit diagram in Figure 1–20. The reader may readily verify that KVL is satisfied around all loops, KCL is satisfied at all nodes, and Ohm's law is satisfied for all resistive branches.

The perceptive reader may have observed other variations in some of the analysis steps that could have been taken in moving through the circuit. Because various techniques are often possible, it is difficult to establish an absolute set of guidelines for recognizing and solving a circuit in this fashion. Some of the success in this process depends on the intuition (and experience) of the person performing the analysis. The more formal methods of the next chapter can often be applied in a "colder" mathematical process, so some people have a tendency to forget

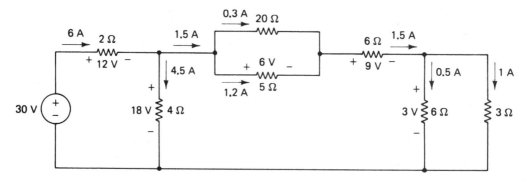

FIGURE 1–20
Circuit for Example 1–7 with all voltages and currents labeled.

about the type of process given here when those methods are learned. However, for circuits in which the step-by-step process can be applied, a quicker solution can often be achieved, and much more can be learned about the circuit in the process.

1–10   DEPENDENT SOURCES

The voltage and current source models introduced in Section 1–3 represent the common independent source models. Recall that independent sources are those whose values are independent of the levels of any voltages or currents in the circuit. The values of all voltage and current sources encountered thus far have either been independent constant values (corresponding to dc sources) or independent time-varying quantities.

Dependent (or controlled) source models are a special class in which the voltage or current is dependent on some other voltage or current within the circuit. Such sources arise from complex physical interactions in many electronic and electrical devices. For example the base current of a bipolar junction transistor causes a much larger collector current to flow, and this process is modeled by a dependent source.

Since the objective here is circuit analysis, no attempt will be made to explain the physical basis for the dependent sources encountered. Throughout the study of electronic circuits and other application areas, dependent source models are introduced to represent the processes involved with various devices. Our goal will be to learn to deal with the dependent sources when they arise in the analysis of circuits.

A dependent source, like an independent source, may be either a voltage source or a current source. However, it may also be controlled by either a voltage or a current. This results in four different combinations for dependent source models, and these are illustrated in Figure 1–21. The symbol used for the dependent source models is in accordance with an increasing portion of the literature, although

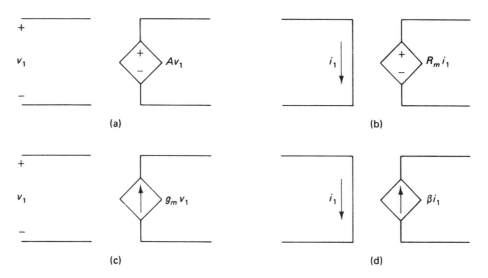

FIGURE 1–21
Four possible models of ideal dependent (or controlled) sources.

earlier texts employ the same symbol (circle) as for independent source models. Each of the four forms will now be discussed briefly.

Voltage-Controlled Voltage Source

The model for the *voltage-controlled voltage source* (VCVS) is shown in Figure 1–21(a). An independent control voltage v_1 is assumed to exist across certain control terminals. This voltage controls a dependent voltage whose value is Av_1. The quantity A is dimensionless and corresponds to a voltage gain when the model represents a voltage amplifier.

Current-Controlled Voltage Source

The model for the *current-controlled voltage source* (ICVS) is shown in Figure 1–21(b). An independent control current i_1 is assumed to be flowing in certain control terminals. This current controls a dependent voltage whose value is $R_m i_1$. The quantity R_m, relating the dependent voltage to the controlling current, has the dimension of ohms and is called the *transresistance*.

Voltage-Controlled Current Source

The model for the *voltage-controlled current source* (VCIS) is shown in Figure 1–21(c). An independent control voltage v_1 is assumed to exist across certain control terminals. This voltage controls a dependent current whose value is $g_m v_1$. The quantity g_m, relating the dependent current to the controlling voltage, has the dimension of siemens and is called the *transconductance*.

Current-Controlled Current Source

The model for the *current-controlled current source* (ICIS) is shown in Figure 1–21(d). An independent control current i_1 is assumed to be flowing in certain control terminals. This current controls a dependent current whose value is βi_1. The quantity β is dimensionless and corresponds to a current gain when the model represents a current amplifier.

EXAMPLE 1–8

The circuit of Figure 1–22 contains an ICIS. Determine I_B and the power delivered by the 24-V source.

Solution

The current through the 6-Ω resistor controls the value of the current source as shown. Since the 24-V source appears across the 6-Ω resistor, the current I_R can be determined directly by Ohm's law as

$$I_R = \frac{24 \text{ V}}{6 \, \Omega} = 4 \text{ A} \tag{1-46}$$

The value of the dependent current source is then readily determined as $2I_R = 2 \times 4 \text{ A} = 8 \text{ A}$. The current I_B is then determined by applying KCL to the upper node. We have

$$I_R + 2I_R - I_B = 0 \tag{1-47a}$$

or

$$4 + 8 - I_B = 0 \tag{1-47b}$$

which results in

$$I_B = 12 \text{ A} \tag{1-48}$$

FIGURE 1–22
Circuit for Example 1–8.

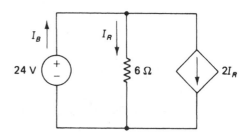

The power P delivered by the 24-V source is

$$P = 24 \times 12 = 288 \text{ W} \qquad \qquad \textbf{(1–49)}$$

The reader is invited to verify that 96 W are absorbed by the 6-Ω resistor, and 192 W are absorbed by the dependent current source.

This dependent source example was chosen to be sufficiently simple so that a step-by-step procedure could be used. In general, however, most problems involving dependent sources require the more sophisticated methods to be developed in the next chapter.

1–11 PSPICE EXAMPLES

Most of the circuits in this chapter are sufficiently simple that they can be solved more quickly with a calculator or by hand than with PSPICE. However, they serve as a good learning tool to develop some initial practice using PSPICE, since the results may be readily checked. In particular, the basic dc analysis capabilities of PSPICE will be illustrated with the examples that follow.

PSPICE EXAMPLE 1–1
Analyze the circuit of Example 1–7 (Figure 1–18) using the .OP command.

Solution
The circuit adapted to the PSPICE format is shown in Figure 1–23(a) and the code is shown in (b). The bus along the bottom is designated as the ground node (0), and all other nodes are numbered as shown. The first line of the code is the title line, which has been identified as PSPICE EXAMPLE 1–1.

The second line is the .OPTIONS statment, which could be omitted if desired. The option NOECHO suppresses the printing of the code again in the output file, which results in some saving of space when that file is to be printed. The portion of the code suppressed is everything that follows this line, so when this option is employed, the .OPTIONS statement should follow the title line. When this particular option is not employed, the .OPTIONS statement can be placed near the end of the program in the control block if desired.

The option NOPAGE suppresses extra header information and generally reduces the number of pages in the output data file. When a graphical output is desired using .PROBE, as will frequently be the case later in the text, neither of these options is usually necessary.

The first line of the circuit description reads

```
VS 1 0 DC 30
```

The letter *V* at the beginning of VS identifies the element as an independent voltage source, which may then be followed by up to seven additional letters or numbers for identification. The numbers *1* and *0* represent the node connections for the

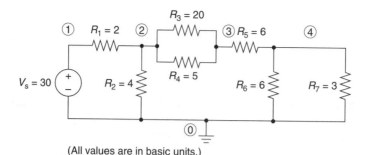

(All values are in basic units.)

```
PSPICE EXAMPLE 1-1
.OPTIONS NOECHO NOPAGE
VS 1 0 DC 30
R1 1 2 2
R2 2 0 4
R3 2 3 20
R4 2 3 5
R5 3 4 6
R6 4 0 6
R7 4 0 3
.OP
.END
```

FIGURE 1-23
Circuit and code for PSPICE Example 1-1.

voltage source, with the first (1) representing the more positive node. The DC designation refers to a dc source, and the value is seen to be 30 V.

The next line of the code reads

```
R1 1 2 2
```

The letter R at the beginning of R1 identifies the element as a resistance, which may be followed by additional letters or numbers. The numbers *1* and *2* following R1 represent the node connections for the resistor. If only node voltages with respect to ground are of interest, the order is unimportant. However, if either branch voltages or branch currents are to be measured, the order is important. For a branch voltage measurement, the first number (1) represents the assumed positive reference node. For a branch current measurement, the assumed positive reference for current flow is from the first node (1) to the second node (2). The final value of 2 is the resistance in ohms.

The next six lines define the connections and values for resistors R2 through R7. These code lines can be readily verified from the circuit diagram.

The analysis in this case is based on the *operating point* control statement .OP, which follows the circuit description. This is one of the simplest commands, since it does not require a separate print statement, but it also provides only limited data, as we will see shortly. Finally, as is required with all programs, an .END statement completes the code.

The output data file, with the compact features produced by the NOECHO and NOPAGE options, is shown in Figure 1–24. The four node voltages are listed (in volts) as 30, 18, 12, and 3, respectively. These values are in perfect agreement with those of Example 1–7.

The .OP command provides the values of all currents through voltage sources, which for this case provides a value of −6 A for the current through VS. The (−) sign may seem puzzling, but it is easily explained by the convention that the positive

```
PSPICE EXAMPLE 1-1


****      CIRCUIT DESCRIPTION


*****************************************************************************


.OPTIONS NOECHO NOPAGE


****      SMALL SIGNAL BIAS SOLUTION        TEMPERATURE =    27.000 DEG C


NODE    VOLTAGE      NODE    VOLTAGE      NODE    VOLTAGE      NODE    VOLTAGE


(   1)    30.0000  (    2)    18.0000  (    3)    12.0000  (    4)     3.0000



       VOLTAGE SOURCE CURRENTS
       NAME           CURRENT

       VS             -6.000E+00

       TOTAL POWER DISSIPATION    1.80E+02   WATTS



****      OPERATING POINT INFORMATION       TEMPERATURE =    27.000 DEG C



       JOB CONCLUDED

       TOTAL JOB TIME                .33
```

FIGURE 1–24
Output data for PSPICE Example 1–1.

direction of current flow for any element is from the first node to the second node in the defining statement. Since the first node for the voltage source is the top node, but the current is actually flowing out of the top node, the program provides a negative value for this variable.

The final data point for the .OP analysis is the total power dissipation, which is 180 W. This value is correct for this particular circuit, but in general, some caution must be exercised in interpreting this quantity. In general, the value provides *the net power delivered to the circuit by voltage sources only* and does *not* provide power associated with *current sources*.

Because of the limited data provided and the possibility of misinterpretation for the total power, we will refrain from using the .OP command in subsequent work. The next example will illustrate a better approach for solving the same circuit using a dc analysis tailored for the variables that are desired.

PSPICE EXAMPLE 1–2

Repeat the analysis of PSPICE Example 1–1 with the .DC command, and obtain values for all node voltages and branch currents.

Solution

The circuit diagram is the same as in PSPICE Example 1–1, so refer to Figure 1–23 for the numbering scheme. The code for the present example is shown in Figure 1–25. All lines through R7 are identical to those in the preceding example and need no further discussion.

The first line that is different reads

```
.DC VS 30 30 1
```

The designation *.DC* indicates a dc analysis, which is assumed to be a "sweep"; that is, it has the capability of assuming a range of values. A source within the circuit must be considered as a "sweep source," and since there is only one source in this case, it must be VS. In general, when there are several sources, one must be arbitrarily selected for this purpose.

```
PSPICE EXAMPLE 1-2
.OPTIONS NOECHO NOPAGE
VS 1 0 DC 30
R1 1 2 2
R2 2 0 4
R3 2 3 20
R4 2 3 5
R5 3 4 6
R6 4 0 6
R7 4 0 3
.DC VS 30 30 1
.PRINT DC V(2) V(3) V(4) I(R1) I(R2) I(R3) I(R4) I(R5) I(R6) I(R7)
.END
```

FIGURE 1–25

Code for PSPICE Example 1–2.

PSPICE EXAMPLE 1-2

****   CIRCUIT DESCRIPTION

****

.OPTIONS NOECHO NOPAGE

****   DC TRANSFER CURVES   TEMPERATURE =   27.000 DEG C

VS	V(2)	V(3)	V(4)	I(R1)	I(R2)
3.000E+01	1.800E+01	1.200E+01	3.000E+00	6.000E+00	4.500E+00

****   DC TRANSFER CURVES   TEMPERATURE =   27.000 DEG C

VS	I(R3)	I(R4)	I(R5)	I(R6)	I(R7)
3.000E+01	3.000E-01	1.200E+00	1.500E+00	5.000E-01	1.000E+00

JOB CONCLUDED

TOTAL JOB TIME   .39

FIGURE 1-26
Output data for PSPICE Example 1-2.

         In general, a dc sweep requires three values to follow the identified source. The first value is the beginning value, and the second is the ending value. The third value represents the step size between successive voltages in the sweep. To force the sweep to assume only one value, the beginning and ending values are chosen to be the same (30 V in this case). The third value can be any nonzero value, so the simplest choice is 1. Thus, we have "fooled" the sweep command and forced it to assume only one value. This is a good example of one of the eccentricities of PSPICE and is related to the original structured format around which the program was developed.

         The next line in the code reads

```
.PRINT DC V(2) V(3) V(4) I(R1) I(R2) I(R3) I(R4) I(R5) I(R6) I(R7)
```

The portion that reads .PRINT DC directs PSPICE to print out a list of the values of the variables that follow. Among these are the node voltages at nodes 2, 3, and 4, which are denoted as V(2), V(3), and V(4), respectively. The next seven entries are the element currents through the seven resistances. Finally, an .END statement is included as required.

The output data file is shown in Figure 1–26. Note that the first printed variable is the "swept" variable VS. This variable is always printed as the first column, and five additional variables are printed on the first line. When more than five variables are required, one or more additional data lines are printed, and VS is repeated at the beginning of the line. In this example, ten variables were requested, so two lines are required.

DRILL PROBLEMS

1–1. A certain electric dryer has a power rating of 4000 W. For 3 hours of continuous operation, determine **(a)** the total consumed energy in joules, **(b)** the corresponding energy in kilowatt-hours (kWh), and **(c)** the cost of operating the heater if the energy rate is 8.5¢ per kWh.

1–2. A certain electric heater uses a total energy of 6 kWh when operated continuously for 5 hours. If the voltage is 120 V, determine **(a)** current and **(b)** resistance of heater.

1–3. The burners on a certain electric stove are each rated at 1 kW, and the oven is rated at 1.5 kW. During the cooking of a certain meal, the following usage is determined:

> Burner #1   15 minutes
> Burner #2   30 minutes
> Oven   45 minutes

If electricity costs 8¢ per kWh, determine the cost of preparing a meal.

1–4. A family decides to leave a 25-W bulb burning continuously near an otherwise dark cellar stairway. If electricity costs 8¢ per kWh, determine the cost of electricity for one full year. (You need not assume a leap year!)

1–5. A certain load is characterized by an instantaneous power that increases linearly from zero to 1 kW over a 3-hour period. Determine the total energy in **(a)** kWh and **(b)** joules.

1–6. The power as a function of time for a certain load is shown in Figure P1–6 for a 12-hour period. Determine the total energy in **(a)** kWh and **(b)** joules.

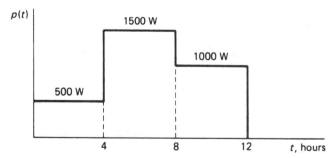

FIGURE P1–6

1–7. For each of the branches shown in Figure P1–7, determine the power and whether it is being delivered or absorbed.

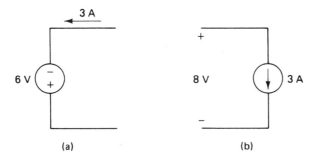

(a)   (b)

FIGURE P1–7

1–8. For each of the branches shown in Figure P1–8, determine the power and whether it is being delivered or absorbed.

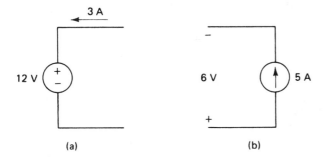

(a)   (b)

FIGURE P1–8

1–9. The dc voltage across a 4.7-kΩ resistor is 12.6 V. Determine the dc current.

1–10. The dc current through a certain resistance is 3 mA when the dc voltage is 120 V. Determine the resistance.

1–11. Calculate the power P dissipated in a resistance for each of the sets of conditions listed below. Voltages and currents are assumed to be dc quantities.
  a. $V = 12$ V, $I = 2$ A
  b. $R = 8\ \Omega$, $I = 3$ A
  c. $V = 24$ V, $R = 6\ \Omega$

1–12. Calculate the "hot" resistance of a light bulb at 100 W and 120 V.

1–13. In the circuit of Figure P1–13, the source voltage and the voltage at the right-hand node with respect to ground are known. Two resistance values are known. Determine V_1, I_1, I_2, and R.

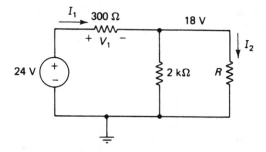

FIGURE P1–13

1–14. In the circuit of Figure P1–14, the two source voltages and the voltage across one resistor are known. Determine I_1, I_2, and R.

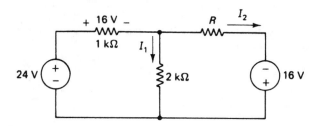

FIGURE P1–14

1–15. For the circuit of Figure P1–15, determine V_0 and R.

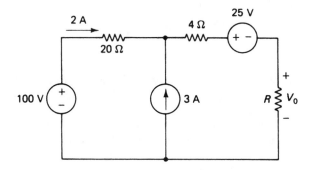

FIGURE P1–15

1–16. For the circuit of Figure P1–16, determine I_1, I_2, and V_0.

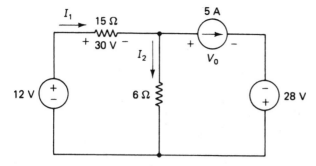

FIGURE P1–16

1–17. Using successive series and parallel reduction, determine the equivalent resistance at the input terminals for the circuit of Figure P1–17.

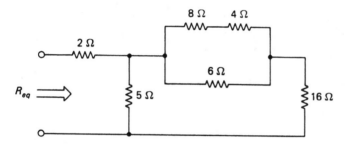

FIGURE P1–17

1–18. Using successive series and parallel reduction, determine the equivalent resistance at the input terminals for the circuit of Figure P1–18.

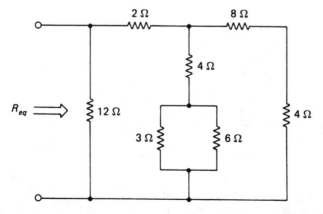

FIGURE P1–18

1-19. Find R_{eq} for the circuit of Figure P1-19.

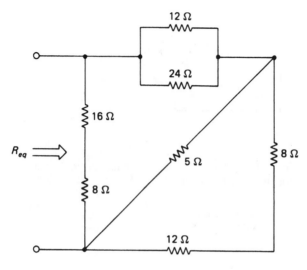

FIGURE P1-19

1-20. Find R_{eq} for the circuit of Figure P1-20.

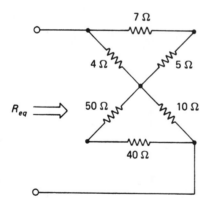

FIGURE P1-20

1-21. In the circuit of Figure P1-21, a dc voltage of 120 V is known to be across the series combination of the three resistors. Determine the three voltages V_1, V_2, and V_3 using the voltage divider rule.

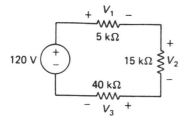

FIGURE P1–21

1–22. In the circuit of Figure P1–22, a dc current of 9 A is known to be flowing into the parallel combination of the three resistors shown. Determine the current I_0 using the current divider rule in two different ways: **(a)** using resistance values and **(b)** using conductance values.

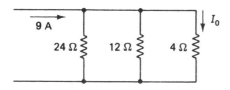

FIGURE P1–22

1–23. Analyze the circuit of Problem 1–21 by the following approach: Determine the equivalent resistance seen by the source, and calculate the current flowing from the source. From a knowledge of the current, determine the voltages across the three resistors.

1–24. Analyze the circuit of Problem 1–22 by the following approach: Determine the equivalent resistance of the three resistors, and calculate the voltage across the combination resulting from the 9-A current. From a knowledge of the voltage, determine the currents through the three resistors.

1–25. Consider the dc resistive circuit of Figure P1–25. The passive part of the circuit is the same as that of Problem 1–17, but a 60-V source has been connected to the input. Using a step-by-step approach as discussed in Section 1–9, determine all branch voltages and currents in the circuit.

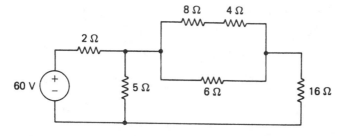

FIGURE P1–25

1–26. Consider the dc resistive circuit of Figure P1–26. The passive part of the circuit is the same as that of Problem 1–18, but a 15-A current source has been connected to the input. Using a step-by-step approach as discussed in Section 1–9, determine all branch voltages and currents in the circuit.

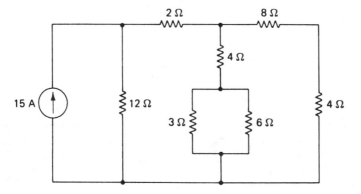

FIGURE P1–26

1–27. The circuit of Figure P1–27 contains an ICIS. Determine I_B and the power delivered by the 12-V source.

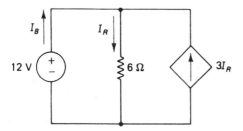

FIGURE P1–27

1–28. The circuit of Figure P1–28 contains a VCVS. Determine **(a)** the voltage V_0 across the current source, **(b)** the power absorbed by the dependent source, and **(c)** the power delivered by the 3-A source.

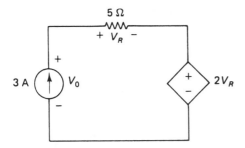

FIGURE P1–28

DERIVATION PROBLEMS

1–29. Consider the series combination of resistors shown in Figure 1–11. Assume a current i flowing clockwise around the loop and a voltage v across the external terminals with the positive terminal at the top. By first expressing all voltage drops across resistors in terms of the resistances and the current, apply KVL to the loop, and *derive* the equivalent resistance of the series connection; i.e., $R_{eq} = v/i$.

1–30. Consider the parallel combination of resistors shown in Figure 1–12. Assume a voltage v across the parallel combination and a current i flowing into the circuit. By first expressing all branch currents in resistors in terms of the resistances and the voltage, apply KCL to one of the nodes, and *derive* the equivalent resistance of the parallel combination. (*Note*: It is easier to determine $1/R_{eq} = i/v$.)

1–31. Consider the voltage divider circuit of Figure 1–15(a). *Derive* the form of the voltage divider rule as given by Equation (1–38) by first determining the equivalent resistance seen by the source, then determining the current flowing through the loop, and finally determining the voltage across R_0 by Ohm's law.

1–32. Consider the current divider circuit of Figure 1–16(a). *Derive* the form of the current divider rule as given by Equation (1–40) by first determining the equivalent resistance seen by the source, then determining the voltage across the parallel combination, and finally determining the current through R_0 by Ohm's law.

APPLICATION PROBLEMS

1–33. A classical approach to designing dc voltmeters is illustrated in Figure P1–33. A basic D'Arsonval galvanometer is used for the indicating device. The series resistances are chosen such that the full-scale current I_{fs} flows when the maximum voltage for a given range is applied. For the circuit illustrated, assume that the four full-scale desired voltages are 3 V, 10 V, 30 V, and 100 V. Assume that $I_{fs} = 100\ \mu A$ and that the meter contains an internal series resistance $R_m = 1\ k\Omega$. Determine the required design values of the four resistances R_1, R_2, R_3, and R_4.

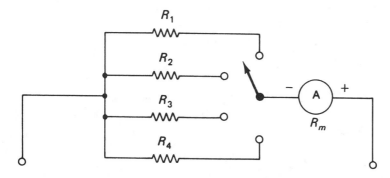

FIGURE P1–33

1–34. Referring to the dc voltmeter concept of Problem 1–33, the sensitivity S of the volunteer is a measure of its equivalent resistance. For a larger sensitivity, the resistance is larger and the loading effect is less. Sensitivity is measured in ohms/volt, and the net voltmeter resistance R_v for a given full-scale voltage V_{fs} is given by

$$R_v = SV_{fs}$$

Show that the sensitivity is given by

$$S = \frac{1}{I_{fs}}$$

1–35. In a certain system shown in Figure P1–35, it is necessary to attenuate a signal v_1 while establishing a certain resistance level. The 3-kΩ load resistance is fixed. The two design criteria are

a. $\dfrac{v_2}{v_1} = \dfrac{1}{10}$ and

b. $R_{eq} = 20$ kΩ

Determine the required values of R_1 and R_2.

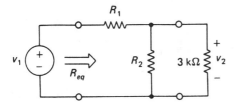

FIGURE P1–35

1–36. The range of an ammeter may be increased by adding a shunt resistance R_s as shown in Figure P1–36. The value of the shunt resistance is chosen to absorb the current in excess of the meter full-scale current. Let I_{fs} represent the full-scale meter current, and assume that it is desired to multiply the meter range by a constant K; that is, the desired full-scale measured current is to be KI_{fs}. If R_m represents the meter internal resistance, show that the required value of R_s is

$$R_s = \frac{R_m}{K - 1}$$

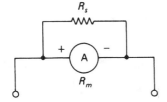

FIGURE P1–36

1–37. Show how to connect four 8-Ω speakers to a single output of an audio amplifier so that the net resistance seen by the amplifier is 8 Ω. If the amplifier is delivering a power of 2 W at a given time, how much power will each speaker receive?

1–38. Assume that there are available a large quantity of 12-V batteries, each of which is rated at 5 A. Design an emergency lighting system capable of providing a load current of 20 A at a voltage of 120 V dc.

2

CIRCUIT ANALYSIS METHODS

OBJECTIVES

After completing this chapter, the reader should be able to:

- Determine the current, voltages, and power in a single-loop circuit.
- Determine the voltage, currents, and power in a single node-pair circuit.
- Apply source transformations to simplify source models.
- Determine appropriate models for realistic voltage and/or current sources.
- Apply mesh current analysis to determine all voltages, currents, and power in a circuit containing several meshes.
- Apply node voltage analysis to determine all voltages, currents, and power in a circuit containing several nodes.
- Determine the Thevenin and Norton equivalent circuits for a given circuit.
- Apply various circuit analysis methods to circuits containing dependent (or controlled) sources.
- Apply the principle of superposition in the analysis of linear circuits.

2–1   SINGLE-LOOP CIRCUIT

The first special circuit that will be analyzed is the single-loop circuit, whose basic form is illustrated in Figure 2–1. Any number of passive components and voltage sources may be in the loop. However, only one ideal current source is allowed, as will be demonstrated later. All elements in the single-loop circuit are said to be connected in *series*.

For analysis purposes, the single-loop circuit will be considered in two categories: (a) voltage sources and passive elements only (i.e., no current source present) and (b) voltage sources, passive elements, and one current source. Each case will be considered separately.

FIGURE 2–1
Basic form of the single-loop circuit.

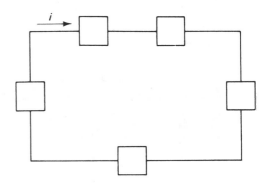

No Current Source Present

Consider first the case when no current source is present. The development here will be restricted to resistive circuits, so assume that an arbitrary number of voltage sources and resistors are in the loop. The voltage sources may or may not be ideal, but any source internal resistances are represented as additional series resistors within the loop.

The procedure for analyzing the single-loop circuit under these conditions is based on the fundamental property that the *current through all elements in a single-loop circuit is the same.* The fact is readily verified by KCL, since at each node within the loop there are only two connecting branches, and the current entering the node must equal the current leaving the same node. The voltage drops across all resistors may be expressed in terms of the single current variable, with the sign convention of Chapter 1 carefully noted, that is, the terminal of a resistor at which the current enters must be assumed to be the most positive. Kirchhoff's voltage law may then be applied to the loop, and the algebraic sum of all the voltages around the loop is equated to zero. As long as there is no more than one unknown quantity in the loop equation, it may be determined from the equation.

As discussed in Chapter 1, lowercase symbols will be used for voltages and currents whenever these quantities *could* be time-varying. When it is desired to emphasize the time-varying nature, symbols such as $v(t)$ and $i(t)$ will be used. However, such terms will be simplified to the forms v and i in many expressions, but the time-varying possibility will be understood.

A common situation in resistive circuits is that the voltage sources and resistances are known, but the current is unknown and must be determined. In this case, an unknown current i is assumed within the loop. With some intuition, it is usually possible to predict the direction in which the current will flow, and it simplifies the "bookkeeping" when the assumed direction is equal to the actual direction. However, it really doesn't matter whether the assumed direction is correct or not, *provided* that all voltage drops across resistors are expressed to be compatible with the assumed current direction.

If the current turns out to be a negative value or function, the direction of the actual current is opposite to the assumed direction. At this point, one of two possible strategies may be taken. The first strategy is to define a new current (say, i'), which has the same magnitude as i but is opposite in direction and thus positive (i.e., $i' = -i$). All voltage drops across resistors are then reversed from the original assumed polarities. The second strategy is to retain i as a negative quantity and keep the same sign conventions as before. As long as all calculations deal with i as a negative quantity or function, the results will still be mathematically correct. Although both strategies are perfectly correct, the second approach tends to encounter more difficulties with sign conventions in subsequent loop calculations.

Example

Formulation of a single-loop equation will be illustrated by the resistive circuit shown in Figure 2–2. The circuit contains three voltage sources and four resistors. All elements in this illustration are labeled with symbols, so no particular quantities will be assumed to be known or unknown. Rather, the goal here is simply to express the KVL equation around the loop in terms of the labeled quantities.

For clarity, the current i is shown at a number of separate points around the circuit. Observe that the positive terminal of the voltage drop across each resistor is assumed at the point where the current enters. However, the signs of the source voltages are a function of the manner in which the sources are connected and have nothing to do with the assumed current direction.

With the circuit properly labeled, we pick a point in the loop at which to start and then we "walk all the way around" to complete a closed loop. Although the equation may be written while completing the loop in either direction, it is preferable to move in the same direction as the assumed current.

Sign Convention

The sign convention employed here and in most places in the book is as follows: In passing through a component, if the potential drops, the voltage is written with a positive sign, but if the potential rises, the voltage is written with a negative sign. This particular convention has the advantage that, in most circuits, the number of positive terms exceeds the number of negative terms. It also has the property that

FIGURE 2–2

Representative single-loop circuit containing dc voltage sources and resistors.

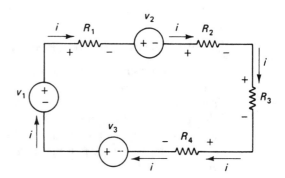

the sign of the term in the KVL equation is the sign of the terminal at which the current enters.

For the circuit of Figure 2–2, we will start at a point just below the negative terminal of the source v_1 and take a clockwise path. Following the convention previously discussed, we obtain the following KVL equation for the loop:

$$-v_1 + R_1 i + v_2 + R_2 i + R_3 i + R_4 i - v_3 = 0 \qquad (2\text{--}1)$$

The reader should carefully study the sign pattern of this equation and relate it to the convention discussed.

Simplified Pattern

An interesting pattern for the single-loop equation results when i is considered as the unknown; an expression for it can be determined in terms of the other parameters. The reader is invited to verify that the result is

$$i = \frac{v_1 - v_2 + v_3}{R_1 + R_2 + R_3 + R_4} \qquad (2\text{--}2)$$

This result may be generalized to the form

$$i = \frac{\text{algebraic sum of voltage source values}}{\text{net loop series resistance}} \qquad (2\text{--}3)$$

In the interpretation of Equation (2–3), voltages of voltage sources are positive when they contribute to current flow in the assumed direction, and negative when they contribute to current flow in the opposite direction. For this form to be valid, *all the voltage sources must be independent sources.*

The reader may prefer a shortcut approach such as Equation (2–3) for a problem such as this, and indeed, the author frequently uses this form. However, one has to exercise caution with this seemingly simpler approach because it is easier to make a sign error in the process. Further, this simpler approach is not valid when there is one or more *dependent voltage sources* in the loop. Therefore, it is recommended that the more fundamental approach of "walking around the loop" be retained as the basic method for analyzing a general single-loop circuit, and that any shortcut procedures be used cautiously and only after a clear understanding of the process is achieved.

Single Current Source Present

Next, we will consider the case where a single ideal current source appears in the loop. *The presence of an ideal current source in a single loop forces the current through all elements in the loop to be the same as that of the source.* Thus, the current is immediately evident if the value of the current source is known. Any number of ideal voltage sources may be in the loop, but they do not affect the current. The various voltage drops across resistors may be determined directly from the known current and the resistances.

One quantity that requires some computation to determine is the voltage across the current source. An unknown voltage (say, v_0) is assumed across the current source. By application of KVL to the loop, v_0 can be determined. The polarity may turn out to have the positive terminal at the point where current flows out of the current source or vice versa, depending on the nature of any voltage sources in the loop.

Thus far, we have assumed only one ideal current source. The assumption of two or more ideal current sources with different currents in series results in a contradictory situation and is impossible. In a real-life situation where two or more current sources with different values are connected in series, the actual models of the current sources containing internal resistances would have to be employed to predict the results, and this would require a more complex model than the one under consideration. If two or more ideal current sources with the same current are connected in series, the result is mathematically equivalent to a single ideal current source. Thus, we need only to consider the presence of a single ideal current source in the single-loop discussion.

EXAMPLE 2–1

For the dc resistive circuit of Figure 2–3(a), determine **(a)** the loop current, **(b)** voltage drops across all resistors, and **(c)** power associated with each element and whether it is absorbed or delivered.

Solution

(a) the circuit as given initially has no labels for the unknown current and voltages desired in order that such labels may be determined as part of the analysis. First, intuition likely tells us that the loop current will be clockwise, since the 24-V source is "more powerful" than the 6-V source, so we will choose a clockwise direction as shown in Figure 2–3(b). (It really doesn't matter which way is chosen, however, as was explained earlier.) Once the current direction is assumed, the various voltage drops across the resistors are chosen in the proper directions as indicated in Figure 2–3(b). For convenience, each of these voltages is also labeled with appropriate defining subscripts. (*Note:* Uppercase symbols are used for the current and all voltages, since this is a dc resistive circuit.)

Starting at a point just below the 24-V source and following a clockwise path, the KVL loop equation reads as follows:

$$-24 + 4I + 6 + 3I + 2I = 0 \qquad \textbf{(2–4)}$$

This simplifies to

$$9I = 18 \qquad \textbf{(2–5)}$$

or

$$I = 2 \text{ A} \qquad \textbf{(2–6)}$$

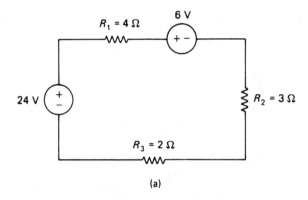

(a)

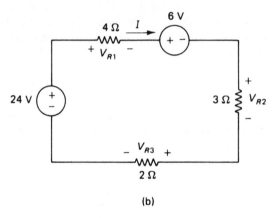

(b)

FIGURE 2–3
Circuit for Example 2–1.

(b) Once the current is known, the voltage drops across the three resistors are readily determined as

$$V_{R1} = R_1I = 4 \times 2 = 8 \text{ V} \qquad (2\text{–}7)$$

$$V_{R2} = R_2I = 3 \times 2 = 6 \text{ V} \qquad (2\text{–}8)$$

$$V_{R3} = R_3I = 2 \times 2 = 4 \text{ V} \qquad (2\text{–}9)$$

(c) All resistive power values represent *absorbed* power. Let P_{R1}, P_{R2}, and P_{R3} represent the three values associated with R_1, R_2, and R_3, respectively. We have

$$P_{R1} = R_1I^2 = 4 \times (2)^2 = 16 \text{ W} \qquad (2\text{–}10)$$

$$P_{R2} = R_2I^2 = 3 \times (2)^2 = 12 \text{ W} \qquad (2\text{–}11)$$

$$P_{R3} = R_3I^2 = 2 \times (2)^2 = 8 \text{ W} \qquad (2\text{–}12)$$

The current is flowing out of the positive terminal of the 24-V source, so that source is *delivering* a power P_{24} given by

$$P_{24} = 24I = 24 \times 2 = 48 \text{ W} \qquad (2\text{–}13)$$

However, the current is flowing into the positive terminal of the 6-V source, so that source is *absorbing* a power P_6 given by

$$P_6 = 6I = 6 \times 2 = 12 \text{ W} \qquad (2\text{--}14)$$

A tabulation of the various values of power absorbed and delivered is useful and is given below. Note that "nature requires a balance," that is, the total power absorbed must equal the total power delivered.

	Power Absorbed	Power Delivered
RESISTORS		
4 Ω	16 W	
3 Ω	12 W	
2 Ω	8 W	
SOURCES		
24 V		48 W
6 V	12 W	
TOTALS	48 W	48 W

EXAMPLE 2–2

For the dc resistive circuit of Figure 2–4(a), determine **(a)** the loop current, **(b)** voltage drops across all resistors, **(c)** voltage across current source, and **(d)** values of power associated with each element and whether the power is absorbed or delivered.

Solution

(a) The circuit, with appropriate labels for various quantities, is shown in Figure 2–4(b). Note that a voltage V_0 is indicated across the current source.

The current I is simply the value of the current source and is

$$I = 3 \text{ A} \qquad (2\text{--}15)$$

(b) The voltage drops across the three resistors are

$$V_{R1} = R_1 I = 2 \times 3 = 6 \text{ V} \qquad (2\text{--}16)$$
$$V_{R2} = R_2 I = 4 \times 3 = 12 \text{ V} \qquad (2\text{--}17)$$
$$V_{R3} = R_3 I = 6 \times 3 = 18 \text{ V} \qquad (2\text{--}18)$$

(c) To determine the voltage V_0, KVL is applied to the loop with V_0 retained as the unknown. Starting at the negative terminal of V_0 and forming a clockwise loop, we have

$$-V_0 + V_{R1} + 4 + V_{R2} - 30 + V_{R3} = 0 \qquad (2\text{--}19)$$

which leads to

$$V_0 = V_{R1} + V_{R2} + V_{R3} - 26 \qquad (2\text{--}20a)$$
$$= 6 + 12 + 18 - 26 = 10 \text{ V} \qquad (2\text{--}20b)$$

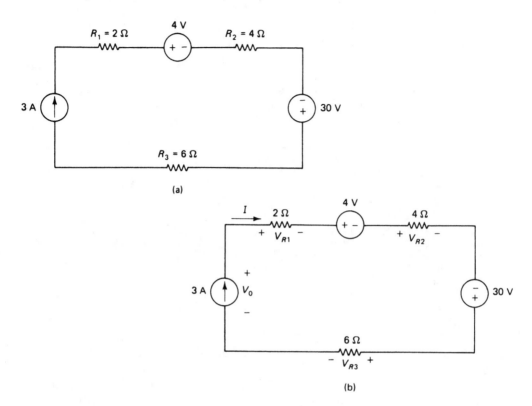

FIGURE 2–4
Circuit for Example 2–2.

(d) Let P_{R1}, P_{R2}, and P_{R3} represent the three values of absorbed power for the three resistors, respectively. These values are

$$P_{R1} = R_1 I^2 = 2 \times (3)^2 = 18 \text{ W} \tag{2-21}$$

$$P_{R2} = R_2 I^2 = 4 \times (3)^2 = 36 \text{ W} \tag{2-22}$$

$$P_{R3} = R_3 I^2 = 6 \times (3)^2 = 54 \text{ W} \tag{2-23}$$

Since the value of V_0 in **(c)** was positive, and since the more positive terminal of the current source was assumed to be the terminal at which current was leaving, the current source is *delivering* power. Let P_3 represent this power, which is

$$P_3 = 3V_0 = 3 \times 10 = 30 \text{ W delivered} \tag{2-24}$$

The 30-V source also delivers power, since current will be flowing out of its more positive terminal. However, the 4-V source will absorb power, since current is flowing into its more positive terminal. Let P_{30} and P_4 represent these two powers, which are

$$P_{30} = 30 \times 3 = 90 \text{ W delivered} \qquad \textbf{(2–25)}$$

and
$$P_4 = 4 \times 3 = 12 \text{ W absorbed} \qquad \textbf{(2–26)}$$

A tabulation of the various power values is given below.

	Power Absorbed	Power Delivered
RESISTORS		
2 Ω	18 W	
4 Ω	36 W	
6 Ω	54 W	
SOURCES		
3 A		30 W
4 V	12 W	
30 V		90 W
TOTALS	120 W	120 W

2–2 SINGLE NODE-PAIR CIRCUIT

The second special circuit that will be analyzed is the single node-pair circuit, whose basic form is illustrated in Figure 2–5. Any number of passive components and current sources may be connected between the pair of nodes. However, only one ideal voltage source is allowed, as will be demonstrated later. All elements in the single node-pair circuit are said to be connected in *parallel*.

For analysis purposes, the single node-pair dc resistive circuit will be considered in two categories: (a) current sources and passive elements only (i.e., no voltage source present) and (b) current sources, passive elements, and one voltage source. Each case will be considered separately.

No Voltage Source Present

Consider first the case when no voltage source is present. The development here will be restricted to resistive circuits, so assume that an arbitrary number of current

FIGURE 2–5
Basic form of the single node-pair circuit.

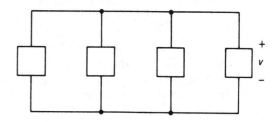

sources and resistors is connected between the pair of nodes. The current sources may or may not be ideal, but any source internal resistances are represented as additional parallel resistors between the pair of nodes.

The procedure for analyzing the single node-pair circuit under these conditions is based on the fundamental property that the *voltage across all elements in a single node-pair circuit is the same.* The currents through all resistors may be expressed in terms of the single voltage variable, with the sign convention of Chapter 1 carefully noted, that is, the current flows into the terminal of a resistor at which the potential is assumed to be the more positive. Kirchhoff's current law may then be applied to one of the nodes (usually the assumed positive node), and the algebraic sum of the currents leaving that node is equated to zero. As long as there is no more than one unknown quantity in the node-pair equation, it may be determined from the equation.

A common situation in resistive circuits is that the current sources and resistances are known, but the voltage is unknown and must be determined. In this case, an unknown voltage v is assumed across the node-pair. With some intuition, it is often possible to predict which terminal will be more positive, and it simplifies the analysis when the assumed polarity is equal to the actual polarity. However, it is immaterial whether the assumed direction is correct or not *provided* that all currents through resistors are expressed to be compatible with the assumed voltage polarity.

If the voltage turns out to be a negative value or function, the polarity of the actual voltage is opposite to the assumed polarity. At this point, one of two possible strategies may be taken. The first strategy is to define a new voltage (say, v') which has the same magnitude as v but is opposite in polarity, and thus positive (i.e., $v' = -v$). All currents through resistors are then reversed from the original assumed directions. The second strategy is to retain v as a negative quantity or function and keep the same sign conventions as before. As long as all calculations deal with v as a negative quantity, the results will still be mathematically correct. Although both strategies are perfectly correct, the second approach is more prone to sign errors in subsequent calculations.

Example

Formulation of a single node-pair equation will be illustrated by the circuit shown in Figure 2–6. The circuit contains three current sources and three resistors. All elements in this illustration are labeled with symbols, so no particular quantities will be assumed to be known or unknown. The goal is to illustrate the formulation of a KCL equation at one of the nodes.

The voltage across the node pair is labeled on the right, but this voltage is the same across all the elements because they are all connected in parallel. Observe that the current through each resistor is assumed to be flowing into the positive terminal. However, the directions of the current sources are a function of the manner in which the sources are connected and have nothing to do with the assumed voltage polarity.

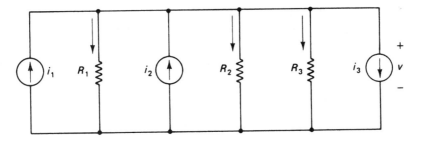

FIGURE 2–6
Representative single node-pair circuit containing current sources and resistances.

With the circuit properly labeled, we select one of the nodes at which to write a KCL equation. It is usually easier to work with the node at which the positive terminal of v is assumed, and thus the upper node in Figure 2–6 will be used. (Incidentally, either node produces the same result, so only one independent equation is possible for a node pair.)

Sign Convention

The sign convention employed here and in most places in the book is as follows: Currents leaving a node are expressed with a positive sign, and currents entering are expressed with a negative sign.

For the circuit of Figure 2–6, an expression of KCL for the upper node is as follows:

$$-i_1 + \frac{v}{R_1} - i_2 + \frac{v}{R_2} + \frac{v}{R_3} + i_3 = 0 \qquad (2\text{--}27)$$

The reader should carefully study the sign pattern of this equation and relate it to the convention discussed.

Simplified Pattern

An interesting pattern for the single node-pair equation results when v is considered as an unknown and an expression for it is determined in terms of the other parameters. The reader is invited to verify that the result is

$$v = \frac{i_1 + i_2 - i_3}{\dfrac{1}{R_1} + \dfrac{1}{R_2} + \dfrac{1}{R_3}} \qquad (2\text{--}28a)$$

$$= R_{eq}(i_1 + i_2 - i_3) \qquad (2\text{--}28b)$$

where R_{eq} is defined as the equivalent parallel resistance of the three resistances, as determined from the equation

$$\frac{1}{R_{eq}} = \frac{1}{R_1} + \frac{1}{R_2} + \frac{1}{R_3}$$

(2–29)

This result can be generalized to the form

$$v = \left(\begin{matrix}\text{net node-pair} \\ \text{parallel resistance}\end{matrix}\right) \times \left(\begin{matrix}\text{algebraic sum of} \\ \text{current source values}\end{matrix}\right)$$

(2–30)

In the interpretation of Equation (2–30), current sources are positive when they contribute to the assumed positive terminal of v, and they are negative when they contribute to the opposite polarity of v. For this form to be valid, *all the current sources must be independent sources.*

The reader may prefer a shortcut approach such as Equation (2–30) for a problem such as this. However, caution must be exercised with this approach because of possible sign errors. Further, this approach is not valid when there is one or more *dependent current sources* across the node pair. Therefore, it is recommended that the more fundamental approach of summing currents be retained as the basic method for analyzing a single node-pair circuit and that any shortcut procedures be used cautiously for applicable circuits.

Single Voltage Source Present

Next, we will consider the case where a single ideal voltage source appears across the node pair. *The presence of an ideal voltage source across a pair of nodes forces the voltage across all elements to be the same as that of the source.* Thus, the voltage is immediately evident if the value of the voltage source is known. Any number of ideal current sources may be connected across the node pair, but they do not affect the voltage. The various currents through resistors may be determined directly from the known voltage and the resistances.

One quantity that requires some computation to determine is the current through the voltage source. An unknown current (say, i_0) is assumed through the voltage source. By application of KCL to a node, i_0 can be determined. The direction may turn out to be either out of or into the positive terminal of the voltage source, depending on the nature of any current sources across the pair of nodes.

Thus far, we have assumed only one ideal voltage source. The assumption of two or more ideal voltage sources with different voltages in parallel results in a contradictory situation and is impossible. In a real-life situation where two or more voltage sources with different voltages are connected in parallel, the actual models of the voltage sources containing internal resistances would have to be employed to predict the results, and this would require a more complex model than the one under consideration. If two or more ideal voltage sources with the same voltage are connected in parallel, the result is mathematically equivalent to a single voltage source, as far as formulation of the circuit equation is concerned. Thus, we need only consider the presence of a single ideal voltage source in the single node-pair discussion.

EXAMPLE 2-3

For the dc resistive circuit of Figure 2–7(a), determine **(a)** the node-pair voltage, **(b)** currents through all resistors, and **(c)** values of power associated with each element and whether the power is absorbed or delivered.

Solution

The circuit as given initially has no labels for the voltage and current unknowns desired in order that such labels may be determined as part of the analysis. First, intuition tells us that the more positive terminal of the voltage is at the top, since the 10-A and 6-A current sources will have more effect than the single 4-A source. With this direction of V chosen, the directions of the three resistive branch currents are then determined in accordance with the standard sign convention as illustrated in Figure 2–7(b). For convenience, each of these currents is labeled with appropriate defining subscripts.

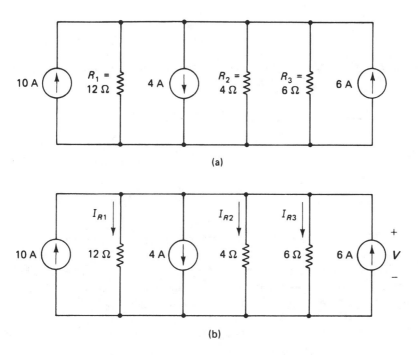

(a)

(b)

FIGURE 2–7
Circuit for Example 2–3.

(a) A KCL equation can be written at the upper node in accordance with the procedure discussed earlier, and it is

$$-10 + \frac{V}{12} + 4 + \frac{V}{4} + \frac{V}{6} - 6 = 0 \qquad \text{(2-31)}$$

This simplifies to

$$\frac{V}{2} = 12 \qquad \text{(2-32)}$$

or
$$V = 24 \text{ V} \qquad \text{(2-33)}$$

(b) Once the voltage is known, the currents through the three resistors are determined as

$$I_{R1} = \frac{V}{R_1} = \frac{24}{12} = 2 \text{ A} \qquad \text{(2-34)}$$

$$I_{R2} = \frac{V}{R_2} = \frac{24}{4} = 6 \text{ A} \qquad \text{(2-35)}$$

$$I_{R3} = \frac{V}{R_3} = \frac{24}{6} = 4 \text{ A} \qquad \text{(2-36)}$$

(c) Let P_{R1}, P_{R2}, and P_{R3} represent the three values of power absorbed for R_1, R_2, and R_3, respectively. We have

$$P_{R1} = \frac{V^2}{R_1} = \frac{(24)^2}{12} = 48 \text{ W} \qquad \text{(2-37)}$$

$$P_{R2} = \frac{V^2}{R_2} = \frac{(24)^2}{4} = 144 \text{ W} \qquad \text{(2-38)}$$

$$P_{R3} = \frac{V^2}{R_3} = \frac{(24)^2}{6} = 96 \text{ W} \qquad \text{(2-39)}$$

Let P_{10}, P_4, and P_6 represent the values of power associated with the 10-A, 4-A, and 6-A current sources, respectively. The 10-A and 6-A sources deliver power, since the current is flowing out of the positive terminal in each case. However, the 4-A source is absorbing power, since current is flowing into its more positive terminal. The three values of power are

$$P_{10} = 24 \times 10 = 240 \text{ W delivered} \qquad \text{(2-40)}$$
$$P_6 = 24 \times 6 = 144 \text{ W delivered} \qquad \text{(2-41)}$$
$$P_4 = 24 \times 4 = 96 \text{ W absorbed} \qquad \text{(2-42)}$$

A tabulation of the various power values is given below.

	Power Absorbed	Power Delivered
RESISTORS		
12 Ω	48 W	
4 Ω	144 W	
6 Ω	96 W	
SOURCES		
10 A		240 W
4 A	96 W	
6 A		144 W
TOTALS	384 W	384 W

EXAMPLE 2–4

For the dc resistive circuit of Figure 2–8(a), determine **(a)** the node-pair voltage, **(b)** currents through all resistors, **(c)** current through voltage source, and **(d)** values of power associated with each element and whether the power is absorbed or delivered.

Solution

(a) The circuit with appropriate labels for various quantities is shown in Figure 2–8(b). Note that a current I_0 is shown through the voltage source.

The voltage V is simply the value of the voltage source and is

$$V = 20 \text{ V} \qquad (2\text{–}43)$$

(b) The currents through the three resistors are

$$I_{R1} = \frac{V}{R_1} = \frac{20}{2} = 10 \text{ A} \qquad (2\text{–}44)$$

$$I_{R2} = \frac{V}{R_2} = \frac{20}{4} = 5 \text{ A} \qquad (2\text{–}45)$$

$$I_{R3} = \frac{V}{R_3} = \frac{20}{5} = 4 \text{ A} \qquad (2\text{–}46)$$

(c) To determine the current I_0, KCL is applied to the upper node, which yields

$$-I_0 + I_{R1} - 7 + I_{R2} + I_{R3} = 0 \qquad (2\text{–}47)$$

This leads to

$$I_0 = I_{R1} + I_{R2} + I_{R3} - 7 \qquad (2\text{–}48a)$$

$$= 10 + 5 + 4 - 7 = 12 \text{ A} \qquad (2\text{–}48b)$$

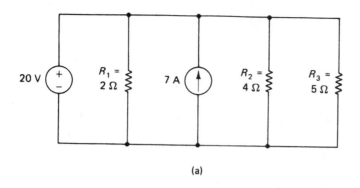

(a)

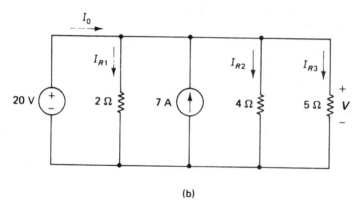

(b)

FIGURE 2–8
Circuit for Example 2–4.

(d) Let P_{R1}, P_{R2}, and P_{R3} represent the three values of absorbed power for the three resistors, respectively. These values are

$$P_{R1} = \frac{V^2}{R_1} = \frac{(20)^2}{2} = 200 \text{ W} \tag{2-49}$$

$$P_{R2} = \frac{V^2}{R_2} = \frac{(20)^2}{4} = 100 \text{ W} \tag{2-50}$$

$$P_{R3} = \frac{V^2}{R_3} = \frac{(20)^2}{5} = 80 \text{ W} \tag{2-51}$$

Let P_{20} and P_7 represent the values of power associated with the 20-V and 7-A sources, respectively. Both sources are delivering power, since the current is flowing out of the more positive terminal in each case. The values of power are:

$$P_{20} = 20 \times 12 = 240 \text{ W delivered} \qquad (2\text{--}52)$$

$$P_7 = 20 \times 7 = 140 \text{ W delivered} \qquad (2\text{--}53)$$

A tabulation of these power values is given below.

	Power Absorbed	Power Delivered
RESISTORS		
2 Ω	200 W	
4 Ω	100 W	
5 Ω	80 W	
SOURCES		
20 V		240 W
7 A		140 W
TOTALS	380 W	380 W

2–3 SOURCE TRANSFORMATIONS

When the effects of inductance and capacitance are negligible, an actual electrical source of energy can be modeled by either an ideal *voltage* source in series with a resistance or an ideal *current* source in parallel with a resistance. As long as the source internal resistance is finite and nonzero, either form may be used to represent a real-life source. Some devices tend to be associated with one particular form. For example, it seems more natural to represent a battery by the voltage source model, and it seems more natural to represent the collector circuit of a transistor by the current source model, but either model could theoretically be used for either case.

As we progress through different methods of circuit analysis, it will be observed that some methods are easier to apply if all sources are voltage sources and other methods are easier to apply if the sources are all current sources. For that reason, the source transformations to be given here are useful in allowing conversion between the types of sources.

Voltage-to-Current Source Conversion

First, consider the voltage source having an instantaneous voltage $v_s(t)$ in series with R_s, shown on the left in Figure 2–9. As far as the external terminals of the source are concerned, the circuit may be replaced by a current source of value $v_s(t)/R_s$ in parallel with a resistance R_s, as shown on the right. Said differently, the

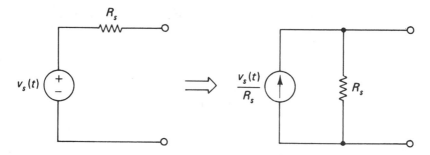

FIGURE 2-9
Transformation of voltage source to current source.

current source model on the right will produce the same *external* effect as the voltage source model on the left.

At the risk of redundancy, it must be stressed that the two circuits are equivalent only as far as effects on the external circuit are concerned. Any internal calculations made in the transformed equivalent circuit may be meaningless.

Current-to-Voltage Source Conversion

Next, consider the current source having an instantaneous current $i_s(t)$ in parallel with R_s, as shown on the left in Figure 2–10. As far as the external terminals of the source are concerned, the circuit may be replaced by a voltage source of value $R_s i_s(t)$ in series with a resistance R_s, as shown on the right. As in the previous case, the new circuit predicts the same external behavior as the circuit it replaces.

To convert a voltage source model to a current source model, a *series* resistance must be present with the voltage source model; and to convert a current source model to a voltage source model, a *parallel* resistance must be present with the current source model. This means that if a source is assumed to be ideal and is standing alone (i.e., not connected to a circuit), it is not possible to perform a

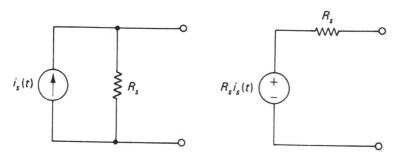

FIGURE 2-10
Transformation of current source to voltage source.

conversion. However, by associating an assumed ideal source with external circuit components, it may still be possible to perform a source conversion if some care is used in interpreting the results.

Readers familiar with Thevenin's and Norton's theorems will recognize that the transformations given here may be considered as special cases of these theorems and may be interpreted as such. A general development of Thevenin's and Norton's theorems will be given in Section 2–7.

EXAMPLE 2–5

In the dc resistive circuit of Figure 2–11(a), it is desired to determine the voltage V_0. Show that the use of a source transformation reduces the circuit to a simplified form in which the desired voltage can be determined with one equation, and then determine V_0.

Solution

The series combination of the 24-V voltage source and the 12-Ω resistor can be converted to a current source of value 24 V/12 Ω = 2 A in parallel with a 12-Ω resistor. When this is performed, the resulting circuit reduces to the form shown

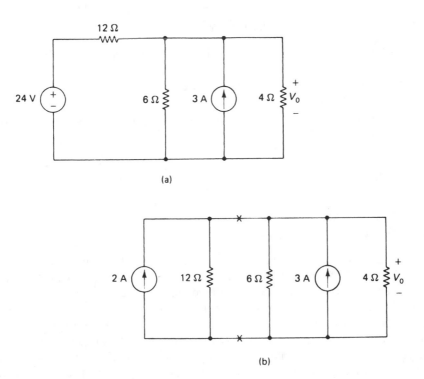

(a)

(b)

FIGURE 2–11
Circuit for Example 2–5.

in Figure 2–11(b). Thus, as far as the desired voltage V_0 is concerned, the circuit is a single node-pair form.

Formulation of a KCL equation at the upper node yields

$$-2 + \frac{V_0}{12} + \frac{V_0}{6} - 3 + \frac{V_0}{4} = 0 \qquad \text{(2–54)}$$

which results in

$$V_0 = 10 \text{ V} \qquad \text{(2–55)}$$

An interesting problem, which will be left as an exercise for the reader (Problem 2–11), is to calculate the total power delivered by the 2-A transformed source and compare it with the total power delivered by the actual 24-V source. The two results are different, which illustrates that only computations *external* to the terminals of the transformed circuit are generally valid.

2–4 REALISTIC SOURCE APPROXIMATIONS

Both ideal and nonideal source models have been considered in the text. In this section, the following three questions will be investigated.

1. Under what conditions can a real source be represented by the ideal voltage source model?
2. Under what conditions can a real source be represented by the ideal current source model?
3. Under what conditions must the internal source resistance be retained in the model?

These questions will be answered by considering two particular circuit conditions, which will be denoted as the low internal resistance case and the high internal resistance case.

Low Internal Resistance

The *low internal resistance* case represents the situation in which the source resistance is small compared with the external load resistance connected to it. It is convenient in this case to use the voltage source model as shown on the left in Figure 2–12(a). The resistance R_s represents the source resistance, and R_L represents the *net* load resistance. Since $R_s \ll R_L$, the voltage drop across R_s is very small compared with the drop across R_L. As a good approximation under many conditions, the circuit may be represented by the model shown on the right. The extent of the approximation depends on the degree of the inequality. As one particular criterion, if $R_s \leq 0.01R_L$, the voltage or current computed in any external resistance will differ by no more than 1% from the true value.

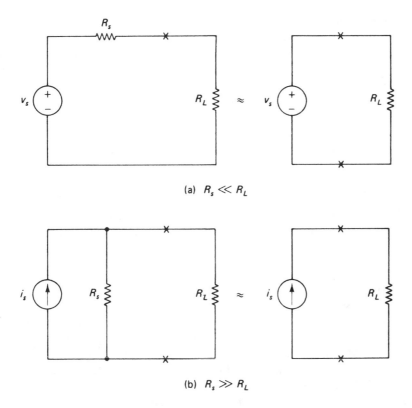

(a) $R_s \ll R_L$

(b) $R_s \gg R_L$

FIGURE 2–12
Realistic source approximations for very low and very high internal resistances.

Some conclusions from this discussion are that (1) the voltage source model is more convenient when the internal source resistance is small compared with the load resistance; (2) if the internal source resistance is sufficiently small, *it may be neglected altogether, and the ideal voltage source model may be used.*

High Internal Resistance

The *high internal resistance* case represents the situation in which the source resistance is large compared with the external load resistance connected to it. It is convenient in this case to use the current source model as shown on the left in Figure 2–12(b). The resistance R_s represents the source resistance, and R_L represents the *net* load resistance. Since $R_s \gg R_L$, the current through R_s is very small compared with the current through R_L. As a good approximation under many conditions, the circuit may be represented by the model shown on the right. The extent of the approximation depends on the degree of the inequality. As one particular criterion, if $R_s \geq 100R_L$, the voltage or current computed in any external resistance will differ by no more than 1% from the true value.

Some conclusions from this discussion are that (1) the current source model is more convenient when the internal source resistance is large compared with the load resistance; (2) if the internal source resistance is sufficiently large, *it may be neglected altogether and the ideal current source model may be used.*

It should be noted that neglecting the internal resistance in the ideal current source model means replacing it with an *open circuit.* Thus, *the internal resistance of an ideal current source is infinite.* This is in sharp contrast to an *ideal voltage source,* in which *the internal resistance is zero.*

When the internal source resistance and the load resistance do not satisfy inequalities of the types just discussed, the internal source resistances must be retained in the model, and the choice of the voltage source in series with a resistance model versus the current source in parallel with a resistance model will depend on a number of factors, particularly the method of analysis employed. The choice will become clearer when we consider mesh and node analysis in the next two sections.

EXAMPLE 2–6

This problem has three objectives: (1) to illustrate the exact equivalence of the models obtained from source transformations, (2) to illustrate why the voltage source model is more convenient for the low-resistance case, and (3) to illustrate that the ideal voltage approximation may be valid for the low-resistance case.

Consider the equivalent circuit of a certain battery shown in Figure 2–13(a) consisting of a 12-V voltage source in series with a 0.1-Ω resistance. **(a)** Assume that the 10-Ω load shown on the right is connected. Calculate the exact current I_L through the load. **(b)** Convert the battery to a current source model and calculate I_L again. **(c)** Using the voltage source model, neglect the internal battery resistance, and calculate the resulting load current I'_L.

Solution

(a) With the 10-Ω load connected and using the voltage source model, the resulting circuit is shown in Figure 2–13(b). The current I_L in this simple circuit is readily determined to be

$$I_L = \frac{12}{10.1} = 1.188 \text{ A} \tag{2–56}$$

(b) Application of the pertinent source transformation of Section 2–3 to the voltage source results in a current source of value 12 V/0.1 Ω = 120 A in parallel with a 0.1-Ω resistor as shown to the left of the terminals in Figure 2–13(c). The current I_L can be determined in this circuit by an application of the current divider rule. We have

$$I_L = \frac{0.1}{0.1 + 10} \times 120 = 1.188 \text{ A} \tag{2–57}$$

which illustrates that both forms produce identical results.

To illustrate how "clumsy" the current source model is in this case, con-

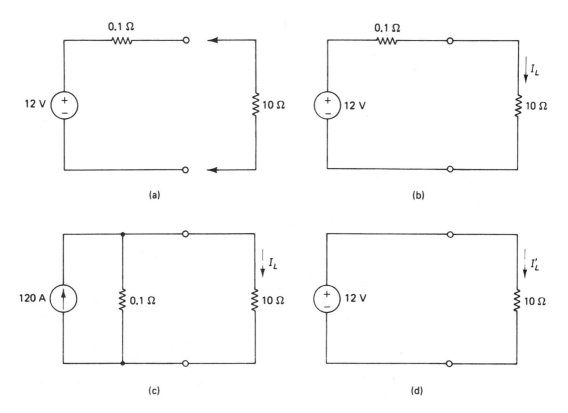

FIGURE 2–13
Circuit models for Example 2–6.

sider how the model would change if the low internal resistance changed from 0.1 Ω to 0.02 Ω. The resulting current source would change from 120 A to 600 A. However, the load current would change only slightly, since the much lower internal resistance would shunt most of the increased current from the source. Thus, the current source model would predict correct terminal results, but working with the very large current and the very low internal resistance is more awkward than with the voltage source model in this case. The opposite situation is true when the internal resistance is very large (see Problem 2–12).

(c) When the internal resistance of the voltage source model is neglected, the equivalent circuit is shown in Figure 2–13(d). The current I'_L is

$$I'_L = \frac{12}{10} = 1.2 \text{ A} \tag{2–58}$$

A comparison of I'_L with I_L indicates that the difference is approximately 1% of the true current. The load resistance of 10 Ω is exactly 100 times the internal resistance of 0.1 Ω, so this approximate level of error is expected.

2-5   MESH CURRENT ANALYSIS

Two of the most widely employed methods for analyzing complex electric circuits are the *mesh current method* and the *node voltage method*. Both methods are rather general in that, in theory, most common circuits can be analyzed by either method. However, many circuits tend to be analyzed more easily with one method than the other, so it is important to establish proficiency in both techniques. Mesh current analysis will be considered in this section, and node voltage analysis will be considered in the next section.

Mesh Currents

To develop the technique of mesh current analysis, consider the network structure illustrated in Figure 2–14. Each branch could contain an arbitrary number of series components and sources, so only the general form is shown. Each of the "openings" represents a *mesh*. A circulating current variable is assigned to each of the meshes. Although the assumed direction is arbitrary, we will follow the fairly common practice of assuming the clockwise direction, since it simplifies the procedure somewhat. Kirchhoff's voltage law is then applied to each mesh by summing voltages around the loop surrounding the mesh. If there are n meshes, the n KVL equations constitute a set of independent simultaneous equations and the n mesh current variables may be determined. Thus, the mesh current method represents a systematic means of expressing a set of KVL equations that may be used to determine currents in the circuit. Once the mesh currents are known, all other circuit variables may be determined by simple one- or two-step manipulations.

Mesh Currents vs. Branch Currents

A few points of explanation about the nature of mesh currents are in order. A given mesh current may or may not represent an actual current that would be measured in the circuit. To illustrate this point, we must distinguish between *mesh* currents and *branch* currents.

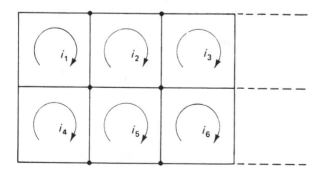

FIGURE 2–14
Electrical circuit structure for mesh current analysis.

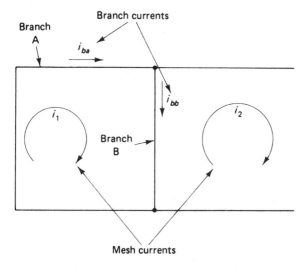

FIGURE 2–15
Comparison of branch and mesh currents.

Consider, for example, two segments of the circuit of Figure 2–14, which are delineated in Figure 2–15. The upper left-hand part of the circuit is designated as Branch A, and its branch current is denoted as i_{ba}. Since there is only one mesh current passing through this branch, clearly $i_{ba} = i_1$. Next, consider the branch designated as Branch B, and let i_{bb} represent its branch current. By a direct application of KCL to the upper junction point, it is noted that $i_{bb} = i_1 - i_2$, which is the current that an ammeter connected in Branch B would measure. Thus, the actual branch current is a linear combination of two mesh currents.

From the preceding discussion, we note that mesh currents in *outer* branches represent true *branch currents,* but mesh currents in *inner* branches must be combined appropriately to determine true branch currents.

Interpreting Direction of Current

In summing voltage drops across resistors, a given voltage drop is expressed as the product of the branch current and the resistance. The branch current can be interpreted as flowing in either direction, but it is usually easier to interpret the branch current as flowing in the direction in which one is "moving" around the loop. To illustrate this point, consider the branch segments shown in Figure 2–16. Assume in (a) that voltages around mesh 1 are being summed in a clockwise fashion, with movement through the branch being downward. The net branch current is $i_1 - i_2$, with the direction being assumed to be the same as for the mesh current i_1. If voltage drops are expressed as positive and rises as negative, the contribution of this branch in the equation of mesh 1 is $+R(i_1 - i_2)$.

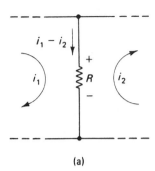 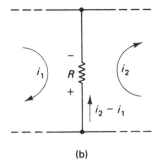

(a)   (b)

FIGURE 2–16
Two interpretations of branch current direction and voltage drop polarity in mesh current analysis.

Next, suppose that the polarity and direction labels of Figure 2–16(a) are retained while the equation of mesh 2 is being formed. In this case, a clockwise path means that movement through the branch is *upward*. With the labels of (a), an apparent voltage *rise* is encountered, and the contribution of this branch in the equation of mesh 2 is $-R(i_1 - i_2)$, which is the negative of the result in mesh 1.

Instead of retaining the polarities and directions of (a), suppose that the branch is interpreted as shown in Figure 2–16(b). This model shows the current going upward, that is, in the same direction as i_2. The value, however, is $i_2 - i_1$. (A current i_x in one direction is equivalent to a current $-i_x$ in the opposite direction.) With the new assumed direction, the positive terminal of the resistive voltage drop must be interpreted as the bottom terminal. In moving upward through the resistor, the contribution will now be expressed as $+R(i_2 - i_1)$. Comparing this value with that of the preceding paragraph, it is readily noted that $+R(i_2 - i_1) = -R(i_1 - i_2)$, meaning that both interpretations produce identical results! In short, it doesn't make any difference which direction is assumed for the unknown branch current, provided that the voltage drop across the resistor and the sign convention are used properly in the equation.

Recommended Assumption

To provide a systematic procedure for the preceding situation, it is recommended that the branch current be assumed in the same direction as the movement through the branch. With this approach, the current for the given mesh will always be the first term in the voltage drop, and the "bookkeeping" will be much easier to manage. Thus, Figure 2–16(a) represents the best interpretation for writing the mesh equation in mesh 1, and Figure 2–16(b) represents the corresponding best interpretation for mesh 2. With some practice, it will not be necessary to label the branch currents and voltage drops as was done for illustration, since frequent erasures would be required.

Because voltages are being summed around the loops, voltage sources contained within a loop are readily incorporated in the mesh equations. Of course, there is no arbitrariness about the polarity of a voltage source, since the manner in which it is connected readily identifies the sign.

Converting Current Sources to Voltage Sources

When a circuit contains a current source in parallel with a resistor, the best way to deal with it in performing mesh analysis is to use a source transformation and convert it to a voltage source in series with a resistor. Although that particular part of the circuit will be modified, after the remainder of the circuit is analyzed, the original circuit can be reconstructed to complete the solution.

Occasionally, a circuit will contain an assumed ideal current source with no simple external parallel resistor, thus eliminating the possibility of a source conversion. There are techniques for applying mesh analysis to this situation, but they are somewhat awkward to apply. Instead, it is recommended that the node voltage method, which will be discussed in the next section, be used.

EXAMPLE 2–7
For the dc resistive circuit of Figure 2–17(a), **(a)** write the three mesh current equations. **(b)** Arrange the results of (a) in standard mathematical form, and solve for the three mesh currents. **(c)** Determine all branch currents. **(d)** Determine values of power associated with each element and whether the power is absorbed or delivered.

Solution
(a) Four of the five active devices in the circuit are voltage sources, so they will be maintained in that form for analysis. However, the 7-A current source in parallel with 7 Ω on the right-hand side of the circuit will be converted to a 49-V voltage source in series with 7 Ω as shown in Figure 2–17(b). The resulting circuit has three meshes, so three clockwise dc mesh variables I_1, I_2, and I_3 are shown on this figure.

The mesh current equations are written by applying KVL to each of the meshes, with drops considered positive and rises negative. The branch currents through the common branches are expressed in terms of the differences between appropriate sets of mesh currents. The mesh numbers correspond to the current subscripts. The mesh current equations are as follows:

Mesh 1

$$-20 + 6I_1 + 5(I_1 - I_2) - 25 = 0 \qquad \textbf{(2–59)}$$

Mesh 2

$$25 + 5(I_2 - I_1) - 32 + 3I_2 + 2(I_2 - I_3) + 6 = 0 \qquad \textbf{(2–60)}$$

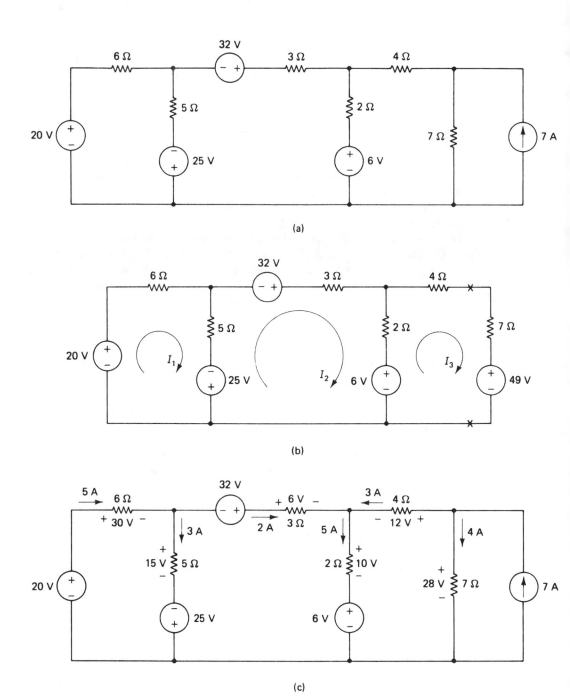

(a)

(b)

(c)

FIGURE 2–17
Circuit for Example 2–7, analyzed by the mesh current method.

Mesh 3

$$-6 + 2(I_3 - I_2) + 4I_3 + 7I_3 + 49 = 0 \qquad \textbf{(2-61)}$$

(b) The standard mathematical form for these simultaneous equations is obtained by combining all I_1, I_2, and I_3 factors together in each equation. All constants are placed on the right-hand sides of the equations. The reader may verify that the results are

$$11I_1 - 5I_2 \qquad = 45 \qquad \textbf{(2-62)}$$

$$-5I_1 + 10I_2 - 2I_3 = 1 \qquad \textbf{(2-63)}$$

$$-2I_2 + 13I_3 = -43 \qquad \textbf{(2-64)}$$

These simultaneous equations may be solved by any of the standard procedures, such as determinants or substitution. For readers requiring some help in this endeavor, Appendix A provides the details of one particular method for solving simultaneous equations.

By whatever method employed, the three currents are determined to be

$$I_1 = 5 \text{ A} \qquad \textbf{(2-65)}$$

$$I_2 = 2 \text{ A} \qquad \textbf{(2-66)}$$

$$I_3 = -3 \text{ A} \qquad \textbf{(2-67)}$$

The negative sign for I_3 indicates that it is actually counterclockwise.

TABLE 2–1
Tabulation of power delivered and absorbed for circuit of Example 2–7

	Power delivered	Power absorbed
SOURCES		
20 V	100 W	
25 V	75 W	
32 V	64 W	
6 V		30 W
7 A	196 W	
RESISTORS		
6 Ω		150 W
5 Ω		45 W
3 Ω		12 W
2 Ω		50 W
4 Ω		36 W
7 Ω		112 W
TOTALS	435 W	435 W

(c) The branch currents are determined by simply identifying the current in each branch as either a mesh current (outer branches) or as a linear combination of two mesh currents (inner branches). Voltages across all resistors can then be calculated using Ohm's law. To obtain information about the original current source and the associated resistor, it is necessary to restore that part of the circuit to the original form. KCL can then be used to determine the actual current through the 7-Ω resistor. All of these currents and voltages are identified on the circuit diagram of Figure 2–17(c). The reader is invited to verify how all of these values are obtained from the mesh currents.

(d) The values of power are readily calculated from all the preceding results, and the sense of power delivered or absorbed is determined from the sign pattern. The results are summarized in Table 2–1.

2–6 NODE VOLTAGE ANALYSIS

The second general circuit analysis technique is the *node voltage method.* To develop the concept of node voltage analysis, consider the network structure of Figure 2–18. Each branch could contain an arbitrary number of parallel components and sources.

Common or Ground Node

In node voltage analysis, one node must be identified as the *common* or *ground* node, and it is often labeled with the ground symbol for clarity. This node may or may not correspond to an actual ground in the corresponding real circuit, although it would certainly be a good choice in many cases. If the circuit contains $n + 1$ nodes, the choice of one ground node results in n additional nodes. A node voltage variable is assigned to each of these nodes, and this voltage is always the potential with respect to the ground node. Kirchhoff's current law is then applied at each of the n nodes having a node voltage variable by summing currents leaving and entering the node. The voltage variables are treated *as if* they are positive with respect to

FIGURE 2–18
Electrical circuit structure for node voltage analysis.

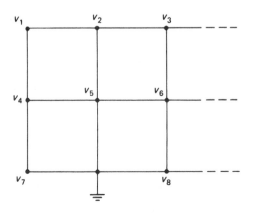

ground when writing the equations, although they may actually be either positive or negative. The n KCL equations constitute a set of independent simultaneous equations, and the n node voltage variables may be determined. Thus, the node voltage method represents a systematic means of expressing a set of KCL equations that may be used to determine voltages in the circuit. Once the node voltages are known, all other circuit variables may be determined by simple one- or two-step manipulations.

Interpreting Voltage Polarity

In summing currents leaving a node, a given current is expressed as the voltage difference across a resistance divided by the resistance. The voltage difference may be interpreted with either polarity, but it is usually easier to *interpret* the polarity at the node of immediate interest as the more positive. To illustrate this point, consider the node segments shown in Figure 2–19. Assume in (a) that currents leaving node 1 are being summed. The net voltage difference across R is $v_1 - v_2$, with the more positive terminal assumed at node 1. If current leaving is expressed as positive and current entering as negative, the contribution of current in this branch to the equation of node 1 is $(v_1 - v_2)/R$.

Next, suppose that the direction of current and polarity labels of Figure 2–19(a) are retained when the equation of node 2 is being formed. In this case, the current through R would be interpreted as *entering* node 2, and the contribution to the equation at node 2 would be expressed as $-(v_1 - v_2)/R$, which is the negative of the result at node 1.

Instead of retaining the direction and polarity of Figure 2–19(a), suppose that the node is interpreted as shown in (b). This model shows the more positive terminal of the voltage across R at node 2, and the value of this voltage difference is $v_2 - v_1$. With the new assumed polarity, the current must be assumed to flow from right to left, that is, out of node 2. The contribution of this current to the equation at node 2 is $(v_2 - v_1)/R$. Comparing this value with that of the preceding paragraph, it is readily noted that $(v_2 - v_1)/R = -(v_1 - v_2)/R$, meaning that both interpretations produce identical results! It doesn't make any difference which direction is assumed for the voltage difference, provided that the current through the resistor and the sign convention are used properly in the equation.

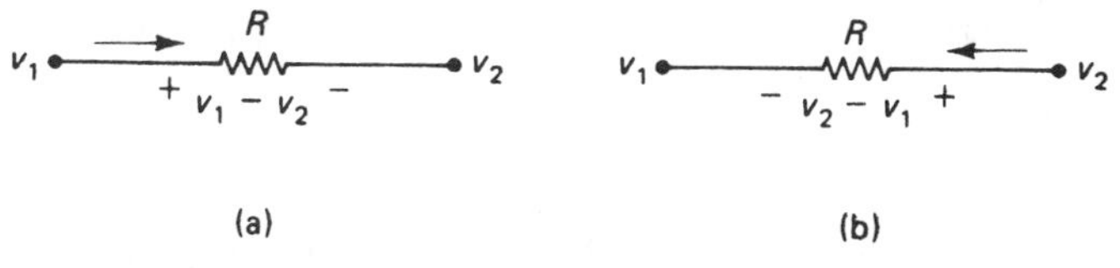

(a)   (b)

FIGURE 2–19
Two interpretations of voltage difference polarity and current direction in node voltage analysis.

Recommended Assumption

To provide a systematic procedure, it is recommended that the node voltage at which the node equation is being written be *interpreted* as the more positive one. With this approach, the current will be interpreted as leaving, and the voltage for that node will be the first one in each term. Thus, Figure 2–19(a) represents the best interpretation for writing the node equation at node 1, and Figure 2–19(b) represents the best interpretation for node 2. With some practice, it will not be necessary to label the voltage differences and currents as done for illustration, since frequent erasures would be required.

Because currents are being summed at nodes, current sources connected to nodes are readily incorporated in the node equations. Of course, there is no arbitrariness about the sign of a current source term, since the manner in which it is connected readily identifies the sign.

Converting Voltage Sources to Current Sources

When a circuit contains a voltage source in series with a resistor, one way to deal with it in performing node analysis is to use a source transformation and convert it to a current source in parallel with a resistor. Although the structure of that particular part of the circuit will be modified, after the remainder of the circuit is analyzed, the original circuit can be reconstructed to complete the solution.

Node with Fixed Voltage

If one side of a given voltage source is connected to the common ground terminal, the other side may be simply treated as a node voltage whose value is already known. This type of node is referred to by some authors as a "dummy variable." In this case, one need not write a node voltage equation at the node whose voltage is known. (Indeed, a conventional node equation cannot easily be written there because the current flowing through the voltage source is not simply expressible.) Rather, all references to the voltage at the given node are made by using the actual value with respect to ground.

Occasionally, a circuit will contain an assumed ideal ungrounded voltage source with no simple external series resistance, thus eliminating the possibility of using either a source transformation or the "dummy variable" concept. There are techniques for applying node analysis to this situation, but they are rather specialized and awkward to apply. Instead, it is recommended that mesh current analysis be used in such a case.

EXAMPLE 2–8

Consider the circuit of Example 2–7, for which a mesh current analysis was performed. **(a)** Using source transformations, convert the circuit to a form appropriate for node voltage analysis. **(b)** Write the node voltage equations. **(c)** Arrange the results of (b) in standard mathematical form and solve for the node voltages.

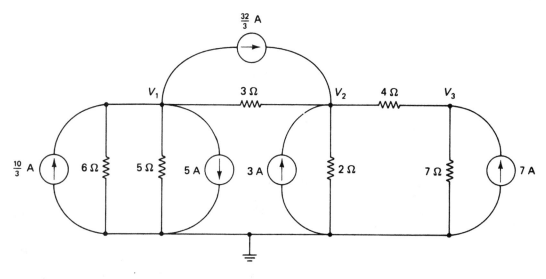

FIGURE 2–20
Circuit for Example 2–8.

Solution

(a) The original circuit was shown in Figure 2–17(a). For convenience in node voltage analysis, the four voltage sources in series with the four separate resistors are all converted to current sources in parallel with the corresonding resistors. However, the original current source is retained in that form. The resulting circuit is shown in Figure 2–20. Observe that the bottom node is established as the ground node, and three dc node voltage variables have been labeled at the other three nodes.

(b) The node voltage equations are obtained by an application of KCL at each of the three upper nodes. At each node, currents leaving are considered positive, and currents entering are considered negative. The code numbers correspond to the voltage subscripts. The node voltage equations are as follows:

Node 1

$$-\frac{10}{3} + \frac{V_1}{6} + \frac{V_1}{5} + 5 + \frac{(V_1 - V_2)}{3} + \frac{32}{3} = 0 \qquad \textbf{(2–68)}$$

Node 2

$$-\frac{32}{3} + \frac{(V_2 - V_1)}{3} - 3 + \frac{V_2}{2} + \frac{(V_2 - V_3)}{4} = 0 \qquad \textbf{(2–69)}$$

Node 3

$$\frac{V_3 - V_2}{4} + \frac{V_3}{7} - 7 = 0 \qquad \textbf{(2–70)}$$

(c) After grouping and combining coefficients in the preceding equations, the results are

$$\frac{7}{10}V_1 - \frac{1}{3}V_2 \qquad\qquad = -\frac{37}{3} \tag{2-71}$$

$$-\frac{1}{3}V_1 + \frac{13}{12}V_2 - \frac{1}{4}V_3 = \frac{41}{3} \tag{2-72}$$

$$-\frac{1}{4}V_2 + \frac{11}{28}V_3 = 7 \tag{2-73}$$

Simultaneous solution of these equations yields

$$V_1 = -10 \text{ V} \tag{2-74}$$
$$V_2 = 16 \text{ V} \tag{2-75}$$
$$V_3 = 28 \text{ V} \tag{2-76}$$

Referring back to Figure 2–17(c), some slight additional analysis on the circuit indicates that the voltages at these nodes are in agreement with those obtained with node analysis, as they should be. However, with the exception of the current source and its parallel resistance on the right and the 4-Ω resistive branch, very little additional information can be inferred from these results without restoring the original form of the circuit. The reason is that all other branches were transformed, and the internal forms of those branches are *not* the same as in the original circuit. Although both methods required three simultaneous equations, the fact that four sources were transformed to allow node analysis, while only one source was transformed for mesh analysis, indicates that mesh analysis was a better choice for this circuit if all branch information were desired. An exception might be if the only variables of interest were the voltages at the nodes, in which case node analysis would be most appropriate. In general, there is no particular "recipe" to determine whether mesh or node analysis is the best method in a particular problem. Whereas the number of voltage sources versus the number of current sources is one criterion, as was the case in the circuit of Examples 2–7 and 2–8, there are other important factors. For example, one method will often result in fewer equations than the other method. Experience plays an important role in the choice, and some persons tend to prefer one method more strongly than the other, even when the other method would be much easier. Our advice to the reader at this point is to learn both mesh and node analysis with equal mastery, because both are important in circuit analysis.

EXAMPLE 2–9

For the dc resistive circuit of Figure 2–21(a), **(a)** write a set of node voltage equations. **(b)** Arrange the results of (a) in standard mathematical form, and solve for the node voltages. **(c)** Determine all branch currents.

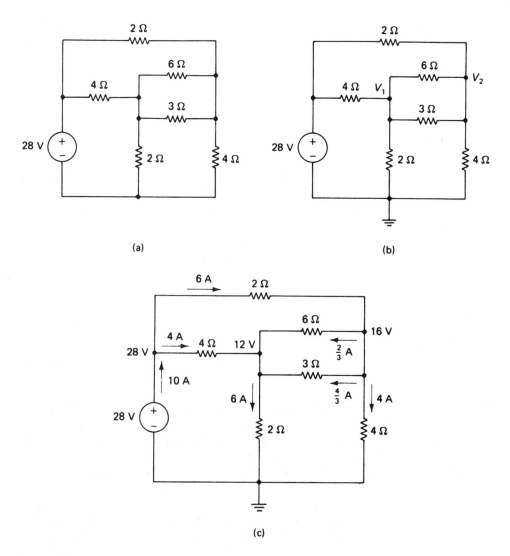

FIGURE 2-21
Circuit for Example 2-9.

Solution

(a) The circuit contains an ideal voltage source, and it is not in series with a simple resistance. Therefore, it cannot be converted to a current source. However, if the bottom node is chosen as the ground node, the potential at the positive terminal of the source will be known (28 V), and this node may be treated as a "dummy variable." Thus, there are only two nodes for which unknown voltages need to be assigned, and they are labeled as V_1 and V_2 in Figure 2-21(b).

Application of KCL to the two nodes labeled as V_1 and V_2 yields the following two equations:

Node 1

$$\frac{V_1 - 28}{4} + \frac{V_1}{2} + \frac{V_1 - V_2}{6} + \frac{V_1 - V_2}{3} = 0 \qquad \textbf{(2-77)}$$

Node 2

$$\frac{V_2 - 28}{2} + \frac{V_2 - V_1}{6} + \frac{V_2 - V_1}{3} + \frac{V_2}{4} = 0 \qquad \textbf{(2-78)}$$

(b) After combining appropriate terms and rearranging, the preceding equations reduce to

$$\frac{5}{4} V_1 - \frac{1}{2} V_2 = 7 \qquad \textbf{(2-79)}$$

$$-\frac{1}{2} V_1 + \frac{5}{4} V_2 = 14 \qquad \textbf{(2-80)}$$

Simultaneous solution of these equations yields

$$V_1 = 12 \text{ V} \qquad \textbf{(2-81)}$$
$$V_2 = 16 \text{ V} \qquad \textbf{(2-82)}$$

(c) All resistive branch currents are readily calculated by determining the difference in potential across each branch and applying Ohm's law. The current flowing through the 28-V source is determined by applying KCL to the node above its positive terminal after the two resistive branch currents connected to it have been determined. The node voltages and branch currents are labeled on Figure 2-21(c).

Incidentally, since there is only one source in this circuit, it must deliver all the power absorbed by the resistors. The value of this total power is 28 V × 10 A = 280 W. The reader is invited to verify that the sum of all the resistive power levels absorbed is equal to this value.

2-7 THEVENIN'S AND NORTON'S THEOREMS

Thevenin's and Norton's theorems are among the most useful of all circuit analysis techniques. They permit the reduction of complex circuits to very simple forms whenever the primary focus of interest is on the external behavior of the complex circuits. A brief statement of each of the theorems as applied to *resistive* circuits will first be made, and explorations of the procedures for determining the equivalent circuits will then be given.

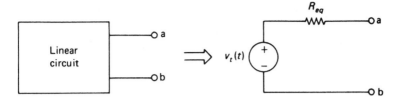

FIGURE 2–22
Thevenin model of an electrical circuit at a set of reference terminals.

Thevenin's Theorem

Thevenin's theorem, as applied to resistive circuits, states that all effects of any linear circuit external to two reference terminals can be completely predicted from a model consisting of a single ideal voltage source in series with a single resistor. The concept is illustrated in Figure 2–22. The voltage $v_t(t)$ will be referred to as the Thevenin voltage, and for reasons that will be seen to be logical shortly, R_{eq} will be referred to simply as the equivalent resistance.

Norton's Theorem

Norton's theorem, as applied to resistive circuits, states that all effects of any linear circuit external to two reference terminals can be completely predicted from a model consisting of a single current source in parallel with a single resistor. The concept is illustrated in Figure 2–23. The current $i_n(t)$ will be referred to as the Norton current, and R_{eq} is the equivalent resistance.

For a given circuit and a given set of terminals, R_{eq} will be the same for either the Thevenin or the Norton model, and for that reason, the terms "Thevenin resistance" and "Norton resistance," which are used in some books, will not be used here. Such terms could give the impression that the resistance is different for the two models, which is not the case.

The reader should recognize that if either the Thevenin or the Norton model is known, the other may be readily determined with one of the source transformations of Section 2–3. For example, if the Thevenin model is known, the Norton

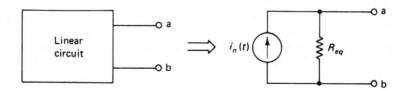

FIGURE 2–23
Norton model of an electrical circuit at a set of reference terminals.

current is readily determined as

$$i_n(t) = \frac{v_t(t)}{R_{eq}} \tag{2-83}$$

Further, if the Norton model is known, the Thevenin voltage is determined as

$$v_t(t) = R_{eq} i_n(t) \tag{2-84}$$

Because of the close similarity between the two forms, it is not necessary to treat them as separate methods; rather, they can be viewed as slightly different perspectives on the same general approach. In simplifying complex circuits, we will frequently use both theorems.

Using the Theorems

At the outset, a fundamental point concerning the external nature of those models must be stressed. When a complex circuit is replaced by a Thevenin or a Norton model, the model predicts correct results *external* to the reference terminals only. The internal action of the original circuit has been "lost," and any internal calculations in the Thevenin or Norton model are generally meaningless. For example, power delivered by the sources v_t or i_n generally is not equivalent to the power delivered by the source or sources in the original circuit. If any internal calculations are required, the original circuit must be reconstructed in the analysis process. This is a limitation frequently misunderstood by beginners in circuit analysis.

The Thevenin and Norton models are most useful when there is a portion of a circuit that remains fixed for which there is little interest in the internal behavior but in which the effect on an external circuit is to be studied under varying conditions. Rather than perform a complete circuit analysis for each load, the static part of the circuit is replaced by its Thevenin or Norton model, and the effect for each load condition is then more easily studied.

Practical Implications

A few comments about the practical nature of Thevenin's and Norton's theorems are in order. Although laboratory experiments to verify these theorems are widely performed in electrical circuits courses, their major utility is in "paper" analysis. Most real-life practical circuits are wired a certain way to perform a job, and occasionally the use of one of these models may suggest a better way to design the circuit. However, the major benefit of the theorems is in analyzing the circuits on paper and in simplifying the calculations. Thus, in using an expression such as "replacing a circuit with a Thevenin equivalent," or the like, we will be referring to the analysis of these circuits rather than their construction. Indeed, some of the operations may be totally impractical or unrealistic when viewed from a practical point of view. For example, reference will be made to determining the "short-circuit current" in certain steps that follow, and this may be totally impractical (or even dangerous) in a realistic situation.

Having established the concepts of Thevenin's and Norton's theorems, we now turn to the necessary task of determining the values of the Thevenin voltage, the Norton current, and the equivalent resistance. The problem is complicated by the fact that the easier methods to apply do not work in the most general cases, and the more general methods tend to be clumsy when applied to the simpler circuits. No doubt these facts contribute to the difficulty that many students apparently encounter with determining Thevenin's and Norton's equivalent circuits.

For the purpose of Thevenin and Norton simplification, it is convenient to divide circuits into two categories: circuits containing only independent sources and circuits containing one or more dependent sources. The second category presents some special difficulties in analysis, and their treatment will be deferred to Section 2–9. Thus, all developments in this section will be restricted to circuits containing only independent sources.

Determining a Thevenin Circuit with Only Independent Sources

A general approach for determining the Thevenin equivalent circuit for a resistive circuit when the circuit contains *only independent sources* is illustrated in Figure 2–24. The procedure consists of the following two steps:

1. Determine the open-circuit voltage v_{oc} across the two reference terminals. The value of this open-circuit voltage is the Thevenin voltage, that is,

$$v_t = v_{oc} \qquad \textbf{(2–85)}$$

2. De-energize all internal sources, and determine the equivalent resistance as viewed from the two reference terminals. This resistance is the equivalent resistance R_{eq}.

To de-energize sources as indicated in step (2), ideal *voltage sources* are replaced by *short circuits,* and ideal *current sources* are replaced by *open circuits.* Since nonideal sources contain resistances, these resistances will remain in the

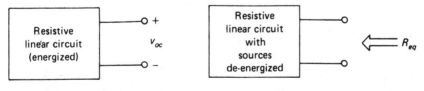

(a) Determine open-circuit voltage

(b) De-energize all internal sources and determine R_{eq}

FIGURE 2–24

General approach for determining Thevenin equivalent of resistive current when circuit contains no dependent sources.

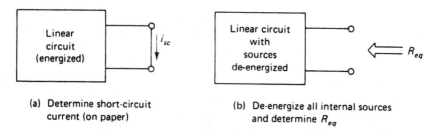

(a) Determine short-circuit current (on paper)

(b) De-energize all internal sources and determine R_{eq}

FIGURE 2–25
General approach for determining Norton equivalent of resistive circuit when circuit contains no dependent sources.

circuit models when the ideal source portions of the actual models are de-energized by the preceding steps.

Although any Thevenin model could be readily replaced by a Norton model by use of a circuit transformation, a separate statement of a general approach for determining the Norton model is needed for completeness. Furthermore, it is easier in many cases to determine the Norton model directly from the given circuit.

Determining a Norton Circuit with Only Independent Sources

A general approach for determining the Norton equivalent circuit for a resistive circuit when the circuit contains *only independent sources* is illustrated in Figure 2–25. The procedure consists of two steps:

1. Assume that a short circuit (on paper) is placed across the two reference terminals. Determine the current i_{sc} that would flow through the short. The value of this short-circuit current is the Norton current i_n, that is,

$$i_n = i_{sc} \qquad\qquad (2\text{–}86)$$

2. The second step is the same as for determining the Thevenin equivalent circuit, namely, de-energize all internal sources, and determine the equivalent resistance R_{eq} as viewed from the two reference terminals.

Relationship to Source Transformations

It can now be stated that the source transformations of Section 2–3 were nothing more than special cases of Thevenin and Norton equivalent circuits. Reviewing Figure 2–9, the value of the current source in the right-hand model represents the current that would flow through a short circuit connected across the terminals of the left-hand model. In Figure 2–10, the value of the voltage source in the right-hand model represents the open-circuit voltage that would be measured in the left-hand model. In all of these cases, the equivalent resistance, $R_{eq} = R_s$, is the resistance that would be viewed from the output terminals when the internal sources are de-energized according to the procedures delineated earlier.

Although these procedures may, in theory, be applied to any circuit containing only independent sources, they are often clumsy and inefficient. Consider, for example, a circuit containing a number of meshes and nodes for which a Thevenin model is desired at a set of terminals on one side of the circuit. To determine the open-circuit voltage or the short-circuit current, the solution of several simultaneous equations may be required. While this may be necessary if all else fails, a simpler step-by-step approach is often possible.

Simplifying Procedures

A large number of circuits that the "average" technical person will need to analyze in detail can be reduced to Thevenin or Norton form by successive application of the source transformations discussed in Section 2–3, coupled with combining resistances by series and parallel equivalents. Although this approach does not work in all cases, it works in enough important and useful cases to justify a major focus of our efforts in that direction. It is difficult to identify a simple "formula" for recognizing such a circuit, but with some practice, an intuitive process will quickly evolve. As a general guideline, if the circuit is one in which an equivalent resistance could be determined by successive series and parallel combinations if all sources were de-energized, the technique under consideration will usually work. However, for bridge-type structures, which are not usually amenable to series and parallel combinations, the simplified procedure will usually be inadequate. The examples that follow this section should help to clarify applicable cases.

EXAMPLE 2–10

Determine the Thevenin equivalent circuit for the dc circuit of Figure 2–26(a) at the terminals x-y two ways: **(a)** by determining V_{oc} and R_{eq} directly and **(b)** by successive internal reduction.

Solution

(a) This particular structure occurs frequently in practical problems, which justifies the decision to use two different approaches for illustration. Reference should be made to Figure 2–26 in this part. First, the open-circuit voltage V_{oc} at the terminals x-y must be determined, and this process is illustrated in Figure 2–26(b). An important point is that the open-circuit voltage across the output is the same as the voltage across the 6-Ω resistor. The reason is that no current flows through the 5-Ω resistor under open-circuit conditions, and there is no voltage drop across it. In determining the open-circuit voltage, imagine that one is using an ideal voltmeter, which would theoretically draw no current from the circuit to which it is connected. The type of situation illustrated here occurs often in performing open-circuit voltage measurements, so it is important to fully understand this concept.

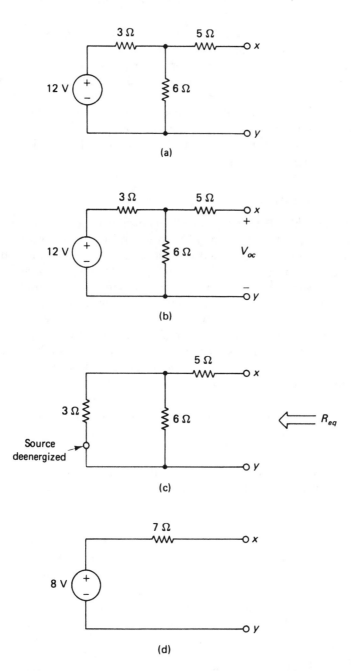

FIGURE 2–26
Circuit for Example 2–10 and the development of the Thevenin equivalent circuit.

Since the open-circuit voltage is the same as the voltage across the 6-Ω resistor, a simple application of the voltage divider rule to the left-hand loop may be used to determine V_{oc}. We have

$$V_{oc} = \frac{6}{6 + 3} \times 12 = 8 \text{ V} \qquad (2\text{--}87)$$

To determine the equivalent resistance, the 12-V independent voltage source is replaced by a short circuit as shown in Figure 2–26(c). The resistance "looking in" from the terminals x-y is then determined. Note that the 3-Ω and 6-Ω resistors turn out to be in parallel as far as R_{eq} is concerned. The parallel equivalent resistance is 2 Ω, which is in series with 5 Ω. The net equivalent resistance is thus

$$R_{eq} = 5 + 2 = 7 \, \Omega \qquad (2\text{--}88)$$

The Thevenin equivalent circuit is shown in Figure 2–26(d).

(b) The various steps involved in a successive reduction process are delineated in Figure 2–27. First, the 12-V voltage source in series with a 3-Ω resistor is converted to 4 A in parallel with 3 Ω as shown in (a). This circuit is then connected to the 6-Ω shunt resistor, and the parallel combination of the two resistors is determined as shown in (b). The resulting 4-A current source in parallel with 2 Ω is converted to 8 V in series with 2 Ω as shown in (c). The additional 5 Ω is connected as shown in (d), and the net series resistance of 7 Ω is determined. The resulting circuit is the Thevenin equivalent circuit.

For this particular circuit form, the author personally prefers the first method because it is more direct. However, the next example will illustrate a circuit in which it would be very difficult to determine the open-circuit voltage directly, but for which successive internal reduction may be readily applied.

EXAMPLE 2–11

Determine the Thevenin and Norton equivalent circuits at the terminals x-y in Figure 2–28.

Solution

The process of determining V_{oc} directly with this circuit would require a somewhat detailed evaluation, so a step-by-step approach using source transformations will be used. Reference should be made to the circuit diagrams of Figure 2–29 in the steps that follow.

First, the circuit to the left of the 11-Ω resistor in Figure 2–28 is reduced to a 20-V source in series with 4 Ω as shown in Figure 2–29(a). The details of this reduction are not shown, but it can be achieved by calculating the open-circuit voltage across the 12-Ω resistor and then determining the parallel combination of the two resistors after the source is de-energized. Next, the 11-Ω resistor and 40-V source are combined with the results of the preceding step, and the circuit is simplified as shown in Figure 2–29(b).

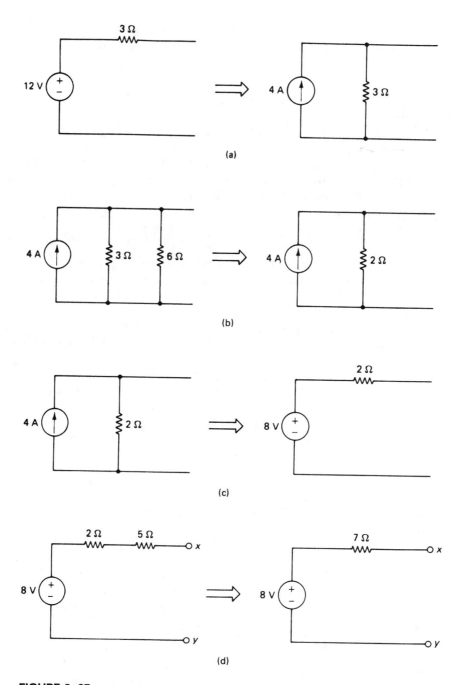

FIGURE 2–27
Alternate development of the Thevenin equivalent circuit in Example 2–10.

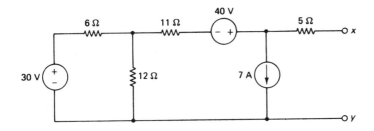

FIGURE 2–28
Circuit for Example 2–11.

In order to combine this circuit with the 7-A shunt current source in Figure 2–29(d), the voltage sources and series resistor are first converted to a current source in parallel with a resistor as shown in (c). Then, when the shunt current source is added, the two current sources may be combined as shown in (d).

The next step consists of converting back to a voltage source form as shown

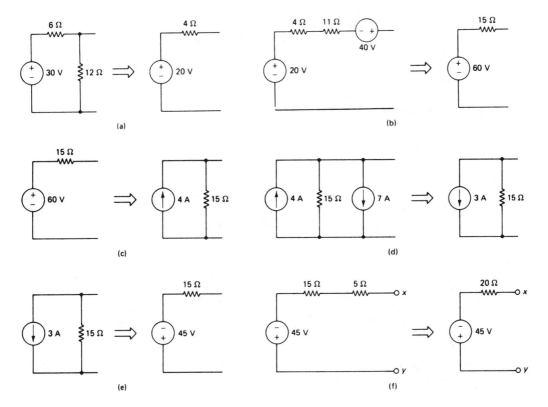

FIGURE 2–29
Successive reduction of the circuit for Example 2–11.

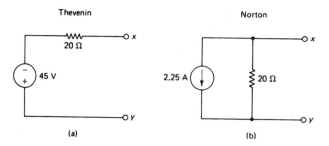

FIGURE 2–30
Thevenin and Norton models for Example 2–11.

in Figure 2–29(e). The additional 5-Ω resistance is included and combined with the 15-Ω resistance as shown in (f), which is the desired Thevenin form.

The final Thevenin model is shown again in Figure 2–30(a). A direct application of a source transformation leads to the Norton model shown in (b).

2-8   CIRCUIT ANALYSIS WITH DEPENDENT SOURCES

When dependent sources appear in circuits, the various circuit analysis equations are written in the normal manner as if the sources were independent. However, the presence of the dependency relationships requires some special attention before a solution can be formulated. A quantity that would be a known value if only independent sources were present may turn out to be a function of one of the dependent variables. Such terms must be moved around in the equations so that the desired variables may be determined uniquely.

The methods for dealing with dependent source models in analysis are best illustrated by representative solutions, and the examples that follow this section were chosen for that purpose. Thevenin's and Norton's theorems with dependent source models warrant a special treatment, and the next section is devoted to that topic.

EXAMPLE 2–12

The single-loop circuit of Figure 2–31 contains a VCVS whose value is $2V_{R1}$, where V_{R1} is the voltage across R_1 in the direction shown. Determine the current I.

Solution

We start by simply "walking around the loop" in the manner established in Section 2–1. The resulting loop equation is

$$-12 + 3I + 6I - 2V_{R1} = 0 \qquad (2\text{–}89)$$

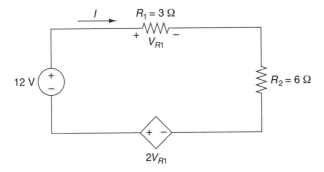

FIGURE 2–31
Circuit for Example 2–12.

Unlike all cases considered in Section 2–1, however, the result of Equation (2–89) is not directly solvable because it contains two unknowns. We must determine a second relationship before a solution is possible.

The key to the problem is to recognize that V_{R1} can be expressed in terms of the other unknown, I. We observe from Figure 2–31 that

$$V_{R1} = 3I \qquad\qquad \textbf{(2–90)}$$

Substitution of Equation (2–90) in Equation (2–89) and subsequent solution yield

$$-12 + 3I + 6I - 2(3I) = 0 \qquad\qquad \textbf{(2–91)}$$

and

$$I = 4 \text{ A} \qquad\qquad \textbf{(2–92)}$$

The resulting value of current means that $V_{R1} = 3 \times 4 = 12$ V and the dependent source has a value of $2 \times 12 = 24$ V.

Note that no simplified Ohm's law–type of approach, such as indicated by Equation (2–3), is possible for this circuit due to the presence of the dependent source. Instead, it is necessary to return to the basic concept of KVL and further establish all constraining mathematical relationships.

EXAMPLE 2–13
The circuit of Figure 2–32 contains a VCIS whose value is $5V_1$. Write a set of node voltage equations, and determine V_1 and V_2.

Solution
Since one side of the 5-V source is grounded, the voltage at the upper left-hand node is known (5 V). Node voltage equations may be written at the nodes labeled

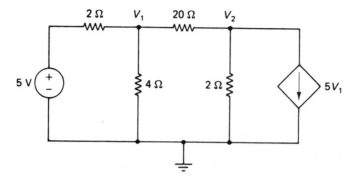

FIGURE 2–32
Circuit for Example 2–13.

V_1 and V_2 by expressing KCL at these nodes. At the node labeled V_1, we have

$$\frac{V_1 - 5}{2} + \frac{V_1}{4} + \frac{V_1 - V_2}{20} = 0 \qquad \textbf{(2–93)}$$

At the node labeled V_2, the equation is

$$\frac{V_2 - V_1}{20} + \frac{V_2}{2} + 5V_1 = 0 \qquad \textbf{(2–94)}$$

These equations were written in the same fashion developed earlier in the chapter, when only independent sources were present. However, the difference is that the term $5V_1$ in Equation (2–94) represents a source term, but it must be grouped with the other V_1 term in that equation. After all terms in both equations are sorted and combined, we have

$$0.8V_1 - 0.05V_2 = 2.5 \qquad \textbf{(2–95)}$$

and

$$4.95V_1 + 0.55V_2 = 0 \qquad \textbf{(2–96)}$$

Simultaneous solution of these equations yields

$$V_1 = 2 \text{ V} \qquad \textbf{(2–97)}$$

and

$$V_2 = -18 \text{ V} \qquad \textbf{(2–98)}$$

2–9 THEVENIN'S AND NORTON'S THEOREMS WITH DEPENDENT SOURCES

In Section 2–7, the procedures for determining Thevenin and Norton equivalent circuits for circuits having only independent sources were developed. Now that circuits containing dependent sources have been introduced, we are ready to consider the extension of Thevenin's and Norton's theorems to such circuits. The need for this procedure arises in determining the gain, input resistance, and output resistance of many active electronic circuits with feedback, for example.

We will present two separate, somewhat general procedures, which will be referred to respectively as the *open circuit–short circuit method* and the *external generator method.* Actually, these methods could be used for the circuits considered earlier containing only independent sources, but this is usually "overkill" for such circuits, and the methods considered earlier are usually much simpler.

Whereas both of the methods to be discussed are general, occasionally an indeterminant form may result with the open circuit–short circuit method. In such a case, the external generator method will usually work. Each of the methods will now be explained. Only *resistive* circuits will be considered at this time.

Open Circuit–Short Circuit Method

Refer to Figure 2–33 for the explanation that follows. With the circuit fully energized, the open-circuit voltage v_{oc} appearing across the output terminals is calculated as shown in (a). Next, a short (on paper) is placed across the terminals, and the short-circuit current i_{sc} is calculated as shown in (b). The relationship of these variables to the Thevenin and Norton circuits is

$$\text{Thevenin voltage: } v_t = v_{oc} \qquad \textbf{(2–99)}$$

$$\text{Norton current: } i_n = i_{sc} \qquad \textbf{(2–100)}$$

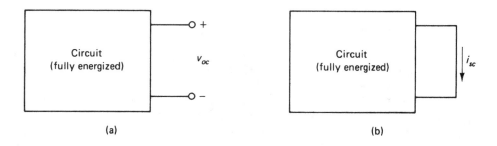

(a)   (b)

FIGURE 2–33
Steps involved in the open circuit–short circuit method for Thevenin's and Norton's theorems.

Assuming a resistive circuit, the equivalent resistance R_{eq} for either circuit is

$$R_{eq} = \frac{v_{oc}}{i_{sc}} \qquad (2\text{--}101)$$

In a sense, this particular procedure is partially a rehash of procedures developed in Section 2–7, but it is the process of determining R_{eq} from Equation (2–101) that gives it more generality, particularly when dependent sources are present.

External Generator Method

First, with the circuit fully excited, *either* the open-circuit voltage *or* the short-circuit current is calculated as in the preceding method. The variable that is calculated could depend on which is easier to determine, or whether the Thevenin or Norton circuit is the form desired. Actually, it is usually not critical which is determined, because eventually a source transformation can be used to change the form after the equivalent resistance is determined.

Where the external generator method changes drastically from the open circuit–short circuit method is in determining the equivalent resistance, and this is illustrated in Figure 2–34. For this purpose, all *independent sources* are de-energized. This means all ideal *independent* voltage sources are replaced by short circuits, and all ideal *independent* current sources are replaced by open circuits. Next, a fictitious external generator is connected to the output terminals. Although the source may be either a voltage source or a current source, we will assume a voltage source v_x for this discussion. The current i_x flowing from this source is calculated. The equivalent resistance is then

$$R_{eq} = \frac{v_x}{i_x} \qquad (2\text{--}102)$$

Alternately, a current source could be chosen and the voltage could be calculated. The relationship of Equation (2–102) is still the same.

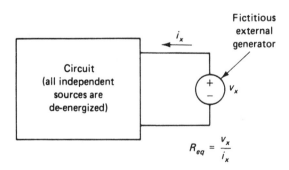

FIGURE 2–34
Application of fictitious external generator to determine Thevenin and Norton equivalent resistances.

It should be stressed that whereas all independent sources are de-energized, any dependent sources that are functions of circuit variables that would not be zero when the external generator is applied must be retained in the analysis.

EXAMPLE 2–14

The circuit of Figure 2–35 represent a field effect transistor (FET) common drain amplifier. The voltages v_1 and v_2 represent the input and output signals, respectively. In the mid-frequency range, where coupling capacitors act essentially as short circuits, a simplified, small-signal equivalent circuit employing the so-called hybrid-pi model for the FET is shown in Figure 2–36(a). Note that the dependent source is a function of both the *independent* input signal voltage v_1 and the *dependent* output signal voltage v_2. Apply Thevenin's and Norton's theorems in the general forms of the preceding section to obtain a new Thevenin equivalent circuit in which the voltage gain and output resistance of the circuit are explicitly delineated.

Solution

The process that will be illustrated here is representative of the analysis involved with electronic circuits having feedback. Readers not having a reasonably good background in electronic circuits may not fully comprehend the significance of the objective. However, the intent here is to apply the procedures developed earlier to a real circuit rather than a fictitious example.

(a) *Open circuit–short circuit method.* To apply this method, the open-circuit output voltage, which will be denoted as v_{2oc}, is measured as illustrated in Figure 2–36(b). Note that since the output voltage under these conditions is denoted

FIGURE 2–35
Common drain FET amplifier circuit for Example 2–14.

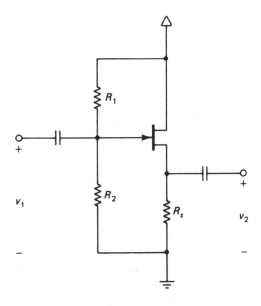

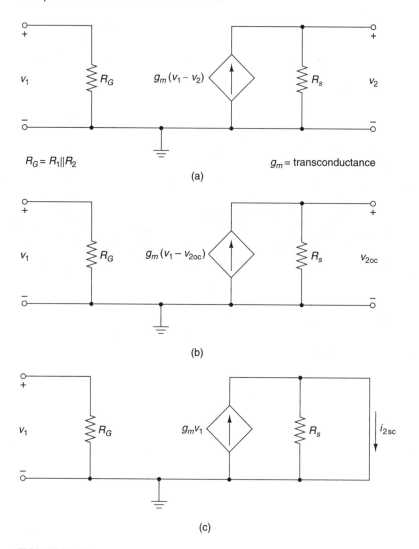

$R_G = R_1 \| R_2$

g_m = transconductance

(a)

(b)

(c)

FIGURE 2–36
Mid-frequency small-signal equivalent circuit of common drain amplifier and steps involved in applying the open circuit–short circuit method.

as v_{2oc}, the v_2 term in the dependent source is also changed to v_{2oc}. This dependent current must flow into R_s, and the open-circuit voltage is

$$v_{2oc} = g_m R_s (v_1 - v_{2oc}) \qquad \textbf{(2–103)}$$

The result of Equation (2–103) may impress the reader as somewhat strange in that v_{2oc} is expressed in terms of itself! In this form, the result is not complete because there is no explicit relationship for v_{2oc} in terms of the independent variable v_1. The problem is circumvented by rearranging the equation so that all v_{2oc} terms appear on the left. This results in

$$v_{2oc}(1 + g_m R_s) = g_m R_s v_1 \qquad (2\text{–}104)$$

The output voltage is then determined to be

$$v_{2oc} = \frac{g_m R_s v_1}{1 + g_m R_s} \qquad (2\text{–}105)$$

The open-circuit voltage is now a function of only the input variable v_1.

The next step in the procedure is to determine the short-circuit output current, which will be denoted as i_{2sc}. The equivalent circuit used to determine this quantity is shown in Figure 2–36(c). Note that since $v_2 = 0$ in this step, the dependent source reduces to $g_m v_1$. In this equivalent circuit, all of the available current from the dependent source flows through the short, and we have

$$i_{2sc} = g_m v_1 \qquad (2\text{–}106)$$

The equivalent resistance R_{eq} is

$$R_{eq} = \frac{v_{2oc}}{i_{2sc}} = \frac{(g_m R_s v_1)/(1 + g_m R_s)}{g_m v_1} = \frac{R_s}{1 + g_m R_s} \qquad (2\text{–}107)$$

Some discussion of the results of Equations (2–105) and (2–107) will be made after a different procedure is used to determine R_{eq}.

(b) *External generator method.* Both v_{2oc} and i_{2sc} have already been determined, so it is necessary only to determine R_{eq}. Referring to Figure 2–37, the independent source v_1 is first de-energized. Either an external fictitious voltage source or a current source could be applied to the output, but since the dependent source is a function of the voltage across the output, it is easier to work with an external voltage generator. This fictitious generator is denoted by v_x here, although a specific value (e.g., 1 V) could be used if desired.

An important point to clarify before proceeding further is the behavior of the dependent current source $g_m(v_1 - v_2)$. Since the independent input source v_1 is de-energized, that term in the dependent source is equated to zero.

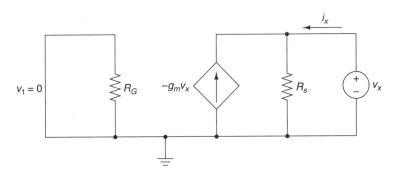

FIGURE 2–37
Application of the external generator method to the common drain equivalent circuit.

However, since $v_2 = v_x$, that portion of the dependent source *cannot* be de-energized. Thus, when de-energizing a circuit to apply the external generator method, it is necessary to carefully inspect all dependent sources to determine those which are zero and those which are not. In this case, the resulting dependent current source is $-g_m v_x$ into the upper node (which is equivalent, of course, to $+g_m v_x$ out of the upper node).

The current in R_s is v_x/R_s flowing downward. By KCL applied to the upper node, we have

$$\frac{v_x}{R_s} - (-g_m v_x) - i_x = 0 \qquad (2\text{--}108)$$

This result leads to

$$i_x = \frac{(1 + g_m R_s)}{R_s} v_x \qquad (2\text{--}109)$$

From Equation (2–109), the equivalent resistance is

$$R_{eq} = \frac{v_x}{i_x} = \frac{R_s}{1 + g_m R_s} \qquad (2\text{--}110)$$

which is obviously in agreement with Equation (2–107).

A final equivalent circuit of the amplifier showing a Thevenin VCVS in series with the equivalent resistance is given in Figure 2–38. The factor of v_1 on the dependent source is the open-circuit voltage gain. Normally, $g_m R_s \gg 1$, so this voltage gain is slightly less than unity. The output resistance $R_s/(1 + g_m R_s)$ is relatively small. These properties are basic to a common drain circuit and are not immediately apparent from the original equivalent circuit of Figure 2–36(a). In fact, someone might erroneously assume that the Thevenin or Norton output resistance is R_s from that circuit. However, the application of the techniques here has eliminated the effects of all dependent variables on the controlled sources, thereby delineating the true effective gain and output resistance.

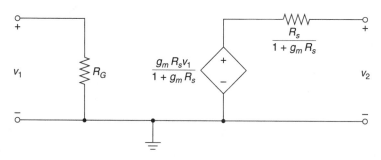

FIGURE 2–38
Final equivalent circuit of common drain amplifier showing voltage gain and output resistance.

2-10   SUPERPOSITION

The principle of superposition states that any voltage or current response in a linear circuit resulting from several voltage and/or current sources may be determined by first considering the response produced by each source individually and then algebraically combining the individual responses. As the effect of each source is considered, all other sources are de-energized. To de-energize an ideal *voltage source,* it is replaced by a *short circuit.* To de-energize an ideal *current source,* it is replaced by an *open circuit.* Any internal resistances assumed in the source models are retained in the circuit when the ideal sources are de-energized.

At this point in the text, the use of superposition will be illustrated by an example. Some rather general properties of superposition will be developed later in the text as applied to linear systems. The major point to stress here is that a circuit must be linear for superposition to apply. All circuits considered thus far in the text were assumed to be linear.

EXAMPLE 2-15
Consider the circuit of Example 2–11 (Figure 2–28), for which Thevenin and Norton equivalent circuits were established at terminals *x-y*. The circuit was simplified in Example 2–11 by using successive source transformations. Use superposition to compute V_{oc} from the original network structure.

Solution
For convenience, the circuit is repeated in Figure 2–39(a), and the desired variable V_{oc} is indicated. Since there are three sources in the circuit, there are three contributions to the voltage. Let

$$V_{oc} = V'_{oc} + V''_{oc} + V'''_{oc} \qquad (2\text{--}111)$$

where V'_{oc} is the response due to the 30-V source, V''_{oc} is the response due to the 40-V source, and V'''_{oc} is the response due to the 7-A source.

To determine V'_{oc}, we de-energize the 40-V source and the 7-A source as shown in Figure 2–39(b). Note that the 40-V source is replaced by a short circuit and the 7-A source is replaced by an open circuit. Since no current flows through the 11-Ω and 5-Ω resistors, V'_{oc} is the voltage across the 12-Ω resistor, and this is determined by the voltage divider rule as follows:

$$V'_{oc} = \frac{12}{12 + 6} \times 30 = 20 \text{ V} \qquad (2\text{--}112)$$

To determine V''_{oc}, we de-energize the 30-V source and the 7-A source as shown in Figure 2–39(c). No current flows in this case, so the open-circuit voltage is the same as the source, that is,

$$V''_{oc} = 40 \text{ V} \qquad (2\text{--}113)$$

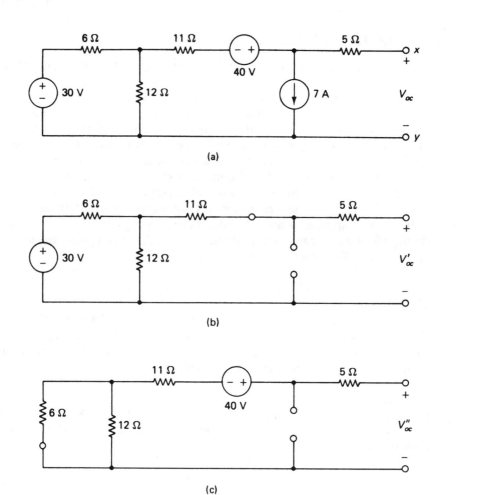

(a)

(b)

(c)

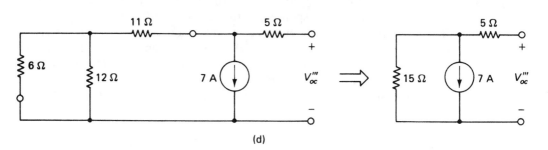

(d)

FIGURE 2–39
Circuit for Example 2–15 used to illustrate superposition.

To determine V_{oc}''', we de-energize the 30-V source and the 40-V voltage as shown on the left in Figure 39(d). To simplify this circuit, the resistance in parallel with the 7-A source is determined to be 15 Ω, as shown on the right. The open-circuit voltage will have its more positive terminal at the bottom, and V_{oc}''' is

$$V_{oc}''' = -7 \times 15 = -105 \text{ V} \tag{2-114}$$

The net voltage from Equation (2–111) is

$$V_{oc} = 20 + 40 - 105 = -45 \text{ V} \tag{2-115}$$

This result is in agreement with the outcome of Example 2–11, as shown in Figures 2–29 and 2–30.

2–11 PSPICE EXAMPLES

The dc analysis capabilities of PSPICE introduced in Chapter 1 will be extended in this chapter. The transfer function command will be used to determine the Thevenin and/or Norton equivalent circuits in several examples, and dependent sources will be introduced.

PSPICE EXAMPLE 2–1

Use PSPICE to solve for various node voltages, branch voltages, and branch currents in the circuit of Example 2–7 (Figure 2–17).

Solution

The circuit adapted to the PSPICE format and the code are shown in Figure 2–40. The options NOECHO and NOPAGE discussed in Chapter 1 are included to reduce the size of the output file. Note again that only code lines following NOECHO are suppressed in the output, so it is desirable to place the options code after the title line when NOECHO is employed.

Most of the circuit block description follows the format discussed in the examples of Chapter 1 and can be easily checked. The only new element is the independent current source, for which the code line reads

```
I1 0 7 DC 7
```

The direction of current flow is from the first node (0) to the second node (7), and the value (7 A) follows the DC designation.

As explained in Chapter 1, a general dc analysis requires that one source be "swept" even when there is a one-point analysis. This was arbitrarily selected as V1 in the .DC command, and its value is forced to be 20 V with the three required values.

The .PRINT statement would be very long if written altogether on one line. One option would be to use more than one .PRINT statement. An equally useful approach is to use a line continuation control, which is achieved by the use of + as the first character of continuation lines. The lines have been arbitrarily organized in groups of five variables, which corresponds to the grouping in the data output.

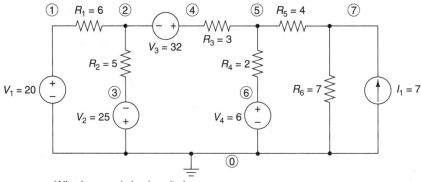

(All values are in basic units.)

```
PSPICE EXAMPLE 2-1
.OPTIONS  NOECHO  NOPAGE
V1  1  0  DC  20
R1  1  2  6
R2  2  3  5
V2  0  3  DC  25
V3  4  2  DC  32
R3  4  5  3
R4  5  6  2
V4  6  0. DC  6
R5  5  7  4
R6  7  0  7
I1  0  7  DC  7
.DC  V1  20  20  1
.PRINT  DC  V(2)  V(4)  V(5)  V(7)  V(R1)
+V(R2)  V(R3)  V(R4)  V(R5)  V(R6)
+I(R1)  I(R2)  I(R3)  I(R4)  I(R5
.END
```

FIGURE 2-40
Circuit and code for PSPICE Example 2-1.

The output data file is shown in Figure 2–41. Note that the "swept" variable V1 repeats at the beginning of each line. All values are in perfect agreement with the results of Example 2–7, when interpreted properly. A negative value, of course, means that the given voltage or current has the opposite sign from the defined positive direction for that variable.

PSPICE EXAMPLE 2-2

Use the transfer function command (.TF) to determine the Thevenin equivalent circuit in Example 2–10 (Figure 2–26).

Solution

The transfer function command is a useful feature for determining input resistance, output resistance, and voltage gain for active circuits having one independent source

PSPICE EXAMPLE 2-1

****   CIRCUIT DESCRIPTION

****

.OPTIONS NOECHO NOPAGE

****   DC TRANSFER CURVES   TEMPERATURE =   27.000 DEG C

V1	V(2)	V(4)	V(5)	V(7)	V(R1)
2.000E+01	-1.000E+01	2.200E+01	1.600E+01	2.800E+01	3.000E+01

****   DC TRANSFER CURVES   TEMPERATURE =   27.000 DEG C

V1	V(R2)	V(R3)	V(R4)	V(R5)	V(R6)
2.000E+01	1.500E+01	6.000E+00	1.000E+01	-1.200E+01	2.800E+01

****   DC TRANSFER CURVES   TEMPERATURE =   27.000 DEG C

V1	I(R1)	I(R2)	I(R3)	I(R4)	I(R5)
2.000E+01	5.000E+00	3.000E+00	2.000E+00	5.000E+00	-3.000E+00

JOB CONCLUDED

TOTAL JOB TIME   .44

FIGURE 2–41
Output data for PSPICE Example 2–1.

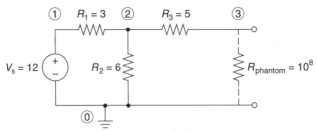

(All values are in basic units.)

```
PSPICE EXAMPLE 2-2
.OPTIONS NOECHO NOPAGE
VS 1 0 DC 12
R1 1 2 3
R2 2 0 6
R3 2 3 5
RPHANTOM 3 0 1E8
.TF V(3) VS
.END
```

FIGURE 2-42
Circuit and code for PSPICE Example 2-2.

and one or more dependent sources when there are *no frequency dependent elements present* (i.e., neither L nor C). (It may also be used, with some caution, for the case of two or more independent sources, as will be demonstrated in PSPICE Example 2-3.)

The circuit adapted to the PSPICE format and code are shown in Figure 2-42. All code lines through R3 follow formats previously discussed, and can be easily checked.

A problem would arise if analysis were attempted with the 5-Ω resistor (R_3) left "dangling" as in the original configuration. With some exceptions that will be noted later in the text, all PSPICE nodes generally require two or more branches at the connection point, and an error message would be generated with the original configuration. The problem is circumvented by connecting a "phantom" resistance (RPHANTOM) between node 3 and ground as shown. The value of this resistance must have negligible effect on the remainder of the circuit, and this is achieved by using a very large value (10^8 Ω in this case). In one sense, this could be considered in the same manner as connecting a voltmeter with a large internal resistance across the terminals.

The command line that directs a transfer function analysis reads

```
.TF V(3) VS
```

The command .TF designates "transfer function," and this is followed by two variables. The first variable is the desired output node voltage, which in this case is V(3). The second variable is the *name* of the assumed input voltage, which is VS in this case. No print statement is required for the transfer function analysis, and the .END statement completes the code.

The output data file is shown in Figure 2–43. It consists of (a) a small-signal bias solution and (b) the small-signal characteristics. The small-signal bias solution is the same form as that obtained from the .OP analysis in Example 1–1. The term

```
PSPICE EXAMPLE 2-2


****     CIRCUIT DESCRIPTION


*****************************************************************************


.OPTIONS NOECHO NOPAGE


****     SMALL SIGNAL BIAS SOLUTION      TEMPERATURE =   27.000 DEG C


NODE   VOLTAGE      NODE   VOLTAGE     NODE   VOLTAGE     NODE   VOLTAGE

(   1)  12.0000  (   2)   8.0000  (   3)   8.0000



    VOLTAGE SOURCE CURRENTS
    NAME           CURRENT

    VS            -1.333E+00

    TOTAL POWER DISSIPATION   1.60E+01  WATTS



****     SMALL-SIGNAL CHARACTERISTICS

    V(3)/VS =  6.667E-01

    INPUT RESISTANCE AT VS =  9.000E+00

    OUTPUT RESISTANCE AT V(3) =  7.000E+00


       JOB CONCLUDED

       TOTAL JOB TIME            .33
```

FIGURE 2–43
Output data for PSPICE Example 2–2.

small signal refers to the type of operation performed with an active circuit having nonlinear characteristics, and need not concern us here. From this analysis, however, we can obtain the value of the open-circuit voltage. Both V(2) and V(3) represent the open-circuit voltage, which is 8 V. Clearly, the large resistance across the output presents no significant loading effect since the voltages at nodes 2 and 3 are the same.

The remaining part of the Thevenin equivalent circuit is obtained from the "small-signal characteristics." Actually, the open-circuit voltage could have been obtained by multiplying the value of V(3)/VS = 0.6667 by the source voltage of 12 V. This approach works when there is only one independent source, as is the case here. However, when there is more than one source, as will be demonstrated in PSPICE Example 2–3, that approach will not work.

The second line of this block represents the input resistance of the circuit, which is seen to be 9 Ω. This value is readily verified from the circuit diagram, but for an active electronic circuit, the value may not be immediately evident, so the result here can be useful.

The third line of this block provides the output resistance looking back from node 3 to ground, which is the Thevenin equivalent resistance and is seen to be 7 Ω, as expected. Again, we see that the shunting effect of the very large resistance across the output has no measurable effect.

PSPICE EXAMPLE 2–3

Use the transfer function command to determine the Thevenin equivalent circuit in Example 2–11 (Figure 2–28).

Solution
The circuit adapted to the PSPICE format and code are shown in Figure 2–44. The forms of all code lines have previously been discussed and can be readily verified. As in the previous example, a large additional resistance has been connected from the output node (5) to ground to prevent an error message. The output voltage in this case is V(5) and the name of the input voltage is VS1.

The output data file is shown in Figure 2–45. The open-circuit output voltage is read from the "small-signal bias solution" as V(5) = V(4) = −45 V. (This means, of course, that the positive terminal of the output voltage is at node 0 or ground.)

Looking at the small-signal characteristics, we note that V(5)/VS1 = 0.6667. When that value is multiplied by VS1 = 30 V, we obtain 20 V, which is clearly not the value of V(5). What is wrong?

The answer lies in the process that is being performed by the transfer function command. When an input independent source is identified, all other independent sources are de-energized, and the resulting transfer ratio represents the gain constant associated only with that particular source. Thus, the contribution of VS1 to V(5) can be deduced as 20 V, but the effects of the other two sources would have to be determined separately. Thus, the small-signal bias solution is the only correct way to determine the output open-circuit voltage when there is more than one independent source.

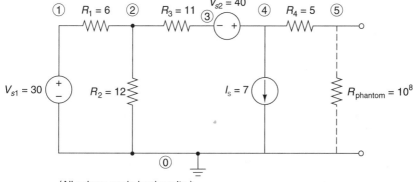

(All values are in basic units.)

```
PSPICE EXAMPLE 2-3
.OPTIONS NOECHO NOPAGE
VS1 1 0 DC 30
R1 1 2 6
R2 2 0 12
R3 2 3 11
VS2 4 3 DC 40
IS 4 0 DC 7
R4 4 5 5
RPHANTOM 5 0 1E8
.TF V(5) VS1
.END
```

FIGURE 2–44
Circuit and code for PSPICE Example 2–3.

The other two values in the small-signal characteristics apply irrespective of the number of independent sources. Thus, the input resistance seen at the VS1 terminals is 18 Ω, and the output (or Thevenin equivalent) resistance is 20 Ω.

PSPICE EXAMPLE 2–4
Use PSPICE to solve for the current in the circuit of Example 2–12 (Figure 2–31).

Solution
The circuit adapted to the PSPICE format and the code are shown in Figure 2–46. The only line representing a new situation is that of the dependent voltage source, which reads

```
E 0 3 1 2 2
```

The symbol E represents a voltage-controlled voltage source (VCVS), and additional letters or numbers could be added if desired. Following the name of the source, five entries are required. The first two numbers (0 3) represent the node connections for the source itself, with the first (0) representing the reference positive

PSPICE EXAMPLE 2-3

**** CIRCUIT DESCRIPTION

****

.OPTIONS NOECHO NOPAGE

**** SMALL SIGNAL BIAS SOLUTION TEMPERATURE = 27.000 DEG C

NODE VOLTAGE NODE VOLTAGE NODE VOLTAGE NODE VOLTAGE

(1) 30.0000 (2) -8.0000 (3) -85.0000 (4) -45.0000

(5) -45.0000

```
    VOLTAGE SOURCE CURRENTS
    NAME            CURRENT

    VS1            -6.333E+00
    VS2            -7.000E+00

    TOTAL POWER DISSIPATION   4.70E+02   WATTS
```

**** SMALL-SIGNAL CHARACTERISTICS

V(5)/VS1 = 6.667E-01

INPUT RESISTANCE AT VS1 = 1.800E+01

OUTPUT RESISTANCE AT V(5) = 2.000E+01

JOB CONCLUDED

TOTAL JOB TIME .33

FIGURE 2-45
Output data for PSPICE Example 2-3.

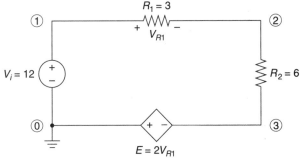

(All values are in basic units.)

```
PSPICE EXAMPLE 2-4
.OPTIONS  NOECHO  NOPAGE
VI  1  0  DC  12
R1  1  2  3
R2  2  3  6
E  0  3  1  2  2
.DC  VI  12  12  1
.PRINT  DC  I (R1)
.END
```

FIGURE 2-46
Circuit and code for PSPICE Example 2-4.

terminal. The next two numbers (1 2) represent the nodes across which the controlling voltage is defined, with the first (1) representing the positive reference for that voltage. The last number (2) represents the gain constant, which has no dimensions for a VCVS.

Since the current in a single-loop circuit is the same everywhere, any one of the branch currents could be used in the .PRINT statement. The current I(R1) was arbitrarily selected.

The output data file is shown in Figure 2–47. The value of 4 A is readily seen to be in agreement with the result of Example 2–12.

PSPICE EXAMPLE 2-5
Use PSPICE to solve for the node voltages in the circuit of Example 2–13 (Figure 2–32).

Solution
The circuit adapted to the PSPICE format and code are shown in Figure 2–48. The only line representing a new situation is that of the dependent current source, which reads

```
PSPICE EXAMPLE 2-4


****      CIRCUIT DESCRIPTION


******************************************************************************


.OPTIONS NOECHO NOPAGE


****      DC TRANSFER CURVES                TEMPERATURE =    27.000 DEG C


  VI           I(R1)

  1.200E+01    4.000E+00




          JOB CONCLUDED

          TOTAL JOB TIME              .38
```

FIGURE 2–47
Output data for PSPICE Example 2–4.

The symbol G represents a voltage-controlled current source (VCIS), and additional letters or numbers could be added if desired. Following the name of the source, five entries are required. The first two numbers (3 0) represent the direction of flow of the current source; that is, the current flows from node 3 to node 0. The next two numbers (2 0) represent the nodes across which the controlling voltage is defined, with the first (2) representing the positive reference for that voltage. The last number (5) represents the transconductance in siemens.

The output data file is shown in Figure 2–49. The voltage V(2) represents V_1 in the original circuit and is 2 V. The voltage V(3) represents V_2 in the original circuit and is −18 V. These values are in perfect agreement with those of Example 2–13.

PSPICE EXAMPLE 2-6
Use PSPICE to model the FET source follower of Example 2–14 with the following parameter values: $R_G = 500$ kΩ, $g_m = 9.5$ mS, $R_s = 2$ kΩ. Use the transfer function command to determine the voltage gain, the input resistance, and the output resistance.

Solution
Whereas a theoretical development using symbols was performed in Example 2–14, it is necessary with PSPICE to employ specific values. The values assumed are typical for the type of circuit under consideration.

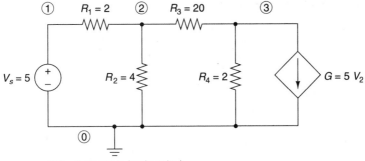

(All values are in basic units.)

```
PSPICE EXAMPLE 2-5
.OPTIONS NOECHO NOPAGE
VS 1 0 DC 5
R1 1 2 2
R2 2 0 4
R3 2 3 20
R4 3 0 2
G 3 0 2 0 5
.DC VS 5 5 1
.PRINT DC V(2) V(3)
.END
```

FIGURE 2-48
Circuit and code for PSPICE Example 2-5.

PSPICE EXAMPLE 2-5

****   CIRCUIT DESCRIPTION

****

.OPTIONS NOECHO NOPAGE

****   DC TRANSFER CURVES   TEMPERATURE =   27.000 DEG C

```
 VS           V(2)         V(3)

  5.000E+00    2.000E+00   -1.800E+01
```

JOB CONCLUDED

TOTAL JOB TIME   .39

FIGURE 2-49
Output data for PSPICE Example 2-5.

117

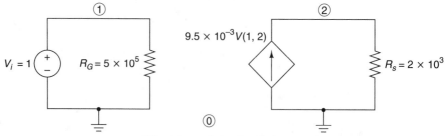

(All values are in basic units.)

```
PSPICE EXAMPLE 2-6
.OPTIONS NOECHO NOPAGE NOBIAS
V1 1 0 DC 1
RG 1 0 5E5
G 0 2 1 2 9.5E-3
RS 2 0 2E3
.TF V (2) V1
.END
```

FIGURE 2-50
Circuit and code for PSPICE Example 2-6.

```
PSPICE EXAMPLE 2-6


****      CIRCUIT DESCRIPTION


*********************************************************************************


.OPTIONS NOECHO NOPAGE NOBIAS



****      SMALL-SIGNAL CHARACTERISTICS


     V(2)/V1 =  9.500E-01

     INPUT RESISTANCE AT V1 =   5.000E+05

     OUTPUT RESISTANCE AT V(2) =   1.000E+02


          JOB CONCLUDED

          TOTAL JOB TIME          .33
```

FIGURE 2-51
Output data for PSPICE Example 2-6.

The circuit diagram adapted to the PSPICE format with the values assumed is shown in Figure 2–50. Since there is only one independent source V_1, and since voltage gain (rather than an actual voltage value) is desired, it was decided to suppress the "small-signal bias solution" in the output data file. This is achieved by adding NOBIAS to the options line.

The code should be fairly evident, since the forms of all entries have been previously considered. One interesting observation is that one of the two nodes representing the VCVS source (2) is also one of the controlling nodes. This is perfectly acceptable, and this type of situation frequently arises in active electronic circuits with feedback. The transfer function line defines V(2) as the output voltage and V1 as the input voltage.

The output data file is shown in Figure 2–51. The voltage gain is read as V(2)/V1 = 0.95. The input and output resistances are 500 kΩ and 100 Ω, respectively. These values can be readily verified from the final equivalent circuit of Figure 2–38 for the given parameter values.

2–12 MATLAB EXAMPLE

MATLAB is a powerful, comprehensive software package for performing mathematical analysis. It is marketed by The Math Works, Inc.* and is available in both professional and student versions. Along with the core program, there are a number of "toolboxes" that provide applications to specific disciplines. It is one of the most popular mathematical software programs available, and is widely used in industry, government, and education.

Because of its wide scope, we can illustrate only a few of its features in this text. The features shown will be those which can be readily used in support of the circuit analysis methods covered. The coverage should, however, be sufficient to expose the reader to the power of the program, and it should disperse any anxieties about using it. Further information can be obtained from the program manuals and other texts devoted to its applications.

The examples in the text were analyzed within the Workspace environment. This is an interactive approach achieved through the process of typing in the various commands and observing the results directly on the screen. For many of the examples in later chapters, the results are transferred to a graph, and a hard copy is obtained with a printer.

All of the MATLAB examples in the text were solved with the *Student Edition of MATLAB,* which is available from Prentice Hall, Inc.

MATLAB EXAMPLE 2–1

Use MATLAB to solve simultaneously the three mesh current equations of Example 2–7 [Equations (2–62), (2–63), and (2–64)].

---

* The Math Works, Inc.
24 Prime Park Way
Natick, MA 01760-1500

Solution

Many readers have calculators that solve equations simultaneously, so the use of MATLAB may be overkill for a situation this simple. However, this is a good warmup exercise, and it will assist in establishing the type of approach used in MATLAB.

The three equations that are to be solved simultaneously are altered slightly and repeated here as follows:

$$11I_1 - 5I_2 + 0 \cdot I_3 = 45 \tag{2-116}$$

$$-5I_1 + 10I_2 - 2I_3 = 1 \tag{2-117}$$

$$0 \cdot I_1 - 2I_2 + 13I_3 = -43 \tag{2-118}$$

Note that I_3 has been added to the first row with a multiplier of 0, and I_1 has been added to the last equation with a multiplier of 0. In other words, it is necessary to fill in all variables in each equation so that the array is complete. (With a little practice, this can be done mentally, without actually writing out all the equations.)

The matrix formulation of the preceding three equations reads as follows:

$$\begin{bmatrix} 11 & -5 & 0 \\ -5 & 10 & -2 \\ 0 & -2 & 13 \end{bmatrix} \begin{bmatrix} I_1 \\ I_2 \\ I_3 \end{bmatrix} = \begin{bmatrix} 45 \\ 1 \\ -43 \end{bmatrix} \tag{2-119}$$

The reader need not necessarily understand matrix theory in order to use the concept.

The array of Equation (2–119) can be expressed in compact matrix notation as

$$[R][I] = [V] \tag{2-120}$$

where $[R]$ represents the square matrix of resistance values; $[I]$ represents the column matrix of unknown current values; and $[V]$ represents the column matrix of voltage values on the right-hand side of the equation.

In matrix theory, the solution of Equation (2–120) is given by

$$[I] = [R]^{-1}[V] \tag{2-121}$$

where $[R]^{-1}$ is called the inverse matrix of $[R]$. We do not need to perform this operation manually, since it can be easily achieved with MATLAB.

Refer to the printout of the MATLAB workspace shown in Figure 2–52. Note that the cursor EDU>> appears on the screen at the beginning of each line that is typed. Thus, you will be able to recognize all lines that are typed by the user by its presence.

First, the matrix $[R]$ is entered by typing

```
R=[11 -5 0;-5 10 -2;0 -2 13]
```

Note that the elements of each row are entered in sequence, with a space between the elements. (A comma may also be used to separate elements in a given row.) Each time a new row is desired, a semicolon (;) is placed on the line. The matrix begins with a left-hand bracket and ends with a right-hand bracket. After this

```
Commands to get started: intro, demo, help help
Commands for more information: help, whatsnew, info, subscribe

EDU» R=[11 -5 0;-5 10 -2;0 -2 13]

R =

    11     -5      0
    -5     10     -2
     0     -2     13

EDU» V=[45;1;-43]

V =

    45
     1
   -43

EDU» I=R\V

I =

    5.0000
    2.0000
   -3.0000

EDU»
```

FIGURE 2–52
Workspace for MATLAB Example 2–1.

operation is entered on the computer, MATLAB returns the entire matrix in rectangular form as noted.

Next, the column matrix $[V]$ is established by typing

```
V=[45;1;-43]
```

Note the necessity in this case to place semicolons between successive entries. Otherwise, the array would be interpreted as a row matrix. After entering this line, MATLAB returns the data as a column matrix.

There are two matrix forms in MATLAB that can be used to determine the unknown current matrix. The one that most closely resembles the form from matrix theory is the command

```
I=inv(R)*V
```

where inv(R) represents the inverse matrix and (*) denotes multiplication. This operation works fine for most functions that will be encountered in this text. However, for certain advanced concepts beyond the scope of this text, the MATLAB manual encourages the use of the following alternate form:

```
I=R\V
```

Note that the reverse slash \ is required. When this operation is used with single numbers, the denominator appears on the left and the numerator appears on the right. Although we are dealing with matrices here, and the process taking place is a matrix inversion, it helps to remember the command by thinking that we "divide R into V" to obtain I.

After this command is entered, the column matrix appears as the output. The three values are readily seen to be the three correct values for the mesh currents.

DRILL PROBLEMS

2–1. For the circuit of Figure P2–1, determine **(a)** the loop current I, **(b)** voltage drops across all resistors, and **(c)** power associated with each element and whether it is absorbed or delivered.

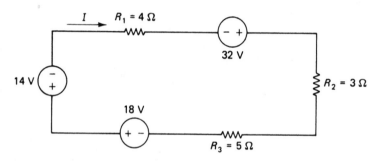

FIGURE P2–1

2–2. For the circuit of Figure P2–2, determine **(a)** the loop current, I, **(b)** voltage drops across all resistors, and **(c)** power associated with each element and whether it is absorbed or delivered.

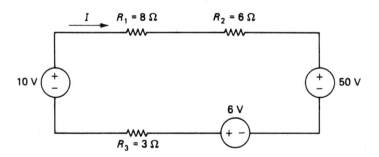

FIGURE P2–2

2–3. For the circuit of Figure P2–3, determine **(a)** the loop current I, **(b)** voltage drops across all resistors, **(c)** voltage V_0 across current source, and **(d)** power associated with each element and whether it is absorbed or delivered.

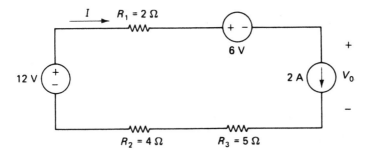

FIGURE P2–3

2–4. For the circuit of Figure P2–4, determine **(a)** the loop current I, **(b)** voltage drops across all resistors, **(c)** voltage V_0 across current source, and **(b)** power associated with each element and whether it is absorbed or delivered.

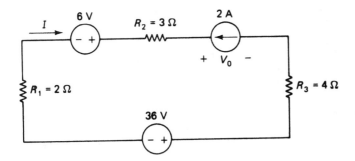

FIGURE P2–4

2–5. For the circuit of Figure P2–5, determine **(a)** the node-pair voltage V, **(b)** currents through all resistors, and **(c)** power associated with each element and whether it is absorbed or delivered.

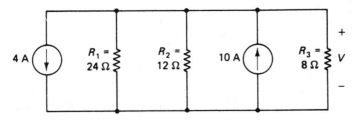

FIGURE P2–5

2–6. For the circuit of Figure P2–6, determine **(a)** the node-pair voltage V, **(b)** currents through all resistors, and **(c)** power associated with each element and whether it is absorbed or delivered.

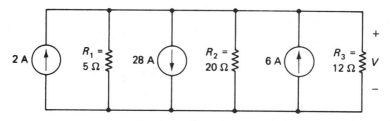

FIGURE P2–6

2–7. For the circuit of Figure P2–7, determine **(a)** the node-pair voltage V, **(b)** currents through all resistors, **(c)** current I_0 through voltage source, and **(d)** power associated with each element and whether it is absorbed or delivered.

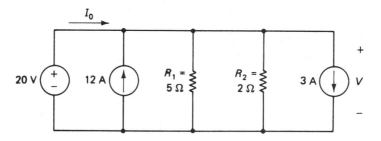

FIGURE P2–7

2–8. For the circuit of Figure P2–8, determine **(a)** the node-pair voltage V, **(b)** currents through all resistors, **(c)** current I_0 through voltage source, and **(d)** power associated with each element and whether it is absorbed or delivered.

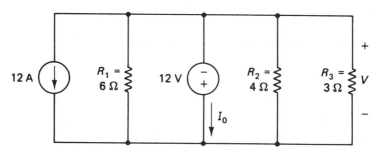

FIGURE P2–8

2–9. The voltage V_0 is to be determined in the circuit of Figure P2–9. Show that by performing a source transformation to the left of *x-y*, the circuit reduces to a single node-pair form as far as the part of the circuit of interest is concerned. Using the modified circuit, determine V_0.

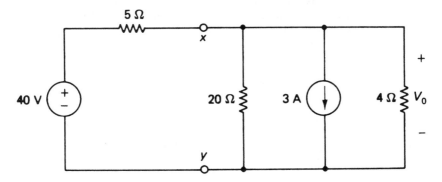

FIGURE P2–9

2–10. The current I_0 is to be determined in the circuit of Figure P2–10. Show that by performing a source transformation to the left of *x-y*, the circuit reduces to a single-loop form as far as the part of the circuit of interest is concerned. Using the modified circuit, determine I_0.

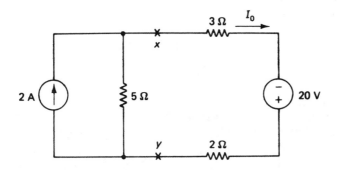

FIGURE P2–10

2–11. In Example 2–5, a source transformation was performed to simplify the computation of V_0. **(a)** From a knowledge of V_0, return to the original circuit of Figure 2–11(a), and compute the power P delivered by the actual 24-V source. **(b)** From the transformed circuit of Figure 2–11(b), compute the power P' delivered by the fictitious 2-A current source and verify that the two power levels are different. (This illustrates that only computations *external* to the transformed circuit are meaningful.)

2–12. This problem has similar objectives to those of Example 2–6, except that the circuit will employ a high-resistance source in which the current source model is more convenient. Consider the circuit of Figure P2–12, consisting of a 2-mA current source in parallel with a 100-kΩ internal source resistance. **(a)** Assume that the 1-kΩ load shown on the right is connected. Calculate the exact voltage V_L across the load, **(b)** convert the source to a voltage source model and calculate V_L again, and **(c)** using the current source model, neglect the internal source resistance (i.e., replace it by an open circuit) and calculate the resulting load voltage V_L'.

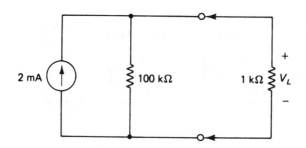

FIGURE P2–12

2–13. For the circuit of Figure P2–13, **(a)** write the mesh current equations, **(b)** arrange the results of (a) in standard mathematical form and solve for the mesh currents, and **(c)** determine all branch currents.

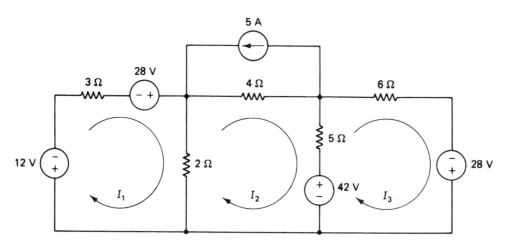

FIGURE P2–13

2–14. For the circuit of Figure P2–14, **(a)** write the mesh current equations, **(b)** arrange the results of (a) in standard mathematical form and solve for the mesh currents, and **(c)** determine all branch currents.

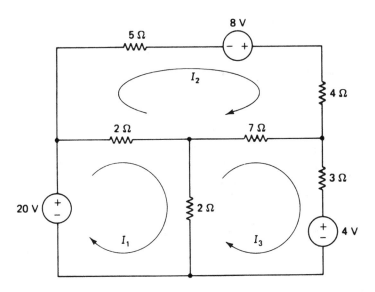

FIGURE P2–14

2–15. For the circuit of Figure P2–15, **(a)** write the node voltage equations, and **(b)** arrange the results of (a) in standard mathematical form and solve for the node voltages.

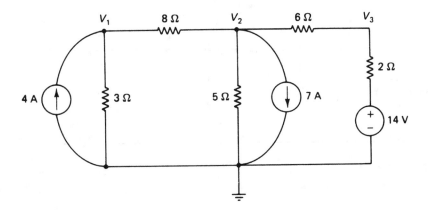

FIGURE P2–15

2–16. For the circuit of Figure P2–16, **(a)** write the node voltage equations, and **(b)** arrange the results of (a) in standard mathematical form and solve for the node voltages.

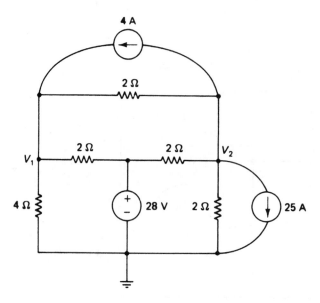

FIGURE P2–16

2–17. Determine the Thevenin and Norton equivalent circuits at the terminals *x-y* in Figure P2–17.

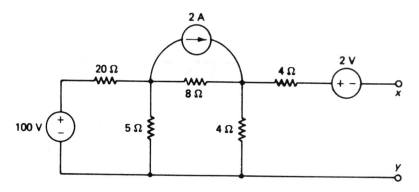

FIGURE P2–17

2–18. Determine the Thevenin and Norton equivalent circuits at the terminals *x-y* in Figure P2–18.

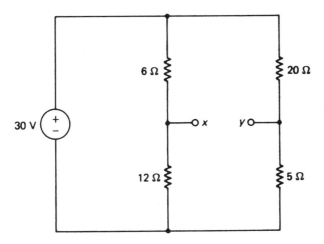

FIGURE P2-18

2-19. For the circuit of Figure P2–19, determine **(a)** the loop current I, **(b)** voltage drops across all resistors, **(c)** voltage of dependent source, and **(d)** power associated with each element and whether it is absorbed or delivered.

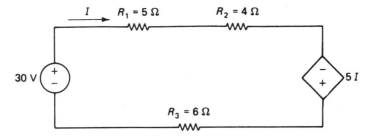

FIGURE P2-19

2-20. For the circuit of Figure P2–20, determine **(a)** the loop current I, **(b)** voltage drops across all resistors, **(c)** voltage of dependent source, and **(d)** voltage V_0 across current source.

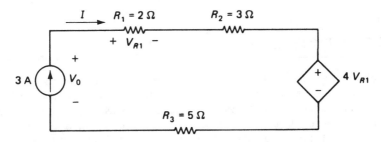

FIGURE P2-20

2–21. For the circuit of Figure P2–21, determine **(c)** the node-pair voltage V, **(b)** currents through all resistors, **(c)** current of dependent source, and **(d)** power associated with each element and whether it is absorbed or delivered.

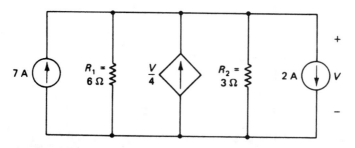

FIGURE P2–21

2–22. For the circuit of Figure P2–22, determine **(a)** the node-pair voltage V, **(b)** currents through all resistors, **(c)** current of dependent source, and **(d)** current I_0 through voltage source.

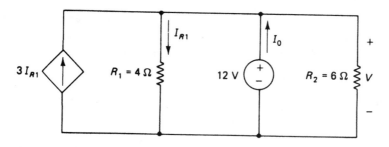

FIGURE P2–22

2–23. Determine the voltage V_0 in Problem 2–9 (Figure P2–9) using the principle of super-position.

2–24. Determine the current I_0 in Problem 2–10 (Figure P2–10) using the principle of super-position.

DERIVATION PROBLEMS

2–25. The objective of this problem is to derive the maximum power transfer theorem for a dc resistive circuit. Consider the circuit of Figure P2–25 containing a source with open-circuit voltage V_s and *internal* resistance R_s. It is assumed that R_s is fixed and cannot be changed. Under that constraint, an *external* resistance R_L, in which the power P_L is of interest, is adjusted to give maximum load power. **(a)** Derive an

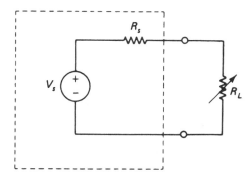

FIGURE P2-25

expression for P_L in terms of R_L, V_S, and R_s. **(b)** Using differential calculus, prove that the value of R_L resulting in maximum load power is $R_L = R_s$. **(c)** Show that the maximum value of the load power under this condition is $P_L(max) = V_s^2/4R_s$. **(d)** Sketch a curve of P_L as a function of R_L. Identify the values of P_L when $R_L = 0$ and $R_L = \infty$.

2-26. Assume that a given dc series source is modeled by a voltage source V_s in series with a resistance R_s as shown in Figure P2-26. A load requiring a current I_L is connected as shown. Assume that the source model is valid for any current requirement in which the load voltage V_L varies from V_s to zero. **(a)** Determine an expression for V_L as a function of I_L, V_s, and R_s, and **(b)** show that the result of (a) is a linear (straight-line) equation. Sketch the form in the first quadrant with I_L as the abscissa and V_L as the ordinate. Identify the slope and the two intercepts.

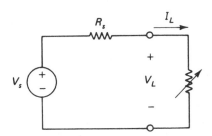

FIGURE P2-26

2-27. This analysis has some similarity to that of Problem 2-26, but the independent variable in this case is the load resistance R_L. Refer to Figure P2-27, and assume the same conditions for the source as in Problem 2-26. **(a)** Determine an equation for V_L as a function of R_L. Is this equation a linear form? **(b)** Sketch the form of the equation with R_L as the abscissa and V_L as the ordinate. Identify on the curve the value of V_L when $R_L = 0$ (vertical intercept) and the value of V_L as $R_L \to \infty$. The latter value of V_L is referred to as a *horizontal asymptote*.

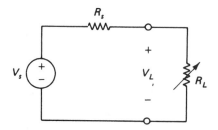

FIGURE P2-27

2-28. The determination of the Thevenin or Norton equivalent circuit of a signal source in the laboratory usually requires a completely different procedure than the "paper" procedures discussed in the chapter. A procedure that works well for small signal sources with moderate to high values of internal resistance is illustated in Figure P2-28. First, the open-circuit voltage v_{oc} is measured, and we know that $v_t = v_{oc}$. Next,

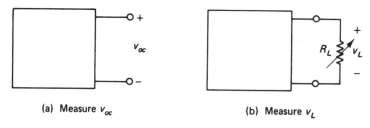

(a) Measure v_{oc} (b) Measure v_L

FIGURE P2-28

a variable resistance R_L is connected across the output, and the resistance is adjusted until the load voltage v_L is

$$v_L = \frac{v_{oc}}{2}$$

At this point, R_L is disconnected and measured. Show that the equivalent resistance R_{eq} is

$$R_{eq} = R_L$$

2-29. The procedure of Problem 2-28 for determining the equivalent resistance of a source is generally not suitable for low internal resistance sources due to the excessive loading when $R_L = R_{eq}$. Referring again to Figure P2-28, an alternate procedure is to adjust R_L for only a moderate decrease in the terminal voltage. For a terminal voltage v_L with load, show that the value of R_{eq} is

$$R_{eq} = \frac{(v_{oc} - v_L)R_L}{v_L}$$

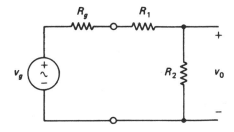

FIGURE P2-30

2-30. A requirement that frequently arises in laboratory testing is to provide a simple voltage divider to attenuate (reduce) the level of a signal without changing the output resistance. Referring to Figure P2–30, assume that a given laboratory generator has an open-circuit voltage v_g and an output resistance R_g. The resistances R_1 and R_2 are added externally, and the following requirements are imposed:

a. $\dfrac{v_0}{v_g} = \alpha \qquad (\alpha < 1)$

b. R_{eq} = Thevenin equivalent resistance looking back from output = R_g. Show that the resistances R_1 and R_2 are

$$R_1 = \frac{(1-\alpha)}{\alpha} R_g$$

$$R_2 = \frac{R_g}{1-\alpha}$$

APPLICATION PROBLEMS

2-31. Using the results of Problem 2–29, assume that measurements are taken on a certain amplifier to determine the Thevenin equivalent resistance. The measured open-circuit signal output voltage is 6 V. When a 120-Ω load is connected across the output, the voltage drops to 5.5 V. Determine the Thevenin equivalent resistance.

2-32. Using the results of Problem 2–30, assume that a given laboratory generator has an output resistance of 50 Ω and that it is desired to reduce the output voltage by a 10 to 1 ratio, i.e., $\alpha = 0.1$. Design the attenuation circuit.

2-33. Repeat Problem 2–32 if $\alpha = 0.01$.

2-34. An unbalanced bridge circuit is an example of a circuit in which the equivalent resistance R_{eq} cannot be determined by simple series-parallel combinations as considered in Chapter 1. Referring to Figure P2–34, one rather general way in which R_{eq} can be determined is from the ratio $R_{eq} = V_s/I_1$, where V_s is an assumed source voltage and I_1 is the current through this source. For this circuit, two other variables, I_2 and I_3, are required to constitute a complete set of mesh current equations. The variables I_2 and I_3 are eliminated, thus allowing I_1 to be expressed directly in terms of V_s, and the desired ratio can then be determined. (*Note:* An alternate way to determine the

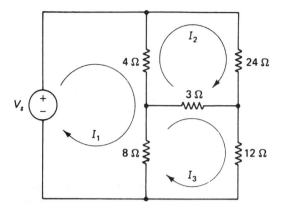

FIGURE P2–34

equivalent resistance of this circuit is with Y-Δ transformations, but that topic is not considered in this text.) For the resistance values in this particular bridge circuit of Figure P2–34, determine R_{eq}.

2–35. A three-terminal variable resistance (also called a potentiometer) is shown in Figure P2–35(a). The adjustable terminal B can be moved all the way from point A to point C. Let R_{AB} represent the resistance between A and B, and let R_{BC} represent the resistance between B and C. If R is the total resistance between A and B, it is evident that $R = R_{AB} + R_{BC}$. Assume that a 10-kΩ potentiometer (i.e., $R = 10^4$ Ω) is connected in the circuit of Figure P2–35(b), in which $V_1 = 100$ V and $V_2 = 10$ V, but in which no load current is drawn. Determine R_{AB}, R_{BC}, and the total power dissipated in the potentiometer.

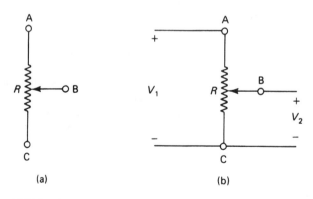

FIGURE P2–35

2–36. The 10-kΩ potentiometer of Problem 2–35 is connected in the circuit of Figure P2–36 in which $V_1 = 100$ V, $V_2 = 10$ V, and the required load current is $I_L = 50$ mA. Determine R_{AB}, R_{BC}, the total power dissipated in the potentiometer, and the total power supplied by the 100-V source.

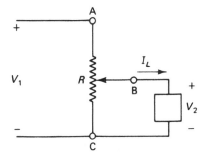

FIGURE P2–36

2–37. When a voltmeter is connected between two points in a circuit to measure the voltage, it presents an additional load to the circuit, which can affect the voltage being measured. As long as the resistance of the voltmeter is very large compared to the Thevenin equivalent resistance looking back from the two points at which the measurement is made, the loading effect is usually negligible. However, this is not always the case, particularly in high-resistance circuits when a classical passive voltmeter based on the concept of Problems 1–33 and 1–34 is used. (Modern electronic voltmeters provide very high input resistances.)

Assume that the measurement of a certain voltage in a relatively high-resistance circuit is attempted with a dc voltmeter having a sensitivity of 10,000 Ω/V. Readings on two scales are determined as follows:

    a. 10-V scale: 5 V
    b. 30-V scale: 6.67 V

Assuming a linear circuit, determine the true value of the desired open-circuit voltage V_{oc} and the equivalent resistance R_{eq} seen from the terminals.

3

CAPACITIVE AND INDUCTIVE TRANSIENTS AND EQUIVALENT CIRCUITS

OBJECTIVES

After completing this chapter, the reader should be able to:

- Apply graphical differentiation to piecewise linear continuous functions.
- Apply graphical integration to piecewise linear functions.
- State and apply the instantaneous voltage–current relationships for a capacitance.
- State and apply the instantaneous voltage–current relationships for an inductance.
- Obtain the equivalent capacitance for a combination of capacitors and the equivalent inductance for a combination of inductors.
- Apply voltage and current divider rules for capacitance or inductance.
- State and apply the voltage–current relationships for mutual inductance.
- State and apply the relationships for ideal transformers.

3–1 GRAPHICAL SIGNIFICANCE OF DIFFERENTIATION

As we prepare to study the instantaneous behavior of capacitance and inductance in electrical circuits, it is necessary to have a firm foundation in the elements of differential and integral calculus. Students learn to perform the operations of differentiation and integration in basic calculus courses, but they seldom grasp the full significance of these operations at that time. It is important to understand how these operations occur in electrical components and how they affect the nature of the waveforms involved.

In this section and the next, the standard mathematical variables x and y will be used so that the development here will correlate well with that given in basic calculus texts. The quantity x will be considered as the *independent variable,* and $y = f(x)$ will be considered as the *dependent variable.* The designation $f(x)$ refers to a "function of x." No particular units or electrical quantities will be associated with these symbols at this time. Later, however, we will see that the differentiation and/or integration processes are frequently applied to voltage, current, power, and energy functions; and time is usually the independent variable of interest.

Piecewise Linear Functions

The operations of differentiation and integration will be closely investigated as they relate to *piecewise linear functions.* A piecewise linear function is one composed of straight-line segments. In the case of differentiation, we will further restrict consideration at this time to functions that are *continuous.* A continuous function is one that does not possess abrupt jumps; that is, all changes in the function take some amount of time to occur.

Examples of two piecewise linear functions are shown in Figure 3–1. The function of (a) is a piecewise linear continuous function, since it has no sudden

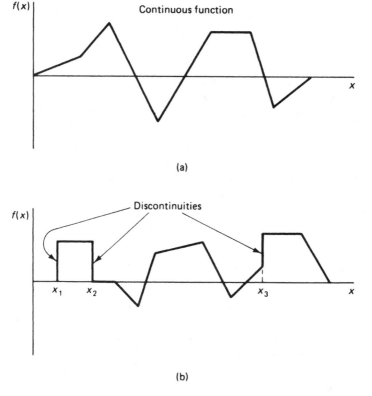

(a)

(b)

FIGURE 3–1
Examples of piecewise linear functions.

jumps. However, the function of (b) has three discontinuities (at x_1, x_2, and x_3), so it is not continuous.

Our interest in piecewise linear functions is twofold. First, many common waveforms arising in electrical circuit applications are of this general type. Second, and of major importance for our immediate purposes, working with piecewise linear functions in basic transient analysis provides some insight into the physical nature of the operations that would be difficult with other functions.

Although it is possible to write equations that completely describe piecewise linear functions (and this will be done in Chapter 10 for special applications), the types of problems considered in this chapter can be solved much more easily by a graphical approach. This approach helps to establish the physical basis for the operations involved. Thus, all waveforms considered in the next several sections will consist of piecewise linear functions for which graphical solutions will be used.

The remainder of this section will be devoted to a review of the basic properties of differentiation and a development of the techniques for differentiating a piecewise linear continuous function by inspection. The corresponding process for integration will be performed in the next section. It is assumed that all readers will have previously studied calculus, since the development here is not intended to be a complete treatment. Rather, it is intended as a practical supplement to the standard developments in basic calculus texts.

Derivative

In the study of calculus, the process of differentiation consists of determining one or more *derivatives* of the function. The first derivative is a new function arising from differentiating (i.e., applying the process of differentiation to) the original function once. The second derivative is the first derivative of the first derivative function, and the third derivative is the first derivative of the second derivative function, and so on. Thus, most general properties of derivatives can be deduced from the properties of the first derivative.

Derivative as Slope

The first derivative of a function is the *slope* of the function at any point. The slope is a measure of how quickly the function is changing. The concept of the slope for some arbitrary function is illustrated in Figure 3–2. At any value of x, assume that a tangent to the curve of $y = f(x)$ is constructed as shown. Assume that the tangent line changes by an amount Δy as x changes by Δx. The slope of the tangent line is defined as $\Delta y / \Delta x$. The first derivative is denoted as either

$$f'(x), \quad \frac{df(x)}{dx}, \quad \text{or} \quad \frac{dy}{dx}$$

The definition of the first derivative is

$$\frac{dy}{dx} = \lim_{\Delta x \to 0} \frac{\Delta y}{\Delta x} \tag{3–1}$$

FIGURE 3–2
Tangent line to a curve leading to the definition of the derivative.

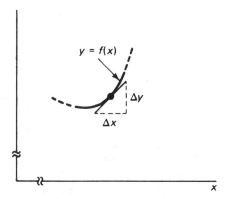

where "lim" means the limit of the quantity to the right as the change becomes infinitesimally small (Δx approaches zero). The result is the slope of the curve at a given point.

The slope of a piecewise linear function is a constant value on a given line segment. Furthermore, the slope of the segment may be readily determined by simply measuring the change in the dependent function y and dividing by the corresponding change in the independent variable x. The sign of the slope is very important. If y increases as x increases, it has a positive slope, but if the function decreases as x increases, the slope is negative. If there is no change in y as x changes (i.e., y is a constant), the slope is zero.

In the strictest mathematical sense, the derivative does not exist at a point where a function is discontinuous. However, by defining a quantity called the *impulse function,* it is possible to provide a meaningful and useful way of dealing with the derivatives of discontinuous functions such as square waves. This concept will be deferred until Chapter 10. At this point in the text, all waveforms for which differentiation is to be performed will be restricted to continuous functions.

Slope of Piecewise Linear Function

Consider the straight-line segment shown in Figure 3–3. This segment starts at $x = x_1$ and ends at $x = x_2$. The corresponding values of the function $y = f(x)$ are $y = y_1$ and $y = y_2$. The first derivative, which is constant everywhere along that segment, is calculated as follows:

$$\frac{dy}{dx} = \frac{y_2 - y_1}{x_2 - x_1} \tag{3–2}$$

Although stated in terms of the beginning and end points, which are usually the most convenient points with which to work, any two points on the given segment could be used, since the slope has a constant value over the entire segment.

The concept will be illustrated by the example that follows.

FIGURE 3–3
A straight-line segment for which the derivative (or slope) may be readily calculated.

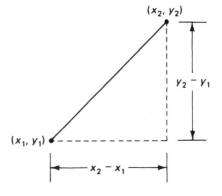

EXAMPLE 3–1

Determine graphically and plot the first derivative with respect to x of the piecewise linear continuous function y shown in Figure 3–4(a).

Solution

Differentiation of this piecewise linear function consists of determining numerically the slope of each segment and plotting the resulting function separately. The differentiated function, which is shown in Figure 3–4(b), employs the same horizontal (independent variable) scale as the original function. However, the derivative of a function has different units than the function and may require a vertical scale with quite a different numerical range than that of the function. In general, *the units of a differentiated function are the units of the dependent variable divided by the units of the independent variable.* For example, the derivative with respect to time of a voltage waveform has the dimension of volts/second. Since y and x are undefined "pure" functions, no units will be used in this example.

For analysis, the function y is divided into four segments.

0 < x < 4   In this segment, we note that $x_1 = 0$, $y_1 = f(0) = 0$, $x_2 = 4$, and $y_2 = f(4) = 20$. We have

$$\frac{dy}{dx} = \frac{y_2 - y_1}{x_2 - x_1} = \frac{20 - 0}{4 - 0} = 5 \qquad (3\text{–}3)$$

4 < x < 8   In this segment, $x_1 = 4$, $y_1 = f(4) = 20$, $x_2 = 8$, and $y_2 = f(8) = -20$. We have

$$\frac{dy}{dx} = \frac{-20 - 20}{8 - 4} = -10 \qquad (3\text{–}4)$$

8 < x < 12   In this segment, y is a constant and does not change. The slope is thus zero, that is,

$$\frac{dy}{dx} = 0 \qquad (3\text{–}5)$$

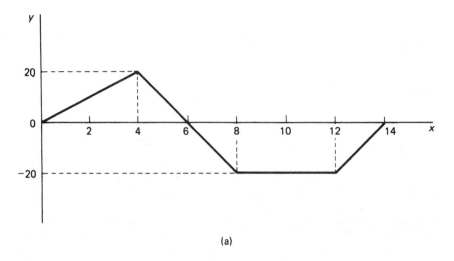

(a)

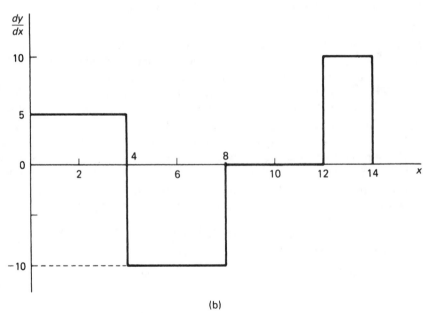

(b)

FIGURE 3–4
A piecewise linear continuous function and its first derivative.

12 < x < 14   In this segment, $x_1 = 12$, $y_1 = -20$, $x_2 = 14$, and $y_2 = 0$. Thus,

$$\frac{dy}{dx} = \frac{0 - (-20)}{14 - 12} = \frac{20}{2} = 10 \qquad \textbf{(3–6)}$$

The reader should carefully compare the plots of the function and its first derivative in Figure 3–4. Observe how to determine when the derivative is zero, positive, or negative in accordance with the direction of the change of the given function. Even when a function is not piecewise linear, one can determine visually whether a slope is positive or negative and, in some sense, the relative size of a derivative. Observe that there is one or more intervals in which both the function and its derivative are positive, both are negative, and both combinations of one positive and the other negative occur.

3–2 GRAPHICAL SIGNIFICANCE OF INTEGRATION

In calculus, the process of integration consists of determining the *integral* (or "anti-derivative") of a given function. Integrals may be classified as *indefinite* or *definite*. Although either type could be used for our purposes, it is much easier to use definite integrals in applying integration to most applications in circuit analysis, and the remainder of this section will be restricted to definite integral forms.

Integral as Area

Refer to Figure 3–5 for the discussion that follows. The definite integral of a function over the interval from $x = a$ to $x = b$ is the area under the curve. This is established in basic calculus by first considering the approximate area as a finite summation of a number of rectangular slabs as shown. The width of a given slab is Δx, the height is $y_k = f(x_k)$, and the corresponding area is $y_k \Delta x$, where k is an integer denoting the particular value of y for that rectangle. The net approximate area can be

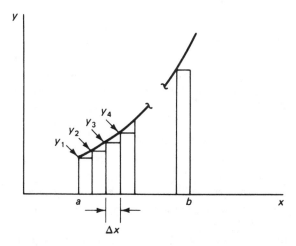

FIGURE 3–5
Area summation leading to the development of the definite integral.

FIGURE 3–6
Evaluation of the definite integral for a straight-line segment.

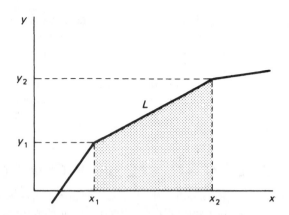

expressed as $\sum_k y_k \Delta x$, where $\sum_k$ indicates the sum of the small areas for all values of k. The definite integral is defined as follows:

$$\int_a^b y \, dx = \lim_{\Delta x \to 0} \sum_k y_k \, \Delta x \qquad (3\text{–}7)$$

The quantity on the left of Equation (3–7) is read as "the definite integral of y from a to b." The quantity $y \, dx$ is defined as the *integrand,* and a and b (which are specific values of x) are called the *limits* of the integral. Alternately, the integrand may be expressed as $f(x) \, dx$. Once the limiting form of the definite integral is established, the area is *exact* rather than an approximation, as was the case of the finite summation. Thus, the definite integral of a function from $x = a$ to $x = b$ may be interpreted as the area under the curve of the function between the limits.

Area of Piecewise Linear Function

For piecewise linear functions, the area between two limits on a line segment of a given slope may be determined by a simple geometric computation. Consider the particular line segment L shown in Figure 3–6, which lies between two segments with different slopes. The segment L starts at $x = x_1$ and ends at $x = x_2$. The definite integral between the limits x_1 and x_2 may be determined by evaluating the area of the trapezoid created by the line segment, and it is

$$\int_{x_1}^{x_2} y \, dx = \frac{1}{2}(y_2 + y_1)(x_2 - x_1) \qquad (3\text{–}8)$$

In case the reader's geometry is "rusty," the first two factors, namely, $(1/2)(y_2 + y_1)$, represent the average height of the trapezoid, and $x_2 - x_1$ is the width of the base. This formula may be interpreted to include a triangle (either y_1 or y_2 is zero) or a rectangle ($y_1 = y_2$) as special cases.

The integral for a given segment may be either positive or negative. In all cases of interest in this text, integration will be performed in the direction of increasing x, that is, from left to right in a conventional coordinate system. In this case, the sign of the integral for a given segment may be determined by the following simple inspection process: If y is *positive* over the range of the segment, the *area* is *positive*; and if y is *negative* for the segment, the *area* is *negative*. If a segment with a given slope crosses the x-axis (i.e., changes sign), it is necessary to divide it into segments, with the x-axis crossing representing the boundary between the segments. One area will be positive, and the other area will be negative.

The net integral over the entire range of x may be determined by first evaluating the individual areas under the curve for the different segments using the concept of Equation (3–8), with the sign of each segment carefully observed. The areas are then summed algebraically to yield the total area or integral. For example, assume that there are three segments defined over the limits of x_1 to x_2, x_2 to x_3, and x_3 to x_4, with increasing subscript number indicating the increasing direction of x. The net integral from x_1 to x_4 may be expressed as

$$\int_{x_1}^{x_4} y \, dx = \int_{x_1}^{x_2} y \, dx + \int_{x_2}^{x_3} y \, dx + \int_{x_3}^{x_4} y \, dx \qquad \textbf{(3–9)}$$

where each of the individual areas could be either positive or negative.

Definite Integral with Variable Upper Limit

Whereas the definite integral over the range of the limits is an important value, another useful concept in circuit analysis is that of predicting the behavior of the integral as the end point of the integration interval varies. Consider, then, an integral of the form

$$\int_{x_1}^{x} y \, dx$$

The beginning point is fixed at $x = x_1$, but the end point x may be anywhere along the interval between x_1 and x_2. What we are attempting to do is determine how the area under the curve accumulates as the right-hand limit of the integral varies. When x reaches the end of the segment, that is, $x = x_2$, the end value of the integral should be the value determined by the trapezoid formula of Equation (3–8).

Definite Integral as Sum of Areas of Slabs

In performing the analysis with x as an upper limit, it is helpful to imagine that the area is divided up into a series of narrow slabs as was done in the formulation of the integral definition. As x increases, the area values for successive slabs will either increase, decrease, or remain at the same value as for the preceding slab. The integral function is the successive accumulation of the areas of all preceding slabs, so the manner in which this function varies depends on the preceding growth pattern of the slabs. For successive slabs either increasing or decreasing in area at a linear rate, the integral function of a straight-line segment will follow a parabolic (second-degree) variation. If successive slabs have equal areas, which means that the function being integrated is simply a constant, the integral function will change at a constant rate.

Comparison of Derivative and Integral

Before closing this section, a few comments about differentiation and integration are in order. The derivative of a function is a measure of the slope or rate of change of the function. Consequently, it can change quickly, and the first derivative of a continuous function may have discontinuities. For example, refer back to Figure 3–4 and note the pronounced change in the derivative when y changes slope.

In contrast, the integral of a function cannot change too quickly. With the exception of impulsive conditions, which will be treated as a special case in Chapter 10, the integral of a function cannot have discontinuities. As area is accumulated, the integral of a function moves either upward or downward, but never with a sudden jump (except with impulsive conditions). If a given area is positive, the integral increases; and if the area is negative, the integral decreases. The movement, however, is from the level previously reached. For example, if the level previously established is positive, negative area in the next segment would cause the integral to decrease from that positive level.

EXAMPLE 3–2
Determine graphically and plot the definite integral as x varies from 0 to 14 of the piecewise linear function shown in Figure 3–7(a). The integral has no previously accumulated value.

Solution
The function desired is

$$\int_0^x y \, dx$$

For evaluation purposes, the total interval may be divided into a number of short intervals in which the function has a constant slope and one sign. The area accumulation for each segment starts at the level attained for the preceding segment. The end point for the new segment is determined by computing the net area for the new area and algebraically adding it to the level previously reached. The variation between the beginning and end points is predicted by noting the manner in which the area accumulates.

The integrated function, which is shown in Figure 3–7(b), employs the same horizontal (independent variable) scale as the original function. However, the integral of a function has different units than the function and may require a vertical scale with quite a different numerical range than that of the function. In general, *the units of an integrated function are the units of the dependent function times the units of the independent variable,* which agrees with the concept of the area property. For example, the integral over time of a current waveform has the dimensions of

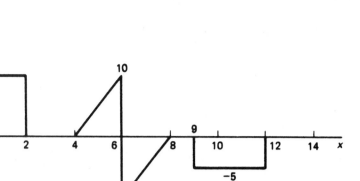

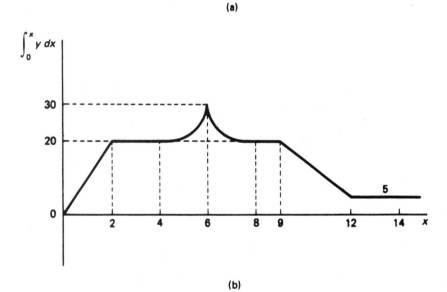

FIGURE 3–7
A piecewise linear function and its definite integral.

amperes times seconds. Since y and x are undefined "pure" functions, no units will be used in this example.

Each specific interval will be considered separately. Reference will be made to Figure 3–7(b) in the steps that follow.

$0 < x < 2$ As stated in the problem, the integral is to begin with zero initial value. Later, we will see how some definite integrals start with an initial value established at some earlier time. In the first segment, y is a constant value, so each slab has the same area as the preceding slab, and the area accumulates at a constant rate. Since the area is positive, the integral will increase. The value at the end of the interval is determined by evaluating the total area for the rectangle. Thus,

$$\int_0^2 y\,dx = 10 \times 2 = 20 \tag{3–10}$$

Thus, the integral increases linearly to a value of 20 at $x = 2$.

$2 < x < 4$ In this interval, $y = 0$, and no area is added. We can state for this interval that

$$\int_2^x y\,dx = 0 \quad \text{for } 2 < x < 4 \tag{3–11}$$

However, since we are plotting the integral starting from $x = 0$, we can say that

$$\int_0^x y\,dx = \int_0^2 y\,dx + \int_2^x y\,dx \quad \text{for } 2 < x < 4 \tag{3–12a}$$
$$= 20 + \int_2^x y\,dx \quad \text{for } 2 < x < 4 \tag{3–12b}$$

Thus, the net integral starts at the level previously established, and any new area (zero in this case) adds to that level. It is important to recognize this fact, because many beginners have a tendency to think that the integral should be zero when the function being integrated is zero. The function, however, remains at the level of 20 for the entire interval.

$4 < x < 6$ At the outset we observe that

$$\int_0^x y\,dx = \int_0^4 y\,dx + \int_4^x y\,dx \quad \text{for } 4 < x < 6 \tag{3–13a}$$
$$= 20 + \int_4^x y\,dx \quad \text{for } 4 < x < 6 \tag{3–13b}$$

Once again, we see that the area at any point in the interval is the area at the beginning plus any additional area accumulated during the interval. Equations of the form of (3–13a) and (3–13b) could be written for any of the intervals, but having established the concept, we will deal with it intuitively for the remainder of this example.

In the interval from 4 to 6, y varies linearly. Since y starts at a value of zero, the area changes slowly at first. However, as x increases, the area begins to increase at a more rapid rate, indicating a second-degree variation of the function. The total area from 4 to 6 is the area of the triangle involved and is

$$\int_4^6 y\,dx = \frac{1}{2} \times 10 \times 2 = 10 \tag{3–14}$$

This value added to the initial value of 20 results in a final value of 30. The parabolic variation is sketched in between the limits as noted in Figure 3–7(b).

6 < x < 8 In this interval, the area is *negative.* This means that the integral function must decrease. The rate of decrease is more significant at first because y has a larger magnitude for x near 6. However, as x approaches 8, the area decrease will be less. The total area from 6 to 8 is the area of the triangle involved and is

$$\int_6^8 y \, dx = -\frac{1}{2} \times 10 \times 2 = -10 \tag{3-15}$$

This value added to the initial value of 30 results in a final value of 20. The parabolic variation is sketched in between the limits as noted in Figure 3–7(b). Note that this particular negative area exactly balances out the positive area of the preceding interval.

8 < x < 9 Since $y = 0$ in this interval, the integral remains fixed at a level of 20.

9 < x < 12 In this interval, the area is negative again, so the integral function must decrease. The function has a constant negative value, so the integral decreases at a linear rate. The net area from 9 to 12 is the area of a rectangle and is

$$\int_9^{12} y \, dx = -5 \times 3 = -15 \tag{3-16}$$

This value added to the initial value of 20 results in a final value of 5. The straight-line variation involved is shown in Figure 3–7(b).

x > 12 For this interval, $y = 0$, and there is no further area. Thus, the integral remains at a level of 5 for all values of x thereafter. Some additional negative area would be required to bring the integral function back to zero.

3–3 CAPACITIVE TRANSIENTS

The physical basis of capacitance is established in basic books in electricity and circuits. A lumped element of capacitance is called a *capacitor.* The symbol for capacitance is C, and the basic unit is the *farad* (abbreviated as F). The farad is an enormous unit, so most practical capacitances are measured in microfarads (μF) and picofarads (pF). However, many simple, unrealistic values of capacitance will be assumed in conceptual problems to reduce the number of manipulations.

Schematic Representation

The schematic representation for a capacitor is shown in Figure 3–8, along with symbols for any voltage and current that may appear with it. Observe that the assumed directions of voltage and current imply power absorbed, which is the same convention as for resistance. Unlike resistance, however, the sign pattern for

FIGURE 3–8
Schematic representation of capacitor
and voltage and current directions.

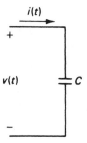

capacitance is more complex, and the power may turn out to be either absorbed
or delivered. Since capacitance is a passive parameter, any power delivered will
have been absorbed at an earlier time.

v–i Relationships

The voltage–current relationship for capacitance is

$$i(t) = C\frac{dv(t)}{dt} \qquad (3\text{–}17)$$

Stated in words, *capacitive current is proportional to the rate of change of the voltage
across the capacitor.* If the voltage is not changing, there is no current. Conversely,
if the voltage is changing rapidly, there may be a very large capacitive current.

    The inverse relationship to Equation (3–17) is obtained by integrating both
sides with respect to time. In general, all previous "history" of the capacitor must
be considered. To indicate this fact symbolically, the lower limit of the integral is
started at $t = -\infty$. The result is

$$v(t) = \frac{1}{C}\int_{-\infty}^{t} i(t)\,dt \qquad (3\text{–}18)$$

    In many practical problems, the time $t = 0$ is established as the beginning of
the interval of interest. The expression of Equation (3–18) can then be written as

$$v(t) = \frac{1}{C}\int_{-\infty}^{0} i(t)\,dt + \frac{1}{C}\int_{0}^{t} i(t)\,dt \quad \text{for } t > 0 \qquad (3\text{–}19)$$

The first term in Equation (3–19) represents the value of the voltage stored on the
capacitor at $t = 0$. Denoting this initial voltage as V_0 and reversing the order in
(3–19), we have

$$v(t) = \frac{1}{C}\int_{0}^{t} i(t)\,dt + V_0 \quad \text{for } t > 0 \qquad (3\text{–}20)$$

This is the form that will be used in most of the book. If $V_0 = 0$, the capacitor is
said to be initially *relaxed* or *uncharged*. If, however, $V_0 \neq 0$, the capacitor is said
to be initially *charged to a voltage V_0*.

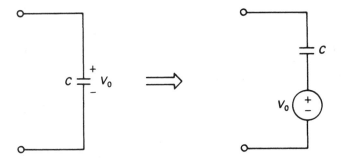

FIGURE 3–9
Equivalent circuit model of charged capacitor.

For a given $i(t)$, the presence of the V_0 term in Equation (3–20) shifts the voltage function up or down according to whether V_0 is positive or negative. The voltage starts to accumulate from a level V_0 as t increases.

Circuit Model of Charged Capacitor

The expression of Equation (3–20) leads to a useful circuit model for an initially charged capacitor. The sum of two voltages can be interpreted as a series connection, so a charged capacitor can be modeled as an uncharged capacitor in series with a dc voltage of value V_0. This equivalent circuit is shown in Figure 3–9. It must be stressed that the external terminals of the capacitor *must include* this dc voltage for the model to be correct.

Capacitive Power

The instantaneous power $p(t)$ for the capacitor can be expressed as

$$p(t) = v(t)\,i(t) = Cv(t)\frac{dv\,(t)}{dt} \tag{3–21}$$

where Equation (3–17) was substituted to yield the second form. This power may be either positive or negative. When the power is *positive*, the capacitor is *absorbing* power and it is *storing energy*. When the power is *negative*, the capacitor is *delivering* power and it is *releasing* this energy back to the circuit.

Energy Stored in Capacitance

The energy stored in a capacitor is a function of the voltage across it. It can be shown that the energy $w_C(t)$ stored in a capacitor at any time t is given by

$$w_C(t) = \frac{1}{2}Cv^2(t) \tag{3–22}$$

where $v(t)$ is the voltage across the capacitor at time t. If a capacitor is charged to a certain final voltage V_f and current ceases to flow, this voltage and an energy given by $CV_f^2/2$ will remain on the capacitor until a new current flow is established. In an ideal capacitor left open-circuited, the voltage and energy would remain stored forever. Practical capacitors, however, always have some leakage in the dielectric, which will eventually discharge the voltage and dissipate the energy. The energy stored in a capacitor represents *potential energy,* since it is a function of charge at rest.

We will next consider some numerical examples of voltage–current relationships for capacitors. The examples considered will employ piecewise linear functions, and the graphical differentiation and integration techniques of the preceding two sections will be used freely. Any piecewise linear voltage waveforms will be continuous, so that differentiation of a discontinuous function can be avoided at this time. However, finite discontinuities in current waveforms will be permitted.

EXAMPLE 3–3

The 0.5-μF capacitor of Figure 3–10 has a voltage waveform $v(t)$ as shown. Determine and plot as functions of time the (a) current, (b) power, and (c) energy.

Solution

(a) The current waveform may be determined by the relationship of Equation (3–17) as

$$i(t) = C\frac{dv(t)}{dt} = 0.5 \times 10^{-6}\frac{dv(t)}{dt} \tag{3-23}$$

Since the voltage waveform is a piecewise linear continuous function, it may be differentiated by determining the slopes of various segments according to the procedure of Section 3–1. The derivative is then multiplied by the constant C to determine the current. The calculations for different segments follow.

$0 < t < 2$ ms

$$\frac{dv}{dt} = \frac{40\text{ V}}{2 \times 10^{-3}\text{ s}} = 20 \times 10^3\text{ V/s} \tag{3-24}$$

$$i = 0.5 \times 10^{-6} \times 20 \times 10^3 = 10\text{ mA} \tag{3-25}$$

2 ms $< t < 4$ ms

$$\frac{dv}{dt} = 0 \tag{3-26}$$

$$i = 0 \tag{3-27}$$

4 ms $< t < 8$ ms

$$\frac{dv}{dt} = -\frac{40\text{ V}}{4 \times 10^{-3}\text{ s}} = -10 \times 10^3\text{ V/s} \tag{3-28}$$

$$i = 0.5 \times 10^{-6} \times (-10 \times 10^3) = -5\text{ mA} \tag{3-29}$$

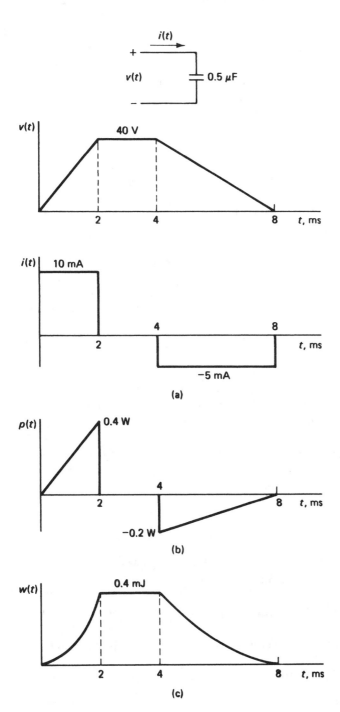

FIGURE 3–10
Capacitor and waveforms for Example 3–3.

A complete plot of $i(t)$ is given in Figure 3–10 (a). Observe that for the capacitor, *current flows when the voltage is changing, but no current flows when the voltage is constant.* Further, the current is *positive* when the voltage is *increasing*, but *negative* when the voltage is *decreasing.* All of these properties may be deduced from the basic relationship of Equation (3–17), but many students have not acquired such deductive logic with calculus. Examples such as this should help.

(b) The power function $p(t)$ is determined by

$$p(t) = v(t)\, i(t) \tag{3–30}$$

It is thus necessary to multiply the voltage by the current at all points. This task is made easier by the fact that the current has a constant level in each interval, and the basic shape of the power function is a function of the voltage waveform in that interval. We will consider each interval separately.

$0 < t < 2$ ms   The current is a constant positive value, so the power function has the same shape as the voltage waveform. The peak reached at the end of the interval is 40 V × 0.01 A = 0.4 W. Thus, the power increases linearly to 0.4 W as shown in Figure 3–10(b). This represents power *absorbed* by the capacitor.

2 ms $< t < 4$ ms   Since $i(t) = 0$ in this interval, $p(t) = 0$.

4 ms $< t < 8$ ms   As in the first interval, the current is constant, so the shape of the power function is related to the shape of the voltage. However, since the current is negative, the power is *inverted* with respect to the voltage. The peak value of this power occurs at the beginning of the interval and is 40 V × (-5×10^{-3} A) = -0.2 W. The power then changes linearly from this initial value to a final value of zero, as shown in Figure 3–10(b). Since the function is negative in this interval, the capacitor is *delivering* power back to the circuit.

(c) The energy function $w(t)$ is

$$w(t) = \int_0^t p(t)\, dt \tag{3–31}$$

Since $p(t)$ is a piecewise linear function, the integral may be evaluated graphically. Each interval will now be considered.

$0 < t < 2$ ms   The power is a first-order linear increasing positive function, so the energy will be a second-order (parabolic) increasing function. The final value reached in the interval is the area of the first triangle and is

$$w(2 \times 10^{-3}) = \frac{1}{2} \times (0.4\text{ W}) \times (2 \times 10^{-3}\text{ s}) = 0.4\text{ mJ} \tag{3–32}$$

2 ms $< t < 4$ ms   The power is zero in this interval, so the energy remains fixed at 0.4 mJ.

4 ms < t < 8 ms In this interval, the power is negative, so the integral decreases. The decrease is more pronounced at first, but tapers off later at a parabolic rate. The total decrease in the interval is determined by evaluating the negative area under the triangle, which is $(1/2) \times (-0.2 \text{ W}) \times (4 \times 10^{-3} \text{ s}) = -0.4$ mJ. This value subtracts from the initial interval value of 0.4 mJ, thus leaving the final value at zero. The energy stored in the capacitor earlier is released during this last interval.

As an additional check point, the formula of Equation (3–22) applied during the interval when the voltage is 40 V yields a value of energy given by $(1/2) \times (0.5 \times 10^{-6}) \times (40)^2 = 0.4$ mJ, which agrees with the result of (c) in Figure 3–10 at the peak level.

The reader should carefully study all the waveforms so that the behavior of each function in each interval is clearly established.

EXAMPLE 3–4

The 0.2-μF capacitor of Figure 3–11 has a current waveform $i(t)$ as shown. Determine and plot the voltage waveform as a function of time. The capacitor is initially uncharged.

Solution

The voltage waveform may be determined by the relationship of Equation (3–20) with $V_0 = 0$. That is,

$$v(t) = \frac{1}{C} \int_0^t i(t)\, dt = \frac{1}{0.2 \times 10^{-6}} \int_0^t i(t)\, dt = 5 \times 10^6 \int_0^t i(t)\, dt \qquad \textbf{(3–33)}$$

Since the current waveform is a piecewise linear continuous function, it may be integrated by evaluating the area under successive segments of the curve. The area is then multiplied by the constant $(1/C)$ to determine the voltage. The calculations for different segments follow.

0 < t < 2 ms The current is positive and increases linearly. The area is thus positive, and it follows a second-degree variation. The net integral from 0 to 2 ms is the area of the triangle involved and is

$$\int_0^{2 \times 10^{-3}} i(t)\, dt = \frac{1}{2} \times (3 \times 10^{-3} \text{ A}) \times (2 \times 10^{-3} \text{ s}) = 3 \times 10^{-6} \text{ A} \cdot \text{s} \qquad \textbf{(3–34)}$$

$$v(2 \times 10^{-3}) = (5 \times 10^6) \times (3 \times 10^{-6}) = 15 \text{ V} \qquad \textbf{(3–35)}$$

Thus, the voltage reaches a level of 15 V at $t = 2$ ms.

2 ms < t < 8 ms The current is positive and is constant in this interval. The area thus increases at a linear rate. The net area accumulated from 2 ms to 8 ms is the area of the rectangle involved and is

$$\int_{2 \times 10^{-3}}^{8 \times 10^{-3}} i(t)\, dt = (3 \times 10^{-3}) \times (6 \times 10^{-3}) = 18 \times 10^{-6} \text{ A} \cdot \text{s} \qquad \textbf{(3–36)}$$

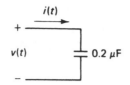

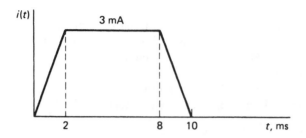

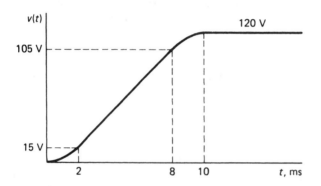

FIGURE 3–11
Capacitor and waveforms for Example 3–4.

The voltage at 8 ms is this area times $(1/C)$ plus the initial interval voltage of 15 V, that is,

$$v(8 \times 10^{-3}) = (5 \times 10^6) \times (18 \times 10^{-6}) + 15 = 105 \text{ V} \qquad (3\text{–}37)$$

The voltage thus increases linearly from 15 V to 105 V in this interval.

8 ms $< t <$ 10 ms   The current is still *positive,* so the area must continue to *increase.* However, since the current is decreasing, the rate of increase of the area is greater at the beginning of the interval. There is a tendency for some students to incorrectly consider the area as decreasing in a situation such as this, but as long as the quantity being integrated is *positive,* the area is *increasing.*

The net area accumulated from 8 ms to 10 ms is the area of the triangle involved and is

$$\int_{8\times10^{-3}}^{10\times10^{-3}} i(t)\,dt = \frac{1}{2} \times (3 \times 10^{-3}\,\text{A}) \times (2 \times 10^{-3}\,\text{s}) = 3 \times 10^{-6}\,\text{A} \cdot \text{s} \qquad \textbf{(3–38)}$$

The voltage at 10 ms is then

$$v(10 \times 10^{-3}) = (5 \times 10^{6}) \times (3 \times 10^{-6}) + 105 = 120\,\text{V} \qquad \textbf{(3–39)}$$

The voltage increases in this interval from 105 V to 120 V.

The complete waveform is shown in Figure 3–11. At the risk of redundancy, note again that since the current is always either positive or zero, the voltage is either increasing or remaining constant. Note how the voltage increase in the range from 8 ms to 10 ms gradually tapers off as the current waveform returns to zero. Note also how the slope changes are gradual near 2 ms and 8 ms, a result of the continuous nature of the function being integrated. Finally, observe that the voltage across the capacitor would theoretically remain charged to 120 V if no further current flows.

EXAMPLE 3–5

For the capacitor and current waveform of Example 3–4, assume that the capacitor is initially charged to a voltage of -50 V at the time the current starts to flow, as shown in Figure 3–12. Plot the voltage waveform as a function of time.

Solution
The relationship of Equation (3–20) is used again, but we now have $V_0 = -50$ V, and so

$$v(t) = 5 \times 10^{6} \int_{0}^{t} i(t)\,dt - 50 \qquad \textbf{(3–40)}$$

Comparing Equations (3–40) and (3–33), the only difference is that the area accumulation starts at the level of -50 V instead of zero. Thus, the required waveform is readily determined by taking the result of Example 3–4 and shifting it downward by 50 V. The result is shown in Figure 3–12.

To avoid creating a false impression, note that this result does not mean that placing a charged capacitor in a circuit, in general, will cause the resulting voltage waveform to be the same as for an uncharged capacitor displaced by the initial voltage. It would be true only if the current remains the same, as was assumed here. Unless the current is being generated by an ideal current source, the effect of an initial voltage on the capacitor will usually cause the current to change, so the situation assumed here should be carefully noted.

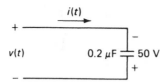

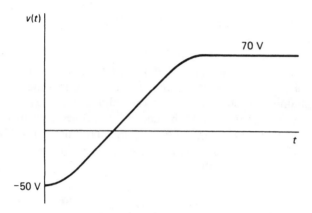

FIGURE 3–12
Capacitor and voltage waveform for Example 3–5.

3–4 INDUCTIVE TRANSIENTS

The physical basis of inductance is established in basic books in electricity and circuits. A lumped element of inductance is called an *inductor.* Depending on the application area, other names for specific inductors are *coil* and *choke.*

Two Forms of Inductance

Actually, there are two forms of inductance: *self-inductance* and *mutual inductance.* The latter phenomenon is associated with coupling between two or more inductors and will be considered as a separate topic in Section 3–6. As a rather common convention, the term *inductance* used alone usually refers to *self-inductance,* and that will be the assumption in this text. The symbol for self-inductance is *L,* and the basic unit is the *henry,* abbreviated H.

Schematic Representation

The schematic representation for an inductor is shown in Figure 3–13, along with symbols for any voltage across and current through it. Observe that the assumed directions of voltage and current imply power absorbed, which is the same convention as for resistance. Like capacitance, however, the sign pattern for inductance

FIGURE 3–13
Schematic representation of inductor and voltage and current directions.

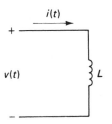

is more complex, and the power may turn out to be either absorbed or delivered. Since inductance is a passive parameter, any power delivered will have been absorbed at an earlier time.

v–i Relationships

The voltage–current relationship for inductance is

$$v(t) = L\frac{di(t)}{dt} \tag{3–41}$$

Stated in words, *inductive voltage is proportional to the rate of change of the current through the inductor.* If the current is not changing, there is no voltage. Conversely, if the current is changing rapidly, there may be a very large voltage.

The inverse relationship to Equation (3–41) is obtained by integrating both sides with respect to time. Symbolically, the integral is started at $t = -\infty$, and we have

$$i(t) = \frac{1}{L}\int_{-\infty}^{t} v(t)\,dt \tag{3–42}$$

In many practical problems, the time $t = 0$ is established as the beginning of the interval of interest. The current of Equation (3–42) can then be expressed as

$$i(t) = \frac{1}{L}\int_{-\infty}^{0} v(t)\,dt + \frac{1}{L}\int_{0}^{t} v(t)\,dt \tag{3–43}$$

The first term in Equation (3–43) represents the value of the current flowing in the inductor at $t = 0$. Denoting this initial current as I_0 and reversing the order in Equation (3–43), we have

$$i(t) = \frac{1}{L}\int_{0}^{t} v(t)\,dt + I_0 \quad \text{for } t > 0 \tag{3–44}$$

If $I_0 = 0$, the inductor is said to be initially *relaxed* or *unfluxed.* If $I_0 \neq 0$, the inductor is said to be initially *fluxed to a current I_0.* This terminology stems from the concept of magnetic flux, which exists in an inductor in which current is flowing. Magnetic flux for an inductor is analogous to electric charge for a capacitor.

For a given $v(t)$, the presence of the I_0 term in Equation (3–44) shifts the current function up or down according to whether I_0 is positive or negative. The current starts to accumulate from a level I_0 as t increases.

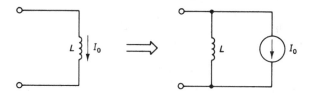

FIGURE 3–14
Equivalent circuit of fluxed inductor.

Circuit Model of Fluxed Inductor

The expression of Equation (3–44) leads to a useful circuit model for an initially fluxed inductor. The sum of two currents can be interpreted as a parallel connection, so a fluxed inductor can be modeled as an unfluxed inductor in a parallel with a dc current source of value I_0. This equivalent circuit is shown in Figure 3–14. It must be stressed that the external terminals of the inductor *must include* this dc current for the model to be valid.

Inductive Power

The instantaneous power $p(t)$ for the inductor can be expressed as

$$p(t) = v(t)i(t) = Li(t)\frac{di(t)}{dt} \tag{3–45}$$

where Equation (3–41) was substituted to yield the second form. As in the case of a capacitor, inductive power may be either positive or negative. When the power is *positive,* the inductor is *absorbing* power, *and it is storing energy.* When the power is *negative,* the inductor is *delivering* power and it is *releasing* this energy back to the circuit.

Energy Stored in Inductance

The energy stored in an inductor is a function of the current through it. It can be shown that the energy $w_L(t)$ stored in an inductor at any time t is given by

$$w_L(t) = \frac{1}{2}Li^2(t) \tag{3–46}$$

where $i(t)$ is the current through the inductor at time t. If a certain final current I_f is established in an inductor and the terminals are then shorted, the current would continue to circulate through the inductor and the short. Under such idealized conditions, an energy given by $LI_f^2/2$ would remain in storage. However, this situation is very difficult to achieve in practice because inductors always possess resistance in the conducting medium, and the energy is soon dissipated. Some experiments with superconducting materials at near absolute zero temperatures have produced conditions where the resistance is essentially zero, and the ideal conditions apply.

The energy stored in an inductor represents a form of *kinetic* energy, since it is a function of charge in motion.

Comparison of Capacitive and Inductive *V–i* Relationships

An interesting comparison of the voltage–current relationship for capacitance and inductance can be made by inspecting equations (3–17), (3–20), (3–41), and (3–44). The relationship for the current as a function of the voltage for a capacitor has the same mathematical form (derivative function) as for the voltage as a function of the current for an inductor, with C being the constant factor for a capacitor and L being the corresponding factor for an inductor. Conversely, the relationship for the voltage as a function of the current for a capacitor has the same mathematical form (integral function) as for the current as a function of the voltage for an inductor, with $1/C$ being the constant factor for a capacitor and $1/L$ being the corresponding factor for an inductor.

In the most basic physical form, a capacitor stores energy in the form of electric charge, and an inductor stores energy in the form of magnetic flux. In the sense of circuit analysis, however, it is more convenient to consider that a *capacitor stores voltage* and an *inductor stores current.*

We will now consider some numerical examples of voltage–current relationships for inductors. As in the last section, the examples considered will employ piecewise linear functions in which graphical techniques will be used. Any piecewise linear current waveforms will be continuous, so that differentiation of a discontinuous function can be avoided at this time. However, finite discontinuities in voltage waveforms will be permitted.

EXAMPLE 3–6

The 250-mH inductor of Figure 3–15 has a current waveform $i(t)$ as shown. Determine and plot as functions of time the **(a)** voltage, **(b)** power, and **(c)** energy.

Solution

(a) The voltage waveform may be determined by the relationship of (3–41), that is,

$$v(t) = L\frac{di(t)}{dt} = 0.25\frac{di(t)}{dt} \tag{3–47}$$

The piecewise linear current waveform may be differentiated by determining the slope of various segments. The derivative is then multiplied by the constant L to determine the voltage. The calculations for different segments follow.

$0 < t < 5$ ms

$$\frac{di}{dt} = \frac{0.4 \text{ A}}{5 \times 10^{-3} \text{ s}} = 80 \text{ A/s} \tag{3–48}$$

$$v = 0.25 \times 80 = 20 \text{ V} \tag{3–49}$$

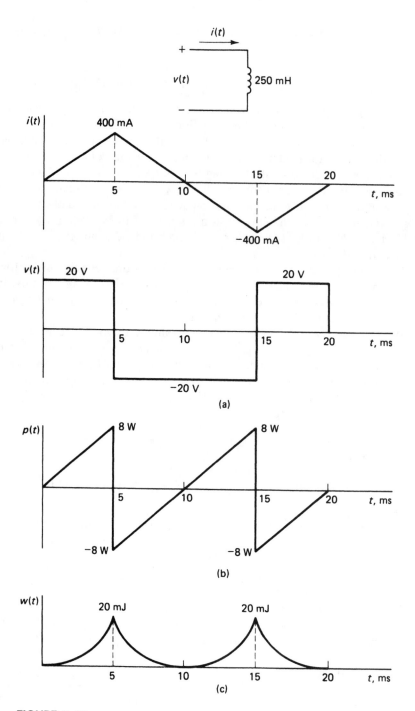

FIGURE 3–15

Inductor and waveforms for Example 3–6.

5 ms < t < 15 ms

$$\frac{di}{dt} = \frac{-0.4 - 0.4}{15 \times 10^{-3} - 5 \times 10^{-3}} = -80 \text{ A/s} \tag{3-50}$$

$$v = 0.25 \times (-80) = -20 \text{ V} \tag{3-51}$$

15 ms < t < 20 ms

$$\frac{di}{dt} = \frac{0 - (-0.4)}{(20 \times 10^{-3}) - (15 \times 10^{-3})} = 80 \text{ A/s} \tag{3-52}$$

$$v = 0.25 \times 80 = 20 \text{ V} \tag{3-53}$$

A complete plot of $v(t)$ is given in (a) of Figure 3–15. Observe that for the inductor, *a voltage appears when the current is changing.* However, *when the current is constant, the voltage across an inductance is zero.* Further, the voltage is *positive* when the current is *increasing,* but the voltage is *negative* when the current is *decreasing.*

(b) The power function $p(t)$ is determined by

$$p(t) = v(t)i(t) \tag{3-54}$$

The product of the voltage waveform and the current waveform must be formed at all points. Since the voltage waveform is composed of constant level segments, the shape of the power function will be primarily dependent on the current waveform. Each segment will be considered separately.

0 < t < 5 ms   The voltage is a constant positive value so the power function has the same shape as the current waveform. The peak reached at the end of the interval is (0.4 A × 20 V) = 8 W. Thus, the power increases linearly to 8 W as shown in Figure 3–15(b). This represents power *absorbed* by the inductor.

5 ms < t < 10 ms   For power computations, it is necessary to divide the original interval from 5 ms to 15 ms into two equal intervals. In the first interval, the current is positive and the voltage is negative, so the power is negative. Thus, energy is being returned to the circuit. The peak occurs at the beginning of the interval and is (0.4 A) × (−20 V) = −8 W. The power returns to zero linearly.

10 ms < t < 15 ms   In this interval, both the current and voltage are negative, so the power is positive. Thus, power is being absorbed again. The peak power level occurs at the end of the interval and is (−0.4 A) × (−20 V) = 8 W.

15 ms < t < 20 ms   In this interval, the current is negative and the voltage is positive, so the power is negative. The peak value occurs at the beginning and is (−0.4 A) × (20 V) = −8 W.

(c) The energy function $w(t)$ is

$$w(t) = \int_0^t p(t)\, dt \qquad\qquad (3\text{--}55)$$

Since $p(t)$ is a piecewise linear function, the integral may be evaluated graphically. Consider first the interval from 0 to 5 ms. The power is increasing linearly, so the energy must increase parabolically. The peak value of the energy occurs at $t = 5$ ms and is the area under the triangle from 0 to 5 ms. This value is

$$w(5 \times 10^{-3}) = \frac{1}{2} \times 8 \text{ W} \times (5 \times 10^{-3}\,\text{s}) = 20 \text{ mJ} \qquad (3\text{--}56)$$

In the interval from 5 ms to 10 ms, the power is negative, so the energy decreases. The decrease is sharpest at the beginning of the interval, since the power has the maximum negative value. Note that the negative area in this interval is equal in magnitude to the area in the first interval, so the energy returns to zero.

From the symmetry of the power function, it can be deduced that the behavior of the energy function in the last two intervals will duplicate that of the first two intervals. The complete function is shown in Figure 3–15(c).

As an additional check point, the formula of Equation (3–46) applied at the peak of the current waveform yields a value of energy given by $(1/2) \times (0.25) \times (0.4)^2 = 20$ mJ, which agrees with the result of Figure 3-15(c) at the peak level.

EXAMPLE 3–7

The 25-mH inductor of Figure 3–16 has a voltage waveform $v(t)$ as shown. Determine and plot the current waveform as a function of time. The inductor is initially unfluxed.

Solution

The current waveform may be determined by the relationship of Equation (3–44) with $I_0 = 0$, that is,

$$i(t) = \frac{1}{L} \int_0^t v(t)\, dt = \frac{1}{25 \times 10^{-3}} \int_0^t v(t)\, dt = 40 \int_0^t v(t)\, dt \qquad (3\text{--}57)$$

The piecewise linear voltage waveform may be integrated by evaluating the area under successive segments, and the integral is multiplied by $(1/L)$ to determine the current. The calculations for different segments follow.

$0 < t < 5\ \mu$s The voltage is positive and constant. The area thus increases linearly. The net integral from 0 to 5 μs is the area of the rectangle involved and is

$$\int_0^{5\times10^{-6}} v(t)\, dt = 20 \text{ V} \times (5 \times 10^{-6}\,\text{s}) = 100 \times 10^{-6} \text{ V} \cdot \text{s} \qquad (3\text{--}58)$$

$$i(5 \times 10^{-6}) = 40 \times (100 \times 10^{-6}) = 4 \text{ mA} \qquad (3\text{--}59)$$

Thus, the current rises to a level of 4 mA at $t = 5\ \mu$s.

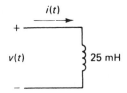

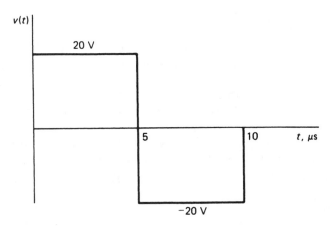

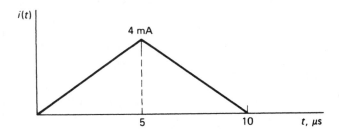

FIGURE 3–16
Inductor and waveforms for Example 3–7.

5 μs < t < 10 μs The voltage is negative and constant, and the area is negative. The net integral thus decreases at a linear rate from the level of 4 mA established at the end of the preceding interval. The net area added in this interval is

$$\int_{5\times10^{-6}}^{10\times10^{-6}} v(t)\, dt = -20\text{ V} \times (5 \times 10^{-6}\text{ s}) = -100 \times 10^{-6}\text{ V} \cdot \text{s} \qquad \textbf{(3–60)}$$

The value of the final current at $t = 10$ μs is

$$\begin{aligned}
i(10 \times 10^{-6}) &= (4 \times 10^{-3}) + 40 \times (-100 \times 10^{-6}) \\
&= (4 \times 10^{-3}) - (4 \times 10^{-3}) \\
&= 0
\end{aligned} \qquad \textbf{(3–61)}$$

This result could have been predicted intuitively, since the negative area is exactly equal in magnitude to the positive area.

3–5 COMBINATIONS OF CAPACITANCE AND INDUCTANCE

A combination of several capacitances may be represented by a single capacitance at a given set of terminals. Voltage and current divider rules may also be developed for capacitors. Similar properties hold for a combination of several inductors. These various relationships for both capacitance and inductance will be discussed in this section.

The focus here will be on stating, interpreting, and applying the relationships. The derivations of most of the concepts are given as guided exercises in the Derivation Problems at the end of the chapter.

The various relationships, along with any required conditions, will first be stated with very few comments. However, at the end of the section, a general overview and comparison of all of the relationships will be made.

Series Capacitors

The series combination of an arbitrary number of capacitors is illustrated in Figure 3–17. The equivalent capacitance C_{eq} is determined from the relationship

$$\frac{1}{C_{eq}} = \frac{1}{C_1} + \frac{1}{C_2} + \cdots + \frac{1}{C_n} \tag{3–62}$$

Parallel Capacitors

The parallel combination of an arbitrary number of capacitors is illustrated in Figure 3–18. The equivalent capacitance C_{eq} is given by

$$C_{eq} = C_1 + C_2 + \cdots + C_n \tag{3–63}$$

FIGURE 3–17
Equivalent capacitance of n capacitors connected in series.

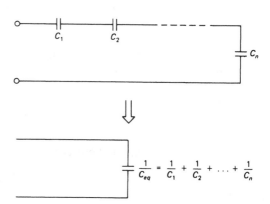

FIGURE 3–18
Equivalent capacitance of *n* capacitors connected in parallel.

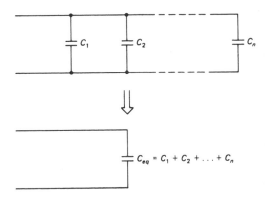

Series Inductors

The series combination of an arbitrary number of inductors is illustrated in Figure
3–19. It will be assumed at this point that there is no mutual coupling between
the inductors. This means that the inductors are either magnetically shielded or
sufficiently separated from each other that the magnetic flux generated in a given
inductor cannot link with a different inductor. With no mutual coupling, the equivalent inductance L_{eq} is

$$L_{eq} = L_1 + L_2 + \cdots + L_n \qquad (3\text{–}64)$$

Parallel Inductors

The parallel combination of an arbitrary number of inductors is illustrated in Figure
3–20. Assuming no mutual coupling, the equivalent inductance L_{eq} is determined
from the relationship

$$\frac{1}{L_{eq}} = \frac{1}{L_1} + \frac{1}{L_2} + \cdots + \frac{1}{L_n} \qquad (3\text{–}65)$$

FIGURE 3–19
Equivalent inductance of *n* inductors connected in series (no mutual coupling).

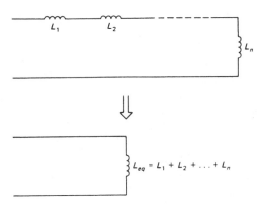

FIGURE 3–20
Equivalent inductance of n inductors connected in parallel (no mutual coupling).

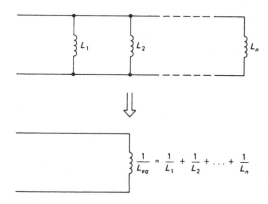

Capacitive Voltage Divider

Consider two capacitors C_1 and C_0 in series as shown in Figure 3–21, and assume a voltage $v_s(t)$ across the input. This voltage may be a source, or it may be a voltage existing within a circuit. It will be assumed that there are no voltages on either capacitor resulting from previous excitations. The voltage $v_0(t)$ across C_0 is directly proportional to $v_s(t)$ according to the following relationship:

$$v_0(t) = \frac{C_1}{C_1 + C_0} v_s(t) \qquad (3\text{--}66)$$

Capacitive Current Divider

Consider two capacitors C_1 and C_0 in parallel as shown in Figure 3–22, and assume a current $i_s(t)$ flowing into the parallel combination. This current may be a source, or it may be a current existing within a circuit. The current $i_0(t)$ flowing into C_0 is directly proportional to $i_s(t)$ according to the following relationship:

$$i_0(t) = \frac{C_0}{C_1 + C_0} i_s(t) \qquad (3\text{--}67)$$

FIGURE 3–21
Circuit used to illustrate capacitive voltage divider rule.

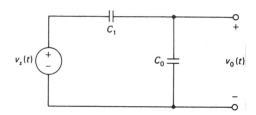

FIGURE 3–22
Circuit used to illustrate capacitive current divider rule.

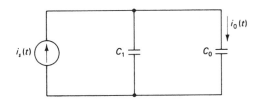

Inductive Voltage Divider

Consider two inductors L_1 and L_0 in series as shown in Figure 3–23 with a voltage $v_s(t)$ across the series combination. Assuming no mutual coupling between the inductors and no initial current, the voltage $v_0(t)$ is proportional to $v_s(t)$ in accordance with the following relationship:

$$v_0(t) = \frac{L_0}{L_0 + L_1} v_s(t) \qquad (3\text{–}68)$$

Inductive Current Divider

Consider two inductors L_1 and L_0 in parallel as shown in Figure 3–24 with current $i_s(t)$ flowing into the parallel combination. Assuming no mutual coupling between the inductors and no initial current, the current $i_0(t)$ is proportional to $i_s(t)$ in accordance with the following relationship:

$$i_0(t) = \frac{L_1}{L_0 + L_1} i_s(t) \qquad (3\text{–}69)$$

It is now appropriate to discuss the various relationships and their similarities and differences. In particular, the results for capacitance and inductance will be compared with those for resistance given in Chapter 1.

Comparison of Various Capacitance and Inductance Relationships

Comparing the formulas for combining series and parallel resistances as given by Equations (1–34) and (1–35) with the formulas for capacitance as given by Equa-

FIGURE 3–23
Circuit used to illustrate inductive voltage divider rule (no mutual coupling).

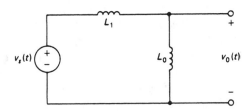

FIGURE 3–24

Circuit used to illustrate inductive current divider rule (no mutual coupling).

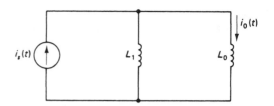

tions (3–62) and (3–63) and with the formulas for inductance as given by Equations (3–64) and (3–65), the following conclusions can be made: *Inductances with no mutual coupling combine in the same manner as resistances.* However, *capacitances combine in a sense opposite to that of resistances;* that is, *capacitances in parallel combine like resistances in series, and capacitances in series combine like resistances in parallel.*

Next, the voltage and current divider rules will be compared. The pertinent resistive relationships are Equations (1–38) and (1–40), and these can be compared with Equations (3–66) and (3–67) for capacitance and Equations (3–68) and (3–69) for inductance. The following conclusions can be made: *Inductive voltage and current divider rules* (where there is no mutual coupling) *have the same forms as the corresponding resistive voltage and current divider rules.* However, *capacitive voltage and current divider rules have opposite forms.* In this sense, the capacitive voltage divider rule has the "other capacitance" in the numerator, which gives it the same appearance as the current divider rule for resistance. When the voltage divider rule for capacitance is applied to more than two capacitors, the "other capacitance," namely, C_1, must represent the equivalent capacitance of all capacitors other than C_0. The same holds for the "other inductance" in the inductive current divider rule, and, as discussed in Chapter 1, a similar requirement holds in the resistive current divider rule.

The combination and divider rules as given here and in Chapter 1 hold only for "pure" circuits, that is, where all components are of the same type. Many circuits may be assumed to be totally resistive in nature, particularly at dc and at low frequencies. Of the two types of energy storage parameters, capacitance is usually the more ideal. Many practical capacitors have almost no losses, and may be modeled as ideal capacitors in many cases. The combination and divider rules given here may then be applied with little error to such circuits.

As a general rule, inductors are the least perfect of the three basic circuit parameters. All inductors contain some series resistance in the coil winding. However, the dc resistance of this wire may represent only a small fraction of the dynamic resistance. When a time-varying current flows through the inductor, various power losses occur, particularly if the core is composed of ferromagnetic material. The manner in which these losses are "charged" to the external circuit is by increasing the effective resistance of the coil. Thus, the dynamic or ac resistance of an inductor may be much greater than the dc resistance of the winding.

In view of these significant losses in many practical inductors, the various combination and divider rules for inductance must be carefully tempered and treated with reasonable caution in applications.

EXAMPLE 3–8

Consider three capacitors connected in the manner shown in Figure 3–25(a). Determine the equivalent capacitance C_{eq} as viewed from the left-hand terminals.

Solution
Starting on the right, the equivalent capacitance of the 0.02-μF and 0.1-μF capacitors in parallel is $0.02 + 0.1 = 0.12$ μF as shown in Figure 3–25(b). When this value is series combined with the 0.06-μF capacitor in accordance with Equation (3–62), the net result is $C_{eq} = 0.04$ μF as shown in Figure 3–25(c).

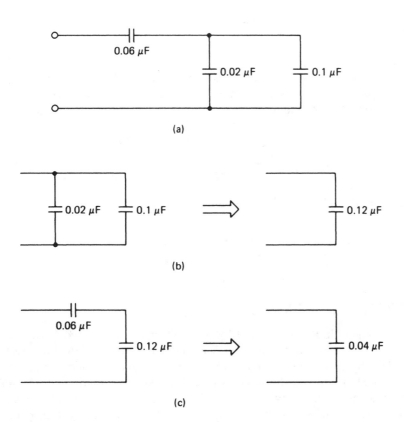

(a)

(b)

(c)

FIGURE 3–25
Circuit for Example 3–8.

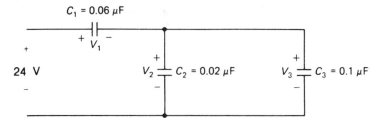

FIGURE 3–26
Circuit for Example 3–9.

EXAMPLE 3–9
Consider the combination of the three capacitors considered in Example 3–8, and assume that the combination has been charged, from an external circuit not shown, to a voltage of 24 V as shown in Figure 3–26. Determine the voltage across each capacitor.

Solution
It is convenient to consider the circuit as the series equivalent of a 0.06-μF capacitor and a 0.12-μF capacitor as was done in Figure 3–25(c). The capacitive voltage divider relationship of Equation (3–66) may then be applied. First, the voltage V_1 is determined as

$$V_1 = \frac{0.12\ \mu F}{0.12\ \mu F + 0.06\ \mu F} \times 24 = \frac{2}{3} \times 24 = 16\ \text{V} \qquad \text{(3–70)}$$

The voltages V_2 and V_3 are the same and can be determined as

$$V_2 = V_3 = \frac{0.06\ \mu F}{0.12\ \mu F + 0.06\ \mu F} \times 24 = \frac{1}{3} \times 24 = 8\ \text{V} \qquad \text{(3–71)}$$

3–6 MUTUAL INDUCTANCE

When the magnetic flux generated by one coil links with a second coil, the two coils are said to be *mutually coupled*. From a circuit point of view, the effects may be described by the concept of *mutual inductance*. The symbol for mutual inductance is M, but the basic unit is the same as for self-inductance, namely, the henry.

Mutual inductance may represent a deliberate and very desirable design property, of which the most notable example is the transformer. On the other hand, undesirable mutual coupling between coils in a circuit may completely disable an otherwise effective circuit.

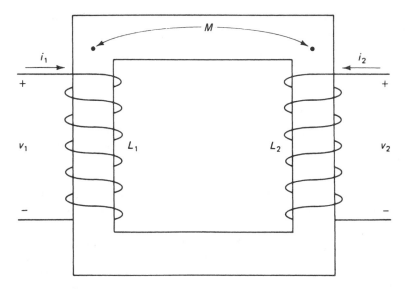

FIGURE 3–27
A transformer structure used to illustrate mutual inductance.

Transformer

To illustrate mutual inductance, consider two coils wound on a common core as shown in Figure 3–27. The structure shown is the basis of a *transformer,* and we will refer to it by that name. (Actually, most references to the term *transformer* tend to assume properties close to certain ideal conditions to be given in Section 3–7, but we need not be concerned about that now.) Each of the coils possesses a self-inductance, and these are denoted as L_1 and L_2 for the coils in Figure 3–27. In addition, there is a mutual inductance M between the coils whose effect will be described shortly. For convenience, we will refer to the left-hand coil (L_1) of the transformer as the *primary* and the right-hand coil (L_2) as the *secondary.*

Voltages Induced by Mutual Inductance Only

Assume first that the secondary is open ($i_2 = 0$) and that a current i_1 is flowing in the primary. If i_1 is changing, there is a voltage v_1 induced across the primary resulting from the primary self-inductance L_1. In addition, the changing magnetic flux linking the secondary induces a voltage v_{2oc} across the secondary given by

$$v_{2oc} = M \frac{di_1}{dt} \tag{3-72}$$

Next, assume that the primary is open ($i_1 = 0$) and that a current i_2 is flowing in the secondary. If i_2 is changing, there is a voltage v_2 induced across the secondary resulting from the secondary self-inductance L_2. In addition, the changing magnetic flux linking the primary induces a voltage v_{1oc} across the primary given by

$$v_{1oc} = M \frac{di_2}{dt} \qquad (3\text{–}73)$$

Both Equations (3–72) and (3–73) describe a voltage across the "other coil" produced by changing current in the reference coil, and this is the basis for mutual inductance. Note that the constant M is the same for both cases, that is, *a given rate of change of current in either coil induces the same voltage in the other coil.*

The polarity of the induced voltage depends on the relative directions of the windings. For the simplified coils in Figure 3–27, a positive rate of change for i_1 results in a positive value of v_{2oc}, and vice versa. If either coil (but not both) were wound in the opposite direction, the induced voltage for a positive rate of change of i_1 would result in a negative value of v_{2oc} and vice versa. This could be accounted for by defining M as either positive or negative, and some texts employ that approach. However, the sign convention can be confusing, and the approach taken here will be to define M as a positive number and to indicate the induced polarity by the method to be discussed in the next several paragraphs.

Dot Convention

The simplified transformer of Figure 3–27 shows only a few turns, and such a transformer is practical only for small inductances in high-frequency applications. Most transformers, particularly those used for power and audio applications, contain hundreds of turns. In such cases, it is not practical to inspect the directions of the windings. Rather, a common means of showing the relative polarities is by means of a *dot convention.* Referring again to Figure 3–27, the dot convention is defined as follows: *If the positive direction of a given current is flowing into the dot of one coil, the voltage induced in the other coil will be positive at that dot.*

Combined Effects of Mutual and Self-Inductances

When current is flowing in both coils, the total voltage induced in each coil will consist of two components: a self-inductance voltage generated by the rate of change of the current through the given coil and a mutual inductance voltage generated by the rate of change of current through the other coil. The sign pattern of these two components can be confusing to the uninitiated, but the procedure to be discussed shortly provides a straightforward means of "bookkeeping" for the circuit.

Procedure for Modeling Combined Effects

When a pair of mutually coupled coils is contained in a circuit to be analyzed, the following procedure may be used to account for the coupling:

1. Assume currents i_1 and i_2 in the two coils. The choice of directions is usually arbitrary.
2. As a result of i_1 in the primary, a fictitious generator $M(di_1/dt)$ is placed in series with the secondary coil. This generator is assumed to be "inside" coil 2. If i_1 is *entering* the dot of coil 1, the assumed direction of the generator is such that the *positive* reference would be measured at the dot of coil 2. The opposite is true, of course, if the current i_1 is assumed to be leaving the dot of coil 1.
3. As a result of i_2 in the secondary, a fictitious generator $M(di_2/dt)$ is placed in series with the primary coil. If i_2 is *entering* the dot of coil 2, the assumed direction of the generator is such that the *positive* reference would be measured at the dot of coil 1, and vice versa.

Including these fictitious generators takes care of the mutual coupling, and appropriate circuit analysis methods may then be applied. We are not yet ready to deal with circuits containing several types of circuit parameters, but these steps may be performed as a prelude to general analysis methods.

Two examples of circuits containing mutual inductance will now be considered. The first (Example 3–10) illustrates how mutual inductance is modeled and how the equations are written, although it is not a complete analysis. The second (Example 3–11) illustrates how the equivalent inductance of two inductors with mutual inductance can be determined.

EXAMPLE 3–10

The circuit of Figure 3–28(a) contains a pair of coils with mutual inductance M. **(a)** Construct an equivalent circuit with mutual coupling generators. **(b)** Write the two mesh current equations.

Solution

(a) Refer to Figure 3–28(b) in the discussion that follows. The current i_1 is assumed to enter the dot of L_1, so the generator $M\,di_1/dt$ must have a polarity such that its positive reference is measured at the dot of L_2. This is achieved by placing the positive reference of the secondary generator as the upper terminal. (Understand that this fictitious generator must be considered to be within the coil terminals so it could be placed either above or below the coil.) The current i_2 is assumed to enter the end of L_2 *opposite* to the dot, so the generator $M\,di_2/dt$ must have a polarity such that its positive reference is measured at the end of L_1 opposite to the dot. This is achieved by placing the positive reference of the primary generator as the lower terminal.

(b) With the equivalent generators shown, the mesh equations may be written using standard conventions. The voltage v_1 could represent a source, or it could represent a portion of some circuit not shown, but this need not concern us for our objective. For each loop, there will be a voltage drop due to the self-inductance plus a voltage resulting from the mutual inductance, the latter of which is determined from the fictitious generator.

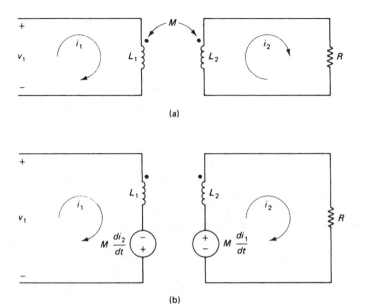

FIGURE 3–28
Circuit for Example 3–10.

The equation for mesh 1 reads

$$-v_1 + L_1 \frac{di_1}{dt} - M \frac{di_2}{dt} = 0 \qquad \textbf{(3–74)}$$

A similar equation for mesh 2 is

$$-M \frac{di_1}{dt} + L_2 \frac{di_2}{dt} + R\, i_2 = 0 \qquad \textbf{(3–75)}$$

EXAMPLE 3–11

Determine the equivalent inductance of two mutually coupled inductors connected in series with *fields aiding* as shown in Figure 3–29(a).

Solution

Since there is mutual coupling, we cannot simply add the two self-inductances. Rather, it is necessary to construct a dynamic circuit model and use the voltage–current relationship for the complete circuit to infer the equivalent inductance.

An equivalent circuit displaying the coupling effects is shown in Figure 3–29(b). A voltage v and a current i are assumed. Note that the same current flows through both coils. Thus, the current i flowing in L_1 results in a fictitious generator $M di/dt$ in series with L_2, while the same current flowing in L_2 results in a similar generator in series with L_1. Observe the signs of the generators in accordance with the previous procedures.

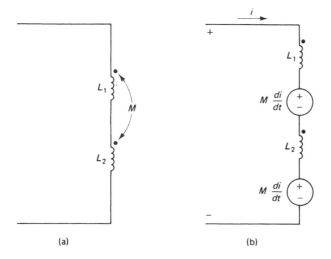

FIGURE 3–29
Circuit for Example 3–11.

A mesh equation for the loop reads

$$-v + L_1 \frac{di}{dt} + M \frac{di}{dt} + L_2 \frac{di}{dt} + M \frac{di}{dt} = 0 \tag{3–76}$$

This equation simplifies to

$$v = (L_1 + L_2 + 2M) \frac{di}{dt} \tag{3–77}$$

Since the circuit is totally inductive, it must satisfy an equation of the form

$$v = L_{eq} \frac{di}{dt} \tag{3–78}$$

where L_{eq} is the equivalent inductance. Comparing Equations (3–77) and (3–78), we obtain

$$L_{eq} = L_1 + L_2 + 2M \tag{3–79}$$

The result of Equation (3–79) tells us that the equivalent inductance of two inductors in series with *fields aiding* is *greater* than the sum of the two self-inductances. This situation would be the case if the two coils were wound in the same directions so that the flux generated by one adds to the flux generated by the other. If the two coils were wound in opposite directions, the flux generated by one would oppose the flux generated by the other, and the net inductance would be less than the sum of the two self-inductances. This case is left as an exercise for the reader (Problem 3–41).

3-7 IDEAL TRANSFORMER

When *all* of the magnetic flux generated by one coil links with a second coil, the mutual inductance M can be shown to satisfy the relationship

$$M = \sqrt{L_1 L_2} \tag{3-80}$$

The extent to which this condition is approached can be specified in terms of a parameter called the *coefficient of coupling*, which will be denoted by k. This quantity is defined as

$$k = \frac{M}{\sqrt{L_1 L_2}} \text{ where } 0 \le k \le 1 \tag{3-81}$$

The coefficient of coupling can be considered as the fraction of the total flux generated by one coil that links with the other coil. When there is no coupling, $k = 0$. Conversely, when all the flux generated by one coil links with the other coil, $k = 1$. This latter condition is referred to as *unity coupling*.

Ideal Transformer Assumptions

We will now define the concept of the ideal transformer. The ideal transformer possesses two major characteristics:

1. Unity coupling exists between the coils, that is, $k = 1$.
2. The self-inductance of the two windings are infinitely large, that is, $L_1 = \infty$ and $L_2 = \infty$.

It turns out that when these two limiting conditions are met, it is no longer necessary to use the somewhat awkward form of the mutual inductance equations and models to predict circuit conditions. Instead, the equations reduce to rather simple, widely used algebraic forms, as we will see shortly.

Although the ideal transformer is a mathematical limit that can never be fully achieved, many practical transformers are sufficiently close that the results to be stated may be applied with very little error. Transformers meeting these requirements usually are wound on ferromagnetic cores constructed to minimize losses. The self-inductances are made to be quite large by using many turns in each winding. Transformers meeting these requirements include, for example, power transformers used at 60 Hz and many audio transformers.

It should be stressed that even a transformer approaching the ideal model may not function properly if used under different conditions (such as frequency range) than those for which it was designed. In particular, it is always assumed that the voltages and currents are varying with time at a sufficiently high rate to ensure satisfactory operation. This is usually specified in terms of the frequency limits of the transformer. Remember that a transformer winding appears to dc as nearly a short circuit, and this can be a disastrous situation! The currents in the windings must be changing with time to generate the time-varying magnetic flux required for satisfactory operation.

FIGURE 3–30
Ideal transformer model and relationships.

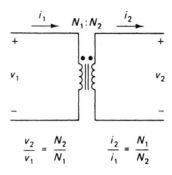

$$\frac{v_2}{v_1} = \frac{N_2}{N_1} \qquad \frac{i_2}{i_1} = \frac{N_1}{N_2}$$

Ideal Transformer Model

The model of the ideal transformer with voltages and currents labeled on both terminals is shown in Figure 3–30. Note that the current i_2 is shown leaving the secondary terminal, which is opposite to the direction assumed on the basic mutual inductance model in Figure 3–27.

Turns Ratio

A most significant parameter for an ideal transformer is the *turns ratio*. It is assumed here that N_1 is the number of primary turns, and N_2 is the number of secondary turns. From an external circuit point of view, the actual number of turns on each winding is usually unimportant (and usually not known anyway), but it is the ratio N_1/N_2 or N_2/N_1 that is significant. For example, a turns ratio could be specified as 4/1, meaning that the number of primary turns is four times the number of secondary turns, but there may be hundreds of actual turns on each winding.

v–i Relationships

The following voltage and current ratios hold for an ideal transformer:

$$\frac{v_2}{v_1} = \frac{N_2}{N_1} \tag{3–82}$$

and

$$\frac{i_2}{i_1} = \frac{N_1}{N_2} \tag{3–83}$$

Thus, the voltage ratio is in direct proportion to the turns ratio in the same sense, whereas the current ratio is opposite in sense. An easy way to remember these relationships is that the side with the larger number of turns has the larger voltage and the smaller current. Any unknown quantities can then be expressed in terms of known quantities by simply organizing the turns ratio in the manner that will ensure these conditions are met. This thought process is recommended rather than rigid memorization of the equations.

Next, the left-hand and right-hand sides of Equations (3–82) and (3–83) will be multiplied together. This process gives

$$\left(\frac{v_2}{v_1}\right)\left(\frac{i_2}{i_1}\right) = \left(\frac{N_2}{N_1}\right)\left(\frac{N_1}{N_2}\right) \qquad \textbf{(3–84a)}$$

or

$$\frac{v_2 i_2}{v_1 i_1} = 1 \qquad \textbf{(3–84b)}$$

The quantities $v_2 i_2$ and $v_1 i_1$ represent the instantaneous output and input power functions, respectively. According to Equation (3–84b), these power functions are equal, meaning that all the power accepted by the ideal transformer must be delivered to the load. Thus, the ideal transformer is characterized by the property that either voltage is increased and current is decreased or voltage is decreased and current increased. However, the power is not changed. In practical transformers, some power is dissipated internally, so the output power is always less than the input power.

Application to Power Transmission

A major application of large power transformers is in the commercial transmission and distribution of power. All transmission lines exhibit some series resistance R_s. If an effective current I is flowing in the resistance, the power loss during transmission is proportional to $I^2 R_s$. In view of the squared relationship, it would be virtually impossible to transmit large amounts of power over any appreciable distance at the high current levels required for large load requirements (e.g., an entire city). To circumvent this problem, the voltage is "stepped up" (increased) at generation stations to very high voltages. As the voltage is increased, the current is decreased in opposite proportion, and the $I^2 R_s$ losses decrease markedly. Such high-voltage lines are used for long-distance transmission and distribution. In the vicinity of customer user areas, the voltage is "stepped down" (decreased) by transformers to the levels required for practical users. This concept will be illustrated in Example 3–13.

Virtually all electronic equipment designed to operate from standard ac voltages requires one or more moderate-size transformers as part of the power supply. These transformers change the standard ac voltage to levels required for the rectifier circuits. The rectifier circuits change the ac to dc and provide proper voltage levels to operate the electronic devices.

Reflecting Secondary to Primary Side

Circuits containing ideal transformers can be analyzed by conventional methods such as mesh and node analysis, but with the additional transformer equations given by Equations (3–82) and (3–83) used to establish constraints between primary and secondary. However, this approach is often awkward to manage because of the extra equations and variables. A more expedient approach for many circuits is achieved by the process of *reflection.* The concept of reflection involves obtaining an equivalent circuit on one side of the transformer that can be used to predict all the results on that particular side. The circuit can be reflected to either the primary or secondary side, and the choice will often be dictated by the side on which most of the analysis is desired. However, once the response is known on one side, it is a straightforward process to return to the original circuit, so the choice is often arbitrary.

The concept of reflection will be developed with the circuit of Figure 3–31(a). A time-varying source $v_s(t)$ in series with a resistance R_1 is connected to the primary side, and a resistance R_2 is connected to the secondary side. Since the secondary has the simpler form, it will be reflected to the primary in this development.

The transformer equations will be stated with v_1 and i_1 as the numerators to assist in subsequent simplification.

We have

$$\frac{v_1}{v_2} = \frac{N_1}{N_2} \tag{3–85}$$

and

$$\frac{i_1}{i_2} = \frac{N_2}{N_1} \tag{3–86}$$

The two sides of Equation (3–85) will next be divided by the corresponding sides of Equation (3–86). After a slight rearrangement, there results

$$\frac{v_1}{i_1} = \left(\frac{N_1}{N_2}\right)^2 \frac{v_2}{i_2} \tag{3–87}$$

The secondary voltage v_2 and the secondary current i_2 are related by Ohm's law applied to R_2, thus:

$$\frac{v_2}{i_2} = R_2 \tag{3–88}$$

Substitution of Equation (3–88) in (3–87) yields

$$\frac{v_1}{i_1} = \left(\frac{N_1}{N_2}\right)^2 R_2 \tag{3–89}$$

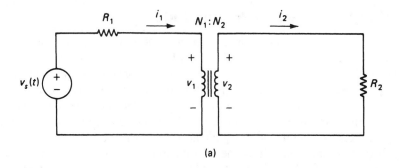

(a)

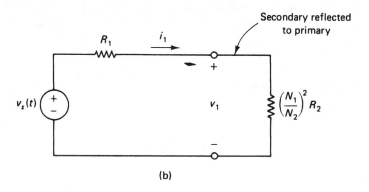

(b)

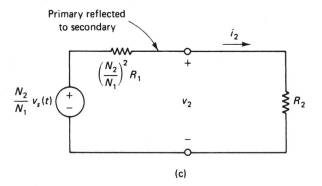

(c)

FIGURE 3–31
Circuit with transformer and reflections to both sides.

The ratio v_1/i_1 in Equation (3–89) describes the voltage-to-current ratio "looking in" to the primary side of the transformer. This ratio is equivalent to an effective input resistance, which will be denoted as R_{in}. Thus,

$$R_{in} = \left(\frac{N_1}{N_2}\right)^2 R_2 \qquad \qquad \text{(3–90)}$$

The implication of the preceding result is that the transformer reflects an equivalent resistance to the primary side given by Equation (3–90). This suggests the equivalent circuit shown in Figure 3–31(b). This equivalent circuit may be used to perform any circuit calculations on the primary side. Any further results needed for the secondary side may be inferred by the appropriate turns ratio applied to the primary results.

Note that whereas voltage and current ratios vary as the first degree of the turns ratio, reflected resistance varies with the square of the turns ratio. This is due to the fact that if the voltage increases, the current decreases, and the ratio increases as a second-degree effect.

Reflecting Primary to Secondary Side

Next, the process of reflecting the circuit to the secondary will be developed. Since the primary contains both a source and a resistance, Thevenin's theorem can be applied for this purpose. Returning to the circuit of Figure 3–31(a), assume that R_2 is momentarily disconnected. In this case, there is no load reflected to the primary ($R_{in} = \infty$), and $v_s(t)$ appears across the primary. The open-circuit secondary voltage will be denoted as v_{2oc}, and it is

$$v_{2oc} = \frac{N_2}{N_1} v_s(t) \qquad \qquad \text{(3–91)}$$

The resistance R_1 reflected to the secondary should have the same form as when R_2 was reflected to the primary, except that the turns ratio is inverted, so this value is $(N_2/N_1)^2 R_1$. The resulting equivalent circuit is shown in Figure 3–31(c).

Although the circuit used in this development is rather simple in form, the concept may be extended to include more complex circuit configurations. In general, the structure of the reflected part of the circuit can be maintained as before, but element values and sources are changed in accordance with the relationships given here. Variables calculated in the part of the circuit remaining unchanged will be correct. On the other hand, any variables calculated in the reflected part of the circuit represent fictitious quantities and must be "reflected back" to the original circuit by using appropriate turns ratio forms to obtain the true values.

Application to Impedance Matching

One classical application for transformers is for *impedance matching* to achieve maximum power transfer. This concept is illustrated by the circuit of Figure 3–32(a). Consider a source with voltage v_s and internal resistance R_s. It is assumed that the

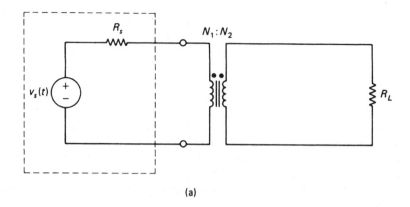

(a)

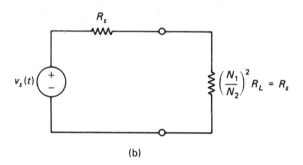

(b)

FIGURE 3–32
Application of transformer for impedance matching.

internal resistance is an inherent part of the source and cannot be changed. Assume that it is necessary to deliver as much power from the source to a load R_L as possible. If $R_s \neq R_L$, not all the available power would be delivered to the load if the source were connected directly to the load. However, including the transformer can accomplish the desired objective, as we will see shortly.

From the maximum power transfer theorem, to extract as much power as possible from the source, the load resistance reflected to the primary should equal the internal source resistance, as indicated in Figure 3–32(b). The following constraint can then be imposed:

$$\left(\frac{N_1}{N_2}\right)^2 R_L = R_s \tag{3–92}$$

This results in

$$\frac{N_1}{N_2} = \sqrt{\frac{R_s}{R_L}} \tag{3–93}$$

The result of Equation (3–93) provides the required turns ratio to achieve maximum power transfer. The source "sees" the optimum resistance for maximum power transfer, and since the ideal transformer consumes no power, the power is transferred to the load. Realistic transformers dissipate some of the available power, of course.

The use of transformers for impedance matching and optimum power transfer applies primarily to sources with very limited available power and with appreciable source resistance levels. Commercial power generation systems, on the other hand, are operated under nearly ideal voltage source conditions for which impedance matching in the sense discussed here is not applicable.

EXAMPLE 3–12

Consider the circuit at the top of Figure 3–33, which contains a sinusoidal source and an ideal transformer. Determine the voltages v_1 and v_2 and the currents i_1 and i_2 in two ways: **(a)** by reflecting the secondary to the primary and **(b)** by reflecting the primary to the secondary.

Solution

(a) To reflect the secondary to the primary, the 48-Ω resistance is multiplied by $(1/2)^2 = 1/4$, which results in an equivalent resistance of 12 Ω on the primary side. The equivalent circuit referred to the primary side is shown in Figure 3–33(a).

The current i_1 is determined as

$$i_1 = \frac{v_s}{8 + 12} = \frac{25 \sin \omega t}{20} = 1.25 \sin \omega t \tag{3–94}$$

The primary voltage v_1 is

$$v_1 = 12i_1 = 15 \sin \omega t \tag{3–95}$$

Now that the transformer primary voltage and current are known, it is an easy process to determine the secondary voltage and current by the basic transformer equations of (3–82) and (3–83). We have

$$v_2 = \frac{N_2}{N_1} v_1 = \frac{2}{1} (15 \sin \omega t) = 30 \sin \omega t \tag{3–96}$$

$$i_2 = \frac{N_1}{N_2} i_1 = \frac{1}{2} (1.25 \sin \omega t) = 0.625 \sin \omega t \tag{3–97}$$

(b) To reflect the primary to the secondary, the voltage source is multiplied by $(2/1)$, and the 8-Ω resistance is multiplied by $(2/1)^2 = 4$. The resulting circuit referred to the secondary side is shown in Figure 3–33(b).

The secondary current i_2 is determined as

$$i_2 = \frac{50 \sin \omega t}{32 + 48} = 0.625 \sin \omega t \tag{3–98}$$

The secondary voltage v_2 is

$$v_2 = 48i_2 = 48(0.625 \sin \omega t) = 30 \sin \omega t \tag{3–99}$$

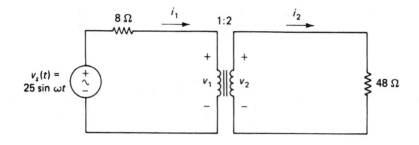

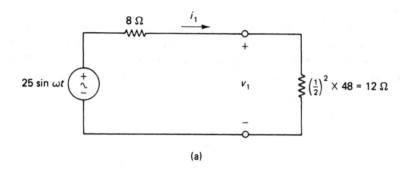

(a)

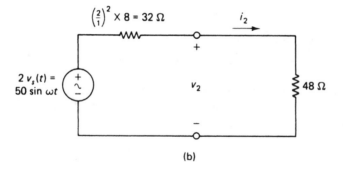

(b)

FIGURE 3–33
Circuit for Example 3–12.

The primary voltage and current are then determined as

$$v_1 = \frac{N_1}{N_2} v_2 = \frac{1}{2} (30 \sin \omega t) = 15 \sin \omega t \qquad \textbf{(3–100)}$$

$$i_1 = \frac{N_2}{N_1} i_2 = 2(0.625 \sin \omega t) = 1.25 \sin \omega t \qquad \textbf{(3–101)}$$

The results obtained from both approaches are obviously in agreement.

Observe that all variables can be determined with either approach. The variables on the side remaining intact are determined first, and from the basic transformer equations, the variables on the other side can be inferred.

EXAMPLE 3–13

This problem has been designed to illustrate how losses can be reduced and efficiency improved by transforming the voltage and current levels in an ac power distribution system. Assume that a certain remote resistive load requires ac *effective values* of voltage and current of 240 V and 100 A, respectively. The concept of *effective value* as applied to ac will be developed later in the text, but for the moment, assume that effective values of voltage and current are treated the same as dc values for calculating average power in a resistive load. Assume that the feeder line from the ac source to the load has a net resistance $R_s = 0.5\ \Omega$. Compare line losses and efficiency under the following two conditions: **(a)** No transformer is used. **(b)** A $1\!:\!10$ "step-up" *ideal* transformer is used at the source end followed by a $10\!:\!1$ "step-down" transformer at the load end. To simplify the analysis, assume that the source voltage is adjusted to compensate for the line voltage drop so that the load voltage is exactly 240 V in each case.

Solution
The load power P_L is

$$P_L = 240\text{ V} \times 100\text{ A} = 24{,}000\text{ W} = 24\text{ kW} \tag{3–102}$$

The power loss P_{loss} in the line is

$$P_{\text{loss}} = I^2 R_s = 0.5 I^2 \tag{3–103}$$

where I is the effective line current. The efficiency η is defined as

$$\eta = \frac{\text{output power}}{\text{input power}} \times 100\% = \frac{P_L}{P_L + P_{\text{loss}}} \times 100\% \tag{3–104}$$

(a) With no transformer, the line current is the same as the load current, and

$$P_{\text{loss}} = 0.5 \times (100)^2 = 5000\text{ W} = 5\text{ kW} \tag{3–105}$$

The efficiency is

$$\eta = \frac{24{,}000}{24{,}000 + 5000} \times 100\% = 82.8\% \tag{3–106}$$

(b) When the line voltage is increased by a factor of 10, the current will be reduced by the same ratio. The primary voltage at the receiving end is now 2400 V, and the line current is now $100/10 = 10$ A. The line loss changes to

$$P_{\text{loss}} = 0.5 \times (10)^2 = 50\text{ W} \tag{3–107}$$

Note that the line loss changes dramatically from 5 kW to only 50 kW. The efficiency is now

$$\eta = \frac{24,000}{24,000 + 50} \times 100\% = 99.8\% \tag{3-108}$$

This problem has assumed ideal transformers, and when realistic transformer losses are considered, the efficiency will be reduced somewhat from the optimistic value obtained in Equation (3–108). However, this problem has illustrated with a specific system the concept of reducing transmission losses by transforming voltage and current ratios.

3–8 PSPICE EXAMPLES

Most of the PSPICE examples of this chapter will utilize transient analysis. With PSPICE, transient analysis involves the determination of the instantaneous behavior of one or more circuit variables as a function of time. Although tabular data may be obtained if desired, most results obtained from a transient analysis are more useful if they are plotted. The PROBE utility of PSPICE permits various circuit variables to be directly plotted by the computer. It also allows various mathematical operations, such as differentiation and integration, to be performed on the waveforms.

PSPICE EXAMPLE 3–1
Use PSPICE to determine the current in Example 3–3 (Figure 3–10), and use PROBE to plot the voltage and current.

Solution
The circuit adapted to the PSPICE format and the code are shown in Figure 3–34. This example is the first of many that will utilize transient analysis, so a detailed explanation is in order.

First, PSPICE offers a number of choices for instantaneous waveforms required in transient analysis, all of which can be tailored to the problem as required. The input waveform is a voltage in this case, which has been called simply *V*, and it is connected between nodes 1 and 0 with 1 indicated as the positive reference. The remainder of the line is the exact description of this voltage, which is a *piecewise linear source* (PWL).

The general format for a PWL source, which of course follows the source name and node numbers, is as follows:

```
PWL(t₁ v₁)(t₂ v₂)-----(tₖ vₖ)
```

where the values of *t* must increase from left to right. The first pair $(t_1 \ v_1)$ indicates that at time t_1, the value of the voltage (or current, if a current source is being described) is v_1. The second pair $(t_2 \ v_2)$ indicates that at time t_2, the value of the

FIGURE 3–34
Circuit and code for PSPICE Example 3–1.

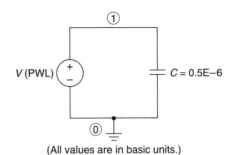

(All values are in basic units.)

```
PSPICE EXAMPLE 3-1
V 1 0 PWL (0 0) (2E-3 40) (4E-3 40) (8E-3 0)
C 1 0 0.5E-6
.TRAN 8E-3 8E-3
.PROBE
.END
```

source is v_2, and so forth. PSPICE then connects straight-line segments between the various points and thus creates a waveform composed of straight-line segments. If a sudden jump is desired (i.e., a pulse function), a very small time difference is chosen so that on the overall time scale, it appears as a jump. Actually, the parentheses are optional and are used at this time to clarify the grouping. Commas may also be inserted between the points in parentheses, but a space will suffice.

Referring back to Figure 3–10 and the code of Figure 3–34, note that four points are required in the description. The first point has the coordinates (0 0); that is, at time $t = 0$, the voltage is 0. The second point has the coordinates (2E-3 40); that is, at time $t = 2$ ms, the voltage is 40 V. The third point has the coordinates (4E-3 40); the voltage is also 40 V at $t = 4$ ms. Finally, the fourth point has coordinates (8E-3 0), which means that at $t = 8$ ms, the voltage is 0 again. The voltage will remain at the level of the last point for all time thereafter.

This is the first time we have encountered a capacitor, but its format is similar to that of a resistor. The first letter of the name must be C, and additional numbers or letters could be added if desired. The two numbers that follow (1 0) represent the node connections. Finally, the value of the capacitance is 0.5E-6 F (0.5 μF). A capacitor could also have an additional data point representing an initial voltage, and this will be considered in PSPICE Example 3–3. The absence of any additional data defaults to an initially uncharged condition.

The instantaneous behavior of the circuit as a function of time is generated by a transient command (.TRAN), which for this example reads

```
.TRAN 8E-3 8E-3
```

Transient analysis is one of the most complex of the PSPICE commands, and for elaborate circuit situations, it may require judgment on the part of the user (and possibly some trial and error). The simplest form of the statement occurs when there are only two data entries following the .TRAN command, which is the case here.

In general, when there are only two data points, they represent, in order, the time step for tabular output and the ending time of analysis. (It always starts at $t = 0$.) The second value is the more obvious here, and it is chosen as 8E-3 (8 ms), which represents the ending time of the waveform. If we desired tabular output, we would want to select the time step between successive points for the first entry. However, our goal is to use PROBE for plotting the curves, and it automatically selects time increments somewhat independent of any time step chosen for tabular data output. Yet the program requires a value to be placed ahead of the final time value. Therefore, when no tabular output is desired and only graphical results are to be displayed, the simplest form of the transient statement is to follow .TRAN with the final time value listed twice. This choice will be made extensively throughout the text, although more complex forms will be encountered later, when the need arises.

It should be pointed out that the time step chosen for tabular data output is not the same as the internal step that PSPICE uses in the numerical solution of the differential equations involved. PSPICE uses an adjustable step size for integration, which depends on the behavior of the time function. If graphical output is desired (which will be the case for many examples throughout the text), a .PROBE statement is required as shown. Finally, the usual .END statement is given.

After the program is executed, a graphical screen is created by PROBE. It is menu driven and can be learned easily, with practice (along with some trial and error). A number of books devoted to PSPICE provide detailed information about the various features.

Plots of the voltage V(1) and the current I(C) as obtained from PROBE are shown in Figures 3–35 and 3–36, respectively. These waveforms are seen to be in agreement with those of parts (a) and (b) of Figure 3–10. We will delay the generation of the power and energy functions (PSPICE Example 3–6) until the reader has more practice with the simpler operations of PROBE.

PSPICE EXAMPLE 3–2

Use PSPICE to determine the voltage of Example 3–4 (Figure 3–11), and use PROBE to plot the current and voltage.

Solution

It would appear that the circuit would consist simply of a current source in parallel with the capacitor, as suggested by Figure 3–11. Unfortunately, however, that approach would lead to an error message indicating that there is a "floating node." PSPICE requires that there be a dc path to ground at each node. As we will see in Chapter 4, a capacitor acts like an open circuit, to pure dc. The current source does not qualify as a path since, when it is deenergized, it acts as an open circuit. There was no difficulty with the capacitor in parallel with the ideal voltage source in PSPICE Example 3–1 since a voltage source, when deenergized, acts as a short circuit.

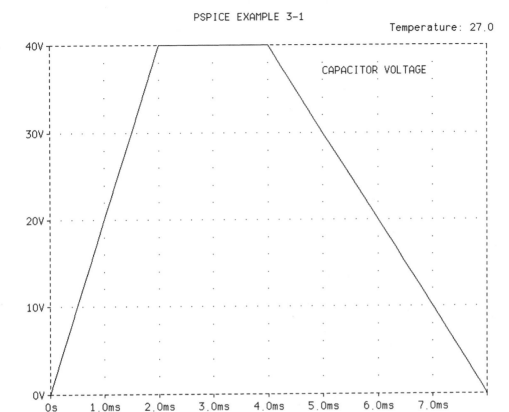

FIGURE 3–35
Capacitor voltage for PSPICE Example 3–1.

The problem can be circumvented by the addition of a very large resistance in parallel with the capacitor. The circuit adapted to the PSPICE format with the added resistor and the code are shown in Figure 3–37. The resistor is denoted as RX, and a value of 10^9 Ω has been assumed.

For the most part, the code follows a pattern similar to that of PSPICE Example 3–1. As in that example, the PWL source with four points is used. In this case, the source is a current source flowing from node 0 to node 1. The transient interval in this example has been chosen to run to 12 ms to demonstrate the pattern after the input current has ceased to flow. Following execution of the program and with the assistance of PROBE, plots of the current and voltage are shown in Figures 3–38 and 3–39, respectively. Note that the voltage starts at zero and reaches a final level of 120 V, at which it remains after current has ceased to flow.

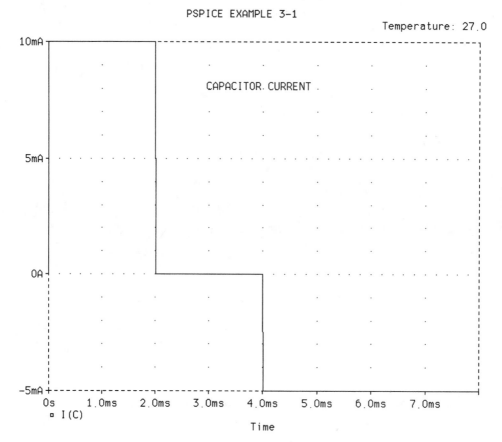

FIGURE 3–36
Capacitor current for PSPICE Example 3–1.

PSPICE EXAMPLE 3–3
Repeat PSPICE Example 3–2 to determine the voltage if the capacitor is initially charged to −50 V as in Example 3–5 (Figure 3–12).

Solution
The circuit diagram is the same as in the previous example and will not be repeated. The code, however, requires two modifications and is shown in Figure 3–40.

The first modification is that an initial condition is placed on the code line for C. Following the value of the capacitance, the initial condition is specified as IC = −50. For a capacitor, this value is interpreted as a voltage, and the (−) sign indicates that it is opposite to the positive reference direction for the branch.

The second modification is that UIC "use initial condition" must be placed as the last entry on the .TRAN control line. Without this latter statement, the

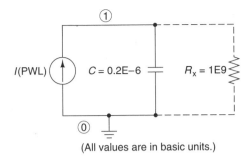

(All values are in basic units.)

PSPICE EXAMPLE 3–2
```
I 0 1 PWL (0 0) (2E–3 3E–3) (8E–3 3E–3) (10E–3 0)
C 1 0 0.2E–6
RX 1 0 1E9
.TRAN 12E–3 12E–3
.PROBE
.END
```

FIGURE 3–37
Circuit and code for PSPICE Example 3–2.

program will ignore the initial condition and treat the capacitor as if it were uncharged.

The capacitor voltage obtained with PROBE is shown in Figure 3–41. In this case, the voltage starts at -50 V and eventually charges to 70 V as in Example 3–5.

PSPICE EXAMPLE 3–4

Use PSPICE to determine the voltage in Example 3–6 (Figure 3–15), and use PROBE to plot the current and voltage.

Solution
The circuit adapted to the PSPICE format and the code are shown in Figure 3–42. The code follows a format similar to that of the preceding examples and can be readily checked. Once again, the PWL source has been used, but in this case, a triangular waveform should be generated. This illustrates the flexibility of the PWL source in a variety of waveforms.

The only new element in this circuit is an inductance, for which the first letter must be L. Additional letters or numbers can be added as desired. The two numbers following the name represent the nodes, as in the case of resistance and capacitance, and the next number represents the inductance value in henries. If an initial current other than zero were desired, an additional label of the type IC = "value" would be required as in the case of the capacitor. For an inductance, this value would be interpreted as a current. The additional label UIC on the .TRAN statement would also be required.

After the program is executed, PROBE can be used to generate the plot. For

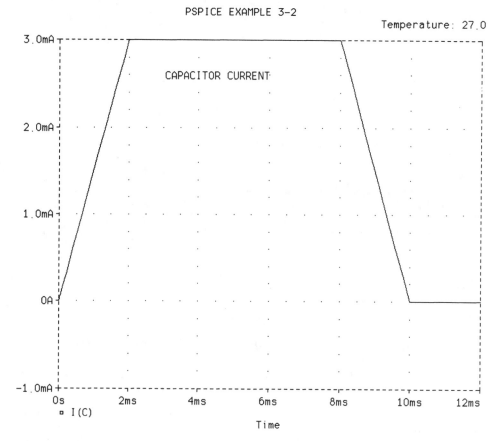

FIGURE 3–38
Capacitor current for PSPICE Example 3–2.

the first time thus far, we will choose to plot both waveforms desired on the same page. This can be achieved by use of the "plot control" in the menu, followed by "add plot." The stacking order is from bottom to top, so the inductor voltage V(1) was plotted first, followed by the current. The results are shown in Figure 3–43.

PSPICE EXAMPLE 3–5

Use PSPICE to determine the current of Example 3–7 (Figure 3–15), and use PROBE to plot the voltage and current.

Solution
It would appear that the circuit would consist simply of a voltage source in parallel with the inductor, as suggested by Figure 3–15. That approach, however, would lead to an error message indicating a "voltage source and/or inductor loop." As

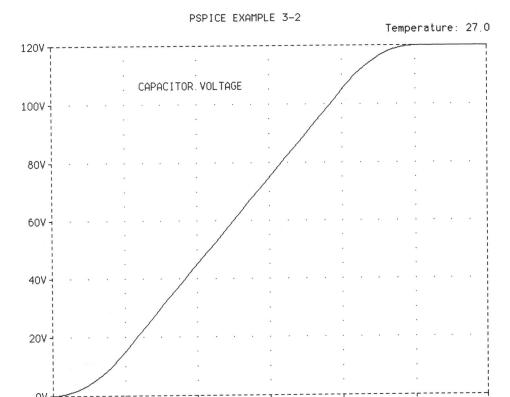

FIGURE 3–39
Capacitor voltage for PSPICE Example 3–2.

we will see in Chapter 4, an inductor acts like a short circuit at dc, and since a deenergized voltage source also acts like a short circuit, an unacceptable loop occurs.

The problem can be circumvented by the addition of a small resistance in series in the loop. The circuit adapted to the PSPICE format with the added resistor and the code are shown in Figure 3–44. The resistor is denoted as RX, and a value of 1E-6 Ω is chosen.

The code is similar to that of previous examples and can be readily checked. The PWL function is chosen once again, but with a slightly different strategy in this case. The point at which the waveform starts is indicated as (0 0), but the next point is indicated as (1E-9 20). This corresponds to a rise time of 1 ns, which is considerably smaller than the width of 5 ns for the first pulse. The third point is (5E-6 20), and the fourth point is (5.001E-6 -20), corresponding to a transition time of 1 ns. A similar transition time is chosen at the end of the negative pulse.

```
PSPICE EXAMPLE 3-3
I 0 1 PWL (0 0) (2E-3 3E-3) (8E-3 3E-3) (10E-3 0)
C 1 0 0.2E-6 IC=-50
RX 1 0 1E9
.TRAN 12E-3 12E-3 UIC
.PROBE
.END
```

FIGURE 3–40

Code for PSPICE Example 3–3.

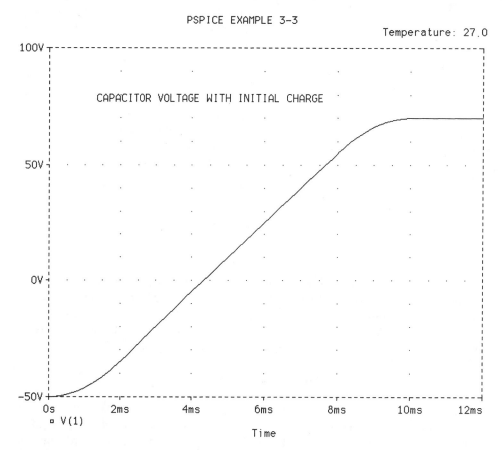

FIGURE 3–41

Capacitor voltage for PSPICE Example 3–3.

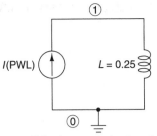

(All values are in basic units.)

```
PSPICE EXAMPLE 3-4
I 0 1 PWL (0 0) (5E-3 0.4) (15E-3 -0.4) (20E-3 0)
L 1 0 0.25
.TRAN 20E-3 20E-3
.PROBE
.END
```

FIGURE 3–42
Circuit and code for PSPICE Example 3–4.

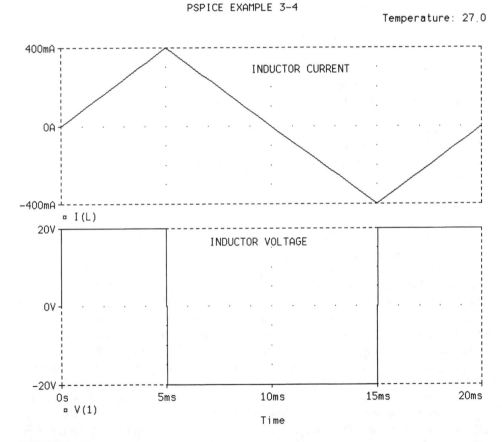

FIGURE 3–43
Waveforms for PSPICE Example 3–4.

197

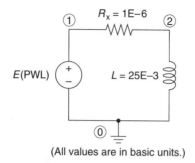

(All values are in basic units.)

```
PSPICE EXAMPLE 3–5
V 1 0 PWL (0 0) (1E–9 20) (5E–6 20) (5.001E–6 –20) (10E–6 –20) (10.001E–6 0)
RX 1 2 1E–6
L 2 0 25E–3
.TRAN 12E–6 12E–6
.PROBE
.END
```

FIGURE 3–44
Circuit and code for PSPICE Example 3–5.

The inductor voltage and current are shown in Figure 3–45. The results are in agreement with those of Example 3–7.

PSPICE EXAMPLE 3–6

Extend the analysis of PSPICE Example 3–1 to include plots of the instantaneous power and the energy.

Solution

The circuit diagram and code are exactly the same as in PSPICE Example 3–1, so refer back to Figure 3–34 as necessary. The additional work in this example will be achieved with PROBE.

In this example, the four plots representing the voltage, current, power, and energy will be made on the same page. This can be achieved with PROBE using the "plot control" option of the menu followed by the "add plot" option. Each time this sequence is followed, the existing plots on the screen shrink in size, and a new coordinate system is added. A new variable can then be plotted on the new coordinate system, and the process can be repeated. Obviously, the figures get smaller after each step, but the presentation is more compact, and the time scale is the same for all quantities plotted.

Refer to Figure 3–46 for the net plots of the four variables. The plots for voltage and current can be made directly from the menu of variables provided, as before. However, the plots for power and energy can be achieved through the mathematical operations available in PROBE.

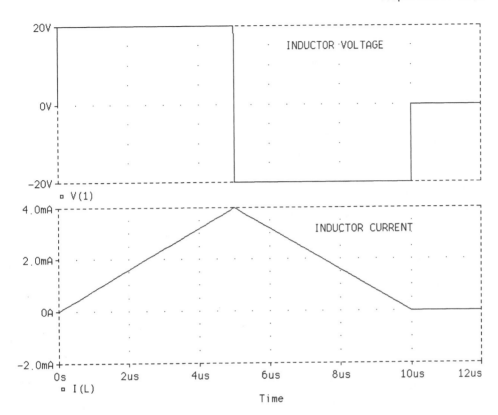

FIGURE 3–45
Waveforms for PSPICE Example 3–5.

The power is obtained by multiplying the capacitor voltage by the capacitor current. The capacitor voltage is V(1) and the capacitor current is I(C). The product of these variables is entered as V(1) * I(C), and this variable represents the instantaneous power.

PROBE will perform an integration of any variable appearing in the menu or defined by an appropriate equation. If X is any variable so identified, S(X) represents its integral up to any time on the screen scale. Since energy represents the integral of power, the proper entry for the instantaneous energy is S(V(1)* I(C)). The four plots of Figure 3–46 are in agreement with those of Figure 3–10. In case the reader attempts to repeat this analysis, it should be noted that the author created the plots in reverse order so that the given input voltage would be at the top and the energy would be at the bottom. Each new plot added by PROBE appears above the previous plot(s), so the first one generated will be the bottom

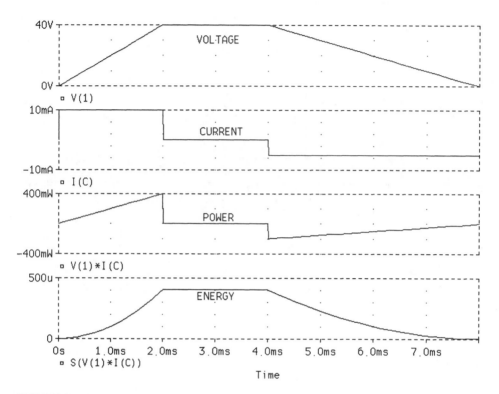

FIGURE 3–46
Waveforms for PSPICE Example 3–6.

one in the stack and the last one will be the top plot. Thus, the author generated S(V(1) * I(C)) first and V(1) last.

PSPICE EXAMPLE 3–7

Use PSPICE to analyze the ideal transformer circuit of Example 3–12 (Figure 3–33).

Solution
Although not specifically emphasized in the text, the transformer is assumed to be operating under steady-state ac conditions, since the source is of a sinusoidal form. Steady-state ac analysis will be developed in detail in Chapter 8, but it will be assumed here that most readers are sufficiently familiar with the concept that we can utilize it for the circuit at hand. (One could utilize a transient analysis and wait

until steady-state conditions are reached, but that approach is not the best one for the present objective.)

In steady-state sinusoidal analysis, a sinusoidal function is represented as a so-called phasor having a magnitude and an angle. The magnitude can be chosen as the peak value of the sinusoid, and the angle is chosen with reference to either a sine function or a cosine function. As will be seen in Chapter 8, the approach in this text will be to use the sine function as the phase reference. Thus, a function of the form $A \sin(\omega t + \theta)$ will be represented as a phasor with magnitude A and an angle θ, and the notation for this is $A \underline{/\theta}$. The complete details will be developed in Chapter 8, so if the concept is not clear at this point, simply accept the notation for the purpose at hand.

A nearly ideal transformer model can be created by two inductors L_1 and L_2, with appropriate additional coupling achieved through the mutual inductance parameter M, or more specifically, through the coefficient of coupling K. The values of L_1 and L_2 must be chosen to be very large, and K can be chosen to be nearly or equal to unity. In determining the size of L_1 and L_2, the smaller of the two must be such that $\omega L = 2\pi f L$ is much greater than the largest resistance in the circuit, where f is the frequency in hertz (Hz). The inductance ratio is chosen to satisfy

$$\frac{L_1}{L_2} = \left(\frac{N_1}{N_2}\right)^2 \tag{3-109}$$

The circuit adapted to the PSPICE format and the code are shown in Figure 3–47. Note that the source $25 \sin \omega t$ is represented as $25 \underline{/0°}$ in accordance with the earlier discussion for ac analysis. Whereas no specific frequency has been specified, in order to relate the value to a possible real-life sitaution, we will select $f = 60$ Hz. The desired turns ratio is $N_1/N_2 = 1/2$, so from Equation (3–109), $L_1/L_2 = 1/4$. As a check, $\omega L_1 = 2\pi \times 60 \times 10 = 3770 \ \Omega$, which is considerably larger than the largest resistance in the circuit.

The source line in the code reads

```
VS 1 0 AC 25 0
```

This code is similar to that for a DC source, except that the type now reads AC instead. For AC, two entries can follow. The first is the phasor magnitude (25 V), and the second is the angle (0°). The last entry can be omitted when the angle is 0°, and the code automatically defaults to that value. We have included it here for clarity.

The primary and secondary of the transformer are denoted by L1 and L2 in the same manner as uncoupled inductances. The coupling is achieved by the next line, which reads

```
K L1 L2 1
```

This is the code for establishing a coefficient of coupling between the inductors, and it must begin with K. Additional letters or numbers may be added to K if desired. Following K, the names of the two inductors coupled together are listed.

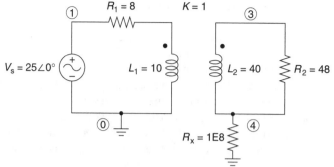

(All values are in basic units.)

```
PSPICE EXAMPLE 3–7
.OPTIONS NOECHO NOPAGE NOBIAS
VS 1 0 AC 25 0
R1 1 2 8
L1 2 0 10
L2 3 4 40
K L1 L2 1
R2 3 4 48
RX 4 0 1E8
.AC LIN 1 60 60
.PRINT AC I(R1) I(R2) V(L1) V(L2)
.END
```

FIGURE 3–47
Circuit and code for PSPICE Example 3–7.

The last data point is the actual value of the coupling coefficient. For an ideal transformer, $K = 1$.

All PSPICE programs must have a ground node, and all nodes must have a dc path to ground. On the primary side, the bottom side of the inductance is connected to node 0 to satisfy the requirement. On the secondary side, the requirement is met by connecting a very large resistance RX from the bottom of the inductance to ground. A value of $10^8 \ \Omega$ was arbitrarily chosen.

Tabular output is desired in this example, so several options were added to minimize the size of the data file. Along with NOECHO and NOPAGE, which have been used extensively, NOBIAS is used in this program. An ac analysis is preceded by an initial bias point analysis, which produces zero values in this example since there are no dc sources. To suppress the superfluous data, this option has been added.

The ac analysis command reads as follows:

```
.AC LIN 1 60 60
```

The first entry .AC commands PSPICE to perform a steady-state ac analysis. The second entry may be one of three, defined as follows:

PSPICE EXAMPLE 3-7

**** CIRCUIT DESCRIPTION

****

.OPTIONS NOECHO NOPAGE NOBIAS

**** AC ANALYSIS   TEMPERATURE = 27.000 DEG C

FREQ	I(R1)	I(R2)	V(L1)	V(L2)
6.000E+01	1.250E+00	6.250E-01	1.500E+01	3.000E+01

JOB CONCLUDED

TOTAL JOB TIME   .38

FIGURE 3-48
Output data for PSPICE Example 3–7.

LIN linear sweep

LOG logarithmic sweep

DEC decade sweep (also a logarithmic type)

The last two are particularly useful when the frequency is to be swept over a very wide range, and they will be used in Chapter 9. The first one is appropriate for a narrow frequency range and is probably the simplest to use when a single frequency is desired, as is the case here. Thus, the second entry is LIN.

Following the type of sweep, three values are required. For the linear sweep, the first entry is the total number of frequencies desired (1). The second entry is the beginning frequency (60 Hz), and the third entry is the ending frequency (also 60 Hz). Once again, we have a structured command for a sweep requiring several entries which must be "rigged" to produce a single point.

The desired output variables are the primary and secondary currents and voltages. The primary and secondary currents are obtained from I(R1) and I(R2), and the primary and secondary voltages are obtained from V(L1) and V(L2).

The output data file is shown in Figure 3–48. The values of the output currents and voltages are the phasor magnitudes, which are the peak values of the sinusoidal functions involved. These values are in perfect agreement with those of Example 3–12.

DRILL PROBLEMS

In all problems in which voltage or current waveforms for a capacitor or inductor are given, assume positive references in accordance with the convention of Figure 3–8 or Figure 3–13.

3–1. Determine graphically and plot the first derivative with respect to x of the piecewise linear continuous function y shown in Figure P3–1.

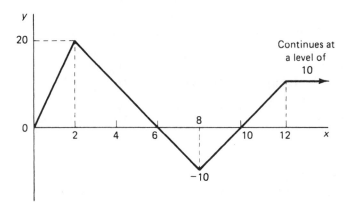

FIGURE P3–1

3–2. Determine graphically and plot the first derivative with respect to x of the piecewise linear continuous function y shown in Figure P3–2.

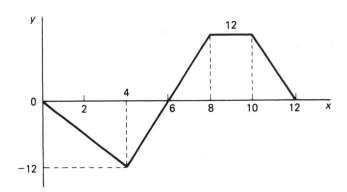

FIGURE P3–2

3–3. Determine graphically and plot the definite integral as x varies from 0 to 12 of the piecewise linear function shown in Figure P3–3. The integral has no previous accumulated value.

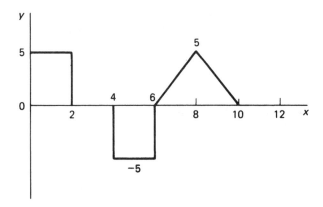

FIGURE P3-3

3-4. Determine graphically and plot the definite integral as x varies from 0 to 12 of the piecewise linear function shown in Figure P3-4. The integral has no previous accumulated value.

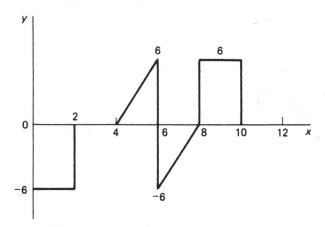

FIGURE P3-4

3-5. A 0.2-F capacitor has a voltage waveform $v(t)$ as shown in Figure P3-5. Determine and plot as functions of time the **(a)** current, **(b)** power, and **(c)** energy.

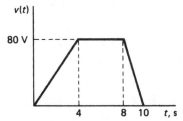

FIGURE P3-5

3–6. A 0.5-F capacitor has a voltage waveform $v(t)$ as shown in Figure P3–6. Determine and plot as functions of time the **(a)** current, **(b)** power, and **(c)** energy.

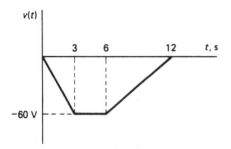

FIGURE P3–6

3–7. A 0.5-μF capacitor has a voltage waveform $v(t)$ as shown in Figure P3–7. Determine and plot the current waveform $i(t)$.

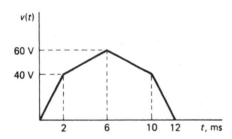

FIGURE P3–7

3–8. A 0.2-μF capacitor has a periodic voltage waveform $v(t)$ as shown in Figure P3–8. Determine and plot the current waveform $i(t)$ over the range of the two cycles shown.

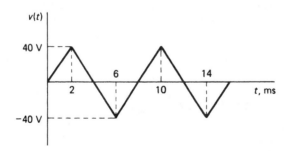

FIGURE P3–8

3–9. A 0.5-F capacitor has a current waveform $i(t)$ as shown in Figure P3–9. Determine and plot the voltage waveform $v(t)$ as a function of time. The capacitor is initially uncharged.

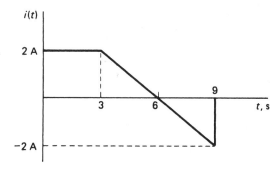

FIGURE P3–9

3–10. A 0.25-F capacitor has a current waveform $i(t)$ as shown in Figure P3–10. Determine and plot the voltage waveform $v(t)$ as a function of time. The capacitor is initially uncharged.

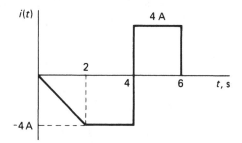

FIGURE P3–10

3–11. For the capacitor and current waveform of Problem 3–9, assume that the capacitor is initially charged to a voltage of +5 V at the time the current starts to flow. Plot the voltage waveform as a function of time.

3–12. For the capacitor and current waveform of Problem 3–10, assume that the capacitor is initially charged to a voltage of +16 V at the time the current starts to flow. Plot the voltage waveform as a function of time.

3–13. A 5-μF capacitor has a current waveform $i(t)$ as shown in Figure P3–13. Determine and plot the voltage waveform $v(t)$ as a function of time. The capacitor is initially uncharged.

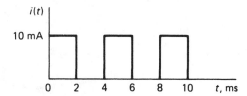

FIGURE P3–13

3-14. A 0.5-μF capacitor has a periodic current waveform $i(t)$ as shown in Figure P3-14. Determine and plot the voltage waveform $v(t)$ over the range of the two cycles shown. The capacitor is initially uncharged.

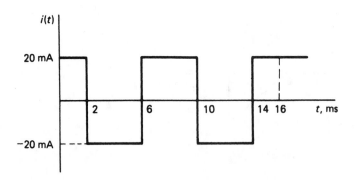

FIGURE P3-14

3-15. A 5-H inductor has a current waveform $i(t)$ as shown in Figure P3-15. Determine and plot as functions of time the **(a)** voltage, **(b)** power, and **(c)** energy.

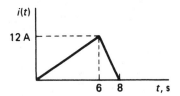

FIGURE P3-15

3-16. A 3-H inductor has a current waveform $i(t)$ as shown in Figure P3-16. Determine and plot as functions of time the **(a)** voltage, **(b)** power, and **(c)** energy.

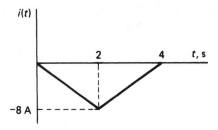

FIGURE P3-16

3-17. A 50-mH inductor has a current waveform $i(t)$ as shown in Figure P3-17. Determine and plot the voltage waveform $v(t)$.

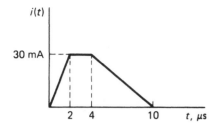

FIGURE P3–17

3–18. A 200-mH inductor has a current waveform $i(t)$ as shown in Figure P3–18. Determine and plot the voltage waveform $v(t)$.

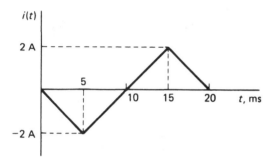

FIGURE P3–18

3–19. A 0.5-H inductor has a voltage waveform $v(t)$ as shown in Figure P3–19. Determine and plot the current waveform $i(t)$ as a function of time. The inductor is initially influxed.

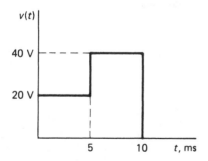

FIGURE P3–19

3–20. A 2-mH inductor has a voltage waveform $v(t)$ as shown in Figure P3–20. Determine and plot the current waveform $i(t)$ as a function of time. The inductor is initially unfluxed.

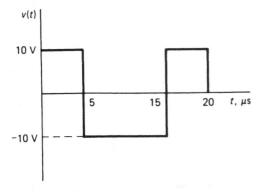

FIGURE P3–20

3–21. Consider the combination of three capacitors connected in the manner shown in Figure P3–21. Determine the equivalent capacitance C_{eq} as viewed from the left-hand terminals.

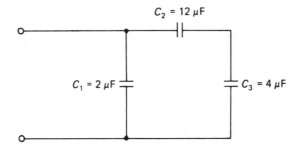

FIGURE P3–21

3–22. Consider the combination of four capacitors connected in the manner shown in Figure P3–22. Determine the equivalent capacitance C_{eq} as viewed from the left-hand terminals.

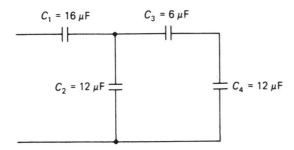

FIGURE P3–22

3–23. Consider the combination of three capacitors considered in Problem 3–21, and assume that the combination has been charged, from an external circuit not shown, to a voltage of 32 V. Determine the voltage across each capacitor.

3-24. Consider the combination of four capacitors considered in Problem 3–22, and assume that the combination has been charged, from an external circuit not shown, to a voltage of 108 V. Determine the voltage across each capacitor.

3-25. Consider the circuit of Example 3–10 (Figure 3–28), but assume that the direction of i_2 is reversed. **(a)** Construct an equivalent circuit with mutual coupling generators. **(b)** Write the two mesh current equations.

3-26. One form of an *autotransformer* structure is shown in Figure P3–26. **(a)** Construct an equivalent circuit with mutual coupling generators. **(b)** Write the two mesh current equations.

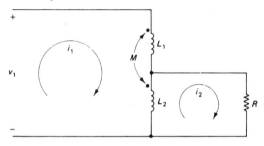

FIGURE P3–26

3-27. Consider the circuit of Figure P3–27 containing a sinusoidal source and an ideal transformer. Determine the voltages v_1 and v_2 and the currents i_1 and i_2 two ways: **(a)** by reflecting the secondary to the primary and **(b)** by reflecting the primary to the secondary.

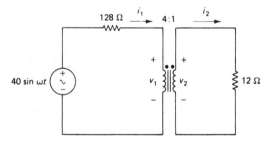

FIGURE P3–27

3-28. Consider the circuit of Figure P3–28 containing a sinusoidal source and an ideal transformer. Determine the currents i_1, i_2, and i_3 and the voltages v_1, v_2, and v_3 two

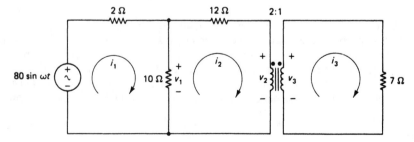

FIGURE P3–28

ways: **(a)** by reflecting the secondary to the primary and **(b)** by reflecting the primary to the secondary.

3-29. Assume that a certain remote resistive load requires ac effective values of voltage and current of 240 V and 150 A respectively. The feeder line has a net resistance of 0.2 Ω. Compare line losses and efficiency under the following conditions: **(a)** No transformer is used. **(b)** A 1:5 ideal transformer is used at the source end followed by a 5:1 transformer at the load end.

3-30. A certain amplifier has an open-circuit effective signal voltage of 60 V (maximum level) and an internal resistance of 150 Ω. It is required to drive a speaker with 8-Ω resistance. **(a)** If the amplifier is connected directly to the speaker, determine the speaker power (maximum level). **(b)** Determine the required turns ratio of a matching transformer. **(c)** Determine the speaker power (maximum level) after matching.

DERIVATION PROBLEMS

3-31. The voltage across a capacitance of value C is given by

$$v(t) = V_p \sin \omega t$$

Derive an expression for the current $i(t)$.

3-32. The current flow in an initially uncharged capacitance of value C is given by

$$i(t) = I_p \cos \omega t$$

for $t > 0$. Derive an expression for the voltage $v(t)$.

3-33. The current through an inductance of value L is given by

$$i(t) = I_p \sin \omega t$$

Derive an expression for the voltage $v(t)$.

3-34. The voltage across an initially unfluxed inductance of value L is given by

$$v(t) = V_p \cos \omega t$$

for $t > 0$. Derive an expression for the current $i(t)$.

3-35. In this problem, the formula for combining capacitances in series will be established. Consider the circuit of Figure 3-17. Assume a current $i(t)$ in the loop and a source voltage $v_s(t)$. Write a loop equation expressing all capacitive voltage drops in integral form in terms of the current. Based on the assumption that there is a single equivalent capacitance C_{eq}, an equation at the terminals may be written as

$$v_s(t) = \frac{1}{C_{eq}} \int_0^t i(t)\, dt$$

Equating this expression to the loop equation, derive Equation (3-62).

3-36. In this problem, the formula for combining capacitances in parallel will be established. Consider the circuit of Figure 3-18. Assume a voltage $v(t)$ across the nodes and a source current $i_s(t)$. Write a node voltage equation expressing all capacitive currents in derivative form in terms of the voltage. Based on the assumption that there is a single equivalent capacitance C_{eq}, an equation at the terminals may be written as

$$i_s(t) = C_{eq} \frac{dv(t)}{dt}$$

Equating this expression to the node equation, derive Equation (3-63).

3–37. In this problem, the formula for combining N inductances without mutual coupling in series will be established. Consider the circuit of Figure 3–19. Assume a current $i(t)$ in the loop and a source voltage $v_s(t)$. Write a loop equation expressing all inductive voltages in derivative form in terms of the current. Based on the assumption that there is a single equivalent inductance L_{eq}, an equation at the terminals may be written as

$$v_s(t) = L_{eq}\frac{di(t)}{dt}$$

Equating this expression to the loop equation, derive Equation (3–64).

3–38. In this problem, the formula for combining N inductances without mutual coupling in parallel will be established. Consider the circuit of Figure 3–20. Assume a voltage $v(t)$ across the nodes and a source current $i_s(t)$. Write a node voltage equation expressing all inductive currents in integral form in terms of the voltage. Based on the assumption that there is a single equivalent inductance L_{eq}, an equation at the terminals may be written as

$$i_s(t) = \frac{1}{L_{eq}}\int_0^t v(t)\, dt$$

Equating this expression to the node equation, derive Equation (3–65).

3–39. In this problem, the capacitive voltage divider rule will be established. Consider the circuit of Figure 3–21. Write a loop equation expressing the two capacitive voltage drops in integral form. Next, write an expression for $v_0(t)$ in terms of the current. From these two equations, derive Equation (3–66).

3–40. In this problem, the capacitive current divider rule will be established. Consider the circuit of Figure 3–22. Write a node voltage equation expressing the two capacitive currents in derivative form. Next, write an expression for $i_0(t)$ in terms of the voltage. From these two equations, derive Equation (3–67).

3–41. Show that the equivalent inductance of two mutually coupled inductors connected in series with *fields opposing* as shown in Figure P3–41 is

$$L_{eq} = L_1 + L_2 - 2\,M$$

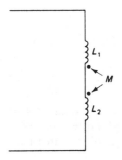

FIGURE P3–41

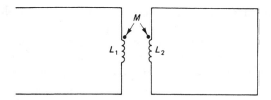

FIGURE P3–42

3–42. Consider a pair of mutually coupled coils with the secondary shorted as shown in Figure P3–42. **(a)** Derive an expression for the equivalent inductance viewed from the primary, and show that it can be expressed as

$$L_{eq} = L_1(1 - k^2)$$

where k is the coefficient of coupling. **(b)** Comment on the relative magnitude of L_{eq} as compared with L_1. What is the significance of the result obtained when there is unity coupling?

APPLICATION PROBLEMS

3–43. It is desired to convert the symmetrical square-wave current waveform of Figure P3–43 to the triangular voltage waveform shown. Specify the type and value of a circuit component that will accomplish the task.

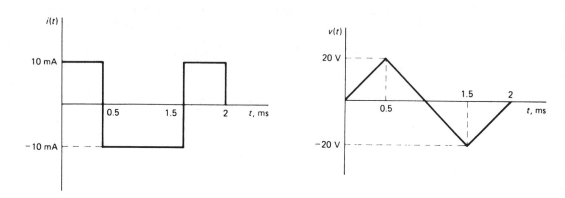

FIGURE P3–43

3–44. A *staircase* voltage waveform can be generated by applying a series of short, positive current pulses to a capacitor as shown in Figure P3–44. For $\tau = 0.1$ ms, $T = 1$ ms, $I_p = 1$ mA, and $C = 0.1$ μF, plot $v(t)$ over the range of the three pulses shown. Assume the capacitor is initially uncharged.

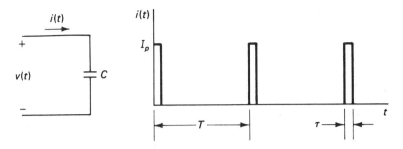

FIGURE P3-44

3-45. When it is necessary to terminate current flow in an inductor, a very large and potentially dangerous negative voltage will be generated unless there is a low resistance path provided to absorb the energy. One circuit used to accomplish this purpose is shown in Figure P3-45. Under conditions in which i is positive and not changing, the

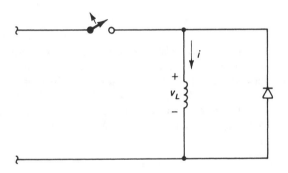

FIGURE P3-45

voltage across an ideal inductance is zero and the diode is reverse biased. If the switch is opened, the negative derivative of the current would result in a large negative voltage across the inductor, but the diode becomes forward biased and limits the voltage.

 To illustrate the concept, assume that a current flow of 5 A is to be terminated and assume that under this condition the diode can be approximated as a resistance of 0.22 Ω. Determine the peak value v_L of the reverse voltage generated.

3-46. The concept of the *single-slope analog-to-digital converter* is illustrated in Figure P3-46. The integrator performs the operation

$$v_o(t) = K_1 \int_0^t v_i(t)\, dt$$

in which the initial value is established as $v_o(0) = 0$ at the beginning of a conversion cycle. Assume that at $t = 0$, a constant reference voltage $v_i(t) = V_{ref}$ is applied to the integrator input. Simultaneously, a standard digital counter is enabled, and it begins producing successive binary numbers. When the output of the integrator reaches the

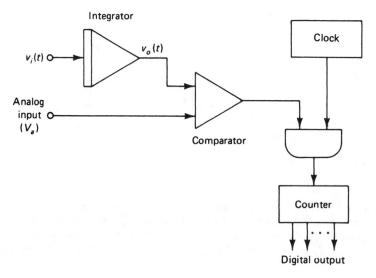

FIGURE P3–46

level of the analog signal V_a, the comparator changes states and inhibits the counter. The value of the binary number in the counter is then the desired digital word.

(a) Let T represent the time required for conversion. Show that T is proportional to the level V_a of the analog signal. **(b)** Let y represent the decimal equivalent of the binary word. Since the binary numbers are produced in sequential order, this value is proportional to the conversion time, i.e., $y = K_2 T$, where K_2 is a constant. Show that

$$y = \frac{K_2 V_a}{K_1 V_{ref}}$$

3–47. The concept of a certain type of function generator that produces square waves and triangular waves simultaneously is shown in Figure P3–47. The integrator circuit performs the operation

$$v_2(t) = -K \int_0^t v_1(t)\, dt + v_2(0)$$

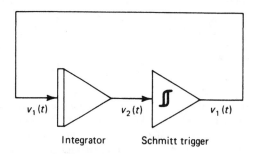

FIGURE P3–47

The Schmitt trigger has two output states, $+V_0$ and $-V_0$, for which the logic is

$$v_1 = +V_0 \quad \text{for } v_2 > +V_{ref}$$
$$= -V_0 \quad \text{for } v_2 < -V_{ref}$$

For $-V_{ref} < v_2 < V_{ref}$, the previous state remains. Assume that at $t = 0^+$ (immediately after a transition), the following conditions exist:

$$v_1(0^+) = -V_0$$
$$v_2(0^+) = -V_{ref}$$

(a) Calculate an expression for $v_2(t)$ which would apply in the interval of time in which $v_2(t) < V_{ref}$.

(b) Let t_1 represent the time at which the comparator changes states. Derive an expression for t_1.

(c) From the symmetry, use the result of (b) to show that the period T is

$$T = \frac{4V_{ref}}{KV_0}$$

(d) Sketch $v_1(t)$ and $v_2(t)$ over two full cycles.

3–48. The result of Problem 3–42 leads to a practical way of determining the coefficient of coupling between two coils. Assume that an inductance bridge is available. First, the self-inductances L_1 and L_2 are measured. (In each measurement, the other coil is left open so that no mutual coupling effects occur.) Next, the external terminals of L_2 are shorted, and the inductance L_{eq} viewed from the primary side is measured. Show that

$$k = \sqrt{1 - \frac{L_{eq}}{L_1}}$$

4

INITIAL, FINAL, AND FIRST-ORDER CIRCUITS

OBJECTIVES

After completing this chapter, the reader should be able to:

- Determine equivalent circuits and predict the voltages and currents in a circuit immediately after an excitation is first applied.
- Determine equivalent circuits and predict the voltages and currents in a circuit after dc steady-state conditions are reached.
- State the mathematical properties for and sketch the exponential function.
- Recognize the form of a first-order circuit with dc excitations.
- Determine a mathematical equation for and sketch the nature of any voltage or current response in a first-order circuit with dc excitations.
- Determine the time required for an exponential response to reach a certain level.

4-1 INITIAL CONDITIONS

The objective of this section is to determine simplified equivalent circuits of a capacitor and an inductor when first excited in an electric circuit. There may have been a previous excitation, in which case initial energy may be stored in one or more capacitors and/or inductors. However, when a sudden change occurs in the circuit, such as a new excitation or the switching of a component in or out of the circuit, the problem may be viewed as a new circuit with initial conditions.

Where there is only one major switching operation of interest in the analysis, it is customary to define the time scale so that $t = 0$ is the beginning of the desired solution. Where there are switching operations occurring at different times, a more elaborate time scale may be necessary, as we will see later.

$t = 0^-$ and 0^+

It is convenient to define $t = 0^-$ and $t = 0^+$ as two variations of the initial time value. The quantity $t = 0^-$ refers to the instant *just before* the switching action occurs, and $t = 0^+$ is the instant *just after* the switching action. Both 0^- and 0^+ are treated as $t = 0$ in mathematical expressions, since the difference is assumed to be negligible. However, the labels convey the descriptions of "just before" and "just after" and are very useful.

Impulsive Conditions

The vast majority of practical circuits are those in which *impulsive conditions* do not appear. Impulsive conditions occur in certain specialized cases, and their treatment will be considered in Chapter 10. This term will be explained further after the initial condition models are developed, but it is necessary to use the term in stating the conditions that follow.

No Initial Energy Case

Consider first the case of *initially relaxed* energy storage components, that is, a capacitor C with no initial voltage and an inductor L with no initial current. Let $v_C(0^+)$ represent the capacitor voltage at $t = 0^+$, and let $i_L(0^+)$ represent the inductor current at $t = 0^+$. Assume a current $i(t)$ for the capacitor and a voltage $v(t)$ for the inductor. We have

$$v_C(0^+) = \frac{1}{C} \int_0^{0^+} i(t)\, dt \qquad \textbf{(4–1)}$$

and

$$i_L(0^+) = \frac{1}{L} \int_0^{0^+} v(t)\, dt \qquad \textbf{(4–2)}$$

For a finite $i(t)$ in Equation (4-1) or a finite $v(t)$ in Equation (4–2), no area can be accumulated in the infinitesimal time from 0 to 0^+. This condition results in the following conditions for initially relaxed components:

$$v_C(0^+) = 0 \qquad \textbf{(4–3)}$$
$$i_L(0^+) = 0 \qquad \textbf{(4–4)}$$

The preceding mathematical formulation tells us that any physical variable resulting from an integration process cannot change instantaneously when the process being integrated has a finite level. In Chapter 10, the concept of the impulse function will be introduced. When impulses are present, it will be shown that it is theoretically possible to produce sudden changes in an integration process (subject to realistic tempering of the results).

The results of Equations (4–3) and (4–4) tell us that when initially relaxed energy-stored elements are first excited, the voltage across a capacitor must initially

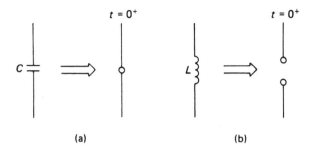

FIGURE 4–1
Initial behavior of uncharged capacitor and unfluxed inductor.

be zero, and the current through an inductor must initially be zero. These conditions are equivalent to the facts that *an uncharged capacitor initially acts like a short circuit, and an unfluxed inductor initially acts like an open circuit.* Models depicting this behavior are shown in Figure 4–1.

Initial Energy Case

Next, consider the case where initial energy is present. It was established in Chapter 3 that a charged capacitor can be represented as an uncharged capacitor in series with a dc voltage source, and a fluxed inductor can be represented as an unfluxed inductor in parallel with a dc current source. The behavior at $t = 0^+$ can then be deduced by combining the effects of the relaxed components in Figure 4–1 with the forms of the dc sources. Alternately, the results of Equations (4–1) and (4–2) can be modified by adding the dc values to the equations given. In either event, the following results are obtained for initial energy storage forms:

$$v_C(0^+) = V_0 \qquad\qquad \text{(4–3)}$$
$$i_L(0^+) = I_0 \qquad\qquad \text{(4–4)}$$

Thus, *a charged capacitor initially acts like a dc voltage source, and a fluxed inductor initially acts like a dc current source.* The forms of the circuit models are shown in Figure 4–2.

In a sense, the models of Figure 4–1 can be thought of as special cases of the models of Figure 4–2. In this sense, a voltage source of zero values reduces to a short circuit, and a current source of zero value reduces to an open circuit. However, the separate sets of models for the two cases have been provided for clarity.

Summary

From these results, the following points should be stressed: For circuits without impulsive conditions, *the voltage across a capacitance cannot be changed instantaneously, and the current through an inductance cannot be changed instantaneously.*

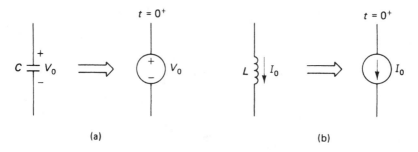

FIGURE 4–2
Initial behavior of charged capacitor and fluxed inductor.

(However, the current flow in a capacitance and the voltage across an inductance can be changed instantaneously.)

Resistance

Thus far, nothing has been said about resistance. Because of the algebraic relationship (Ohm's law) between resistive voltage and current, the form is always the same. Thus, the model for a resistor at $t = 0^+$ is the same as for all time.

EXAMPLE 4–1

The initially relaxed circuit of Figure 4–3(a) is excited at $t = 0$ by the source shown. Determine all element voltages and currents at $t = 0^+$.

Solution
Since there is only one time of interest, the functional notation will be omitted in both the analysis that follows and on the circuit diagram for brevity. However, it is understood that all quantities are evaluated only at $t = 0^+$.

The equivalent circuit at $t = 0^+$ is shown in Figure 4–3(b). The initially uncharged capacitor is replaced by a short circuit, and the initially unfluxed inductor is replaced by an open circuit. Note that it is unnecessary to specify actual values for C and L in this analysis. Labels for all element voltages and currents have been made in Figure 4–3(b), even though some are zero and some are redundant.

The circuit at $t = 0^+$ reduces to a simple single-loop configuration for which a single current i may be used to characterize the solution. We have

$$i = \frac{20 \text{ V}}{4\,\Omega + 6\,\Omega} = 2 \text{ A} \tag{4–5}$$

It is then readily recognized that

$$i_{R1} = i_C = i_{R2} = i = 2 \text{ A} \tag{4–6}$$

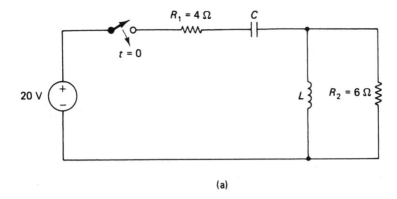

(a)

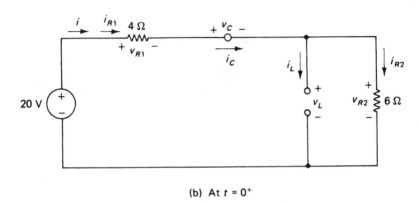

(b) At $t = 0^+$

FIGURE 4-3
Circuit for Example 4-1 and equivalent circuit at $t = 0^+$.

The initial capacitor voltage and the initial inductive current are obviously zero, that is,

$$v_C = 0 \qquad \text{(4-7)}$$
$$i_L = 0 \qquad \text{(4-8)}$$

Finally, the two resistive voltages and v_L are determined as

$$v_{R1} = 2\,\text{A} \times 4\,\Omega = 8\,\text{V} \qquad \text{(4-9)}$$
$$v_L = v_{R2} = 2\,\text{A} \times 6\,\Omega = 12\,\text{V} \qquad \text{(4-10)}$$

It should be stressed again that these values are correct only for a brief instant after the switch is closed. The circuit variables will change as time passes, and we are not yet ready to predict the behavior for increasing time.

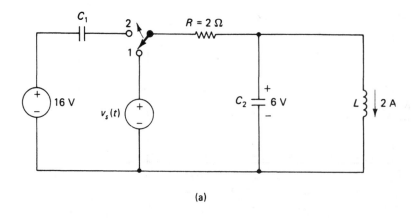

(a)

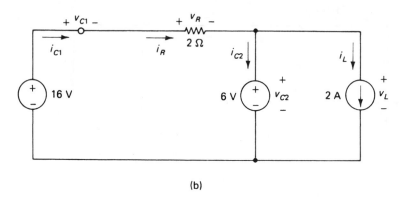

(b)

FIGURE 4–4
Circuit for Example 4–2 and equivalent circuit at $t = 0^+$.

EXAMPLE 4–2

The switch in the circuit of Figure 4–4(a) is initially in position 1. As a result of some excitation $v_s(t)$, whose form is not important for our analysis, the initial values of the capacitor voltage and inductor current are as shown. At $t = 0$, the switch is changed to position 2. Determine all element voltages and currents at $t = 0^+$.

Solution

The equivalent circuit at $t = 0^+$ is shown in Figure 4–4(b). As in the previous problem, the functional notation will be omitted, but the time $t = 0^+$ is understood. The charged capacitor C_2 acts initially like a 6-V dc voltage source, and the fluxed inductor acts initially like a 2-A current source. However, C_1 is initially uncharged, so it acts like a short circuit.

Although several methods could be used to solve this dc circuit, the simplest approach is a step-by-step process. We first note that

$$v_{C1} = 0 \qquad\qquad\qquad \text{(4–11)}$$
$$v_{C2} = v_L = 6 \text{ V} \qquad\qquad \text{(4–12)}$$
$$i_L = 2 \text{ A} \qquad\qquad\qquad \text{(4–13)}$$

The resistive voltage is

$$v_R = 16 \text{ V} - 6 \text{ V} = 10 \text{ V} \qquad\qquad \text{(4–14)}$$

The current in C_1 is the same as that in R and is

$$i_{C1} = i_R = \frac{10 \text{ V}}{2 \text{ }\Omega} = 5 \text{ A} \qquad\qquad \text{(4–15)}$$

The current in C_2 is determined by application of KCL to the upper right-hand node, that is,

$$i_{C2} = i_R - i_L = 5 - 2 = 3 \text{ A} \qquad\qquad \text{(4–16)}$$

4–2   FINAL CONDITIONS

The objective of this section is to determine equivalent circuits of a capacitor and an inductor under conditions in which all voltages and currents in the circuit have ceased to vary with time and have assumed constant values. This condition is referred to as a *steady-state dc circuit*. It is customary to use $t = \infty$ as a reference in many problems for this situation. Thus, a voltage $v(\infty)$ and a current $i(\infty)$ refer to circuit variables under these limiting conditions. It should be noted that $t = \infty$ is only a mathematical way of indicating that a sufficiently long time has passed to ensure the assumed conditions.

Circuits in which dc steady-state conditions are valid include those having dc sources, initial conditions (i.e., charged capacitors and fluxed inductors), and certain types of bounded inputs (e.g., single-pulse input or a decaying exponential input). The primary criterion is that all voltages and currents must eventually assume constant values. The concept would be invalid for circuits with voltages and currents that continue to vary with time (e.g., sinusoidal sources).

Let $i_C(t)$ represent the current flow in a capacitor C, and let $v_L(t)$ represent the voltage across an inductor L. Assume a voltage $v(t)$ across the capacitor and a current $i(t)$ in the inductor. In general, the relationships are

$$i_C(t) = C \frac{dv(t)}{dt} \qquad\qquad \text{(4–17)}$$

$$v_L(t) = L \frac{di(t)}{dt} \qquad\qquad \text{(4–18)}$$

dc Steady-State Behavior for L and C

In the steady-state dc case, all voltages and currents have settled down to constant values and there are no changes. Thus, all derivatives have zero values, and we have

$$i_C(\infty) = 0 \qquad\qquad (4\text{--}19)$$
$$v_L(\infty) = 0 \qquad\qquad (4\text{--}20)$$

The results of Equations (4–19) and (4–20) tell us in the dc steady-state that the current flow in a capacitor is zero and the voltage across an inductance is zero. These conditions are equivalent to the facts that, *in the dc steady-state, a capacitor acts like an open circuit and an inductor acts like a short circuit.* Models illustrating these phenomena are shown in Figure 4–5.

$t = \infty$ or 0^-

In many problems, the time $t = \infty$ will be used to indicate the reference for steady-state conditions, as already indicated. In some cases, however, a steady-state condition for one part of a circuit will serve as the basis for the initial condition of a different part of a circuit. In this type of situation, if $t = 0$ is used as a switching time, $t = 0^-$ is an appropriate symbol for the steady-state condition just prior to switching.

Whenever a steady-state condition existing in one part of a circuit is used as a basis for initial conditions in a switching operation, the two most important variables to be determined prior to the switching operation are *capacitor voltages* and *inductor currents*. From the work of Section 4–1, it is clear that these are the variables that cannot change instantaneously, so they remain the same through a nonimpulsive instantaneous switching operation.

EXAMPLE 4–3

Determine the steady-state voltage and current for each element in the circuit of Figure 4–6(a).

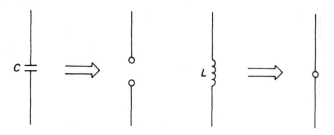

FIGURE 4–5
Steady-state dc behavior of capacitor and inductor.

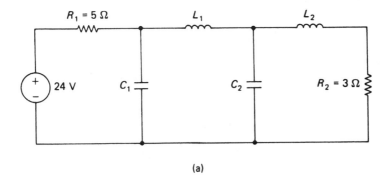

(a)

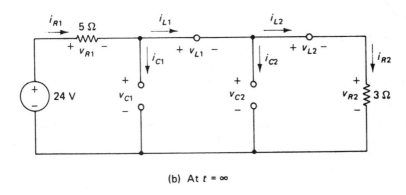

(b) At $t = \infty$

FIGURE 4–6
Circuit for Example 4–3 and equivalent circuit at $t = \infty$.

Solution
Since only steady-state voltages and currents are of interest, functional notation will be omitted as was done in the preceding section. However, it is understood that all values are the steady-state quantities.

The equivalent circuit at $t = \infty$ based on steady-state dc concepts is shown in Figure 4–6(b). The two capacitors are replaced by open circuits, and the two inductors are replaced by short circuits.

By inspection, it is immediately evident that

$$v_{L1} = 0 \qquad\qquad\qquad (4\text{–}21)$$
$$v_{L2} = 0 \qquad\qquad\qquad (4\text{–}22)$$
$$i_{C1} = 0 \qquad\qquad\qquad (4\text{–}23)$$
$$i_{C2} = 0 \qquad\qquad\qquad (4\text{–}24)$$

The circuit form is that of a single-loop circuit, and the same current flows through the two inductors and the two resistors. Thus,

$$i_{R1} = i_{L1} = i_{L2} = i_{R2} = \frac{24\text{ V}}{5\,\Omega + 3\,\Omega} = 3\text{ A} \qquad (4\text{–}25)$$

The two resistor voltages are

$$v_{R1} = 3 \text{ A} \times 5 \, \Omega = 15 \text{ V} \tag{4-26}$$
$$v_{R2} = 3 \text{ A} \times 3 \, \Omega = 9 \text{ V} \tag{4-27}$$

The two capacitor voltages are equal and are the same as the voltage across R_2. Thus

$$v_{C1} = v_{C2} = v_{R2} = 9 \text{ V} \tag{4-28}$$

The capacitors are thus both charged to 9 V, and the inductors are fluxed to 3 A in the dc steady state.

EXAMPLE 4-4

The circuit of Figure 4–7(a) is initially in a steady-state condition in position 1. At $t = 0$, the switch is moved to position 2 instantaneously. **(a)** Determine all element voltages and currents immediately following the switching operation. **(b)** Determine all element voltages and currents after steady-state conditions are again established.

Solution

This problem represents a common practical situation in which initial conditions are not given explicitly, but in which such conditions are determined by analyzing the initial circuit form. The variables to be determined prior to the switching operation are the two capacitor voltages and the inductor current.

The steady-state dc circuit prior to switching is shown in Figure 4–7(b). However, the time is indicated as $t = 0^-$, since this is just before the switching operation. Since steady-state conditions exist, the two capacitors are represented as open circuits, and the inductor is represented as a short circuit. The inductive current is readily determined as

$$i_L(0^-) = \frac{10 \text{ V}}{2 \, \Omega + 3 \, \Omega} = 2 \text{ A} \tag{4-29}$$

The voltage across C_1 is the same as the voltage across the 3-Ω resistor and is

$$v_{C1}(0^-) = 3 \, \Omega \times 2 \text{ A} = 6 \text{ V} \tag{4-30}$$

However, since a short appears across C_2, its initial voltage is

$$v_{C2}(0^-) = 0 \tag{4-31}$$

(a) The equivalent circuit just after the switching operation is shown in Figure 4–7(c). The inductor acts like a dc current source of 2 A, capacitor C_1 acts like a 6-V source, and capacitor C_2 acts like a short circuit. The input source voltage is now the 15-V source.

The circuit can be readily analyzed with a step-by-step approach. The voltage across R_1 is

$$v_{R1}(0^+) = 15 \text{ V} - 6 \text{ V} = 9 \text{ V} \tag{4-32}$$

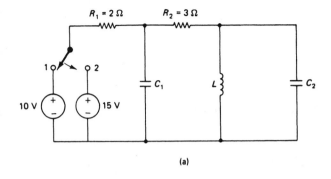

(a)

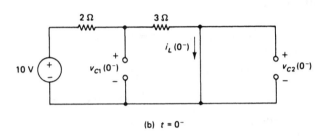

(b) $t = 0^-$

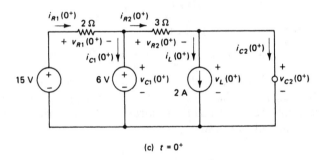

(c) $t = 0^+$

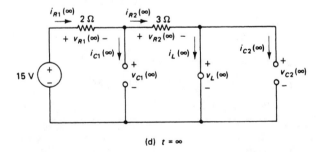

(d) $t = \infty$

FIGURE 4–7
Circuit for Example 4–4 and various models used in analysis.

The current through this resistor is

$$i_{R1}(0^+) = \frac{9\text{ V}}{2\text{ }\Omega} = 4.5\text{ A} \tag{4-33}$$

Because of the short on the right,

$$v_L(0^+) = v_{C2}(0^+) = 0 \tag{4-34}$$

and

$$v_{R2}(0^+) = v_{C1}(0^+) = 6\text{ V} \tag{4-35}$$

The current through R_2 is

$$i_{R2}(0^+) = \frac{6\text{ V}}{3\text{ }\Omega} = 2\text{ A} \tag{4-36}$$

By KCL, the current flow in C_1 is

$$i_{C1}(0^+) = i_{R1}(0^+) - i_{R2}(0^+) = 4.5\text{ A} - 2\text{ A} = 2.5\text{ A} \tag{4-37}$$

The current in the inductance is

$$i_L(0^+) = 2\text{ A} \tag{4-38}$$

By KCL again, the current flow in C_2 is

$$i_{C2}(0^+) = i_{R2}(0^+) - i_L(0^+) = 2 - 2 = 0 \tag{4-39}$$

(b) The steady-state circuit based on the new source voltage is shown in Figure 4–7(d). The two capacitors have been replaced by open circuits once again, and the inductor has been replaced by a short circuit. The various voltages and currents are determined as follows:

$$i_{R1}(\infty) = i_{R2}(\infty) = i_L(\infty) = \frac{15\text{ V}}{2\text{ }\Omega + 3\text{ }\Omega} = 3\text{ A} \tag{4-40}$$

$$i_{C1}(\infty) = 0 \tag{4-41}$$

$$i_{C2}(\infty) = 0 \tag{4-42}$$

$$v_{R1}(\infty) = 3\text{ A} \times 2\text{ }\Omega = 6\text{ V} \tag{4-43}$$

$$v_{C1}(\infty) = v_{R2}(\infty) = 3\text{ A} \times 3\text{ }\Omega = 9\text{ V} \tag{4-44}$$

$$v_L(\infty) = v_{C2}(\infty) = 0 \tag{4-45}$$

4–3 EXPONENTIAL FUNCTION

One of the most important mathematical functions is the exponential function. The most natural and basic form of the exponential function is denoted by e^x, where x is the independent variable and e is the base of the natural logarithmic system. The

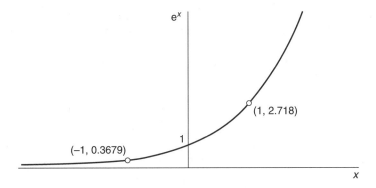

FIGURE 4–8
General form of the exponential function.

value of this base to 4 significant figures is $e = 2.718$. Virtually all modern scientific calculators have the e^x function provided. The general form of the exponential function for both positive and negative x is shown in Figure 4–8.

Basic Differential Equation

The exponential function arises in the solution of many differential equations. There are many physical and numerical processes in which the rate of change of a variable y is proportional to the level of the same variable. This is stated mathematically as

$$\frac{dy}{dx} = ky \tag{4-46}$$

where k is a constant. The solution of this differential equation is of the form

$$y = De^{kx} \tag{4-47}$$

where D is a constant that depends on the initial value of y. Processes of the type given in Equation (4–47) include such widely varied areas as population growth, compounding of interest, heat flow, and, of course, electrical circuits.

Having introduced the form of the exponential function in general terms, we will now change the form and notation slightly to correspond to electrical circuit usage. First, the form of the function arising more often in practical circuits is the portion shown for negative x in Figure 4–8. Second, the independent variable is time.

Decaying Exponential Function

Let $f(t)$ represent any exponential function of interest in the discussion that follows. The *decaying exponential function* is defined as

$$f(t) = e^{-\alpha t} \tag{4-48}$$

where α is a positive constant known as the *damping factor.* Dimensionally, it has the unit of (seconds)$^{-1}$. The function of Equation (4–48) is normally of interest for $t > 0$.

An alternate form of the decaying exponential function is

$$f(t) = e^{-t/\tau} \qquad\qquad \textbf{(4–49)}$$

where τ is a positive constant known as the *time constant.* Comparing Equations (4–48) and (4–49), we note that

$$\tau = \frac{1}{\alpha} \qquad\qquad \textbf{(4–50)}$$

The form of the decaying exponential for $t > 0$ is shown in Figure 4–9. The horizontal scale is the *normalized* variable $t/\tau = \alpha t$, so the curve can be adapted to any particular time constant or damping factor. It is noted on the curve that if the exponential continued to decrease at its initial rate, it would reach a value of zero at $t = \tau$.

The reader may find it instructive to employ the e^x function on a calculator to support the discussion that follows. Reference can also be made to Figure 4–9. The value of the exponential function at $t = 0$ is simply $e^0 = 1$. When $t = \tau$, $t/\tau = 1$, and $e^{-t/\tau} = e^{-1} = 0.36788$ to five significant figures. When $t = 2\tau$, $t/\tau = 2$, and $e^{-t/\tau} = e^{-2} = 0.13534$. The preceding values and several others are summarized thus:

t	$e^{-t/\tau}$
τ	0.36788
2τ	0.13534
3τ	0.04979
4τ	0.01832
5τ	0.00674

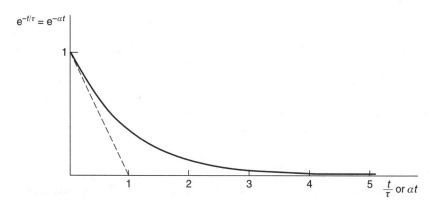

FIGURE 4–9
Form of the decaying exponential function.

From the preceding results, note that the exponential decays to about 37% of its initial value in a time equal to one time constant. In the next interval equal to one time constant, it decays to about 37% of the value at the beginning of that interval, which means it is now at a level of about 13.5% of the initial value, and so on.

Rule of Thumb

Theoretically, the exponential never quite reaches zero. For all practical purposes, however, it does reach zero. Observe from the table that between $t = 4\tau$ and $t = 5\tau$, the function reaches 1% of the initial level. A convenient rule of thumb is that *the exponential function has reached a limiting value of zero in a time interval 5τ.* Like most rules of thumb, however, this should be interpreted as an approximation rather than an exact result. This author has observed some practical precision measurement systems in which a longer time interval was required.

Derivative and Integral of Exponential Function

An important property of the exponential function is that the derivative or integral of an exponential function is an exponential function with the same damping factor or time constant. Let

$$y = e^{-\alpha t} \quad \text{for } t > 0 \tag{4–51}$$

The derivative of Equation (4–51) for $t > 0$ is

$$\frac{dy}{dt} = -\alpha e^{-\alpha t} \quad \text{for } t > 0 \tag{4–52}$$

The indefinite integral of Equation (4–51) is

$$\int e^{-\alpha t}\, dt = \frac{-e^{-\alpha t}}{\alpha} + C \tag{4–53}$$

where C is a constant of integration. The exponential forms of the derivative and integral of the exponential function are immediately evident.

Power Series Form

By a mathematical process called *power series expansion,* the exponential function e^x may be represented as

$$e^x = 1 + x + \frac{x^2}{2!} + \frac{x^3}{3!} + \cdots \tag{4–54a}$$

$$= \sum_{n=0}^{\infty} \frac{x^n}{n!} \tag{4–54b}$$

where $n!$ (read as "n factorial") is defined as

$$n! = n(n-1)(n-2) \cdots (2)(1) \quad \text{for } n > 0 \qquad \text{(4–55a)}$$

and

$$0! = 1 \text{ by definition} \qquad \text{(4–55b)}$$

The power series representation indicates that an infinite number of terms would theoretically be required to determine the exponential function in this manner. Practically speaking, however, the series converges quite rapidly, so that only a finite number of terms is needed to closely represent the function in Equation (4–54a). Indeed, the software in many scientific calculators uses the series form for computing the exponential function.

When x is replaced by $-\alpha t$ in Equation (4–54a), the form of the decaying exponential expressed as a power series expansion is

$$e^{-\alpha t} = 1 - \alpha t + \frac{(\alpha t)^2}{2!} - \frac{(\alpha t)^3}{3!} + \cdots \qquad \text{(4–56a)}$$

$$= \sum_{n=0}^{\infty} \frac{(-1)^n (\alpha t)^n}{n!} \qquad \text{(4–56b)}$$

Approximation for Short Time Interval

In some applications of the exponential function, operation is confined to a short time interval in which $\alpha t \ll 1$ or $t \ll \tau$. In this case, all terms in Equation (4–56a) of second degree or higher are very small and may be neglected. Under such conditions, the decaying exponential may be closely approximated as

$$e^{-\alpha t} \simeq 1 - \alpha t = 1 - \frac{t}{\tau} \quad \text{for } t \ll \tau \qquad \text{(4–57)}$$

This equation indicates that the decaying exponential function behaves approximately like a straight-line equation with a slope of $-\alpha$ or $-1/\tau$ in a time interval very small compared with one time constant.

EXAMPLE 4–5

This example is designed to impart some practical appreciation of the power series representation of the exponential function. To follow the computations, the reader should have available a scientific calculator with an e^x function.

Consider the function $y = e^{-t/\tau}$ and determine how closely power series approximations containing 2 terms, 3 terms, and 4 terms represent the function as t varies from 0 to τ.

Solution

The "true" value of the function will be denoted as y, and its value will be calculated using the e^x function on the calculator. (Actually, the calculator value is an approximation, but it is usually accurate to a large number of decimal places.) Let y_1, y_2, and y_3 represent the approximations containing powers through the first, second, and third as given by Equation (4–56a). Using the time constant form, we have

$$y_1 = 1 - \frac{t}{\tau} \tag{4-58}$$

$$y_2 = 1 - \frac{t}{\tau} + \frac{1}{2}\left(\frac{t}{\tau}\right)^2 = y_1 + \frac{1}{2}\left(\frac{t}{\tau}\right)^2 \tag{4-59}$$

$$y_3 = 1 - \frac{t}{\tau} + \frac{1}{2}\left(\frac{t}{\tau}\right)^2 - \frac{1}{6}\left(\frac{t}{\tau}\right)^3 = y_2 - \frac{1}{6}\left(\frac{t}{\tau}\right)^3 \tag{4-60}$$

Note that each function can be computed from the previous function by the addition of one term.

To illustrate one particular case, consider $t = 0.1\tau$. The quantity $t/\tau = 0.1\tau/\tau = 0.1$, and $e^{-0.1} = 0.904837$ to six significant figures as obtained with the e^x calculator function. The approximations are

$$y_1 = 1 - 0.1 = 0.9 \tag{4-61}$$

$$y_2 = 0.9 + \frac{1}{2}(0.1)^2 = 0.905000 \tag{4-62}$$

$$y_3 = 0.905 - \frac{1}{6}(0.1)^3 = 0.904833 \tag{4-63}$$

The successive approximations get better as the number of terms is increased. By including powers through the third, the results agree through the first five digits.

The preceding and some additional values are given in the short table that follows.

t	y	y_1	y_2	y_3
0	1	1	1	1
0.1τ	0.904837	0.900000	0.905000	0.904833
0.25τ	0.778801	0.750000	0.781250	0.778646
0.5τ	0.606531	0.500000	0.625000	0.604167
0.75τ	0.472367	0.250000	0.531250	0.460938
τ	0.367879	0	0.500000	0.333333

As can be seen, the linear approximation (y_1) is valid only for small t (generally $t < 0.1\tau$ or so). The higher-order approximations are closer for larger t, but for $t = \tau$, even y_3 differs by nearly 10% from the true value. At that point, more terms in the series would be required to closely represent the function.

4–4   FIRST-ORDER CIRCUITS

The complete response of first-order circuits with certain types of input excitations will be considered in this section. First-order circuits are those that can be reduced to one of the following two forms:

1. *RC* One equivalent *R* and one equivalent *C*, *or*
2. *RL* One equivalent *R* and one equivalent *L*

The implication is that there may be more than one component of a particular type initially, provided that the circuit can be reduced to one equivalent value. Note that *L* and *C* will not both be permitted in the same circuit.

The term *first order* is associated with the fact that a differential equation describing the circuit will be a first-order form. If both *L* and *C* were present, the differential equation would be at least a second-order form. The term *single time constant circuit* is also used as a description of a first-order circuit.

As a further restriction in this chapter, we will consider only dc sources and initial conditions. First-order circuits excited by arbitrary sources will be deferred to Chapter 6, after Laplace transforms are introduced.

Why do we wish to put so much emphasis on first-order circuits with dc sources, since this appears to be a rather specialized case? First, many important electronic timing circuits and oscillators use such circuits to control time delays and oscillation periods. Second, this represents the first point in the text at which the solution of a complete circuit with more than one element and with time-varying responses will be considered. It will thus serve to illustrate how complete circuit solutions can be obtained.

Differential Equation Formulation

To illustrate how the differential equation for first-order circuits can be formulated, consider the *RC* circuit shown in Figure 4–10. A dc voltage source of value V_s is applied at $t = 0$, and the capacitor is assumed to be initially uncharged. Assume that the variable of interest is the voltage across the capacitor $v_C(t)$. A node voltage equation at the node representing $v_C(t)$ can be obtained by expressing KCL at that

FIGURE 4–10
RC circuit used to illustrate formulation of differential equation.

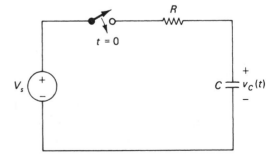

FIGURE 4–11
RL circuit used to illustrate formulation of differential equation.

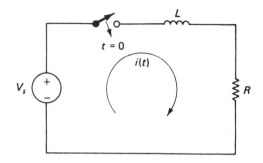

node. We have

$$C\frac{dv_C}{dt} + \frac{v_C - V_s}{R} = 0 \tag{4–64}$$

This can be rearranged as

$$\frac{dv_C}{dt} + \frac{1}{RC}v_C = \frac{V_s}{RC} \tag{4–65}$$

As a second illustration, consider the *RL* circuit of Figure 4–11. A dc voltage source of value V_s is applied at $t = 0$, and the inductor is assumed to be initially unfluxed. Assume that the variable of interest in this case is the current $i(t)$. A mesh current equation can be obtained by applying KVL, and we have

$$-V_s + L\frac{di}{dt} + Ri = 0 \tag{4–66}$$

This can be rearranged as

$$\frac{di}{dt} + \frac{R}{L}i = \frac{V_s}{L} \tag{4–67}$$

Both Equation (4–65) and Equation (4–67) are examples of first-order differential equations with constant excitation sources. All equations of this type could be solved by standard mathematical procedures. However, the solutions are of the same general form. By employing some basic circuit properties familiar to us, we can write the solution directly by a straightforward inspection process, and that will be the basis of our approach.

General form for Solution

Let $y(t)$ represent any voltage or current response of interest. The general response of a first-order circuit with dc excitations will always be of the form

$$y(t) = A + Be^{-t/\tau} \tag{4–68}$$

where A and B are constants dependent on the circuit configuration and source magnitudes, and τ is the time constant.

The constants A and B can be determined from evaluating the initial response by the procedure of Section 4–1, and the final response by the procedure of Section 4–2. Let $y(0^+)$ represent the initial response, and let $y(\infty)$ represent the final response. Setting first $t = 0$ and then $t = \infty$ in Equation (4–68), the following two simultaneous equations are obtained.

$$y(0^+) = A + B \tag{4–69}$$
$$y(\infty) = A \tag{4–70}$$

The values of A and B are then readily determined as

$$A = y(\infty) \tag{4–71}$$
$$B = y(0^+) - y(\infty) \tag{4–72}$$

The form of $y(t)$ is then given by

$$y(t) = y(\infty) + [y(0^+) - y(\infty)]e^{-t/\tau} \tag{4–73}$$

The value of the time constant is always equal to one of the following forms:

a. *RC Circuit:* $\qquad\qquad\qquad\qquad \tau = R_{eq}C_{eq} \tag{4–74}$

b. *RL Circuit:* $\qquad\qquad\qquad\qquad \tau = \dfrac{L_{eq}}{R_{eq}} \tag{4–75}$

where R_{eq}, C_{eq}, and L_{eq} represent the equivalent values when the circuit has been reduced to one resistor and one capacitor or one resistor and one inductor.

The form of Equation (4–73) represents the simplest response equation to use, since one need only plug in the initial value, the final value, and the time constant. However, some people prefer to remember the simpler form of Equation (4–68) and work out the A and B constants in each case, so it is a matter of preference. In the remainder of this section, the form of Equation (4–73) will be used.

Shape of Response

The complete response $y(t)$ changes from $y(0^+)$ to $y(\infty)$ as time changes, and the behavior between is exponential in character. The function is always concave toward the final value, as illustrated by two cases in Figure 4–12. In part (a), $y(\infty) > y(0^+)$, and the exponential function is concave upward. In part (b), $y(\infty) < y(0^+)$, and the function is concave downward.

Simple *RC* Circuit Form

To illustrate some complete response forms, consider first the form of the *RC* circuit considered earlier, which is repeated in Figure 4–13(a). Both the current $i(t)$ and

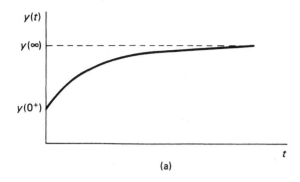

(a)

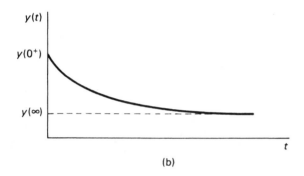

(b)

FIGURE 4–12
Forms of exponentially (a) increasing and (b) decreasing responses.

the capacitor voltage $v_C(t)$ are assumed to be of interest. The initial circuit is shown in Figure 4–13(b). We note that

$$i(0^+) = \frac{V_s}{R} \tag{4–76}$$

$$v_C(0^+) = 0 \tag{4–77}$$

The final circuit is shown in Figure 4–13(c), and we see that

$$i(\infty) = 0 \tag{4–78}$$
$$v_C(\infty) = V_s \tag{4–79}$$

The time constant in this circuit is simply

$$\tau = RC \tag{4–80}$$

Substituting pertinent values of current and voltage in Equation (4–73), we obtain

$$i(t) = \frac{V_s}{R} e^{-t/RC} \tag{4–81}$$

FIGURE 4–13
RC circuit and models used in developing total response.

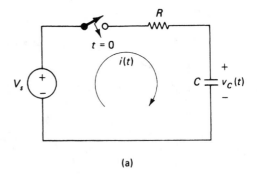

(a)

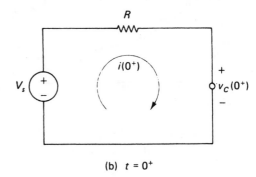

(b) $t = 0^+$

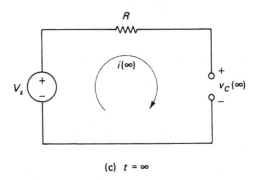

(c) $t = \infty$

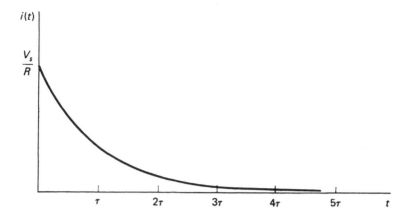

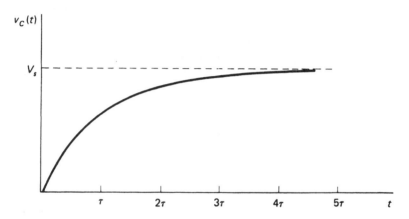

FIGURE 4–14
Response functions in *RC* circuit of Figure 4–13.

and

$$v_C(t) = V_s - V_s e^{-t/RC} = V_s(1 - e^{-t/RC}) \qquad (4\text{–}82)$$

The functions $i(t)$ and $v_C(t)$ are shown in Figure 4–14. The current starts at a value V_s/R and exponentially decreases toward zero. The capacitor voltage, however, starts at zero and exponentially increases toward V_s. For most practical purposes, the current will reach zero, and the capacitor voltage will reach V_s, in an approximate time interval $5\tau = 5RC$.

Simple *RL* Circuit Form

Consider next the form of an *RL* circuit discussed earlier, which is repeated in Figure 4–15(a) (with slight rearrangement). Both the current $i(t)$ and the inductor voltage $v_L(t)$ are assumed to be of interest. The initial circuit is shown in Figure

FIGURE 4–15
RL circuit and models used in developing total response.

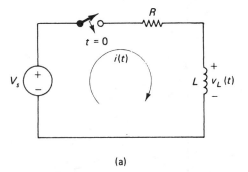

(a)

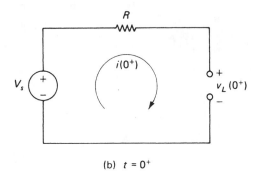

(b) $t = 0^+$

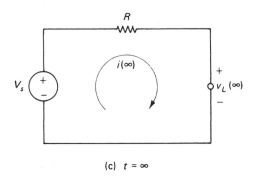

(c) $t = \infty$

4–15(b). We can readily determine that

$$i(0^+) = 0 \qquad\qquad \textbf{(4–83)}$$

$$v_L(0^+) = V_s \qquad\qquad \textbf{(4–84)}$$

The final circuit is shown in Figure 4–15(c), and we note that

$$i(\infty) = \frac{V_s}{R} \qquad\qquad \textbf{(4–85)}$$

$$v_L(\infty) = 0 \qquad\qquad \textbf{(4–86)}$$

The time constant in this circuit is

$$\tau = \frac{L}{R} \tag{4-87}$$

Substituting the values of voltage and current in Equation (4–73), we obtain

$$i(t) = \frac{V_s}{R} - \frac{V_s}{R} e^{-Rt/L} \tag{4-88}$$

or

$$i(t) = \frac{V_s}{R} (1 - e^{-Rt/L}) \tag{4-89}$$

and

$$v_L(t) = V_s e^{-Rt/L} \tag{4-90}$$

The functions $i(t)$ and $v_L(t)$ are shown in Figure 4–16. The current starts at a value of zero and exponentially increases toward V_s/R. The inductor voltage starts at V_s and exponentially decreases toward zero.

Comparing Figures 4–14 and 4–16, we observe that the process of charging a capacitor has the same mathematical form as the process of establishing current through an inductor.

EXAMPLE 4–6

The 1-μF capacitor in the circuit of Figure 4–17(a) is initially charged to 5 V in the direction shown. At $t = 0$, a 10-V source is connected in the direction shown. Determine expressions for $i(t)$ and $v_C(t)$ for $t > 0$, and sketch the functions.

Solution
The equivalent circuit at $t = 0^+$ is shown in Figure 4–17(b). At that time, the capacitor acts like a voltage source of 5 V, with the more positive terminal downward. The initial value of the capacitor voltage is thus

$$v_C(0^+) = -5 \text{ V} \tag{4-91}$$

The initial current is

$$i(0^+) = \frac{10 \text{ V} + 5 \text{ V}}{10 \text{ k}\Omega} = 1.5 \text{ mA} \tag{4-92}$$

The equivalent circuit at $t = \infty$ is shown in Figure 4–17(c). The capacitor now acts like an open circuit, and we have

$$v_C(\infty) = 10 \text{ V} \tag{4-93}$$
$$i(\infty) = 0 \tag{4-94}$$

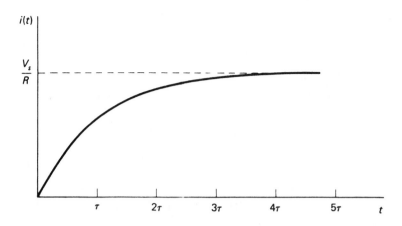

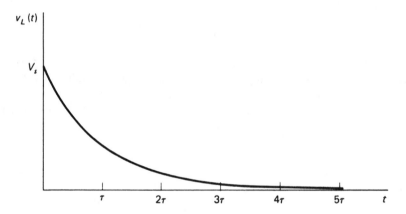

FIGURE 4–16
Response functions in *RL* circuit of Figure 4–15.

The time constant is

$$\tau = 10^4 \ \Omega \times 10^{-6} \ \text{F} = 0.01 \ \text{s} \qquad \textbf{(4–95)}$$

Expressions for $v_C(t)$ and $i(t)$ can be written using Equation (4–73). The results are

$$v_C(t) = 10 + (-5 - 10)e^{-t/0.01} \qquad \textbf{(4–96a)}$$
$$= 10 - 15e^{-100t} \qquad \textbf{(4–96b)}$$

and

$$i(t) = 1.5 \times 10^{-3}e^{-100t} \qquad \textbf{(4–97)}$$

The two functions are shown in Figure 4–18. Note that $v_C(t)$ changes exponentially from an initial negative voltage to a final positive voltage. Some additional analysis of this function will be made later, in Example 4–13.

FIGURE 4–17
Circuit for Example 4–6 and models used in analysis.

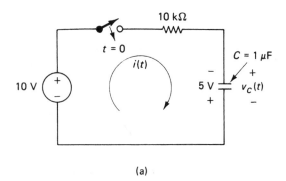

(a)

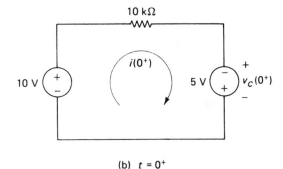

(b) $t = 0^+$

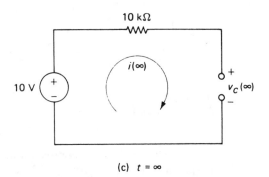

(c) $t = \infty$

EXAMPLE 4–7

Steady-state conditions exist in the circuit of Figure 4–19(a) at $t = 0^-$. The 3-Ω resistor is connected across the capacitor at $t = 0$. Determine an expression for $v_C(t)$ for $t > 0$ and sketch the function.

Solution

The equivalent circuit at $t = 0^-$ is shown in Figure 4–19(b). The only quantity of interest at that time is the capacitor voltage, and it is

$$v_C(0^-) = 12 \text{ V} = v_C(0^+) \tag{4–98}$$

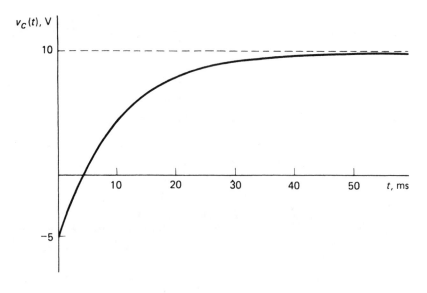

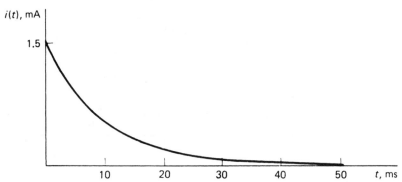

FIGURE 4–18
Waveforms for Example 4–6.

After the 3-Ω resistor is connected in the circuit, it is helpful to redraw the circuit as shown in Figure 4–19(c), with the resistor on the left. A Thevenin equivalent circuit can then be determined to the left of the capacitor, and it is shown in Figure 4–19(d). Thus, the equivalent resistance is readily seen to be 2 Ω. The time constant is then

$$\tau = 2\,\Omega \times \frac{1}{4}\text{F} = 0.5\,\text{s} \qquad\qquad \textbf{(4–99)}$$

The equivalent circuit at $t = \infty$ is shown in Figure 4–19(e). The final capacitor voltage is

$$v_C(\infty) = 4\,\text{V} \qquad\qquad \textbf{(4–100)}$$

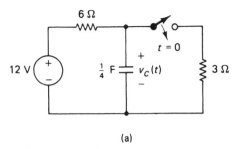

(a)

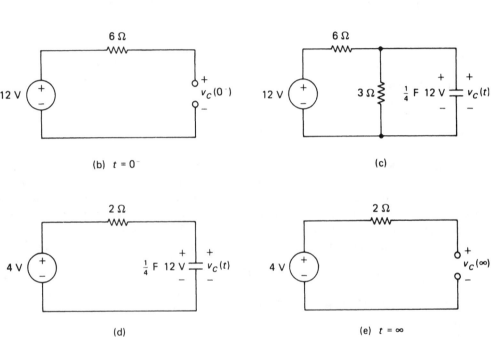

FIGURE 4–19
Circuit for Example 4–7 and models used in analysis.

The complete voltage is then determined as

$$v_C(t) = 4 + (12 - 4)e^{-t/0.5} \qquad \textbf{(4–101a)}$$
$$= 4 + 8e^{-2t} \qquad \textbf{(4–101b)}$$

This function is shown in Figure 4–20.

EXAMPLE 4–8
The two relaxed inductors in the circuit of Figure 4–21(a) are connected at $t = 0$. Determine an expression for $i(t)$ for $t > 0$ and sketch it. There is no mutual coupling between the inductors.

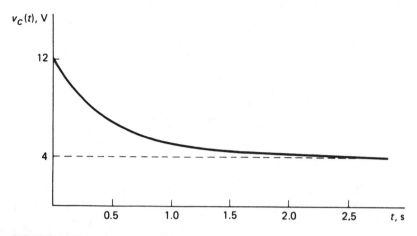

FIGURE 4–20
Waveform for Example 4–7.

Solution
Although the circuit contains two sources, three resistors, and two inductors, it is still a first-order circuit, since it can be reduced to the basic form. First, the Thevenin equivalent circuit to the left of the switch can be determined. The details will not be shown here, but the reader is invited to verify the validity of the result shown in Figure 4–21(b). The equivalent inductance is 20 mH + 30 mH = 50 mH.

The equivalent circuit at $t = 0^+$ is shown in Figure 4–21(c), and we note that

$$i(0^+) = 0 \qquad\qquad (4\text{–}102)$$

The equivalent circuit at $t = \infty$ is shown in Figure 4–21(d), and we determine that

$$i(\infty) = \frac{50\,\text{V}}{25\,\Omega} = 2\,\text{A} \qquad\qquad (4\text{–}103)$$

The time constant is

$$\tau = \frac{0.05\,\text{H}}{25\,\Omega} = 2 \times 10^{-3}\,\text{s} = 2\,\text{ms} \qquad\qquad (4\text{–}104)$$

The current is

$$i(t) = 2 - 2e^{-t/2\times10^{-3}} = 2(1 - e^{-500t}) \qquad\qquad (4\text{–}105)$$

The form of this function is shown in Figure 4–22.

EXAMPLE 4–9
The initially relaxed shunt-C *RC* circuit of Figure 4–23(a) is excited by a voltage pulse of height V_p and width T, as shown in (b). Determine the output voltage $v_2(t)$.

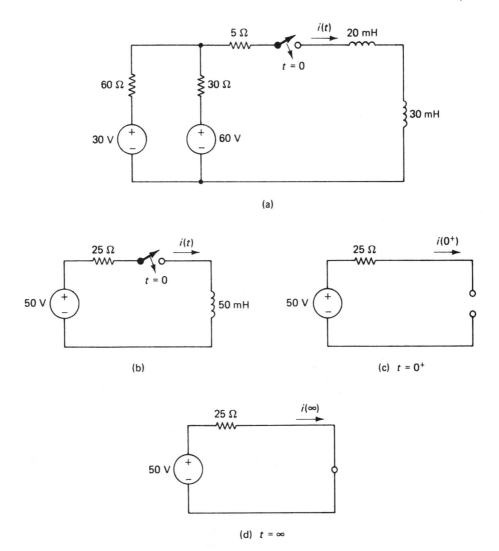

FIGURE 4-21
Circuit for Example 4–8 and models used in the analysis.

Solution
The excitation is not exactly a dc function like that of other examples considered thus far. However, the response in the interval $0 < t < T$ should be the same form *as if* the source were dc. For $t > T$, the solution then consists of determining the response to an initially charged capacitor based on the level reached at $t = T$. Thus, the solution will be considered in two separate intervals.

$0 < t < T$ In this interval, the form of the solution is based on the initial value $v_2(0^+) = 0$ and the assumed final value $v_2(\infty) = V_p$. Note that the final

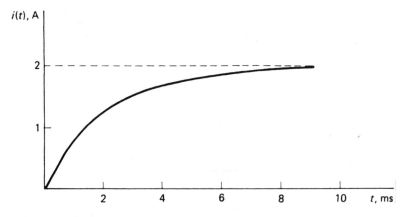

FIGURE 4-22
Waveform for Example 4-8.

condition is determined *as if* the response continued on with a dc input. The response is thus

$$v_2(t) = V_p(1 - e^{-t/RC}) \quad \text{for } 0 < t < T \qquad \textbf{(4-106)}$$

$t > T$ At $t = T$, the source voltage drops to zero, and the equivalent circuit is of the form shown in Figure 4-24. The quantity V_c represents the voltage on the capacitor at the beginning of the second interval, and it is determined by evaluating Equation (4-106) at $t = T$. We have

$$V_c = V_p(1 - e^{-T/RC}) \qquad \textbf{(4-107)}$$

This value represents the initial value of the output voltage during the second interval. The capacitor must eventually discharge so that the final value is zero. It

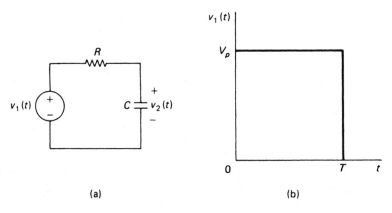

(a)

(b)

FIGURE 4-23
Shunt-C *RC* circuit and excitation for Example 4-9.

FIGURE 4–24
Equivalent circuit for Example 4–9 when input pulse drops to zero.

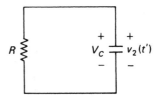

is convenient to define a new time variable $t' = t - T$, measured from the beginning of the second interval. We have

$$v_2(t') = V_c e^{-t'/RC} \quad \text{for } t > T \quad (t' = t - T) \tag{4–108}$$

Some response forms and interpretations of this problem will be given in the next example.

EXAMPLE 4–10

This problem is an extension of Example 4–9. Let $\tau = RC$ represent the time constant of the circuit. Investigate the form of the output voltage for each of the following conditions: **(a)** $\tau \ll T$, **(b)** $\tau = T$, and **(c)** $\tau \gg T$.

Solution
The functions of Equations (4–106) and (4–108) can be expressed in terms of the time constant $\tau = RC$ as

$$v_2(t) = V_p(1 - e^{-t/\tau}) \quad \text{for } 0 < t < T \tag{4–109}$$

and

$$v_2(t') = V_c e^{-t'/\tau} \quad \text{for } t > T \quad (t' = t - T) \tag{4–110}$$

Reference can be made to the appropriate portions of Figure 4–25 in the steps that follow. The input pulse is repeated at the top of the figure.

(a) $\tau \ll T$ The capacitor voltage charges to the final level in a very short time compared with a pulse width, as shown in Figure 4–25(a). For all practical purposes, the capacitor voltage is $V_c = V_p$ at the beginning of the second interval. The discharge time is also very short for $t > T$. This case represents a desirable situation for pulse transmission when very little distortion can be tolerated. There is some slight rounding at the pulse edges.

(b) $\tau = T$ During the first interval, the capacitor charges to a level $v_2(T) = 0.632V_p$, as observed in Figure 4–25(b). In the second interval, the capacitor discharges from that level toward zero.

(c) $\tau \gg T$ During the first interval, only a small fraction of the input voltage level can be reached, in view of the large time constant. Some insight can be

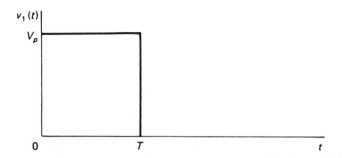

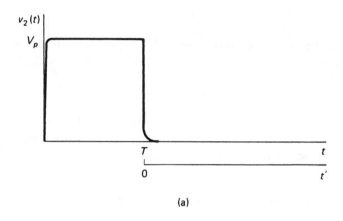

(a)

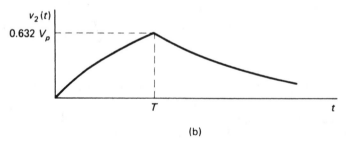

(b)

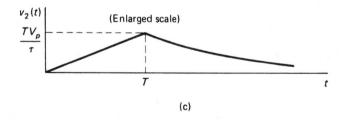

(c)

FIGURE 4–25
Waveforms for Example 4–10.

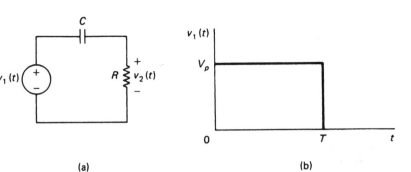

FIGURE 4–26
Series-C *RC* circuit and excitation for Example 4–11.

obtained by employing the approximation of Equation (4–57), which, when applied to Equation (4–109), results in

$$v_2(t) \approx V_p \left[1 - \left(1 - \frac{t}{\tau} \right) \right] = \frac{V_p t}{\tau} \qquad \text{(4–111)}$$

for $0 < t < T$ and $\tau \gg T$.

The preceding result indicates that the output voltage changes approximately linearly for a pulse input voltage. This circuit is referred to as an *integrator circuit* when the time constant is large compared with the pulse width. The logic behind this definition is based on the fact that the integral of a constant level is a linearly increasing function. The resulting function is shown in Figure 4–25(c). (The vertical scale is assumed to be enlarged in this case.)

For $t > T$, the function simply discharges exponentially from a level V_c toward zero. Note that for the single pulse, the discharge no longer approximates a straight-line equation. However, when the input is a repetitive pulse train, all segments of the output approximate straight-line forms for the time constant inequality assumed here.

EXAMPLE 4–11
The initially relaxed series-C *RC* circuit of Figure 4–26(a) is excited by a voltage pulse of height V_p and width T as shown in (b). Determine the output voltage $v_2(t)$.

Solution
The approach is the same as that for Example 4–9 and involves establishing two separate intervals.

$0 < t < T$ In the first interval, note that the initial value is $v_2(0^+) = V_p$, and the final assumed value is $v_2(\infty) = 0$. The response is

$$v_2(t) = V_p e^{-t/RC} \quad \text{for } 0 < t < T \qquad \text{(4–112)}$$

FIGURE 4–27
Equivalent circuit for Example 4–11
when input pulse drops to zero.

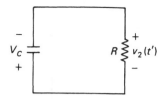

t > T Since the capacitor voltage cannot change instantaneously, it is necessary first to determine the value of that voltage at the point when the excitation drops to zero. Let V_c represent the value of the capacitor voltage at $t = T$, with the positive reference defined on the left. In view of the similarity of this circuit to that of Example 4–9, the expression of Equation (4–107) is the capacitor voltage at $t = T$, that is,

$$V_c = V_p(1 - e^{-T/RC}) \qquad (4\text{–}113)$$

The equivalent circuit used in establishing the response for $t > T$ is shown in Figure 4–27. Note that the positive polarity of V_c now appears at the negative reference terminal of the output. As in Example 4–9, it is convenient to define a time variable $t' = t - T$ measured from the beginning of the second interval. The output voltage must eventually discharge to zero, and we have

$$v_2(t') = -V_c e^{-t'/RC} \quad \text{for } t > T \quad (t' = t - T) \qquad (4\text{–}114)$$

Some response forms and interpretations of this problem will be given in the next example.

This same problem will be analyzed in Example 10–9 in Chapter 10, using a different approach.

EXAMPLE 4–12
This problem is an extension of Example 4–11. Investigate the form of the output voltage for each of the following conditions: **(a)** $\tau \ll T$, **(b)** $\tau = T$, and **(c)** $\tau \gg T$.

Solution
The functions of Equations (4–112) and (4–114) can be expressed in terms of the time constant $\tau = RC$ as

$$v_2(t) = V_p e^{-t/\tau} \quad \text{for } 0 < t < T \qquad (4\text{–}115)$$

and

$$v_2(t') = -V_c e^{-t'/\tau} \quad \text{for } t > T \quad (t' = t - T) \qquad (4\text{–}116)$$

Reference can be made to various portions of Figure 4–28 in the steps that follow. The input pulse is repeated at the top of the figure.

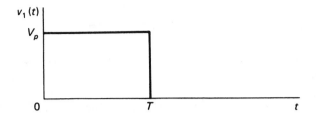

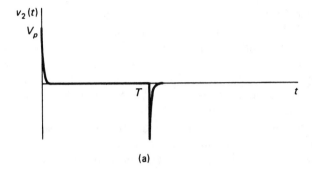

(a)

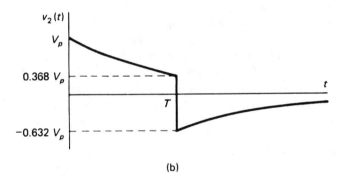

(b)

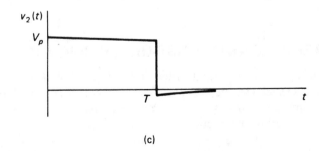

(c)

FIGURE 4–28
Waveforms for Example 4–12.

(a) $\tau \ll T$ The exponential decays very quickly to zero in the interval $0 < t < T$, as shown in Figure 4–28(a). At $t = T$, the output changes suddenly to $-V_p$ and then decays quickly back to zero.

This circuit is referred to as a *differentiator circuit* when the time constant is small compared with the pulse width. Under this condition, the output due to a pulse type input is a series of "spikes" whose polarities represent the directions of the pulse transitions. There are numerous applications of this concept in trigger circuits.

(b) $\tau = T$ The exponential decays to a level $0.368V_p$ at $t = T^-$, as shown in Figure 4–28(b). At the point when the input signal drops to zero, the output drops to a level of $-0.632V_p$. This represents a change of $-V_p$ at the output. The output then decays exponentially toward zero according to Equation (4–116).

(c) $\tau \gg T$ The time constant is so large compared with the pulse width in this case that only a slight "sag" in output has occurred at $t = T^-$. When the input drops to zero, the output changes by $-V_p$, but the result is only a slight negative excursion. Quantity V_c in Equation (4–116) is very small in this case, so the exponential decay for $t > T$ is relatively inconspicuous compared with the pulse. This case represents a desirable situation where it is necessary to pass pulses through an RC coupling circuit with minimum pulse distortion.

In all three cases, note that the output change was $-V_p$ at $t = T$ as the input dropped from V_p to zero. Generalizing this concept somewhat, when the voltage with respect to ground on one side of a capacitor *changes* instantaneously by ΔV, the voltage on the other side of the capacitor with respect to ground must also change instantaneously by ΔV. This constraint results from the fact that the voltage across the capacitor cannot change instantaneously, so a sudden change in voltage on one side must be accompanied by a sudden change of the same value on the other side to maintain the same capacitor voltage at that moment. This concept is very useful, and it arises in a number of trigger and timing circuits. In the present example, the voltage on the left side of the capacitor changes from V_p to zero at $t = T$, which represents a change of $-V_p$. Thus, the output for each case changed by $-V_p$ at $t = T$.

4–5 TIMING INTERVALS WITH FIRST-ORDER CIRCUITS

Many useful timing circuits employ first-order circuits to control the length of timing intervals. In analyzing such circuits, it is necessary to solve an exponential equation to determine the time interval at which the response reaches a certain level. This process will be explained in this section.

Consider an exponential response of the form

$$y(t) = A + Be^{-t/\tau} \tag{4–117}$$

Assume that it is desired to determine a specific time t_1 at which the corresponding response $y(t_1) = y_1$. Substitution of y_1 and t_1 in Equation (4–117) results in

$$y_1 = A + Be^{-t_1/\tau} \qquad \text{(4–118)}$$

for which it is desired to determine t_1.

Procedure

A procedure for determining t_1 can be formulated as follows:

1. Using basic algebraic operations, manipulate the function until the exponential $e^{-t_1/\tau}$ stands alone on one side of the equation, that is,

$$e^{-t_1/\tau} = \frac{y_1 - A}{B} \qquad \text{(4–119)}$$

2. Take the reciprocal of both sides of Equation (4–119). This step converts the exponential to a form with positive argument, which is easier to simplify.

$$e^{t_1/\tau} = \frac{B}{y_1 - A} \qquad \text{(4–120)}$$

3. Take the natural logarithm of both sides of Equation (4–120). Since $\ln e^x = x$, this operation yields

$$\frac{t_1}{\tau} = \ln\left(\frac{B}{y_1 - A}\right) \qquad \text{(4–121)}$$

for which

$$t_1 = \tau \ln\left(\frac{B}{y_1 - A}\right) \qquad \text{(4–122)}$$

A significant point to note is that the interval t_1 is proportional to the time constant τ. The actual expression for t_1 is more complex and contains an additional multiplicative factor, but the basic property that the timing interval is proportional to τ always applies. The timing interval can always be controlled by controlling the time constant. For example, if it is desired to double the timing interval, this can be achieved simply by doubling the time constant.

EXAMPLE 4–13
Consider the voltage response $v_C(t)$ in Example 4–6, which changed exponentially from -5 V to 10 V. Determine the time t_1 at which the voltage reaches the level of 0 V.

Solution

The function $v_C(t)$ was shown in Figure 4–18, and an expression was given in Equation (4–96b). This function is repeated here for convenience.

$$v_C(t) = 10 - 15e^{-100t} \tag{4–123}$$

We first set $t = t_1$ and $v_C(t_1) = 0$.

$$0 = 10 - 15e^{-100t_1} \tag{4–124}$$

Solving for e^{-100t_1}, we obtain

$$e^{-100t} = \tfrac{2}{3} \tag{4–125}$$

or

$$e^{100t} = 1.5 \tag{4–126}$$

Taking the natural logarithm of both sides results in

$$100t_1 = \ln 1.5 = 0.4055 \tag{4–127}$$

Finally,

$$t_1 = 0.004055 \text{ s} = 4.055 \text{ ms} \tag{4–128}$$

EXAMPLE 4–14

The timing circuit of the popular 555 integrated circuit timer in the astable mode is shown in simplified form in Figure 4–29(a). Assume that at $t = 0^+$, the value of the capacitor voltage is $v_C(0^+) = V_b/3$, and the switch is open. When v_C charges to a level of $2V_b/3$, the switch will close, which isolates the voltage V_b from the capacitor. When the capacitor discharges back to a level $V_b/3$, the switch will open again. (The switch is controlled through comparators, which are not shown here.) Determine expressions for the two timing intervals in a cycle.

Solution

In addition to the circuit in Figure 4–29, refer to the waveform of capacitor voltage in Figure 4–30 for the discussion that follows. The first timing interval corresponds to the time required for the capacitor to charge from $V_b/3$ to $2V_b/3$, and this time will be denoted as T_1. The equivalent circuit applicable during this interval is shown in Figure 4–29(b). Without showing the initial and final circuits, we can readily deduce that

$$v_C(0^+) = V_b/3 \tag{4–129}$$

and

$$v_C(\infty) = V_b \tag{4–130}$$

(Actually, the capacitor will never reach V_b due to switching, but it charges *toward* V_b *as if* it would reach that level.) The time constant during this interval will be

FIGURE 4–29
Simplified timing circuit of 555 timer and
models used in Example 4–14.

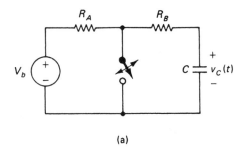

(a)

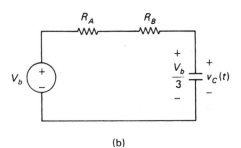

(b)

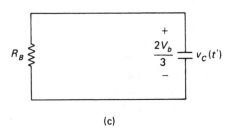

(c)

denoted as τ_1, and it is

$$\tau_1 = (R_A + R_B)C \qquad (4\text{–}131)$$

An expression for $v_C(t)$ is

$$v_C(t) = V_b - \tfrac{2}{3}V_b e^{-t/\tau_1} \qquad (4\text{–}132)$$

The interval T_1 is determined by setting $t = T_1$ and $v_C(T_1) = 2V_b/3$ in Equation
(4–132). The steps involved in determining T_1 follow.

$$\tfrac{2}{3}V_b = V_b - \tfrac{2}{3}V_b e^{-T_1/\tau}$$
$$e^{-T_1/\tau} = \tfrac{1}{2}$$
$$e^{T_1/\tau} = 2$$
$$\frac{T_1}{\tau} = \ln 2 \qquad (4\text{–}133)$$

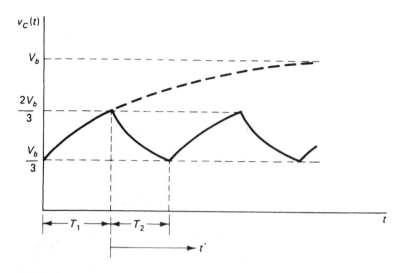

FIGURE 4–30
Capacitor voltage in Example 4–14.

Finally, we obtain

$$T_1 = \tau_1 \ln 2 = (R_A + R_B)C \ln 2 = 0.693(R_A + R_B)C \qquad \textbf{(4–134)}$$

After the voltage reaches a level $2V_b/3$ and the switch closes, the circuit of interest is as shown in Figure 4–29(c). It is convenient to define a new time scale t' representing the beginning of the discharge time interval. For the circuit and the new time scale, we have

$$v_C(0^+) = 2V_b/3 \qquad \textbf{(4–135)}$$
$$v_C(\infty) = 0 \qquad \textbf{(4–136)}$$

where the final value of the capacitor voltage would be zero if it were allowed to discharge fully. The time constant τ_2 in this interval is

$$\tau_2 = R_B C \qquad \textbf{(4–137)}$$

An expression for $v_C(t')$ is

$$v_C(t') = \tfrac{2}{3}V_b e^{-t'/\tau_2} \qquad \textbf{(4–138)}$$

The interval T_2 is determined by setting $t' = T_2$ and $v_C(T_2) = V_b/3$ in Equation (4–138). The steps involved in determining T_2 follow.

$$\tfrac{1}{3}V_b = \tfrac{2}{3}V_b e^{-T_2/\tau_2}$$
$$e^{-T_2/\tau_2} = \tfrac{1}{2}$$
$$e^{T_2/\tau_2} = 2 \qquad \textbf{(4–139)}$$
$$\frac{T_2}{\tau_2} = \ln 2$$

Finally, there results

$$T_2 = \tau_2 \ln 2 = R_B C \ln 2 = 0.693 R_B C \qquad \textbf{(4–140)}$$

Observe that $T_1 > T_2$ since $\tau_1 > \tau_2$.

4–6 PSPICE EXAMPLES

The transient analysis introduced in Chapter 3 will be continued in this chapter. Several examples of first-order circuits will be presented, and the PSPICE switch function will be introduced.

PSPICE EXAMPLE 4–1

Use PSPICE to analyze a simple RC circuit of the form shown in Figure 4–10 with $V = 10$ V, $R = 1$ kΩ, and $C = 1$ μF. Assume that the capacitor is initially uncharged, and obtain plots of the loop current and capacitor voltage.

Solution

The circuit adapted to the PSPICE format with element values labeled and the code are shown in Figure 4–31. The process of switching on the dc voltage at $t = 0$ is most easily achieved by using a PWL source and assuming a small rise time. (Some alternate ways will be discussed at the end of the example.) The PWL source must start with zero voltage at $t = 0$ and reach 10 V in a very short time. To determine a suitable time, we first determine the time constant, which is $RC = 10^3 \, \Omega \times 10^{-6}$ F $= 10^{-3}$ s $= 1$ ms. The rise time must be much shorter than the time constant, and a value of 1 μs is arbitrarily chosen. At this point, the parentheses have been omitted to simplify the notation.

FIGURE 4–31
Circuit and code for PSPICE Example 4–1.

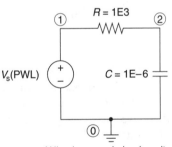

(All values are in basic units.)

```
PSPICE EXAMPLE 4–1
VS 1 0 PWL 0 0 1E–6 10
R 1 2 1E3
C 2 0 1E–6
.TRAN 5E–3 5E–3
.PROBE
.END
```

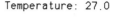

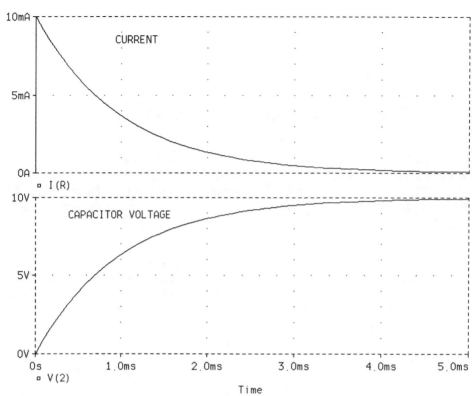

FIGURE 4–32
Waveforms for PSPICE Example 4–1.

The remainder of the code is straightforward and can be readily checked. The transient has been programmed to run to 5 ms, which represents 5 time constants.

Plots of the current and capacitor voltage obtained with PROBE are shown in Figure 4–32. The plots are in agreement with the mathematically predicted curves of Figure 4–14.

Whereas the use of the PWL function is probably the most straightforward way to produce the required switching at $t = 0$, there are several other ways to accomplish this process. One way is to use a dc input voltage and the PSPICE switch function, whose application will be explored in a more complex situation in PSPICE Example 4–5. However, the switch function is a bit of overkill for a simple transient problem like this and should be avoided when a simpler approach can be found.

One other fairly simple way will be mentioned, but before it is discussed, let us suppose that the input was replaced by a pure dc voltage of 10 V without a switch, but with the remainder of the code as in Figure 4–31. The reader is invited to experiment with this approach, and the result may come as a surprise. Indeed, there will be no transient at all, and the circuit will reach steady-state conditions immediately! The reason is that PSPICE will perform an initial analysis to determine initial conditions in the circuit prior to beginning the analysis designated by the code. The result will be equivalent to the process of analyzing the circuit at $t = 0$, which, for the circuit with a pure dc voltage input, will result in zero current and an initial voltage of 10 V across the capacitor.

A way to circumvent this problem, and yet use a pure dc voltage input, is to force the capacitor voltage to be zero at the point where the transient begins. This could be achieved by adding IC = 0 to the capacitor line and adding UIC to the transient line. An alternate method is to add an .IC code line in the control block, which would read

```
.IC V(2)=0
```

This latter method effectively clamps the node voltage above the capacitor to a value of 0 prior to the start of the transient.

These alternate methods may seem confusing, so it is suggested that for a simple source switching at $t = 0$, the use of a pulse-type input starting with a value of 0 at $t = 0$ is the simplest choice.

PSPICE EXAMPLE 4–2

Use PSPICE to analyze a simple RL circuit of the form shown in Figure 4–11 with $V = 10$ V, $R = 1$ kΩ, and $L = 1$ H. Assume that the inductor is initially unfluxed, and obtain plots of the loop current and inductor voltage.

Solution
The circuit adapted to the PSPICE format with element values labeled and the code are shown in Figure 4–33. The PWL function starting with zero voltage at $t = 0$ and reaching a level of 10 V in a very short time will be employed. The time constant is $\tau = L/R = 1$ H$/10^3\,\Omega = 1$ ms. The rise time is selected as 1 μs. The transient analysis is run to 5 ms, representing 5 time constants.

Plots of the current and inductor voltage are shown in Figure 4–34. The plots are in agreement with the predicted curves of Figure 4–16.

PSPICE EXAMPLE 4–3

Use PSPICE to simulate the circuit of Example 4–6 (Figure 4–17), and obtain a plot of the capacitor voltage.

Solution
The circuit adapted to the PSPICE format and the code are shown in Figure 4–35. The time constant is $\tau = 10^4\,\Omega \times 1 \times 10^{-6}$ F $= 10$ ms, and a rise time of 1 μs has

FIGURE 4-33

Circuit and code for PSPICE Example 4-2.

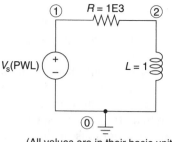

(All values are in their basic units.)

PSPICE EXAMPLE 4-2
VS 1 0 PWL 0 0 1E-6 10
R 1 2 1E3
L 2 0 1
.TRAN 5E-3 5E-3
.PROBE
.END

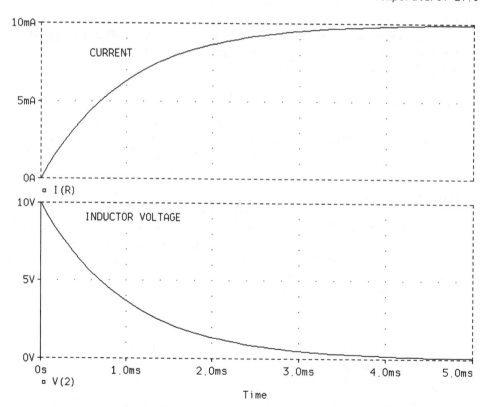

PSPICE EXAMPLE 4-2

Temperature: 27.0

FIGURE 4-34

Waveforms for PSPICE Example 4-2.

FIGURE 4–35
Circuit and code for PSPICE Example 4–3.

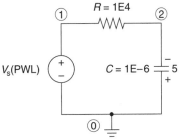

(All values are in basic units.)

```
PSPICE EXAMPLE 4–3
VS  1  0  PWL  0  0  1E–6  10
R  1  2  1E4
C  2  0  1E–6  IC=–5
.TRAN  50E–3  50E–3  UIC
.PROBE
.END
```

been chosen for the PWL input function. Note that an initial voltage of -5 V is placed on the capacitor line, and the entry UIC is placed on the transient command line.

A plot of the capacitor voltage obtained with PROBE is shown in Figure 4–36. This curve is in agreement with the predicted curve of Figure 4–18.

PSPICE EXAMPLE 4–4

In Example 4–13, the time at which the voltage reaches zero was determined mathematically. This circuit was simulated with PSPICE in PSPICE Example 4–3, and a curve was shown in Figure 4–36. Run an additional simulation over a shorter time period, and use the *cursor* of PROBE to determine this point experimentally.

Solution

The code is essentially the same as in PSPICE Example 4–3 and will not be repeated here. (Refer back to Figure 4–35 if necessary.) However, the transient line will be modified for this run. First, we note that the zero crossing will occur below 10 ms, so it will be limited to a time interval from 0 to 8 ms to provide better resolution. Second, we will now demonstrate how the transient command can be modified to provide more resolution and, possibly, more accuracy in some applications.

The transient command extended to four entries has the following format:

.TRAN t_s t_f t_i t_m

The first two entries, t_s and t_f, have the same meaning as before. Since t_s represents a time step for tabular output, if only a PROBE display is desired, this value may be chosen to be the same as the final time, t_f. The value t_i represents the first point at which any type of output (either tabular or graphical) is desired. (It does not

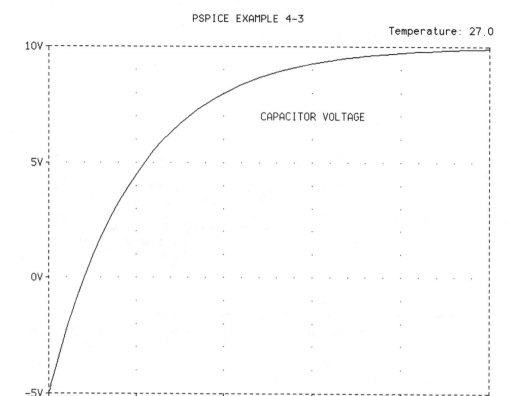

PSPICE EXAMPLE 4-3

Temperature: 27.0

CAPACITOR VOLTAGE

□ V(2)

Time

FIGURE 4–36
Waveform for PSPICE Example 4–3.

represent the beginning of the computation, which is always $t = 0$.) The value t_m represents a *maximum step size* used in the actual computation. By specifying this value, we can often create a higher resolution or a smoother presentation.

For the example under consideration, the transient command was modified to the following:

```
.TRAN 8E-3 8E-3 0 1E-6
```

The first two values are self-explanatory based on the earlier practice, but with the interval limited to 8 ms. The third entry, 0, means that we wish to begin observation at $t = 0$. This value represents no change in previous practices, but in order to use the desired fourth entry, the field requires a value here. The fourth entry, 1E-6, means that no time step greater than 1 μs will be used in the integration process.

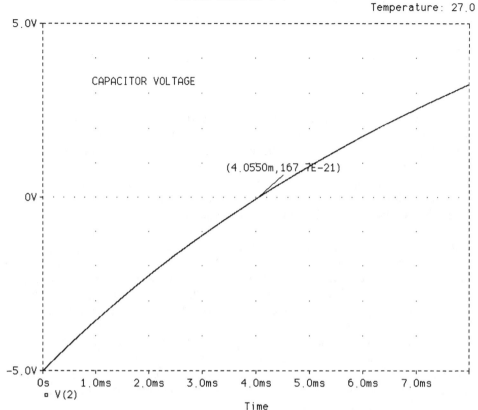

FIGURE 4–37
Waveform for PSPICE Example 4–4.

The capacitor voltage is shown in Figure 4–37. With the use of the cursor, the zero crossing point is identified. (Note that the closest value to zero is about 168×10^{-21}, which is infinitesimally close to zero.) The value of time at which this occurs is 4.055 ms, which agrees with the earlier analysis.

PSPICE EXAMPLE 4–5
Use PSPICE to simulate the circuit of Example 4–7 (including the use of a switch function), and obtain a plot of the capacitor voltage.

Solution
This is an example of a circuit in which an actual switching operation is warranted. PSPICE contains both a voltage-controlled switch and a current-controlled switch. We will choose to use the voltage-controlled switch in this example.

The general format for a voltage-controlled switch is as follows:

```
S  i j k l name
```

Reference must then be made to the model library in the following format:

```
.MODEL name VSWITCH
```

S is the first letter of a switch and may be followed by additional numbers or letters in the title. The integers i and j represent the node numbers that are to be switched. The integers k and l represent the nodes across which the controlling voltage appears, with k being the positive reference. The title *name* represents any arbitrary name that you wish to designate, which will identify the particular switch model.

The .MODEL line has as its second entry the "name" referred to in the switch line. This is followed by VSWITCH, which is the title of the voltage switch model.

The basic strategy in the control mechanism is as follows: When the controlling voltage is zero or negative, the switch is a very high resistance, with a default value of $10^6 \, \Omega$. When the controlling voltage is 1 V or greater, the switch is a very low resistance, with a default value of $1 \, \Omega$. The switch resistance varies between the control limits. (In reality, it is a voltage-controlled resistor.) Thus, it is necessary to create an abrupt change in the controlling voltage to force the switch to change states quickly.

One other problem occurs with this particular circuit when the default parameters are used. Whereas $1 \, \Omega$ is small compared with resistance levels in most typical electronic circuits, it is *not* small compared with the resistances in this particular circuit. Thus, it will be necessary to modify the switch model to create a very small resistance relative to those in the circuit.

A final consideration that must be addressed is the necessity to identify a controlling voltage for the switch. Unless there is a source within the circuit demonstrating the desired controlling nature, it will be necessary to add an additional source external to the circuit to accomplish this purpose.

The circuit adapted to the PSPICE format and the code are shown in Figure 4–38. Since steady-state conditions are desired prior to closing the switch, a pure dc voltage of 12 V is placed at the input. Note that the capacitor is connected between nodes 2 and 0, and R_2 is connected between nodes 3 and 0, thus leaving a space for the switch.

The three lines that constitute the switch and the controller are grouped together (although they need not be), and they read as follows:

```
S 2 3 4 0 SCONTROL
.MODEL SCONTROL VSWITCH RON=1E-6 ROFF=1E6
VCONTROL 4 0 PWL 0 0 1E-3 1
```

Referring back to the earlier general description, the title of the switch is chosen simply as S, and it is connected between nodes 2 and 3. The controlling voltage

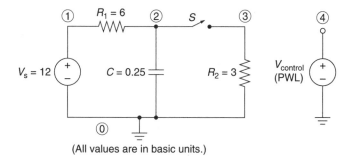

(All values are in basic units.)

```
PSPICE EXAMPLE 4–5
VS 1 0 DC 12
R1 1 2 6
C 2 0 0.25
R2 3 0 3
S 2 3 4 0 SCONTROL
.MODEL SCONTROL VSWITCH RON=1E-6 ROFF=1E6
VCONTROL 4 0 PWL 0 0 1E-3 1
.TRAN 3 3
.PROBE
.END
```

FIGURE 4–38
Circuit and code for PSPICE Example 4–5.

will be connected between nodes 4 and 0, and the name chosen for the model is SCONTROL.

The .MODEL line has the name SCONTROL followed by the required model element VSWITCH. If the default parameters of 1 Ω and 10^6 Ω were acceptable, the line could be terminated at this point. However, the ON resistance is modified by the entry RON=1E-6, which creates an ON resistance of 1 $\mu\Omega$. It would be unnecessary in this example to modify the OFF resistance, but to demonstrate to the reader how it is done, the entry ROFF=1E6 has been included.

The controlling voltage VCONTROL is connected between an extra node 4 and 0, and it is a PWL function. This voltage switches from a value of 0 to 1 V in a time of 1 ms, which should be adequate for the switch.

The capacitor voltage obtained from PROBE is shown in Figure 4–39. As expected, the voltage starts with an initial voltage of 12 V and decays exponentially to a final level of 4 V.

There are other ways that this circuit could have been simulated without the use of the switch. For example, the switch could have been placed on the capacitor. The method employed here demonstrates the simulation in a manner more consistent with the actual behavior and avoids the necessity to perform the analysis at $t = 0$. Of significant importance, however, is that the switch function has been introduced as a viable circuit element.

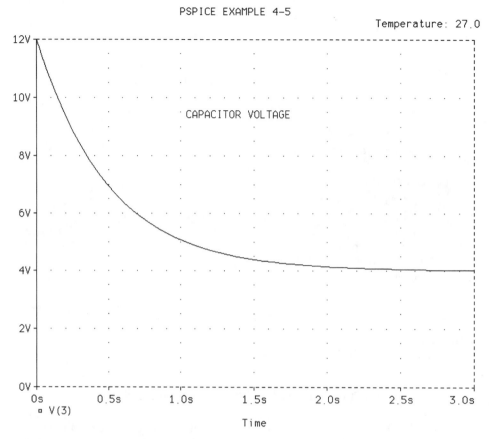

FIGURE 4–39
Waveform for PSPICE Example 4–5.

4–7 MATLAB EXAMPLE

The example that follows will illustrate how to use MATLAB to plot a curve based on a defining mathematical equation.

MATLAB EXAMPLE 4–1

Use MATLAB to plot and label the curve representing the voltage across the capacitor in the solution of Example 4–6 (Equation 4–96b).

Solution

Whereas the equation representing the voltage was obtained through an inspection process, we are using the example to illustrate some of MATLAB's capabilities in defining and plotting equations. The equation that will be plotted is given by

$$v_C(t) = 10 - 15e^{-100t} \tag{4–141}$$

The best way to perform the task is first to define a row matrix representing all values of the independent variable time at which points are to be determined. This could be done manually by actually typing the time values and then enclosing them by brackets [ ]. This would be satisfactory for a few points, and it might be the best way when the values of time are somewhat random in nature. However, the nature of the problem suggests a fairly large number of points with uniform spacing.

MATLAB offers a number of different commands, along with several variations, that can be used for this purpose. For equal spacing, the following format can be used:

$$t = t_1 : \Delta t : t_2$$

where t_1 is the first value, Δt is the time step, and t_2 is the last value. This process generates a number of points equal to $(t_2 - t_1)/\Delta t + 1$, assuming that the last point coincides with t_2.

A second command that achieves the same process is "linspace" (linear spacing), whose format reads as follows:

$$t = \text{linspace}(t_1, t_2, n)$$

where t_1 and t_2 have the same meaning as in the first case and n is the total number of points in the matrix. One can also generate a logarithmic spacing by the command "logspace," and this will be considered later in the text.

Refer to Figure 4–40 in the discussion that follows. First, we will illustrate the process in which all computed points are listed on the screen by selecting a modest number. The first of the two forms for generating the time array is arbitrarily selected, and the following command is entered:

$$t = 0{:}5e{-}3{:}50e{-}3$$

This generates an array of time values ranging from $t = 0$ to $t = 50$ ms in steps of 5 ms. In general, the number of points n based on this format will be

$$n = \frac{t_2 - t_1}{\Delta t} + 1 \qquad \qquad \textbf{(4–142)}$$

Thus, 11 values of time are generated and immediately displayed. An alternate command that could have achieved the same result is

$$t = \text{linspace}(0,50e{-}3,11)$$

The designation for the exponential function in MATLAB is "exp," and multiplication is indicated by an asterisk (*). The capacitor voltage is denoted as vc, and the equation expressed in the workspace is

$$vc=10-15*exp(-100*t)$$

When this expression is entered, MATLAB automatically performs the calculation at all values of time and immediately displays the values on the screen as shown.

```
EDU» t=0:5e-3:50e-3

t =

  Columns 1 through 7

         0    0.0050    0.0100    0.0150    0.0200    0.0250    0.0300

  Columns 8 through 11

    0.0350    0.0400    0.0450    0.0500

EDU» vc=10-15*exp(-100*t)

vc =

  Columns 1 through 7

   -5.0000    0.9020    4.4818    6.6530    7.9700    8.7687    9.2532

  Columns 8 through 11

    9.5470    9.7253    9.8334    9.8989

EDU»
```

FIGURE 4–40
Workspace for MATLAB Example 4–1.

Note that the values of vc appear in the same relative positions in the array as the corresponding values of time.

Although we could obtain a rough plot with the values generated, the number of points is a bit sparse for generating a smooth curve. On the other hand, if we used a very small time with the format used, we would get a large number of points on the screen, and a printout of the workspace would be somewhat clumsy for our purposes.

We will now repeat the analysis with a much finer step size, but suppress the listing of all the values on the screen. The listing of results from any computation can be suppressed by placing a semicolon (;) at the end of the command.

Refer now to Figure 4–41. The first two commands are identical to those of Figure 4–40, except for a smaller time step and the semicolon at the end of each of the commands. The time step has been reduced to 1 ms, and for the time interval used, 51 values of time and the corresponding values of vc will be calculated. Subsequent commands do not directly involve calculations and do not need the semicolons.

To plot the dependent variable vc as a function of t, we need only enter the command

```
plot(t,vc)
```

where t becomes the horizontal (x) axis and vc becomes the vertical (y) axis.

```
EDU» t=0:1e-3:50e-3;
EDU» vc=10-15*exp(-100*t);
EDU» plot(t,vc)
EDU» xlabel('time, seconds')
EDU» ylabel('voltage, volts')
EDU» title('Capacitor Voltage of Example 4-6')
EDU»
```

FIGURE 4–41
Workspace for MATLAB Example 4–1 with screen listing suppressed and additional plot commands.

The curve is now plotted, and MATLAB even extrapolates between points. (Wasn't that much easier than a manual plot?) The remaining work consists of labeling and adding other identifications to the plot. MATLAB has many options available for this purpose, including markers on actual calculated points, and the like. We will leave the curve in smooth form and illustrate a few of the types of labels.

First, the horizontal axis will be labeled. The command used is

```
xlabel('time, seconds')
```

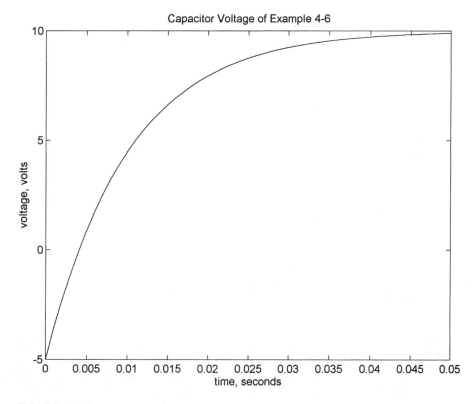

FIGURE 4–42
Waveform for MATLAB Example 4–1 without grid.

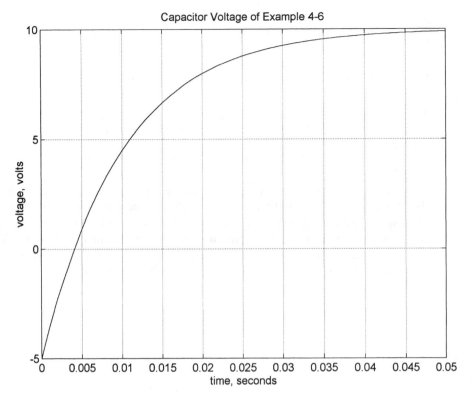

FIGURE 4–43

Waveform for MATLAB Example 4–1 with grid.

where the actual labeling is between the ' ' markers. Next, the vertical axis is labeled by the command

```
ylabel('voltage, volts')
```

Finally, the curve is given a title by the command

```
title('Capacitor Voltage of Example 4-6')
```

The result is shown in Figure 4–42.

To illustrate one additional option, a grid structure will be superimposed on the curve. This is achieved by the simple command

```
grid
```

The resulting plot is shown in Figure 4–43. It is exactly the same as in Figure 4–42, except for the added grid lines.

We have touched on only a few of the many equation and plotting operations available with MATLAB. The interested reader should refer to a MATLAB reference manual or to a textbook devoted to the subject for further information.

DRILL PROBLEMS

4-1. The initially relaxed circuit of Figure P4–1 is excited at $t = 0$ by the source shown. Determine all element voltages and currents at $t = 0^+$.

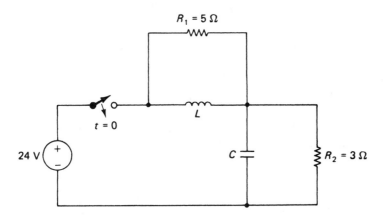

FIGURE P4–1

4-2. The initially relaxed circuit of Figure P4–2 is excited at $t = 0$ by the source shown. Determine all element voltages and currents at $t = 0^+$.

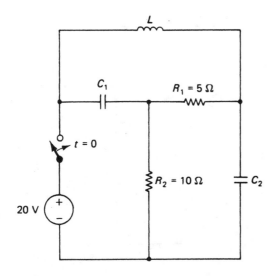

FIGURE P4–2

4–3. The switch in the circuit of Figure P4–3 is initially in position 1. As a result of some excitation $v_s(t)$, the initial values of the capacitor voltage and inductor current are as shown. At $t = 0$, the switch is changed to position 2. Determine all element voltages and currents at $t = 0^+$.

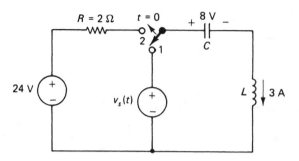

FIGURE P4–3

4–4. The switch in the circuit of Figure P4–4 is initially in position 1. As a result of some excitation $v_s(t)$, the initial values of the capacitor voltage and inductor current are as shown. At $t = 0$, the switch is changed to position 2. Determine all element voltages and currents at $t = 0^+$. (*Note:* Although the new source varies with time, it has a constant value at a specific value of time.)

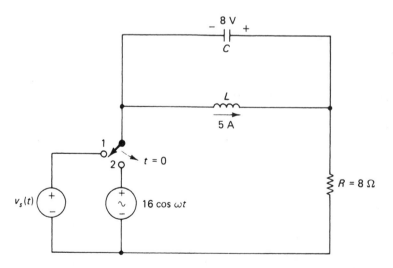

FIGURE P4–4

4–5. Determine the steady-state voltage and current for each element in the circuit of Figure P4–5.

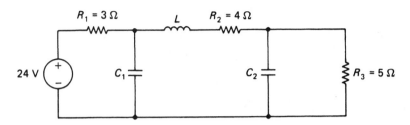

FIGURE P4–5

4–6. Determine the steady-state voltage and current for each element in the circuit of Figure P4–6.

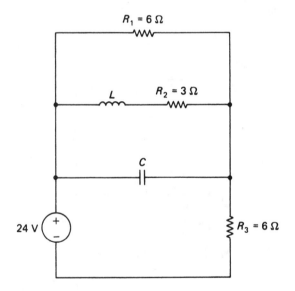

FIGURE P4–6

4–7. The circuit of Figure P4–7 is initially in a steady-state condition in position 1. At $t = 0$, the switch is moved to position 2 instantaneously. **(a)** Determine all element voltages and currents immediately following the switching operation. **(b)** Determine all element voltages and currents after steady-state conditions are again established.

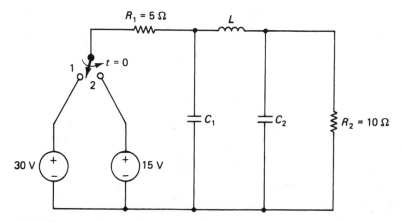

FIGURE P4–7

4–8. The initially relaxed circuit of Figure P4–8 is excited at $t = 0$ by the dc source shown. Determine all element voltages and currents at **(a)** $t = 0^+$ and **(b)** $t = \infty$. Note the reference directions of all variables.

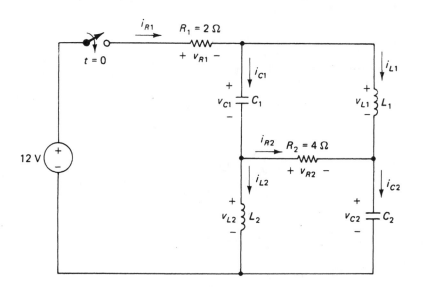

FIGURE P4–8

4–9. An RC circuit of the form of Figure 4–13(a) has $V_s = 100$ V, $R = 100$ kΩ, and $C = 100$ μF.

    (a) Determine the time constant τ.

    (b) Write expressions for $i(t)$ and $v_C(t)$, and sketch the functions.

    (c) Using the exponential function on your calculator, evaluate $i(t)$ and $v_C(t)$ for each of the following values of t: 0, τ, 2τ, 5τ, 10τ.

4–10. Repeat the calculations of Problem 4–9 if $R = 1$ kΩ and $C = 0.01$ μF. The voltage source is still $V_s = 100$ V.

4–11. An RL circuit of the form of Figure 4–15(a) has $V_s = 100$ V, $R = 1$ kΩ, and $L = 10$ H.

    (a) Determine the time constant τ.

    (b) Write expressions for $i(t)$ and $v_L(t)$, and sketch the functions.

    (c) Using the exponential function on your calculator, evaluate $i(t)$ and $v_L(t)$ for each of the following values of t: 0, τ, 2τ, 5τ, 10τ.

4–12. Repeat the calculations of Problem 4–11 if $R = 10$ kΩ and $L = 10$ mH. The voltage source is still $V_s = 100$ V.

4–13. The 0.05-F capacitor in the circuit of Figure P4–13 is initially charged to 8 V in the direction shown. At $t = 0$, a 20-V source is connected as shown. Determine expressions for $i(t)$ and $v_C(t)$ for $t > 0$, and sketch the functions.

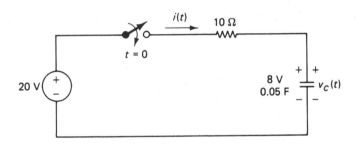

FIGURE P4–13

4–14. The 0.2-F capacitor in the circuit of Figure P4–14 is initially charged to 8 V in the direction shown. At $t = 0$, a 12-V source is connected as shown. Determine expressions for $i(t)$ and $v_C(t)$ for $t > 0$, and sketch the functions.

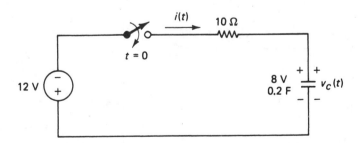

FIGURE P4–14

4–15. Steady-state conditions exist in the circuit of Figure P4–15 at $t = 0^-$. The 5-Ω resistor is connected across the capacitor at $t = 0$. Determine an expression for $v_C(t)$ for $t > 0$, and sketch the function.

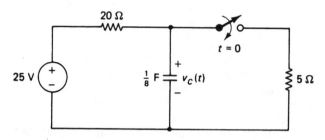

FIGURE P4–15

4–16. Steady-state conditions exist in the circuit of Figure P4–16 at $t = 0^-$. The 12-Ω resistor is disconnected at $t = 0$. Determine an expression for $v_C(t)$ for $t > 0$ and sketch the function.

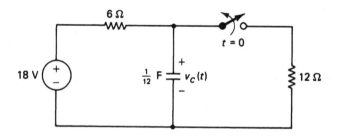

FIGURE P4–16

4–17. The switch in the circuit of Figure P4–17 is closed at $t = 0$, and the capacitor is initially uncharged. Determine an expression for $v_C(t)$ and sketch the function.

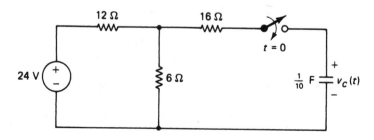

FIGURE P4–17

4–18. The switch in the circuit of Figure P4–18 is closed at $t = 0$, and the capacitors are initially uncharged. Determine expressions for $v_1(t)$ and $v_2(t)$.

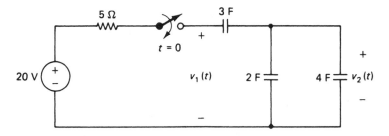

FIGURE P4–18

4–19. Steady-state conditions exist in the circuit of Figure P4–19 at $t = 0^-$. The switch is *closed* at $t = 0$. Determine expressions for $i(t)$ and $v_L(t)$ for $t > 0$, and sketch the functions. On the curves, show also the values at $t = 0^-$.

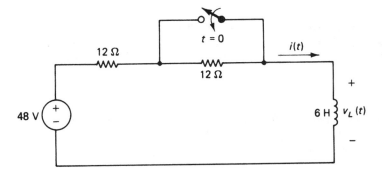

FIGURE P4–19

4–20. Steady-state conditions exist in the circuit of Figure P4–20 at $t = 0^-$. The switch is *opened* at $t = 0$. Determine expressions for $i(t)$ and $v_L(t)$ for $t > 0$, and sketch the functions. On the curves, show also the values at $t = 0^-$.

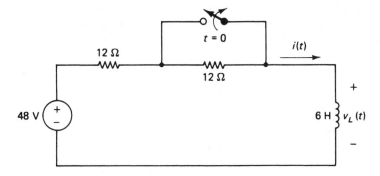

FIGURE P4–20

4–21. Consider a first-order circuit in which a voltage $v(t)$ changes exponentially from -10 V to $+10$ V with a time constant $\tau = 1$ ms. Determine the time t_1 required to reach the level of 0 V.

4–22. Consider a first-order circuit in which a voltage $v(t)$ changes exponentially from -5 V to $+20$ V with a time constant $\tau = 2$ ms. Determine the time t_1 required to reach the level of $+10$ V.

DERIVATION PROBLEMS

4–23. For any first-order circuit with dc excitations, show that the time t_1 required to reach a level y_1 is given by:

$$t_1 = \tau \ln \left[\frac{1 - \dfrac{y(0^+)}{y(\infty)}}{1 - \dfrac{y_1}{y(\infty)}} \right]$$

4–24. The simple RC circuit of Figure 4–23(a) will be investigated as it relates to a possible *integrator circuit,* as in Example 4–10. **(a)** By writing a node voltage equation at the node above the capacitor, show that the following differential equation is obtained:

$$\frac{dv_2}{dt} + \frac{1}{RC} v_2 = \frac{1}{RC} v_1$$

(b) Under certain conditions in which RC is large, the second term on the left in the above equation is small compared with the first term. Under these conditions, show that the approximate response $\hat{v}_2(t)$ is:

$$\hat{v}_2(t) = \frac{1}{RC} \int_0^t v_1(t)\, dt$$

where it is assumed that $\hat{v}_2(0) = 0$. **(c)** Assume that $v_1(t) = 100$ V and $RC = 1$ s. Calculate the approximate response $\hat{v}_2(t)$ for the following values of t (in seconds): $0.001, 0.01, 0.1, 1, 10$. **(d)** For the conditions given in (c), determine the exact response $v_2(t)$ and compare the results. What can you conclude?

4–25. The simple RC circuit of Figure 4–26(a) will be investigated as it relates to a possible *differentiator circuit* as discussed in Example 4–12. **(a)** By writing a node voltage equation at the node above the resistor, show that the following differential equation is obtained:

$$v_2 + RC \frac{dv_2}{dt} = RC \frac{dv_1}{dt}$$

(b) Under certain conditions in which RC is small, the second term on the left in the above equation is small compared with the first term. Under these conditions, show that the approximate response $\hat{v}_2(t)$ is

$$\hat{v}_2(t) = RC \frac{dv_1(t)}{dt}$$

4–26. The effect of shunt capacitance or other frequency-limiting effects is to produce a *rise time* for any dc or pulse-type input to a transmission system. To minimize errors and uncertainties in initial and final levels, the IEEE definition of rise time T_R is

$$T_R = t_{90} - t_{10}$$

where t_{90} represents the time required to reach 90% of the final level and t_{10} is the time required to reach 10% of the final level. For the basic RC circuit of Figure 4–13(a) with a dc input, show that the rise time is approximately

$$T_R \simeq 2.2RC$$

4–27. Consider the simple series RC circuit with a dc source shown back in Figure 4–13(a), and assume that the capacitor is initially uncharged. By determining instantaneous power functions and integrating these functions from $t = 0$ to $t = \infty$ to determine energy, verify the following results:

$$\text{Total energy dissipated in } R = \tfrac{1}{2}CV_s^2$$
$$\text{Final energy stored in } C = \tfrac{1}{2}CV_s^2$$
$$\text{Total energy supplied by } V_s = CV_s^2$$

4–28. Consider a simple RL circuit with a dc source shown back in Figure 4–15(a), and assume the inductor is initially unfluxed. By determining the instantaneous inductive power $p_L(t)$ and integrating this function from $t = 0$ to $t = \infty$, show that the total energy W_L absorbed by the inductor is:

$$W_L = \tfrac{1}{2}LI_f^2$$

where $I_f = V_s/R$ is the final value of the inductive current.

APPLICATION PROBLEMS

4–29. Many practical connection links between system components can be approximated by the shunt-C RC circuit shown in Figure 4–23(a). R represents the effective source resistance, and C represents the net shunt capacitance such as cable capacitance and input capacitance to the next stage. Assume that in a given interconnection system, the net capacitance is 500 pF. Pulses are to be transmitted through the system, and an arbitrary criterion is imposed that the rise times of these pulses must not exceed 10% of the pulse width T. (See Problem 4–26 for the rise time relationship.) **(a)** If the effective source resistance is 1000 Ω, determine the minimum pulse width $T_{\min}$ that can be transmitted. **(b)** If the source is changed to a lower resistance source of 50 Ω, repeat the calculation of (a).

4–30. A shunt-C RC circuit of the type shown in Figure 4–23(a) is to be designed to approximate an integrator. The type of signal to be integrated is a series of pulses whose maximum widths are 1 ms. A decision is made to establish $\tau = 10$ ms. Using the ideal integrator form as the reference, determine the maximum percent deviation from the ideal expected with the approximate integrator. (See Problem 4–24 for the ideal RC integrator relationship.)

4–31. A series-C RC circuit of the type shown in Figure 4–26(a) is to be designed to generate short trigger pulses in an electronic timing circuit. The actual pulse train from which the pulses are to be obtained has pulses not less than 1 ms wide. An arbitrary criterion is imposed that the trigger pulses obtained from the differentiator circuit must not exceed a level greater than 2% of the input pulse level at any time greater than 5% of the pulse width. Determine the maximum value of τ that meets these requirements for the circuit.

4–32. A series-C RC circuit of the type shown in Figure 4–26(a) occurs in many coupling circuits between amplifiers to remove the dc level. Assume that in a given application, it is necessary to couple pulses through such a circuit. The maximum pulse width is 1 ms. An arbitrary criterion is imposed that the sag of any pulse must not exceed 5%. (The percentage sag is referred to the ideal pulse level.) If R is fixed at 100 kΩ as the input resistance for the next stage, determine the minimum value of C required.

4–33. A certain type of operational amplifier oscillator circuit is shown in Figure P4–33(a). The timing circuit in simplified form is shown in Figure P4–33(b). Assume that at $t = 0^+$, $v_C(0^+) = -V_1$ where

$$V_1 = \frac{R_2}{R_1 + R_2} V_{\text{sat}}$$

and V_{sat} is the op-amp saturation voltage. The capacitor will charge toward V_{sat}. However, when the capacitor voltage reaches a level V_1, the circuit changes state and one-half cycle is completed. Let t_1 represent the time for one-half cycle. Show that

$$t_1 = RC \ln \left(1 + \frac{2R_2}{R_1} \right)$$

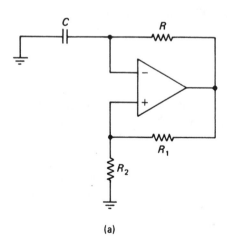

(a)

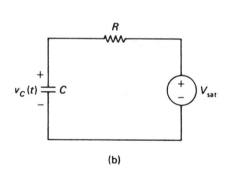

(b)

FIGURE P4–33

5

LAPLACE TRANSFORMS

OBJECTIVES

After completing this chapter, the reader should be able to:

- Define and explain the purposes of the Laplace transform as applied to circuit analysis.
- State the Laplace transforms for the most common functions encountered in circuit analysis.
- State the forms for several of the most common Laplace transform operations.
- Determine the Laplace transform of a given time function.
- Determine the inverse transform of a given s-domain function.

5–1 GENERAL APPROACH

The Laplace transform is a so-called "operational" method for solving differential equations by algebraic methods. Developed by the French mathematician Pierre-Simon de Laplace (1749–1827), this transform is an important part of the study of applied mathematics and is covered from a mathematical point of view in undergraduate courses in differential equations, applied mathematics, and related areas.

Flow Chart for Mathematical Procedure

The manner in which the Laplace transform is generally applied to solve differential equations is illustrated by the flowchart of Figure 5–1. The differential equation to be solved is first converted to an algebraic transform equation using the Laplace transformation. The resulting equation can be completely manipulated by basic algebraic methods, and the desired variable, expressed in transform terms, is deter-

FIGURE 5–1

Steps involved in Laplace transform solution of differential equation.

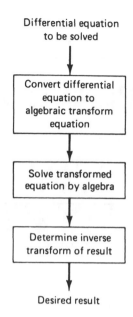

mined. The inverse transform is then found, and the result is the desired solution of the given differential equation.

This approach is the way in which the Laplace transform is used in mathematics courses, and it is the basic method whenever the differential equation is given. However, electrical engineers and technologists tend to think in terms of circuit models such as impedances, rather than differential equation forms. Consequently, methods have been developed for obtaining the transform circuit response more directly by using circuit models rather than by writing the differential equation first. Because of the applied nature of this book, this type of approach will be emphasized.

Flow Chart for Circuit Procedure

The approach that will be emphasized in this book is illustrated by the flowchart of Figure 5–2. It is assumed that a circuit is given for which a desired voltage or current response is to be determined. The first step in the process is to convert the circuit to an equivalent Laplace transform operational form. The original circuit is referred to as the *time domain* form, and the transferred circuit is referred to as the *s-domain* (or *Laplace domain*) form. The symbol *s* refers to the Laplace variable, as will be seen shortly.

Once the *s*-domain circuit model is established, all voltages and currents may be determined by algebraic methods. The desired voltages and currents are solved for using any standard circuit analysis methods. Any variable determined in the *s*-domain is then converted back to the time domain to yield the final desired response. This latter process is referred to as *inverse Laplace transformation*.

FIGURE 5-2
Steps involved in *s*-domain circuit analysis process using Laplace transforms.

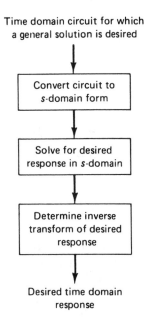

Time domain circuit for which a general solution is desired

Convert circuit to *s*-domain form

Solve for desired response in *s*-domain

Determine inverse transform of desired response

Desired time domain response

This method has the advantage that no differential equation need be written in the process. Further, the various theorems of circuit analysis may be used to determine any desired response.

The emphasis in this chapter will be directed toward determining the Laplace transforms of time functions and determining the inverse transforms of *s*-domain functions. These operations represent the most tedious, and perhaps the least interesting, aspects of the first and third blocks of Figure 5–2, respectively. The circuit modeling aspects, which represent some of the practical and interesting interpretations of these operations, will be considered in Chapter 6. Thus, the reader should patiently study and accept the manipulative aspects of transforms in this chapter as a necessary prerequisite to the useful end results in the next chapter.

5-2   LAPLACE TRANSFORMS

Most time functions of interest in circuit analysis are voltages or currents. To provide some generality, the symbol $f(t)$ will be used in many developments, but remember that in most cases of interest in this text, the function will be either a voltage or a current.

The transform of a time function $f(t)$ is a new function $F(s)$, where s is the Laplace transform variable or operator. (Don't worry about what s is at this point. Simply accept it as a variable that must be kept in all equations when it arises.) The process of transformation is indicated symbolically as

$$\mathcal{L}[f(t)] = F(s) \tag{5-1}$$

This equation is read as "the Laplace transform of $f(t)$ is $F(s)$." The process of inverse transformation is indicated symbolically as

$$\mathscr{L}^{-1}[F(s)] = f(t) \tag{5-2}$$

This latter equation is read as "the inverse Laplace transform of $F(s)$ is $f(t)$."

Definition of Laplace Transform

The actual mathematical definition of the Laplace transform is

$$F(s) = \int_0^\infty f(t)e^{-st}\,dt \tag{5-3}$$

This equation indicates that $f(t)$ is first multiplied by e^{-st}, and the result is integrated over positive time to produce the Laplace transform. Although a few transforms will be developed for illustration, most transforms of interest have been derived and are tabulated in numerous references. Most practical problems requiring Laplace transformation involve using standard tabulated results to determine the transforms.

The most common Laplace transforms of interest in circuit analysis are listed in Table 5–1. Not all of the time functions listed in the table have been used up

TABLE 5–1
Laplace Transform Pairs Encountered in Circuit Analysis

$f(t)$	$F(s) = \mathscr{L}[f(t)]$	
1 or $u(t)$	$\dfrac{1}{s}$	(T–1)
$e^{-\alpha t}$	$\dfrac{1}{s + \alpha}$	(T–2)
$\sin \omega t$	$\dfrac{\omega}{s^2 + \omega^2}$	(T–3)
$\cos \omega t$	$\dfrac{s}{s^2 + \omega^2}$	(T–4)
$e^{-\alpha t} \sin \omega t$	$\dfrac{\omega}{(s + \alpha)^2 + \omega^2}$	(T–5)*
$e^{-\alpha t} \cos \omega t$	$\dfrac{s + \alpha}{(s + \alpha)^2 + \omega^2}$	(T–6)*
t	$\dfrac{1}{s^2}$	(T–7)
t^n	$\dfrac{n!}{s^{n+1}}$	(T–8)
$e^{-\alpha t} t^n$	$\dfrac{n!}{(s + \alpha)^{n+1}}$	(T–9)
$\delta(t)$	1	(T–10)

* Use when roots are complex.

to this point in the book. A brief discussion of most of the functions not yet encountered in the text will be given shortly. This table is also repeated in Appendix C at the end of the text.

The column designated as $f(t)$ in Table 5–1 indicates the basic form of the time function, and the column designated as $F(s)$ indicates the corresponding form of the Laplace transform.

Basic Theorems of Linearity

In order to use these basic results, two simple but important results must be given. These results apply to all transform functions considered in the text. First, let K represent an arbitrary constant. We have

$$\mathscr{L}[Kf(t)] = K \; \mathscr{L}[f(t)] = KF(s) \tag{5-4}$$

Stated in words, the Laplace transform of a constant times a function is the constant times the transform of the function.

Next, let $f_1(t)$ and $f_2(t)$ represent any arbitrary functions, and let $F_1(s)$ and $F_2(s)$ represent their respective transforms. We have

$$\mathscr{L}[f_1(t) + f_2(t)] = \mathscr{L}[f_1(t)] + \mathscr{L}[f_2(t)] = F_1(s) + F_2(s) \tag{5-5}$$

Stated in words, the Laplace transform of the sum of two functions is the sum of the respective Laplace transforms.

The results of Equations (5–4) and (5–5) constitute results that might be "obvious," but they must be stated and understood in order to justify the decomposition of transforms into simpler forms. One form often incorrectly used by beginners is the Laplace transform of a product such as $f_1(t)f_2(t)$. Although it is possible to determine the transform of the product from the separate transforms by a process called *convolution*, that technique is not within the scope of this book. The transform of the product is *not* the product of the transforms, that is, $\mathscr{L}[f_1(t)f_2(t)] \neq F_1(s)F_2(s)$. The only way to deal with the product in this text is to look at the product as a new function directly.

Unless otherwise stated, it is assumed that all time functions for which Laplace transforms are determined start at $t = 0$. In some cases, it is necessary to call attention to this fact for proper interpretation. In Chapter 10, the concept of functions starting at different times will be introduced. Several of the functions listed in Table 5–1 will now be discussed.

Step Function

When it is desirable to describe the process of "turning on" a dc level at $t = 0$, the unit step function $u(t)$ may be used. This function, which is shown in Figure 5–3, is defined as

$$u(t) = 0 \quad \text{for } t < 0$$
$$= 1 \quad \text{for } t > 0$$

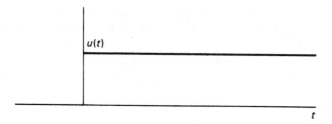

FIGURE 5–3
Form of the unit step function.

Any arbitrary constant level may be expressed by multiplying the unit step function by an appropriate constant. For example, the process of switching on a 5-A dc current $i(t)$ at $t = 0$ can be expressed as $i(t) = 5u(t)$.

The Laplace transform of the step function is given by (T–1) of Table 5–1, which reads

$$\mathscr{L}[u(t)] = \frac{1}{s} \tag{T–1}$$

From Equation (5–4), the Laplace transform of the current referred to in the preceding paragraph is $I(s) = 5/s$.

At this point it is not necessary to be so formal about notation, since most functions of current interest start at $t = 0$. Thus, $u(t)$ will simply be interpreted as a dc level of unity, and the Laplace transform of a dc level K will be expressed as K/s. However, some elaborate waveform decomposition techniques will be developed in Chapter 10, at which point the use of elaborate step function factors will be essential.

Exponential Function

The exponential function $e^{-\alpha t}$ was discussed thoroughly in Chapter 4 and requires no further discussion here. The Laplace transform is

$$\mathscr{L}[e^{-\alpha t}] = \frac{1}{s + \alpha} = \frac{1}{s + \dfrac{1}{\tau}} \tag{T–2}$$

where $\tau = 1/\alpha$.

Sine and Cosine Functions

We expect that most readers will be familiar with the sine and cosine functions, but a brief review is in order. The forms of $\sin \omega t$ and $\cos \omega t$ for $t > 0$ are shown in Figure 5–4. Both sine and cosine are referred to as *sinusoidal functions*, or simply *sinusoids*.

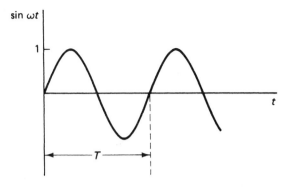

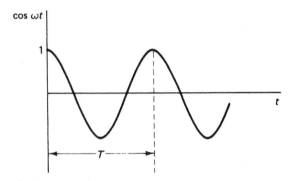

FIGURE 5–4
Forms of sine and cosine functions.

The following definitions relating to the sinusoidal function are important:

ω = angular frequency in radians/second (rad/s) = $2\pi f = 2\pi/T$

f = cyclic frequency in hertz (Hz) = $1/T$

T = period in seconds (s)

The basic sinusoidal functions have an *amplitude* (*peak value*) of unity. When a different amplitude is required, the function is multiplied by that factor. For example, the voltage $v(t) = 20 \sin 1000t$ describes a sinusoid with an amplitude or peak value of 20 V, an angular frequency $\omega = 1000$ rad/s, a cyclic frequency $f = 1000/2\pi = 159.15$ Hz, and a period $T = 1/159.15 = 6.283$ ms.

The Laplace transforms of the two functions $\sin \omega t$ and $\cos \omega t$ are given by pairs (T–3) and (T–4) (see Table 5–1).

Damped Sinusoidal Functions

The two functions in the $f(t)$ column corresponding to (T–5) and (T–6) are *damped sinusoidal functions*. They consist of sinusoidal oscillations in which the peak value decreases exponentially with time. A typical example is shown in Figure 5–5. The

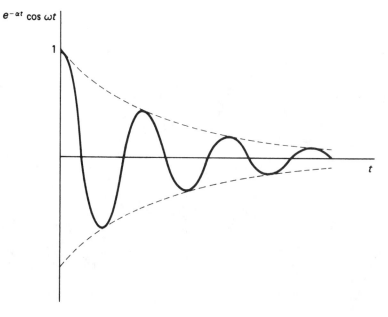

FIGURE 5–5
Typical damped sinusoidal function.

exponential envelope establishes the bound for the function, and the oscillation period is established by the sinusoidal factor. The time constant $\tau = 1/\alpha$ establishes the time that it takes the oscillations to diminish in accordance with the exponential development in Chapter 4.

The Laplace transforms of the damped sine and cosine functions are given by pairs (T–5) and (T–6).

The remaining functions will be discussed as the need arises. Pairs (T–7), (T–8), and (T–9) represent certain basic functions and combinations that arise in some applications. The function $\delta(t)$ of (T–10) represents a quantity called the *impulse function*, whose full treatment will be delayed to Chapter 10. A passing note on this "mysterious" function is that its Laplace transform is simply a constant value of 1!

The derivation of a representative transform and the use of Table 5–1 will now be illustrated with some examples.

EXAMPLE 5–1
Derive the Laplace transform of $f(t) = e^{-\alpha t}$, i.e., (T–2) of Table 5–1.

Solution
The exponential function is substituted in the defining relationship for $F(s)$ as given by Equation (5–3), and we have

$$F(s) = \int_0^\infty e^{-\alpha t} e^{-st}\, dt = \int_0^\infty e^{-(\alpha+s)t}\, dt \qquad (5–6)$$

The Laplace variable s is treated as a constant in the derivation of the transform, since integration is with respect to time. The integral of the exponential function in Equation (5–6) is

$$F(s) = \frac{e^{-(\alpha+s)t}}{-(\alpha + s)} \Bigg]_0^\infty \tag{5–7}$$

In evaluating the upper limit in Equation (5–7), the factor $\alpha + s$ is assumed to be positive, so that the exponential approaches zero as t increases without limit. The function $F(s)$ then becomes

$$F(s) = 0 - \frac{1}{-(\alpha + s)} = \frac{1}{s + \alpha} \tag{5–8}$$

as given in Table 5–1.

EXAMPLE 5–2
Using results provided in Table 5–1, determine the Laplace transform of

$$f(t) = 10 + 5e^{-4t} + 12 \sin 3t + 4e^{-2t} \cos 5t \tag{5–9}$$

Solution
Fundamental to this process are the results of Equations (5–4) and (5–5), allowing the constant multipliers and additive properties to be used. The pertinent transform pairs of Table 5–1 are (T–1), (T–2), (T–3), and (T–6).

$$F(s) = \frac{10}{s} + \frac{5}{s + 4} + \frac{12(3)}{s^2 + (3)^2} + \frac{4(s + 2)}{(s + 2)^2 + (5)^2} \tag{5–10a}$$

$$= \frac{10}{s} + \frac{5}{s + 4} + \frac{36}{s^2 + 9} + \frac{4(s + 2)}{s^2 + 4s + 29} \tag{5–10b}$$

5–3 LAPLACE TRANSFORM OPERATIONS

Having introduced some of the most common Laplace transform function pairs, the next step is to investigate the transform *operation pairs*. For a given functional operation in the time domain (e.g., differentiation), the operation pair indicates the corresponding operation that is performed in the s-domain. A few of the most common operation pairs of interest in circuit analysis are listed in Table 5–2. Not all of these operations will be defined or used at this point in the book, but the table is provided now for subsequent discussion and reference. This table is also repeated in Appendix C at the end of the text.

The column designated as $f(t)$ in Table 5–2 indicates an operation that can be performed in the time domain, and the column designated as $F(s)$ indicates the corresponding operation that must be performed on the Laplace transform of the original function. At this point in the text, only the first three pairs, that is, (O–1),

TABLE 5–2

Laplace Transform Operations Encountered in
Circuit Analysis

$f(t)$	$F(s)$	
$f'(t)$	$sF(s) - f(0)$	(O–1)
$\int_0^t f(t)\, dt$	$\dfrac{F(s)}{s}$	(O–2)
$e^{-\alpha t} f(t)$	$F(s + \alpha)$	(O–3)
$f(t - T)u(t - T)$	$e^{-sT}F(s)$	(O–4)
$f(0)$	$\lim_{s \to \infty} sF(s)$	(O–5)
$\lim_{t \to \infty} f(t)$	$\lim_{s \to 0} sF(s)*$	(O–6)

* Poles of $sF(s)$ must be in left-hand half-plane.

(O–2), and (O–3), will be considered. Each of these three will be discussed in the remainder of this section.

- *Derivative of a Time Function*

$$\mathcal{L}[f'(t)] = sF(s) - f(0) \qquad \textbf{(O–1)}$$

This operation indicates that to find the Laplace transform of the derivative of a function, the transform of the original function is multiplied by s, and the initial value of the time function is subtracted. If we momentarily consider time functions with an initial value of zero, that is, $f(0) = 0$, we can say that *differentiation in the time domain corresponds to multiplication by s in the s-domain.* This means that the differentiation process is replaced by the simpler process of multiplication by s.

- *Integral of a Time Function*

$$\mathcal{L}\left[\int_0^t f(t)\, dt\right] = \frac{F(s)}{s} \qquad \textbf{(O–2)}$$

This operation indicates that to find the Laplace transform of the definite integral of a time function (from 0 to an arbitrary time t), the transform of the original function is divided by s. We can thus say that *integration in the time domain corresponds to division by s in the s-domain.* Again, a process requiring calculus in the time domain is replaced by a simpler process (division in this case) in the s-domain.

- *Multiplication by $e^{-\alpha t}$*

$$\mathcal{L}[e^{-\alpha t}f(t)] = F(s + \alpha) \qquad \textbf{(O–3)}$$

This theorem states that if the original function is multiplied by $e^{-\alpha t}$, the transform of the product function is obtained by replacing s in the transform of the original function by $s + \alpha$. The major benefit of this theorem is to simplify the correlation between transforms of functions with and without exponential factors. Referring back to Table 5–1, first compare pair (T–3) with pair (T–5), and then compare pair (T–4) with pair (T–6). In each case, when the time function is multiplied by $e^{-\alpha t}$, the s-domain function can be determined by application of this transform operation.

EXAMPLE 5–3

Starting with the Laplace transform of the sine function as given by (T–3), derive the Laplace transform of the cosine function by employing the differentiation transform operation of (O–1).

Solution

We start with

$$f(t) = \sin \omega t \tag{5–11}$$

whose Laplace transform is

$$F(s) = \frac{\omega}{s^2 + \omega^2} \tag{5–12}$$

Next, differentiate both sides of Equation (5–11), which yields

$$f'(t) = \omega \cos \omega t \tag{5–13}$$

By means of operation (O–1), the Laplace transform of $f'(t)$ can be expressed as

$$\mathcal{L}[f'(t)] = sF(s) - f(0) = \frac{s\omega}{s^2 + \omega^2} - 0 \tag{5–14}$$

However, from Equation (5–13), we have

$$\mathcal{L}[f'(t)] = \mathcal{L}[\omega \cos \omega t] = \omega \, \mathcal{L}[\cos \omega t] \tag{5–15}$$

Equating Equations (5–14) and (5–15) results in

$$\mathcal{L}[\cos \omega t] = \frac{s}{s^2 + \omega^2} \tag{5–16}$$

which agrees with (T–4).

EXAMPLE 5–4

Starting with the Laplace transform of the constant 1 as given by (T–1), derive the Laplace transform of the linear function t making use of the integration transform operation of (O–2).

Solution

We start with

$$f(t) = 1 \qquad \text{(5-17)}$$

whose Laplace transform is

$$F(s) = \frac{1}{s} \qquad \text{(5-18)}$$

Next, integrate both sides of Equation (5–17), which yields

$$\int_0^t f(t)\, dt = \int_0^t (1)\, dt = t \qquad \text{(5-19)}$$

By means of operation (O–2), the Laplace transform of the integral function can be expressed as

$$\mathscr{L}\left[\int_0^t f(t)\, dt \right] = \frac{F(s)}{s} = \frac{1}{s}\left(\frac{1}{s}\right) = \frac{1}{s^2} \qquad \text{(5-20)}$$

However, from Equation (5–19), we have

$$\mathscr{L}\left[\int_0^t f(t)\, dt \right] = \mathscr{L}[t] \qquad \text{(5-21)}$$

Equating Equations (5–20) and (5–21) results in

$$\mathscr{L}[t] = \frac{1}{s^2} \qquad \text{(5-22)}$$

This result agrees with (T–7) of Table 5–1.

5–4 PREPARATION FOR INVERSE TRANSFORMATION

The process of inverse Laplace transformation consists of finding the time function corresponding to a given s-domain function. We have seen that to determine the Laplace transforms of common time functions, tabulated results such as those given in Tables 5–1 and 5–2 may be readily employed. In a few simple cases, these tables can also be used to determine inverse Laplace transforms directly. In the majority of cases, however, transforms usually appear in forms in which these tables may not be directly applied without considerable reduction.

To illustrate the preceding point, consider the result of Example 5–2, which is given in Equation (5–10b). If the transform were given in this form, one could use pertinent transform pairs from Table 5–1 and quickly determine the original time function as given in Equation (5–9). However, suppose this function appeared in a form with all terms arranged over a common denominator, which turns out to be

$$F(s) = \frac{19s^5 + 160s^4 + 1086s^3 + 3896s^2 + 9819s + 10440}{s(s + 4)(s^2 + 9)(s^2 + 4s + 29)} \qquad (5\text{–}23)$$

This form represents an example of the most common way in which a transform appears as the outcome of a circuit analysis problem. The process of converting this form to the expanded form of Equation (5–10b) to allow use of simpler transform pairs is not an easy task!

To prepare for inverse transformation, a few definitions are in order. Most transforms of interest in circuit analysis turn out to be expressible as ratios of polynomials in the Laplace variable s. We will define a transform function $F(s)$ as

$$F(s) = \frac{N(s)}{D(s)} \qquad (5\text{–}24)$$

where $N(s)$ is a numerator polynomial of the form

$$N(s) = a_n s^n + a_{n-1} s^{n-1} + \cdots + a_0 \qquad (5\text{–}25)$$

and $D(s)$ is a denominator polynomial of the form

$$D(s) = b_m s^m + b_{m-1} s^{m-1} + \cdots + b_0 \qquad (5\text{–}26)$$

Degree or Order

The integer n is the *degree* or *order* of the numerator polynomial, and m is the corresponding degree or order of the denominator polynomial. For all the work in this chapter, we will assume that $m > n$, that is, the degree of the denominator polynomial is greater than that of the numerator. (Later in the book, this restriction will be removed for special cases.)

A fundamental theorem of algebra states that the number of roots of a polynomial is equal to the degree of the polynomial. This means that $N(s)$ is assumed to have n roots, and $D(s)$ is assumed to have m roots.

Zeros and Poles

The roots of the numerator polynomial $N(s)$ are called the *zeros* of $F(s)$, and the roots of the denominator polynomial $D(s)$ are called the *poles* of $F(s)$. The significance of these terms relates to the fact that the value of a polynomial is zero when the variable s assumes the value of any one of its roots. Since $N(s)$ is in the numerator, $F(s) = 0$ when $N(s) = 0$. However, $D(s)$ is in the denominator, and the function $F(s)$ increases without limit (i.e., approaches "∞") when $D(s) = 0$. The behavior of $F(s)$ in the latter case can loosely (and perhaps in an amusing sense) be related to that of a "pole."

Although both $N(s)$ and $D(s)$ can be factored, our major interest at this time is in the poles, as will be seen shortly. Let $s_1, s_2, \ldots, s_m$ represent the m roots of $D(s)$, which are the poles of $F(s)$. This polynomial can be written in factored form as

$$D(s) = b_m(s - s_1)(s - s_2) \cdots (s - s_m) \qquad \text{(5–27)}$$

Comparing Equations (5–26) and (5–27), it can be deduced that *a polynomial can be completely specified by its roots except for a constant multiplier.* However, this constant multiplier (b_m) must be included when expressing the polynomial in factored form. In some cases, partially factored forms may appear. For example, refer to the denominator of Equation (5–23).

Classification of Poles

The poles of a transform function may be classified in two particular ways. The first way is to classify a given pole as either *real*, *imaginary*, or *complex*. A real pole is an ordinary real number (either positive or negative). An imaginary pole is one preceded by $j = \sqrt{-1}$, the basis for imaginary numbers, for example, $j5$. A complex number is one having both a real and an imaginary part, for example, $-3 + j4$. An imaginary or complex pole is always accompanied by its complex *conjugate*.

The *conjugate* of a complex or imaginary number is another complex number formed from the original by retaining the same real part (which is zero in the case of an imaginary number) and reversing the sign of the imaginary part. For example, the complex conjugate of $-2 + j3$ is $-2 - j3$; and if the first value is a root of a given polynomial, the second value is also a root. Similarly, the conjugate of $j5$ is $-j5$, and if one is a root of a given polynomial, so is the other one.

The second way of classifying poles is according to their *order*, which is the number of times a root is repeated in the denominator polynomial. The most common case is that of the *first-order* (or simple-order) root, in which the root appears only once. Higher-order roots are referred to as *multiple-order* roots. A multiple-order root may be further designated by its particular order.

The determination of the roots of a polynomial involves equating the polynomial to zero and finding the specific values of s that satisfy the equation. For a first-order polynomial, the one root is readily determined by inspection. For a second-order polynomial, the quadratic formula is used, or, in a few cases, the polynomial can be factored by inspection. The roots of second-order equations may be either real or complex. For third- and higher-degree equations, numerical methods must often be used. Computer and calculator programs are available for factoring such polynomials.

The examples that follow will deal with the determination of inverse transforms by inspection and with the classification of poles.

EXAMPLE 5–5

Determine the inverse transform of

$$F(s) = \frac{10}{s} + \frac{15}{s^2} + \frac{20}{s + 3} \qquad \text{(5–28)}$$

Solution

Inspection of Table 5–1 reveals that the three terms are of the forms of (T–1), (T–7), and (T–2), respectively (except, of course, for constant multipliers). Thus, $f(t)$ is readily written as

$$f(t) = 10 + 15t + 20e^{-3t} \qquad (5\text{–}29)$$

This illustrates one of the simplest types of functions in which to perform an inverse transformation, since each additive term is recognizable as one of the basic tabulated transforms.

EXAMPLE 5–6

Determine the inverse transform of

$$F(s) = \frac{8s + 30}{s^2 + 25} \qquad (5\text{–}30)$$

Solution

Inspection of Table 5–1 reveals that the denominator is of the form of (T–3) and (T–4), and since the numerator represents both a term containing s and a constant, these two transform pairs should work. First, we recognize that $\omega^2 = 25$ and $\omega = 5$. Note that (T–3) requires ω in the numerator. This is achieved by multiplying and dividing the constant term by 5. We then write Equation (5–30) in the form

$$F(s) = 8 \left[\frac{s}{s^2 + (5)^2} \right] + 6 \left[\frac{5}{s^2 + (5)^2} \right] \qquad (5\text{–}31)$$

The inverse transform is then expressed as

$$f(t) = 8 \cos 5t + 6 \sin 5t \qquad (5\text{–}32)$$

EXAMPLE 5–7

Determine the inverse transform of

$$F(s) = \frac{2s + 26}{s^2 + 6s + 34} \qquad (5\text{–}33)$$

Solution

The denominator is a second-degree polynomial, which appears to be of the same form as the denominator of (T–5) and (T–6) in Table 5–1. However, as noted by the footnote in Table 5–1, (T–5) *and* (T–6) *should be used only if the roots of the quadratic are complex.* The procedure for dealing with real roots of a quadratic will be considered in the next section.

The roots of the quadratic are determined by the quadratic formula to be $-3 \pm j5$, and since these values are complex, (T–5) and (T–6) may be used. The first step is to write the denominator polynomial in the form $(s + \alpha)^2 + \omega^2$, which is accomplished by the process of completing the square. Half of the constant factor for the linear term ($6/2 = 3$) is determined, and this number is squared (9). This value is added and subtracted as follows:

$$s^2 + 6s + 9 + 34 - 9 = (s + 3)^2 + 25 = (s + 3)^2 + (5)^2 \qquad \text{(5–34)}$$

where the first three terms constitute a perfect square. We see that $\alpha = 3$ and $\omega = 5$. Note that the roots of the quadratic are $-3 \pm j5$, so the roots are directly related to the values of α and ω. More will be said about this relationship in the next section.

The next step is to adjust the numerator so that the numerator forms of (T–5) an (T–6) are recognized. Note that (T–6) requires the presence of $s + \alpha = s + 3$ in the numerator. However, since the factor of s is 2, we must force the numerator of one term to be $2(s + 3)$. The additional term contains whatever constant is required to make the function equal to the initial function given. These steps are illustrated as follows:

$$\begin{aligned} F(s) &= \frac{2s + 26}{(s + 3)^2 + (5)^2} = \frac{2(s + 3)}{(s + 3)^2 + (5)^2} + \frac{20}{(s + 3)^2 + (5)^2} \\ &= \frac{2(s + 3)}{(s + 3)^2 + (5)^2} + 4\left[\frac{5}{(s + 3)^2 + (5)^2}\right] \end{aligned} \qquad \text{(5–35)}$$

Inverse transformation using (T–5) and (T–6) results in

$$f(t) = 2e^{-3t}\cos 5t + 4e^{-3t}\sin 5t \qquad \text{(5–36)}$$

EXAMPLE 5–8
The partially factored form of a certain Laplace transform is

$$F(s) = \frac{N(s)}{s(s^2 + 3s + 2)(s^2 + 16)(s^2 + 6s + 34)(s^2 + 8s + 16)} \qquad \text{(5–37)}$$

where $N(s)$ is the numerator polynomial, which is not specified. It is known, however, that no roots of $N(s)$ coincide with those of $D(s)$. Determine all poles of $F(s)$, and classify them according to the two methods discussed in this section.

Solution
The reason that $N(s)$ is not specified is to delineate our major interest at this point in the poles of $F(s)$ rather than the zeros. The additional stipulation that no numerator roots coincide with denominator roots ensures that no cancellations between numerator and denominator factors occur. Fortunately, the denominator polynomial is given in partially factored form. Imagine the complexity of the problem if the denominator had been given as a ninth degree single polynomial!

The nine poles of $F(s)$ will be designated as $s_1, s_2, \ldots, s_9$. Starting on the left, the single factor s means that $s_1 = 0$ is the value of one root. (If desired, the factor s may be interpreted as $s - 0$.) The first quadratic encountered is $s^2 + 3s + 2$, which may be readily factored as $(s + 1)(s + 2)$, which means that the roots correspond to $s_2 = -1$ and $s_3 = -2$. Alternately, the quadratic formula may be used to determine the roots directly.

Note that a given real root s_k must be interpreted in factored form as $(s - s_k)$. Thus, when a factor such as $(s + \alpha)$, with $\alpha > 0$, appears, it must be interpreted as $s - (-\alpha)$ so that the value of the pole is $s_k = -\alpha$.

Continuing to the right, the polynomial $s^2 + 16$ has the purely imaginary roots $s_4, s_5 = \pm j4$. Note that the absence of the first-degree term in the quadratic results in purely imaginary roots when both of the other terms have positive signs.

The polynomial $s^2 + 6s + 34$ cannot be readily factored by inspection. However, application of the quadratic formula leads to the complex roots $s_6, s_7 = -3 \pm j5$. If desired, this polynomial could be expressed in factored form as

$$[s - (-3 + j5)][s - (-3 - j5)] = (s + 3 - j5)(s + 3 + j5) \qquad \textbf{(5–38)}$$

However, this form for complex roots will not be used often in this text.

The final polynomial $(s^2 + 8s + 16)$ is factorable into $(s + 4)^2$. Thus, $s = -4$ is the root, and it occurs twice. This means that $s_8 = s_9 = -4$.

Summarizing all the preceding information, the following are the poles appropriately classified:

- Real, first-order (3 poles): $0, -1, -2$
- Imaginary, first-order (2 poles): $+j4, - j4$ (usually written as $\pm j4$)
- Complex, first-order (2 poles): $-3 + j5, -3 - j5$ (usually written as $-3 \pm j5$)
- Real, second-order (2 poles): $-4, -4$
  (*Note*: The second-order pole is counted twice to yield the nine poles.)

5–5 INVERSE TRANSFORMS OF FIRST-ORDER POLES

For the purpose of inverse transformation, poles will be classified in four categories:

1. First-order real poles
2. First-order complex poles
3. Multiple-order real poles
4. Multiple-order complex poles

Purely imaginary poles will be considered as a special case of complex poles with zero real part.

First-order poles are much more common in practical problems than multiple-order poles, so our major efforts will be directed toward them. This section will be devoted to the consideration of transforms with first-order poles, both real and complex, and the procedures for obtaining the inverse transforms. The procedure for working with multiple-order poles will be considered in the next section.

First Step (Check Poles)

Before developing general procedures, a specific function will be used to illustrate the type of approach required. Consider the particular function

$$F(s) = \frac{50s + 75}{(s^2 + 3s + 2)(s^2 + 4s + 20)} \qquad \textbf{(5–39)}$$

The denominator of this function is in partially factored form with two quadratics. Roots for quadratics can be either real or complex, so a further check is necessary. The quadratic $s^2 + 3s + 2$ has the real roots -1 and -2, so it can be expressed as $(s + 1)(s + 2)$. However, the quadratic $s^2 + 4s + 20$ has the complex roots $-2 \pm j4$, so it will be retained in quadratic form. Thus, there are two first-order real poles and two first-order complex poles.

With the factored form for the first polynomial, $F(s)$ can be expressed as

$$F(s) = \frac{50s + 75}{(s + 1)(s + 2)(s^2 + 4s + 20)} \tag{5-40}$$

This first step has been taken to alert the reader to the necessity to check all denominator quadratics to determine the nature of the roots. If the roots of a given quadratic are real, express the quadratic in factored form. If the roots are complex, leave the quadratic in its original form, but make a note of the roots for later reference. As we will see shortly, the nature of the time function is strongly dependent on the types of poles present.

Partial Fraction Expansion

In mathematics books, the next step normally taken is to perform a *partial fraction expansion*. The function $F(s)$ is expressed as

$$\frac{50s + 75}{(s + 1)(s + 2)(s^2 + 4s + 20)} = \frac{A_1}{s + 1} + \frac{A_2}{s + 2} + \frac{B_1 s + B_2}{s^2 + 4s + 20} \tag{5-41}$$

where A_1, A_2, B_1, and B_2 are constants to be determined. Note that the single-pole denominator terms require only a constant in the numerator, but the quadratic term requires a constant plus a term proportional to s. Various procedures exist for determining the constants, but the results can always be checked by combining back over a common denominator if necessary to see if the original function is obtained.

The various terms in the partial fraction expansion can be inverted with the basic transform pairs of Table 5–1. The first two terms can be inverted with (T–2), and they represent exponential terms in the time domain. The last term in Equation (5–41) can be represented as the sum of a term fitting the form of (T–5) plus a term fitting the form of (T–6). The result in the time domain will be a damped sine function plus a damped cosine function. (Refer back to Example 5–7 to see an example of this form.) It will be shown later that the sum of a damped sine function plus a damped cosine function can be expressed as a damped sine function with an additional phase shift. The inverse transform of Equation (5–41) can then be expressed in the form

$$f(t) = A_1 e^{-t} + A_2 e^{-2t} + B e^{-2t} \sin(4t + \theta) \tag{5-42}$$

where B and θ are determined from the constants B_1 and B_2 by techniques to be discussed later.

The inversion is by no means complete because A_1, A_2, B, and θ are not known. However, we have taken an important step in that direction. With very little computation, we have written down the general form of the inverse transform in which the damping factors and frequencies have been identified. This process is important in establishing a physical "feeling" for the mathematics involved.

Relationship Between Poles and Time Response

The following properties are evident from this example: (1) A first-order real pole corresponds to an exponential time response term. (2) A quadratic factor with complex poles corresponds to a damped sinusoidal term. Said differently, a pair of complex conjugate poles corresponds to a damped sinusoidal time response term.

Each first-order real pole factor can be thought of as producing an exponential function. If the pole is $-\alpha$, then α is the damping factor in the decaying exponential. Each pair of first-order complex conjugate poles can be thought of as producing a damped sinusoidal function. If the poles are $-\alpha \pm j\omega$, the damping factor of the decaying exponential multiplier is α, and the radian frequency of the sinusoid is ω.

General Algorithm for Inversion of Transform

We are now ready to present a general algorithm for inverting a transform containing an arbitrary number of first-order real and complex poles. Assume that $F(s)$ contains K real poles of first-order and R pairs of complex conjugate first-order poles. The degree of the denominator is then $K + 2R$ when expressed in polynomial form. Assume that all poles are known and that all real poles are expressed in first-order factored form. However, complex poles are retained in second-degree quadratic form. Thus, $F(s)$ will be expressed as

$$F(s) = \frac{N(s)}{(s + \alpha_1)(s + \alpha_2) \cdots (s + \alpha_K)(s^2 + a_1s + b_1)(s^2 + a_2s + b_2) \cdots (s^2 + a_Rs + b_R)}$$

(5–43)

Note that the poles of the linear factors are $-\alpha_1$, $-\alpha_2$, . . . , $-\alpha_K$. The poles of the quadratic $s^2 + a_1s + b_1$ are assumed to be $-\alpha_1 + j\omega_1$; the poles of the second quadratic are assumed to be $-\alpha_2 \pm j\omega_2$, etc.

The procedure for inverting $F(s)$ is as follows:

1. Consider that $f(t)$ can be expressed as

$$f(t) = f_{e1}(t) + f_{e2}(t) + \cdots + f_{eK}(t) + f_{s1}(t) + f_{s2}(t) + \cdots + f_{sR}(t) \quad (5–44)$$

The subscript e refers to "exponential," and these K terms can be considered to arise from the K real-pole factors. (A pole of value $s = 0$, which results in a constant term in the term domain, is considered as a limiting case of an exponential with $\alpha = 0$.) The subscript s refers to "sinusoidal," and these R terms can be considered to arise from the R quadratic factors with complex roots.

2. Any arbitrary exponential term $f_{ek}(t)$ can be expressed as

$$f_{ek}(t) = A_k e^{-\alpha_k t} \tag{5-45}$$

where A_k is given by

$$A_k = (s + \alpha_k)F(s)]_{s=-\alpha_k} \tag{5-46}$$

When $F(s)$ is multiplied by $(s + \alpha_k)$, this factor cancels the corresponding denominator factor, and the result contains everything in the original $F(s)$ except that denominator factor.

3. Any arbitrary damped sinusoidal term $f_{sk}(t)$ can be expressed as

$$f_{sk}(t) = B_k e^{-\alpha_k t} \sin(\omega_k t + \theta_k) \tag{5-47}$$

where the poles for the given quadratic are $-\alpha_k \pm j\omega_k$, and B_k and θ_k are determined from the relationship

$$B_k e^{j\theta_k} = B_k \underline{/\theta_k} = \frac{1}{\omega_k}[(s^2 + a_k s + b_k)F(s)]_{s=-\alpha_k+j\omega_k} \tag{5-48}$$

This formula indicates that after the quadratic is cancelled, the complex number $-\alpha_k + j\omega_k$ is substituted for s, and the resulting complex quantity is expressed as a magnitude and an angle. The magnitude is B_k, and the angle is θ_k.

When the poles are purely imaginary,

$$\alpha = 0 \text{ and } e^{-\alpha t} = 1$$

In this case, Equation (5-47) reduces to a sinusoid with no damping. The preceding techniques will be illustrated by a number of examples.

The procedure of Equation (5-48) requires some basic arithmetic manipulations with complex numbers. Readers with little or no background in this area may refer to Appendix B for assistance.

EXAMPLE 5-9

Determine the inverse transform of

$$F(s) = \frac{6s + 42}{s^2 + 7s + 10} \tag{5-49}$$

Solution

The denominator of this function was given as a quadratic to emphasize the point again that all quadratics in the denominator should be checked to see whether the roots are real or complex. The roots of the denominator in this case are readily determined to be -2 and -5, and $F(s)$ will be expressed with the denominator in factored form as

$$F(s) = \frac{6(s+7)}{(s+2)(s+5)} \tag{5–50}$$

(A constant multiplier has also been factored out in the numerator for convenience, but this is not really necessary.)

There are no complex poles in this example, and reference to Equation (5–44) indicates that $f(t)$ can be expressed as

$$f(t) = f_{e1}(t) + f_{e2}(t) \tag{5–51}$$

for two exponential terms. The two terms are of the forms

$$f_{e1}(t) = A_1 e^{-2t} \tag{5–52}$$

and

$$f_{e2}(t) = A_2 e^{-5t} \tag{5–53}$$

The constants A_1 and A_2 are determined from Equation (5–46) as

$$A_1 = (s+2)F(s)]_{s=-2} = \frac{6(s+7)}{(s+5)}\bigg]_{s=-2} = \frac{6(-2+7)}{(-2+5)} = 10 \tag{5–54}$$

$$A_2 = (s+5)F(s)]_{s=-5} = \frac{6(s+7)}{s+2}\bigg]_{s=-5} = \frac{6(-5+7)}{(-5+2)} = -4 \tag{5–55}$$

Observe how the $(s + \alpha_k)$ terms cancel when $F(s)$ is multiplied by $(s + \alpha_k)$. The resulting function can be expressed as

$$f(t) = 10e^{-2t} - 4e^{-5t} \tag{5–56}$$

With some practice, it is not necessary to write out all of the steps in the expanded fashion given here.

EXAMPLE 5–10

Determine the inverse transform of

$$F(s) = \frac{10s^2 + 42s + 24}{s^3 + 4s^2 + 3s} \tag{5–57}$$

Solution

Although the denominator is a third-degree equation, there is no constant term, and the denominator may be expressed as $s(s^2 + 4s + 3)$. The quadratic $s^2 + 4s + 3$ has the two real roots, -1 and -3. The poles are all real, and the function may then be written with the denominator in factored form as

$$F(s) = \frac{10s^2 + 42s + 24}{s(s+1)(s+3)} \tag{5–58}$$

The factor s may be interpreted as $s - 0$ for purposes of applying the inversion procedure, that is, $\alpha = 0$.

Shortening the procedure slightly, $f(t)$ may be expressed as

$$f(t) = A_1 + A_2 e^{-t} + A_3 e^{-3t} \tag{5-59}$$

where $e^{-\alpha_1 t} = 1$ since $\alpha_1 = 0$. The values of these constants are

$$A_1 = \frac{10s^2 + 42s + 24}{(s + 1)(s + 3)}\bigg]_{s=0} = \frac{24}{(1)(3)} = 8 \tag{5-60}$$

$$A_2 = \frac{10s^2 + 42s + 24}{s(s + 3)}\bigg]_{s=-1} = \frac{10 - 42 + 24}{(-1)(2)} = 4 \tag{5-61}$$

$$A_3 = \frac{10s^2 + 42s + 24}{s(s + 1)}\bigg]_{s=-3} = \frac{90 - 126 + 24}{(-3)(-2)} = -2 \tag{5-62}$$

Substitution of these constants in Equation (5–59) yields

$$f(t) = 8 + 4e^{-t} - 2e^{-3t} \tag{5-63}$$

EXAMPLE 5–11

Determine the inverse transform of

$$F(s) = \frac{20(s + 2)}{s(s + 1)(s^2 + 2s + 5)} \tag{5-64}$$

Solution

The two real poles 0 and -1 are readily apparent, but a further check of the quadratic reveals that its roots are $-1 \pm j2$. Thus, there are two real poles and a pair of complex conjugate poles. The composite function may be expressed as

$$f(t) = f_{e1}(t) + f_{e2}(t) + f_s(t) \tag{5-65}$$

where $f_{e1}(t)$ and $f_{e2}(t)$ are due to the real poles and $f_s(t)$ is due to the pair of complex roots.

The three terms are of the forms

$$f_{e1}(t) = A_1 \text{ (since } \alpha_1 = 0) \tag{5-66}$$

$$f_{e2}(t) = A_2 e^{-t} \tag{5-67}$$

$$f_s(t) = Be^{-t}\sin(2t + \theta) \tag{5-68}$$

The constants A_1 and A_2 are determined as

$$A_1 = \frac{20(s + 2)}{(s + 1)(s^2 + 2s + 5)}\bigg]_{s=0} = \frac{20(2)}{(1)(5)} = 8 \tag{5-69}$$

$$A_2 = \frac{20(s + 2)}{s(s^2 + 2s + 5)}\bigg]_{s=-1} = \frac{20(1)}{(-1)(4)} = -5 \tag{5-70}$$

The constants B and θ are determined from the procedure of Equation (5–48) as follows:

$$B\underline{/\theta} = \frac{1}{2}[(s^2 + 2s + 5)F(s)]_{s=-1+j2} = \frac{1}{2}\left[\frac{20(s + 2)}{s(s + 1)}\right]_{s=-1+j2}$$

$$= \frac{10(1 + j2)}{(-1 + j2)(j2)} = \frac{10(2.2361\underline{/63.435°})}{(2.2361\underline{/116.565°})(2\underline{/90°})}$$

$$= 5\underline{/-143.13°} \tag{5–71}$$

Thus, $B = 5$ and $\theta = -143.13°$. The function $f_s(t)$ can be expressed as

$$f_s(t) = 5e^{-t}\sin(2t - 143.13°) \tag{5–72}$$

Combining the preceding results, the total time function can be expressed as

$$f_s(t) = 8 - 5e^{-t} + 5e^{-t}\sin(2t - 143.13°) \tag{5–73}$$

EXAMPLE 5–12
Determine the inverse transform of

$$F(s) = \frac{100s}{(s^2 + 4)(s^2 + 2s + 10)} \tag{5–74}$$

Solution
The quadratic $(s^2 + 4)$ has the purely imaginary roots $\pm j2$, and the quadratic $(s^2 + 2s + 10)$ has the complex roots $-1 \pm j3$. Thus, there is one pair of purely imaginary poles and one pair of complex poles. The time function will then be of the form

$$f(t) = f_{s1}(t) + f_{s2}(t) \tag{5–75}$$

where

$$f_{s1}(t) = B_1 e^{-0t}\sin(2t + \theta_1) \tag{5–76a}$$

$$= B_1 \sin(2t + \theta_1) \tag{5–76b}$$

and

$$f_{s2}(t) = B_2 e^{-t}\sin(3t + \theta_2) \tag{5–77}$$

The constants B_1 and θ_1 are determined from the procedure of Equation(5–48) as follows:

$$B_1\underline{/\theta_1} = \frac{1}{2}[(s^2 + 4)F(s)]_{s=j2} = \frac{1}{2}\left[\frac{100s}{s^2 + 2s + 10}\right]_{s=j2}$$

$$= \frac{50(j2)}{-4 + 10 + j4} = \frac{100\underline{/90°}}{6 + j4} = \frac{100\underline{/90°}}{7.21110\underline{/33.6901°}}$$

$$= 13.8675\underline{/56.3099°} \tag{5–78}$$

Thus, $B_1 = 13.8675$ and $\theta_1 = 56.3099°$.

The constants B_2 and θ_2 are determined as follows:

$$B_2 \,\underline{/\theta_2} = \frac{1}{3}[(s^2 + 2s + 10)F(s)]_{s=-1+j3} = \frac{1}{3}\left[\frac{100s}{s^2 + 4}\right]_{s=-1+j3}$$

$$= \frac{33.33(-1 + j3)}{(-1 + j3)^2 + 4} = \frac{33.33(-1 + j3)}{-4 - j6}$$

$$= \frac{33.33(3.1623 \,\underline{/108.435°})}{(7.2111 \,\underline{/-123.690°})} = 14.6176 \,\underline{/232.125°}$$

$$= 14.6176 \,\underline{/-127.875°} \qquad \textbf{(5–79)}$$

Thus, $B_2 = 14.6176$ and $\theta_2 = -127.875°$. Combining all the preceding results, the net inverse transform is

$$f(t) = 13.8675 \sin(2t + 56.3099°) + 14.6176e^{-t} \sin(3t - 127.875°) \quad \textbf{(5–80)}$$

5–6   INVERSE TRANSFORMS OF MULTIPLE-ORDER POLES

With the possible exception of second-order real poles, multiple-order poles do not arise very often in the types of practical problems for which transforms are best suited. Some general procedures for dealing with multiple-order poles will be presented without proof in this section. These procedures are somewhat unwieldy to apply for all but the simplest cases, so most people (including the author) have to review the formulas each time they are used. The only part of this section that is required in the remainder of the book is the case of the second-order real pole, which is stated as a separate procedure in Equations (5–87) through (5–90).

Before considering the procedures for multiple-order poles, understand that there may also be present any arbitrary number of first-order real and complex poles. The procedures of the last section are used to determine all $f_{ek}(t)$ and $f_{sk}(t)$ functions corresponding to first-order real and complex poles, respectively. In this section, we will deal specifically with the procedures for the multiple-order poles only, which result in additional additive terms in the time domain.

Algorithm for Multiple-Order Real Poles

Consider first that there is a real pole of value $s = -\alpha$ and of order i. This means that the denominator of $F(s)$ has the factor $(s + \alpha)^i$, and $F(s)$ can be expressed as

$$F(s) = \frac{Q(s)}{(s + \alpha)^i} \qquad \textbf{(5–81)}$$

where

$$Q(s) = (s + \alpha)^i F(s) \qquad \textbf{(5–82)}$$

Note that multiplication by $(s + \alpha)^i$ in Equation (5–82) results in a cancellation of the $(s + \alpha)^i$ factor in the denominator of $F(s)$, so $Q(s)$ represents all parts of $F(s)$ except the multiple-order pole factor under consideration.

Let $f_m(t)$ represent the portion of the total time function due to the pole of order i with a value $-\alpha$. This function will be of the form

$$f_m(t) = \left[\frac{C_1 t^{i-1}}{(i-1)!} + \frac{C_2 t^{i-2}}{(i-2)!} + \cdots + \frac{C_k t^{i-k}}{(i-k)!} + \cdots + C_i \right] e^{-\alpha t} \qquad \textbf{(5–83)}$$

A given coefficient C_k can be determined from the expression

$$C_k = \frac{1}{(k-1)!} \frac{d^{k-1}}{ds^{k-1}} Q(s) \Bigg]_{s=-\alpha} \qquad \textbf{(5–84)}$$

where the "!" symbol denotes the factorial operation. (The value of $n!$ is $n! = 1 \cdot 2 \cdot 3 \cdot \cdots n$, and by definition, $0! = 1$.) According to the expression of Equation (5–84), the $Q(s)$ function is successively differentiated with respect to s a total of $i - 1$ times. The function and the first $i - 1$ derivatives evaluated at $s = -\alpha$ and multiplied by suitable constants yield the set of i coefficients.

It should be noted that for a first-order pole (i.e., $i = 1$), the time function reduces to

$$f_m(t) = \frac{C_1 t^0}{0!} e^{-\alpha t} = C_1 e^{-\alpha t} \qquad \textbf{(5–85)}$$

with

$$C_1 = \frac{1}{0!} Q(s) \Bigg]_{s=-\alpha} = (s + \alpha) F(s)]_{s=-\alpha} \qquad \textbf{(5–86)}$$

These results are the same (with slightly different notation) as those given in the preceding section for a first-order pole.

Second-Order Real Poles

The most important case of interest is the second-order real pole. Thus, assume $F(s)$ has a denominator factor $(s + \alpha)^2$ and define

$$Q(s) = (s + \alpha)^2 F(s) \qquad \textbf{(5–87)}$$

The time function $f_m(t)$ produced by this second-order pole will be

$$f_m(t) = (C_1 t + C_2) e^{-\alpha t} \qquad \textbf{(5–88)}$$

where

$$C_1 = Q(s)]_{s=-\alpha} \qquad \textbf{(5–89)}$$

and

$$C_2 = \frac{dQ(s)}{ds} \Bigg]_{s=-\alpha} \qquad \textbf{(5–90)}$$

If a function contains more than one value for multiple-order poles, the procedure discussed earlier is applied separately to each factor. There will thus be a separate $Q(s)$ function for each phase of the analysis, and there will be more than one series of terms of the form of Equation (5–83).

Multiple-Order Complex Poles

Multiple-order complex poles are best inverted by first expressing a quadratic in factored form in terms of the complex roots. The procedure of this section is then applied to each of the multiple-order pole factors. However, complex numbers appear in the process, so $s = -\alpha$ in Equation (5–84) must be interpreted as $s = -\alpha + j\omega$ for one set and $s = -\alpha - j\omega$ in the other set. The resulting sets of values for C_k will be complex numbers, and the exponential terms will be of the forms $e^{-\alpha t}e^{j\omega t}$ and $e^{-\alpha t}e^{-j\omega t}$. By suitable manipulations and the definitions of sine and cosine in terms of complex exponentials, real time functions will ultimately be obtained.

The exponential definitions of cosine and sine are

$$\cos \omega t = \frac{e^{j\omega t} + e^{-j\omega t}}{2} \tag{5–91}$$

and

$$\sin \omega t = \frac{e^{j\omega t} - e^{-j\omega t}}{2j} \tag{5–92}$$

The relative infrequency of this situation in a practical problem decrees no additional consideration at this point.

EXAMPLE 5–13

Determine the inverse transform of

$$F(s) = \frac{20}{s(s^2 + 4s + 4)} \tag{5–93}$$

Solution

There is one first-order real pole of value $s = 0$. A check of the quadratic reveals that it has two equal real roots of value $s = -2$. Thus, $F(s)$ can be expressed as

$$F(s) = \frac{20}{s(s + 2)^2} \tag{5–94}$$

The first-order pole produces a time function $f_e(t)$ of the form

$$f_e(t) = A_1 e^{0t} = A_1 \tag{5–95}$$

where

$$A_1 = \frac{20}{(s + 2)^2}\bigg]_{s=0} = 5 \tag{5–96}$$

The time function corresponding to the second-order pole is

$$f_m(t) = (C_1 t + C_2)e^{-2t} \tag{5–97}$$

The $Q(s)$ function is

$$Q(s) = (s + 2)^2 F(s) = \frac{20}{s} \tag{5–98}$$

Its derivative is

$$\frac{dQ(s)}{ds} = \frac{-20}{s^2} \tag{5–99}$$

The values of C_1 and C_2 are then determined as

$$C_1 = Q(s)]_{s=-2} = \frac{20}{-2} = -10 \tag{5–100}$$

$$C_2 = \frac{dQ(s)}{ds}\bigg]_{s=-2} = \frac{-20}{(-2)^2} = -5 \tag{5–101}$$

Substituting Equations (5–100) and (5–101) in Equation (5–97) and combining that function with $f_e(t)$, we have

$$f(t) = 5 - (10t + 5)e^{-2t} \tag{5–102}$$

5–7   MATLAB EXAMPLES

In this section, some of the capabilities of MATLAB for manipulating Laplace transforms will be investigated. The first example will illustrate how to determine the instantaneous behavior of an inverse Laplace transform as a plot. The second example will show how a symbolic inverse Laplace transform can be determined.

MATLAB EXAMPLE 5–1

Determine numerically the inverse Laplace transform of the function of Example 5–9 [Equation (5–49)], and obtain a plot as a function of time.

Solution
The pertinent function is repeated here for convenience and is

$$F(s) = \frac{6s + 42}{s^2 + 7s + 10} \tag{5–103}$$

In working with a Laplace transform where numerical results are desired, the numerator and denominator polynomial coefficients are defined as row vectors. Although it is possible to define shorter vectors when some powers are missing, it is this author's opinion that there is less chance of confusion if the numerator and denominator polynomials are thought of as having the same degree by adding appropriate zeros. To that end, we think of $F(s)$ in the form

$$F(s) = \frac{0 \cdot s^2 + 6s + 42}{s^2 + 7s + 10} \qquad (5\text{--}104)$$

With a little practice, this process can be performed mentally without the need to actually write it down.

Refer now to the MATLAB workspace of Figure 5–6. The numerator row vector *n* is defined as

```
n = [0 6 42]
```

The denominator row vector *d* is defined as

```
d = [1 7 10]
```

These values are immediately displayed on the screen as indicated on the workspace. Since an inverse transform is to be performed numerically, a time range over which the operation is to be performed will be defined. We will choose to utilize 151 points over a time range from $t = 0$ to $t = 3$, and the linear spacing command will be employed. The command is

```
t = linspace(0,3,151);
```

where printing of the values was suppressed by the semicolon.

The most direct way of obtaining the inverse Laplace transform numerically is through the *impulse command*. The theory behind this will be developed later,

```
EDU» n=[0 6 42]

n =

        0     6    42

EDU» d=[1 7 10]

d =

        1     7    10

EDU» t=linspace(0,3,151);
EDU» f=impulse(n,d,t);
EDU» plot(t,f)
EDU» xlabel('time')
EDU» ylabel('function')
EDU» title('INVERSE LAPLACE TRANSFORM')
EDU»
```

FIGURE 5–6
Workspace for MATLAB Example 5–1.

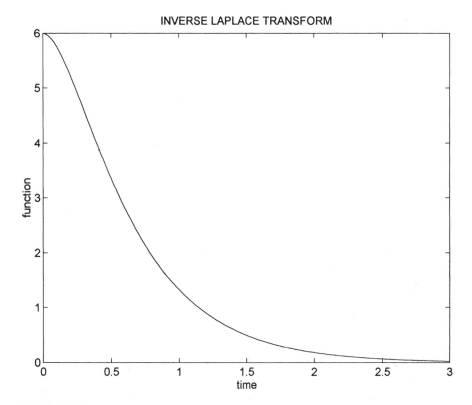

FIGURE 5–7
Plot of inverse transform for MATLAB Example 5–1.

but the impulse command always produces the inverse transform of a given Laplace function without any modifications. The inverse transform will be denoted as *f*, and the command is written as

```
f = impulse(n,d,t);
```

where *n* is the numerator polynomial row vector, *d* is the denominator polynomial row vector, and *t* is the time array. Again, printing of the values on the screen has been suppressed.

The last three lines of the workspace involve labeling of the graph, and they follow the format discussed in MATLAB Example 4–1. The resulting plot is shown in Figure 5–7.

MATLAB EXAMPLE 5–2
Use MATLAB to perform *symbolically* the inverse Laplace transform of Equation (5–49).

FIGURE 5–8
Workspace for MATLAB Example 5–2.

```
EDU»  F='(6*s+42)/(s^2+7*s+10)'

F =

(6*s+42)/(s^2+7*s+10)

EDU» .f=invlaplace(F)

f =

10*exp(-2*t)-4*exp(-5*t)

EDU»
```

Solution
This approach differs from the last example in that we desire an *expression* for the inverse transform, rather than a set of values or a curve. To achieve this goal, the function is entered in *symbolic* form. This is achieved by beginning and ending the equation with the apostrophe (').

Refer to the MATLAB workspace of Figure 5–8. The function of Equation (5–49) is entered as

```
F='(6*s+42)/(s^2+7*s+10)'
```

After entering this expression, MATLAB repeats the operation without the ' indicators.

Let *f* represent the inverse Laplace transform in symbolic form. The command for the inversion is

```
f=invlaplace(F)
```

MATLAB then prints

```
f=10*exp(-2*t)-4*exp(-5*t)
```

This corresponds to the expression

$$f(t) = 10e^{-2t} - 4e^{-5t} \qquad (5\text{–}105)$$

which is identical with Equation (5–56).

DRILL PROBLEMS

Use Table 5–1 to determine the Laplace transforms in Problems 5–1 through 5–10.

5–1. $f(t) = 6$
5–2. $f(t) = 4t$
5–3. $v(t) = 8 \sin 4t$
5–4. $v(t) = 6 \cos 3t$

5–5. $i(t) = 7e^{-2t} \cos 3t$
5–6. $i(t) = 5e^{-3t} \sin 6t$
5–7. $f(t) = 8e^{-2t}$
5–8. $v(t) = 3t^2$
5–9. $i(t) = 2\delta(t) + 3$
5–10. $f(t) = e^{-3t}t^2$

Use Table 5–1 to determine the inverse Laplace transforms in Problems 5–11 through 5–20.

5–11. $F(s) = \dfrac{6}{s+4}$

5–12. $F(s) = \dfrac{8}{s}$

5–13. $V(s) = \dfrac{20}{s^2 + 16}$

5–14. $V(s) = \dfrac{4s}{s^2 + 9}$

5–15. $I(s) = \dfrac{16}{s^2 + 2s + 5}$

5–16. $I(s) = \dfrac{9(s+1)}{s^2 + 2s + 5}$

5–17. $V(s) = \dfrac{5s + 22}{s^2 + 4s + 13}$

5–18. $I(s) = \dfrac{4s - 3}{s^2 + 6s + 25}$

5–19. $F(s) = 3 + \dfrac{4}{s^3}$

5–20. $V(s) = \dfrac{30}{(s+2)^4}$

Use the procedures of Sections 5–5 and 5–6 to determine the inverse Laplace transforms in Problems 5–21 through 5–38.

5–21. $F(s) = \dfrac{6s + 14}{s^2 + 4s + 3}$

5–22. $F(s) = \dfrac{5s + 16}{s^2 + 6s + 8}$

5–23. $V(s) = \dfrac{s^2 + 8s + 4}{s^3 + 3s^2 + 2s}$

5–24. $V(s) = \dfrac{6s + 24}{s^3 + 5s^2 + 6s}$

5–25. $I(s) = \dfrac{18}{s(s+1)(s+2)(s+3)}$

5–26. $I(s) = \dfrac{18}{(s+1)(s+2)(s+3)}$

5-27. $V(s) = \dfrac{10s + 3 \times 10^4}{s^2 + 3 \times 10^3 s + 2 \times 10^6}$

5-28. $I(s) = \dfrac{10^4}{s^2 + 3 \times 10^6 s + 2 \times 10^{12}}$

5-29. $F(s) = \dfrac{100}{s(s^2 + 4s + 13)}$

5-30. $F(s) = \dfrac{100(s + 2)}{s(s^2 + 4s + 13)}$

5-31. $V(s) = \dfrac{100}{s(s + 1)(s^2 + 4s + 13)}$

5-32. $V(s) = \dfrac{100(s + 2)}{s(s + 1)(s^2 + 4s + 13)}$

5-33. $I(s) = \dfrac{20}{s(s^2 + 4)}$

5-34. $I(s) = \dfrac{20(s + 2)}{s(s^2 + 4)}$

5-35. $F(s) = \dfrac{50}{(s^2 + 4)(s^2 + 4s + 13)}$

5-36. $F(s) = \dfrac{50}{(s^2 + 2s + 5)(s^2 + 4s + 13)}$

5-37. $V(s) = \dfrac{20}{s^2(s + 1)}$

5-38. $I(s) = \dfrac{20}{(s + 1)(s^2 + 4s + 4)}$

DERIVATION PROBLEMS

5-39. By employing the basic definition, derive the Laplace transform of $f(t) = 1$, i.e., (T-1) of Table 5-1. You may assume that s is positive in the upper limit of the integral.

5-40. Derive the Laplace transforms of $\sin \omega t$ and $\cos \omega t$ by the following process: Use the Laplace transform of the exponential function, i.e., (T-2) of Table 5-1, to determine the transform of $e^{j\omega t}$. (This means that $j\omega$ replaces $-\alpha$ in the transform.) Next, expand the transform into real and imaginary parts by rationalizing the denominator. Further, the Euler formula reads

$$e^{j\omega t} = \cos \omega t + j \sin \omega t$$

Thus, $\qquad \mathcal{L}[e^{j\omega t}] = \mathcal{L}[\cos \omega t] + \mathcal{L}[j \sin \omega t]$

Finally, equate real and imaginary parts of the equation above to the respective parts of the transform to obtain two transforms in one step!

5-41. Starting with the Laplace transform of the cosine function as given by (T-4), derive the Laplace transform of the sine function by employing the integration transform operation of (O-2).

5–42. Starting with the Laplace transform of the function t as given by (T–7), derive the Laplace transform of the constant 1 by employing the differentiation transform operation of (O–1).

5–43. Starting with the Laplace transform of sin ωt as given by (T–3), derive the Laplace transform of the damped sine function, i.e., (T–5), by employing operation (O–3).

5–44. Starting with the Laplace transform of cos ωt as given by (T–4), derive the Laplace transform of the damped cosine function, i.e., (T–6), by employing operation (O–3).

APPLICATION PROBLEMS

Application problems with Laplace transforms of a type suitable for this book are inappropriate at this point. Much of the work of the next several chapters will be oriented toward establishing applications of the Laplace transform.

6

CIRCUIT ANALYSIS WITH LAPLACE TRANSFORMS

OBJECTIVES

After completing this chapter, the reader should be able to:

- Define transform impedance and admittance and determine these quantities for a given element value.
- Represent initial conditions for capacitors and inductors in terms of s-domain Thevenin and Norton models.
- Determine the complete s-domain model for a given circuit.
- Apply various circuit analysis methods to s-domain circuit models.
- Identify the natural and forced responses and determine when these represent transient and steady-state responses.
- Apply Laplace transform methods to obtain complete solutions for first-order circuits with arbitrary excitations.
- Apply Laplace transform methods to obtain complete solutions for second-order circuits.

6–1 TRANSFORM IMPEDANCES

The major objective in this section is to determine s-domain forms for each of the three basic circuit parameters. As we will see, the use of Laplace transform operations allows the derivative and integral relationships to be converted to algebraic relationships. This is achieved by transforming voltages and currents to the s-domain.

Before considering any specific relationships, the general approach is to represent the effects of passive components as s-domain or Laplace impedances. Refer to the block shown in Figure 6–1(a), which is assumed to contain only passive circuit components. The instantaneous voltage across the circuit is $v(t)$ and the current flowing into the circuit is $i(t)$. With the exception of a resistive circuit, there

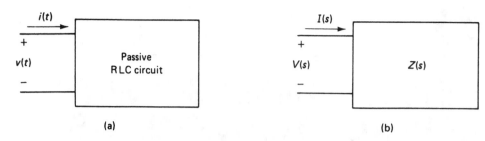

(a)   (b)

FIGURE 6-1
Passive *RLC* circuit and its transform or *s*-domain impedance.

is, in general, no simple algebraic relationship between the voltage and current because of the derivative/integral relationships involved.

Transform Impedance and Admittance

Let $V(s) = \mathcal{L}[v(t)]$ represent the Laplace transform of the voltage, and let $I(s) = \mathcal{L}[i(t)]$ represent the Laplace transform of the current. A function $Z(s)$ called the *s-domain* or *transform impedance* will be defined as

$$Z(s) = \frac{V(s)}{I(s)} \qquad (6\text{-}1)$$

The reciprocal of transform impedance is called *transform admittance $Y(s)$, and it* is defined as

$$Y(s) = \frac{1}{Z(s)} = \frac{I(s)}{V(s)} \qquad (6\text{-}2)$$

The use of transform impedances (and admittances) permits many of the basic circuit analysis techniques to be extended to complex circuits with time-varying voltages and currents through the transform process. Various network reduction techniques and simplifications may be applied, with transform impedances treated in much the same manner as resistances in dc circuits. Of course, all s values must be carefully maintained in the expressions.

The transform impedances for each of the three basic circuit parameters will now be derived. References will be made to Figure 6-2 in the developments that follow. The time-domain form is shown on the left for each parameter, and the corresponding *s*-domain form is shown on the right.

Resistance

The time-domain voltage–current relationship for a resistance is

$$v(t) = Ri(t) \qquad (6\text{-}3)$$

The Laplace transformation is applied to both sides of Equation (6–3), and there results

$$V(s) = RI(s) \qquad\qquad (6\text{–}4)$$

The transform impedance for the resistance is

$$Z(s) = \frac{V(s)}{I(s)} = R \qquad\qquad (6\text{–}5)$$

The time-domain and s-domain forms for resistance are shown along the top row of Figure 6–2.

These results show that the s-domain impedance of a resistance is simply the

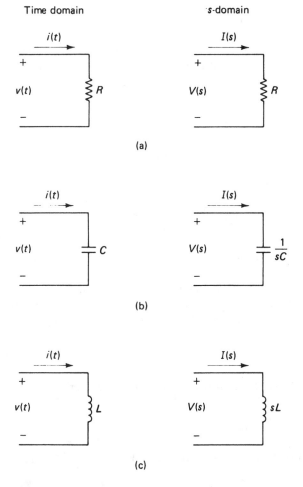

FIGURE 6–2
Conversion to s-domain models for circuit parameters.

value of the resistance, so no change is required in converting a resistance to the s-domain. This equality is a result of the simple algebraic time-domain voltage–current relationship for a resistor, and the voltage and current waveforms are identical.

Capacitance

Consider an *uncharged* capacitor C, in which the voltage–current relationship is

$$v(t) = \frac{1}{C} \int_0^t i(t) \, dt \qquad (6\text{–}6)$$

The Laplace transformation is applied to both sides of Equation (6–6), and we have

$$\mathcal{L}[v(t)] = \mathcal{L}\left[\frac{1}{C} \int_0^t i(t) \, dt\right] = \frac{1}{C} \mathcal{L}\left[\int_0^t i(t) \, dt\right] \qquad (6\text{–}7)$$

The Laplace transform of the integral in Equation (6–7) is expressed in terms of operation (O–2) of Table 5–2. With appropriate s-domain substitutions, this equation becomes

$$V(s) = \frac{1}{sC} I(s) \qquad (6\text{–}8)$$

The integral relationship in the time domain thus becomes an algebraic equation in the s-domain.

The impedance $Z(s)$ for the capacitor is

$$Z(s) = \frac{V(s)}{I(s)} = \frac{1}{sC} \qquad (6\text{–}9)$$

The time-domain and s-domain forms for the capacitance are shown in the middle row of Figure 6–2.

These results show that the s-domain impedance of a capacitance is a quantity $1/sC$. The voltage–current relationship is thus expressed in algebraic form, provided that transform voltage and current variables are used.

Inductance

Consider an *unfluxed* inductance L, in which the voltage–current relationship is

$$v(t) = L \frac{di(t)}{dt} \qquad (6\text{–}10)$$

The Laplace transformation is applied to both sides of Equation (6–10), and we have

$$\mathcal{L}[v(t)] = \mathcal{L}\left[L \frac{di(t)}{dt}\right] = L \mathcal{L}\left[\frac{di(t)}{dt}\right] \qquad (6\text{–}11)$$

The Laplace transform of the derivative in Equation (6–11) is expressed in terms of operation (O–1) of Table 5–2, with $i(0) = 0$. With appropriate s-domain substitutions, this equation becomes

$$V(s) = sLI(s) \tag{6–12}$$

The impedance $Z(s)$ for the inductance is

$$Z(s) = \frac{V(s)}{I(s)} = sL \tag{6–13}$$

The time-domain and s-domain forms for the inductance are shown in the bottom row of Figure 6–2.

These results show that the s-domain impedance of an inductance is a quantity sL. The relationship between the transform variables is thus algebraic in nature.

EXAMPLE 6–1

The following components are shown on the left in Figure 6–3: (a) a 1-kΩ resistor, (b) a 0.5-μF uncharged capacitor, and (c) a 30-mH unfluxed inductor. Show the corresponding s-domain impedance models.

Solution

The s-domain models are quickly established using the forms given in Figure 6–2, and the results are shown on the right in Figure 6–3. Observe that the basic units

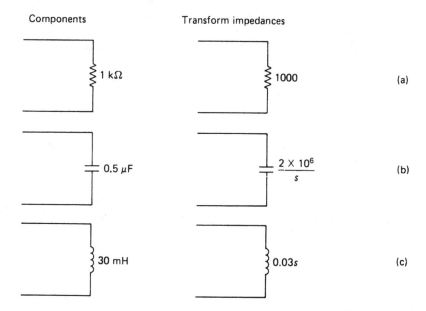

FIGURE 6–3
Components of Example 6–1 and their transform impedances.

of the component values are used. Because of the volume of labeling required in s-domain circuits, the units for the impedances will usually be omitted in the transform circuit forms. However, it will be established and understood throughout the book that basic units for element values are used in impedance functions.

6-2 INITIAL CONDITION MODELS

In the preceding section, s-domain impedance models for the passive circuit parameters were derived. In the developments, the capacitor was assumed to be uncharged, and the inductor was assumed to be unfluxed. We will now extend these models to include the effects of initial energy storage.

The method that will be used is based on some models developed in Chapter 3 for initial energy storage. Refer back to Figure 3–9 for the case of a capacitor, in which the effect of an initial voltage V_0 is represented by a dc voltage source of value V_0 in *series* with an uncharged capacitor. Refer back to Figure 3–14 for the case of an inductor, in which the effect of an initial current I_0 is represented by a dc current source of value I_0 in *parallel* with an unfluxed inductor. Keep these models in mind in the development that follows.

Models for Initially Charged Capacitor

An initially charged capacitor is shown in Figure 6–4(a). By using the model just discussed, the capacitor is represented by an impedance $1/sC$, and the fictitious series voltage source V_0 is converted to its s-domain form, which is a transform voltage V_0/s. (Recall that the Laplace transform of a dc value or constant value is that value divided by s.) The resulting circuit is shown in Figure 6–4(b). Thus, the equivalent circuit given in Chapter 3 leads to a natural form for the s-domain charged capacitor model. Note that this model is in a Thevenin form.

If a Norton form for the charged capacitor is desired, a source transformation may be readily applied to the voltage source in series with the impedance using s-

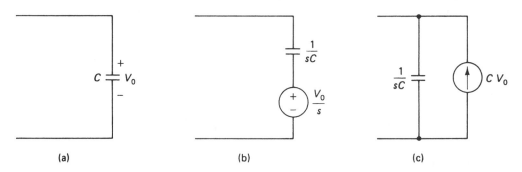

(a)   (b)   (c)

FIGURE 6–4
Capacitor with initial voltage and the two s-domain models.

domain concepts. The value of the current source is $(V_0/s)/(1/sC) = CV_0$, and the resulting model is shown in Figure 6–4(c).

A comment concerning the significance of this current source value is in order. The s-domain current is a constant value CV_0. Referring back to Table 5–1, transform pair (T–10) indicates that a constant value in the s-domain corresponds to an impulse function $\delta(t)$ in the time domain. We have not yet defined or explained the concept of an impulse function, and its full significance will not be encountered until Chapter 10. However, at this point, it is arising indirectly through a source transformation in the s-domain. The mathematical results are telling us that, to represent a charged capacitor as a current source model, an impulse current is required. However, we need not refer to the impulse source explicitly, since the mathematical manipulations in the s-domain take care of the required "book-keeping."

Before continuing, make sure that the concept of a constant value in the s-domain is carefully distinguished from the much more common concept of a constant value in the time domain. A constant value in the time domain is simply a dc value, but its particular representation in the s-domain is a constant divided by s. However, a constant in the s-domain is a totally different quantity, and its time-domain form would require introduction of an impulse function. At this point, we will deal with impulses only through the s-domain interpretation.

Models for Initially Fluxed Inductor

Next, consider the initially fluxed inductor shown in Figure 6–5(a). By using the model of Chapter 3, which was discussed earlier in this section, the inductor is represented by an impedance sL, and the fictitious parallel current source I_0 is converted to its s-domain form, which is a transform current I_0/s. The resulting circuit is shown in Figure 6–5(b). Note that this model is in a *Norton form*.

If a Thevenin form for the fluxed inductor is desired, a source transformation may be readily applied to the current source in parallel with the impedance using s-domain concepts. The value of the voltage source is $(I_0/s) \times sL = LI_0$, and the

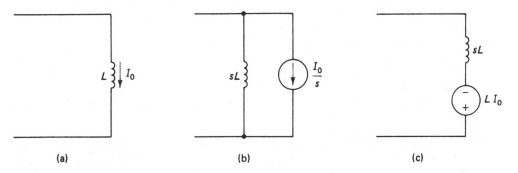

(a)   (b)   (c)

FIGURE 6–5
Inductor with initial current and the two s-domain models.

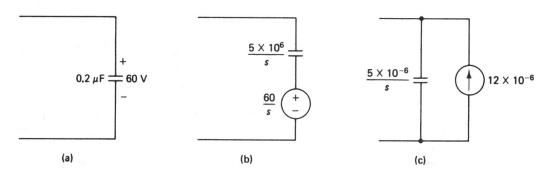

FIGURE 6–6
Charged capacitor for Example 6–2 and the s-domain models.

resulting model is shown in Figure 6–5(c). Once again, a constant value of a function in the s-domain is encountered, and the corresponding time-domain form would be an impulse.

The various s-domain forms were given in Figures 6–4 and 6–5, and these results can always be looked up if necessary. However, to assist in remembering these forms, it is recommended that the original equivalent circuit of the charged capacitor in series with a dc voltage and the equivalent circuit of the fluxed inductor in parallel with a dc current be retained as the most basic forms. When these are converted to the s-domain, the sources become the dc values divided by s, one of the most common of all transform forms. The alternate forms can be determined by source transformations when they are required, since they are, in a sense, less natural.

EXAMPLE 6–2

Show two s-domain models for the charged capacitor shown in Figure 6–6(a).

Solution

As recommended in the text, the Thevenin form is first established as shown in Figure 6–6(b). The Norton model is then determined as shown in (c). Note that the value of the s-domain current source is $(60/s)/(5 \times 10^6/s) = 12 \times 10^{-6}$. Alternately, it can be determined simply as $CV_0 = 0.2 \times 10^{-6} \times 60 = 12 \times 10^{-6}$.

EXAMPLE 6–3

Show two s-domain models for the fluxed inductor shown in Figure 6–7(a).

Solution

In this case, the Norton form is first established, as shown in Figure 6–7(b). The Thevenin model is then determined, as shown in (c). Note that the value of the s-domain voltage source is $(0.05s) \times (0.4/s) = 0.02$. Alternately, it can be determined as $LI_0 = 0.05 \times 0.4 = 0.02$.

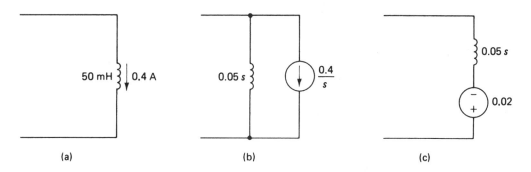

FIGURE 6–7
Fluxed inductor for Example 6–3 and the *s*-domain models.

6–3 COMPLETE CIRCUIT MODELS

We can now bring together all the various transform concepts and models for the purpose of obtaining complete circuit solutions. A general procedure will be formulated in the steps that follow.

1. Transform the complete circuit from the time domain to the *s*-domain. All quantities in the time domain are replaced by *s*-domain quantities as follows:
   a. Any voltage variable $v(t)$ is replaced by a transform variable $V(s)$.
   b. Any current variable $i(t)$ is replaced by a transform variable $I(s)$.
   c. Sources are replaced by their Laplace transforms.
   d. Passive parameters are replaced by their *s*-domain impedances as discussed in Section 6–1.
   e. Initial energy storage in capacitors and inductors is represented by the models developed in Section 6–2. The choice of a voltage or current model would be strongly influenced by the method of analysis planned.

2. Solve for any desired voltage or current variables using the *s*-domain circuit model. In general, standard circuit analysis methods may be employed, and all mathematical manipulations are algebraic in form. However, care must be taken not to lose any *s* variables that appear in various expressions.

3. Using the inverse Laplace transform techniques of Chapter 5, determine the corresponding time-domain forms for the voltages and currents of interest.

The examples that follow will illustrate the process of converting to the *s*-domain and formulating circuit equations.

EXAMPLE 6–4

The switch in the circuit of Figure 6–8(a) is closed at $t = 0$. One of the capacitors is initially charged, and one of the inductors is initially fluxed. The initial conditions shown are the values at $t = 0$. Determine a complete *s*-domain model that could

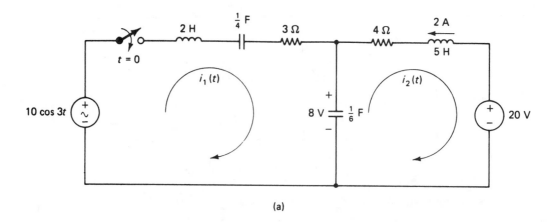

(a)

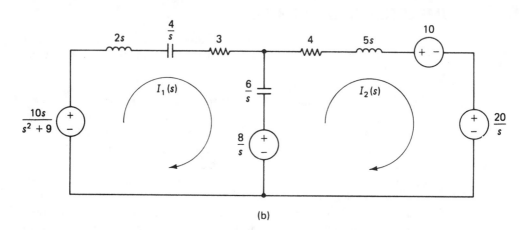

(b)

FIGURE 6-8
Circuit for Example 6-4 and its s-domain model.

be used for solving any circuit response for $t > 0$, and write a pair of simultaneous s-domain mesh current equations.

Solution
Refer to the s-domain circuit of Figure 6–8(b) in the discussion that follows. The time-domain mesh current variables $i_1(t)$ and $i_2(t)$ are replaced by transform current variables $I_1(s)$ and $I_2(s)$. The uncharged $\frac{1}{4}$-F capacitor is represented by an impedance $4/s$, and the unfluxed 2-H inductor is represented by an impedance $2s$. Since mesh analysis to be performed, Thevenin models for the two components with initial energy storage are desired. The $\frac{1}{6}$-F capacitor charged to 8 V is represented by an impedance $6/s$ in series with a voltage $8/s$. However, the 5-H inductor fluxed to 2 A is represented by an impedance $5s$ in series with an s-domain voltage source of

value $5 \times 2 = 10$. The source $10 \cos 3t$ is transformed to $10s/[s^2 + (3)^2] = 10s/(s^2 + 9)$, and the 20-V source is transformed to $20/s$.

The s-domain mesh current equations may now be written by the basic process discussed in Chapter 2. The equations are obtained as follows:

Mesh 1

$$-\frac{10s}{s^2 + 9} + 2sI_1(s) + \frac{4}{s}I_1(s) + 3I_1(s) + \frac{6}{s}[I_1(s) - I_2(s)] + \frac{8}{s} = 0 \qquad \textbf{(6–14)}$$

Mesh 2

$$-\frac{8}{s} + \frac{6}{s}[I_2(s) - I_1(s)] + 4I_2(s) + 5sI_2(s) + 10 + \frac{20}{s} = 0 \qquad \textbf{(6–15)}$$

These two equations may be regrouped in forms more applicable for solution. They are

$$\left(2s + 3 + \frac{10}{s}\right)I_1(s) - \frac{6}{s}I_2(s) = \frac{10s}{s^2 + 9} - \frac{8}{s} \qquad \textbf{(6–16)}$$

$$-\frac{6}{s}I_1(s) + \left(5s + \frac{6}{s} + 4\right)I_2(s) = -10 - \frac{12}{s} \qquad \textbf{(6–17)}$$

These equations could be solved algebraically for either $I_1(s)$ or $I_2(s)$. Our interest has been in establishing the model and formulating the equations, so this circuit will not be pursued further.

EXAMPLE 6–5
The switch in the circuit of Figure 6–9(a) is closed at $t = 0$. One of the capacitors is initially charged, and one of the inductors is initially fluxed. The initial conditions shown are the values at $t = 0$. Determine a complete s-domain model that could be used for solving any circuit response for $t > 0$, and write a pair of simultaneous s-domain node voltage equations.

Solution
Refer to the s-domain circuit of Figure 6–9(b) in the discussion that follows. The time-domain node voltage variables $v_1(t)$ and $v_2(t)$ are replaced by transform voltage variables $V_1(s)$ and $V_2(s)$. The uncharged 1/6-F capacitor is represented by an impedance $6/s$, and the unfluxed 5-H inductor is represented by an impedance $5s$. Since node analysis is to be performed, Norton models for the two components with initial energy storage are desired. The 2-H inductor fluxed to 4 A is represented by an impedance $2s$ in parallel with a current source $4/s$. The 1/2-F capacitor charged to 6 V is represented by an impedance $2/s$ in parallel with an s-domain current source of value $(1/2) \times 6 = 3$. The 7-A source is transformed to $7/s$, and the source $5 \sin 4t$ is transformed to $5 \times 4/[s^2 + (4)^2] = 20/(s^2 + 16)$.

The s-domain node voltage equations may now be written by the basic process discussed in Chapter 2. The equations are obtained as follows:

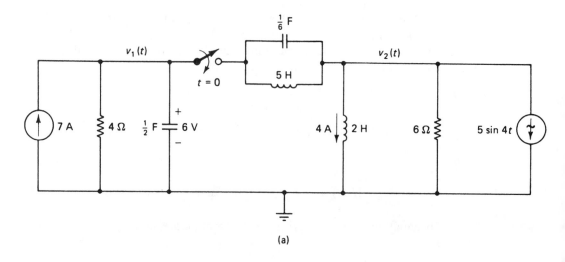

(a)

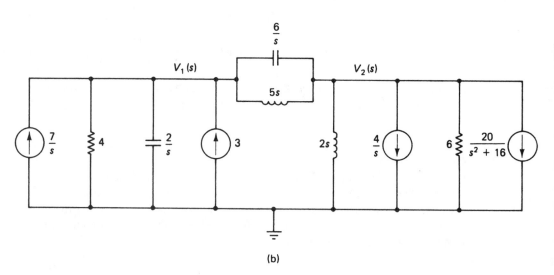

(b)

FIGURE 6–9
Circuit for Example 6–5 and its *s*-domain model.

Node 1

$$-\frac{7}{s} + \frac{V_1(s)}{4} + \frac{V_1(s)}{2/s} - 3 + \frac{V_1(s) - V_2(s)}{5s} + \frac{V_1(s) - V_2(s)}{6/s} = 0 \qquad \textbf{(6–18)}$$

Node 2

$$\frac{V_2(s) - V_1(s)}{5s} + \frac{V_2(s) - V_1(s)}{6/s} + \frac{V_2(s)}{2s} + \frac{4}{s} + \frac{V_2(s)}{6} + \frac{20}{s^2 + 16} = 0 \qquad \textbf{(6–19)}$$

These two equations may be regrouped as follows:

$$\left(\frac{2}{3}s + \frac{1}{4} + \frac{1}{5s}\right) V_1(s) - \left(\frac{s}{6} + \frac{1}{5s}\right) V_2(s) = 3 + \frac{7}{s} \tag{6-20}$$

$$-\left(\frac{s}{6} + \frac{1}{5s}\right) V_1(s) + \left(\frac{s}{6} + \frac{1}{6} + \frac{7}{10s}\right) V_2(s) = -\frac{4}{s} - \frac{20}{s^2 + 16} \tag{6-21}$$

These equations could be solved algebraically for either $V_1(s)$ or $V_2(s)$.

6-4 GENERAL FORMS FOR SOLUTIONS

From the mathematical concepts of the Laplace transform in Chapter 5 and the circuit models given earlier in this chapter, we have seen how a complete circuit analysis could be performed in the s-domain. When a given transform voltage or current is converted back to the time domain, the function obtained is a mathematical description of that variable as a function of time. Depending on the desired end result, a function could be plotted or certain properties could be deduced directly from the form of the function.

Although Laplace transform methods can theoretically be used to obtain general responses for a linear circuit of any order, a practical limitation is imposed by the unwieldy nature of the transform mathematics involved. Digital computer programs such as PSPICE are available for modeling complex circuits, in which numerical solutions to the differential equations are closely approximated by special algorithms. The results provide circuit responses on a point-by-point basis, rather than as mathematical functions.

Transform methods are best suited to circuits of moderate complexity, in which the transform polynomials either are directly factorable, or in which the results automatically appear in partially factored forms. However, the concepts and insights obtained from the transform models are applicable to circuits of any complexity, so many aspects of the transform method have rather general applicability.

In the remainder of this chapter, complete solutions for a number of circuits will be performed using transform techniques. The purpose is twofold. First, it will provide additional practice and skill in transform methods. Second, the general properties of some common circuits can be established. Thus, the reader should be alert to observe how different response forms can be obtained with different circuits.

Natural and Forced Responses

Let us now consider some properties that apply to all circuit responses. Let $y(t)$ represent some arbitrary general circuit response (either a voltage or a current). When the circuit is excited by one or more sources, a general response $y(t)$ may be represented by the sum of two responses as follows:

$$y(t) = y_n(t) + y_f(t) \tag{6-22}$$

The component $y_n(t)$ is called the *natural response,* and $y_f(t)$ is called the *forced response.* The Laplace transform approach provides the total resonse $y(t)$. We will investigate means by which the two components can be determined from a general response and the significance of each.

Form of Natural Response

The form of the natural response is a function of the actual circuit parameters, and it displays the natural time constants and frequencies of the circuit. For example, if the circuit has a time constant of 2 seconds, corresponding to an exponential $e^{-t/2}$, such a term will appear in the response when the circuit is excited by any type of source. The magnitude and sign of the exponential will depend on the type of excitation and its magnitude, but the exponential will always appear in the solution.

The form of the forced response, on the other hand, is a function of the type or types of sources exciting the circuit. For example, if a circuit is excited by a sinusoid having a frequency of 5 rad/s, the general response will always contain a sinusoid with a frequency of 5 rad/s.

To summarize, *the form of the natural response is determined by the circuit parameters, and the form of the forced response is determined by the nature of the excitation source or sources.* The magnitudes and signs of each, however, are dependent on interactions between the different parameters and sources.

Transient and Steady-State Responses

In many common circuits, the natural response term will eventually approach zero, whereas the forced response continues indefinitely. In this situation, it is convenient to redefine the general response $y(t)$ in the following manner:

$$y(t) = y_t(t) + y_{ss}(t) \tag{6–23}$$

The function $y_t(t)$ is called the *transient response,* and it is the same as the natural response in this case. The function $y_{ss}(t)$ is the *steady-state response,* and it is the forced response in this case.

Frequently in this usage, transient response and natural response are considered to be equivalent, and steady-state response and forced response are considered to be equivalent. Strictly speaking, however, the terms *natural* and *forced* are more general, and deal with the basic mathematical forms. The terms *transient* and *steady-state* relate to the common case where the natural response is transient in nature and eventually vanishes, whereas the forced response continues as a steady-state condition indefinitely.

EXAMPLE 6–6

A particular circuit is excited by the single voltage source $v_s(t) = 10 \sin 1000t$. The general solution for a particular current $i(t)$ of interest is

$$i(t) = 2e^{-t} - 3e^{-2t} + 4 \sin (1000t + 30°) \tag{6–24}$$

(a) Determine the natural response and the forced response, and (b) comment on the possible use of the terms *transient response* and *steady-state response,* and define these terms, if appropriate.

Solution

(a) Let $i_n(t)$ and $i_f(t)$ represent the natural and forced responses respectively. Since we are given that the only source is a sinusoid, it is easier to determine the forced response first, since it must have the same form. By inspection,

$$i_f(t) = 4\sin(1000t + 30°) \tag{6-25}$$

The remaining portion of the total response must represent the natural response, and it is

$$i_n(t) = 2e^{-t} - 3e^{-2t} \tag{6-26}$$

(b) Both terms in the natural response approach zero with increasing time, so the natural response is transient in nature. Simultaneously, the forced response continues indefinitely, so it is a steady-state form. It would be appropriate for this example, then, to define a transient response $i_t(t)$ and a steady-state response $i_{ss}(t)$ as

$$i_t(t) = i_n(t) = 2e^{-t} - 3e^{-2t} \tag{6-27}$$
$$i_{ss}(t) = i_f(t) = 4\sin(1000t + 30°) \tag{6-28}$$

6-5   FIRST-ORDER CIRCUITS REVISITED

First-order circuits with dc or step inputs were considered in Chapter 4. Since the voltage or current response in that case is always an exponential term plus a constant, the form of the solution could always be predicted by a simple inspection process. In the context of the discussion of Section 6–4, *the exponential term represents the natural response, and the constant represents the forced response.* Solution of a first-order circuit by inspection is generally limited to the case of dc excitations (plus initial conditions).

First-Order Circuit with Arbitrary Input

We are now in a position to consider the solution of first-order circuits with arbitrary source inputs. Such circuits arise frequently in practical problems, and the determination of voltage and current responses with standard signal sources is generally straightforward. The types of circuit responses will be illustrated with the examples that follow this discussion. For first-order circuits, the natural response $y_n(t)$ will always be an exponential term of the form

$$y_n(t) = Ke^{-t/\tau} = Ke^{-\alpha t} \tag{6-29}$$

where K is a constant, τ is the time constant for the particular circuit, and $\alpha = 1/\tau$ is the damping factor. Since the exponential term approaches zero as time increases, it is proper to designate the natural response as a transient response whenever the forced response continues indefinitely.

EXAMPLE 6–7

The switch in the circuit of Figure 6–10(a) is closed at $t = 0$, and the capacitor is initially uncharged. Using Laplace transform techniques, determine the current $i(t)$ and the voltage $v_C(t)$ for $t > 0$.

Solution

The presence of one resistor and one capacitor clearly establishes the circuit as a first-order circuit. Since the source voltage is a sinusoid, however, the simple inspection procedure of Chapter 4 is not adequate to solve the problem. The transform procedure works quite nicely, as we are about to see.

Using procedures developed in earlier sections, the s-domain model is established as shown in Figure 6–10(b). The circuit is a single-loop form, and a loop

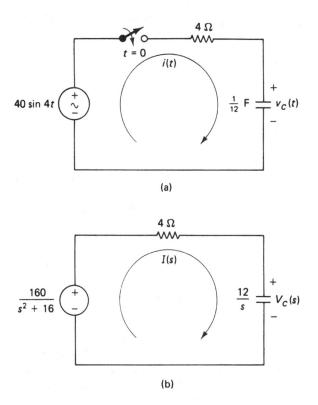

(a)

(b)

FIGURE 6–10

Circuit for Example 6–7 and its s-domain model.

equation may be readily expressed as

$$-\frac{160}{s^2 + 16} + 4I(s) + \frac{12}{s}I(s) = 0 \qquad \textbf{(6-30)}$$

Solution for $I(s)$ and some simplification result in

$$I(s) = \frac{160s}{(4s + 12)(s^2 + 16)} \qquad \textbf{(6-31)}$$

For transform inversion, it is best to force the coefficient of the highest degree s-term in each factor to be unity. This is achieved by factoring out 4 from the $(4s + 12)$ factor and cancelling it with the numerator. This results in

$$I(s) = \frac{40s}{(s + 3)(s^2 + 16)} \qquad \textbf{(6-32)}$$

The function $I(s)$ can be inverted by the procedures established in Chapter 5. The first-order real pole -3 results in an exponential term, and the pair of first-order purely imaginary poles $\pm j4$ results in a sinusoidal term. The reader is invited to show that the inverse transform of Equation (6–32) is

$$i(t) = -4.8e^{-3t} + 8\sin(4t + 36.87°) \qquad \textbf{(6-33)}$$

Before commenting on this solution, the capacitor voltage will be determined. Working again in the s-domain, the capacitor voltage $V_C(s)$ is

$$V_C(s) = \frac{12}{s} \times I(s) = \frac{480}{(s + 3)(s^2 + 16)} \qquad \textbf{(6-34)}$$

where $I(s)$ from Equation (6–32) was used in determining Equation (6–34). The function $V_C(s)$ may be inverted in the same manner as $I(s)$, and the reader is invited to show that the result is

$$v_C(t) = 19.2e^{-3t} + 24\sin(4t - 53.13°) \qquad \textbf{(6-35)}$$

Once the time-domain solution for the current was determined as given by Equation (6–33), an alternate way to determine $v_C(t)$ is by the time-domain relationship, but the transform approach has been emphasized here for convenience.

Compare now the form of $i(t)$ in Equation (6–33) and the form of $v_C(t)$ in Equation (6–35). Each function has an exponential term with a damping factor $\alpha = 3$, which corresponds to a time constant $\tau = 1/3$ s. Observation of the original circuit certainly indicates that the time constant is $\tau = 4 \times (1/12) = 1/3$ s. The exponential term for each function is the *natural* response, which could also be defined as a *transient* response in each case.

Each function also has a sinusoidal term with a frequency of 4 rad/s, which is a result of the fact that the circuit is excited with a sinusoid of the same frequency. The phase angles of the voltage and current differ from that of the source, but the general sinusoidal form and frequency are preserved. The sinusoidal term for each function is the *forced* response, which could also be defined as a *steady-state* response in each case.

An interesting deduction can be made by evaluating $i(t)$ and $v_C(t)$ at $t = 0$. From Equations (6–33) and (6–35), we have

$$i(0) = -4.8 + 8 \sin 36.87° = -4.8 + 4.8 = 0 \qquad \text{(6–36)}$$
$$v_C(0) = 19.2 + 24 \sin(-53.13°) = 19.2 - 19.2 = 0 \qquad \text{(6–37)}$$

In this example, both the current and capacitor voltage are zero at the instant after the switch is closed. Do these results make sense? The answer is yes, because the sinusoidal source has a value of zero volts at the instant the switch is closed, and the capacitor is initially uncharged.

EXAMPLE 6–8

Consider the circuit of Figure 6–11(a), which is the same as that of Figure 6–10(a) as analyzed in Example 6–7, except that the capacitor is initially charged to 16 V with the polarity shown at $t = 0$. Determine $i(t)$ and $v_C(t)$ in this case.

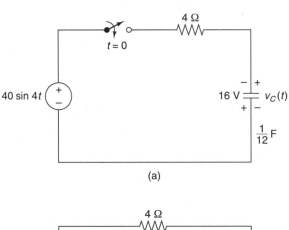

(a)

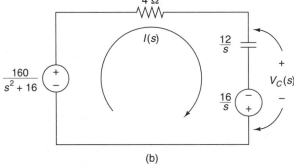

(b)

FIGURE 6–11
Circuit for Example 6–8 and its *s*-domain model.

Solution

The *s*-domain model for the case is shown in Figure 6–11(b). Observe the additional source model representing the initial voltage on the capacitor. Observe also that this fictitious source must be included in the equivalent capacitor terminals in calculating the capacitor voltage as indicated.

For this circuit, the loop equation becomes

$$-\frac{160}{s^2 + 16} + 4I(s) + \frac{12}{s} I(s) - \frac{16}{s} = 0 \tag{6-38}$$

Solution for $I(s)$ and several simplifications yield

$$I(s) = \frac{40s}{(s + 3)(s^2 + 16)} + \frac{4}{s + 3} \tag{6-39}$$

In this form, the first term is recognized as the same as Equation (6–32), which was the case when the capacitor was initially uncharged. The second term then represents an additional component of current produced by the initial capacitor voltage, and its inverse transform is simply $4e^{-3t}$. In fact, the form of Equation (6–39) is the same as would be obtained by using the superposition principle and considering separately the effects of the two sources. One could combine all terms over the same denominator before inversion if desired, but there is no advantage in doing that. Combining the additional exponential term with Equation (6–33), the complete current solution is

$$i(t) = -0.8e^{-3t} + 8 \sin (4t + 36.78°) \tag{6-40}$$

The transform capacitor voltage $V_C(s)$ is

$$V_C(s) = \frac{12}{s} \times I(s) - \frac{16}{s} = \frac{480}{(s + 3)(s^2 + 16)} + \frac{48}{s(s + 3)} - \frac{16}{s} \tag{6-41}$$

in which the current $I(s)$ of Equation (6–39) was substituted. The first term in Equation (6–41) is the same form as Equation (6–34), which corresponded to an uncharged capacitor. Let $V_{C1}(s)$ represent the middle term in Equation (6–41), which is

$$V_{C1}(s) = \frac{48}{s(s + 3)} \tag{6-42}$$

This function has two first-order real poles, (including $s = 0$), and the inverse transform can be readily shown to be

$$v_{C1}(t) = 16 - 16e^{-3t} \tag{6-43}$$

The last term in Equation (6–41) inverts to the constant value -16, which will cancel with the value 16 in Equation (6–43) when all terms are combined. The complete voltage is obtained from the combination of Equations (6–35) and (6–43), and the additional value of -16. This function is

$$v_C(t) = 3.2e^{-3t} + 24 \sin (4t - 53.13°) \tag{6-44}$$

Let us now evaluate the current and capacitor voltage at $t = 0$. From Equations (6–40) and (6–44), we have

$$i(0) = -0.8 + 8 \sin 36.87° = -0.8 + 4.8 = 4 \text{ A} \tag{6-45}$$
$$v_C(0) = 3.2 + 24 \sin (-53.13°) = 3.2 - 19.2 = -16 \text{ V} \tag{6-46}$$

Do the results now seem reasonable?

Since the voltage on the capacitor cannot change instantaneously, the initial voltage of -16 V is certainly proper. The source voltage initially has zero voltage, and the initial current should be 16 V/4 Ω = 4 A, which is the case.

EXAMPLE 6–9

The switch in the circuit of Figure 6–12(a) is closed at $t = 0$, and the inductor is initially unfluxed. Using Laplace transform techniques, determine the current $i(t)$ for $t > 0$.

Solution

The s-domain model for the circuit is shown in Figure 6–12(b). Although a loop equation could be readily written, we will choose in this example to use the impedance concept. The series impedance $Z(s)$ "seen" by the source is

$$Z(s) = 3s + 6 \tag{6-47}$$

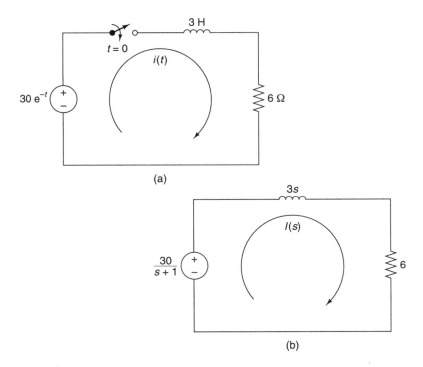

(a)

(b)

FIGURE 6–12
Circuit for Example 6–9 and its s-domain model.

The transform current $I(s)$ is determined by dividing the transform of the source voltage by the impedance. Thus,

$$I(s) = \frac{30/(s+1)}{Z(s)} = \frac{30/(s+1)}{3s+6} \quad\quad (6\text{--}48)$$

$$= \frac{30}{(3s+6)(s+1)} = \frac{10}{(s+1)(s+2)}$$

where the factor of one s term was factored out and cancelled with the numerator in the last step.

The time domain current $i(t)$ consists of two exponential terms, and it can be determined to be

$$i(t) = 10e^{-t} - 10e^{-2t} \quad\quad (6\text{--}49)$$

The term containing e^{-2t} represents the natural response, since it arises from the time constant of the circuit, that is, $\tau = L/R = 3/6 = 1/2$ s. The term containing e^{-t}, however, represents the forced response, since it is of the form of the source. Although the natural response term is transient in nature, the forced response also approaches zero with increasing time, so the terms *transient* and *steady-state* will not be used here. Actually, the complete response is of a transient nature.

EXAMPLE 6—10
Consider the circuit of Figure 6–13, which is the same as that of Figure 6–12 as analyzed in Example 6–9, except for one parameter. The damping factor for the exponential source is now 2 (corresponding to a time constant of 0.5 s). Repeat the analysis of Example 6–9.

Solution
At first glance, this problem may appear to be just another version of Example 6–9, but the one parameter change has a significant effect on the final form of the solution. The transform circuit model is shown in Figure 6–13(b). The current $I(s)$ in this case is

$$I(s) = \frac{30/(s+2)}{3s+6} = \frac{10}{(s+2)^2} \qu\quad (6\text{--}50)$$

The denominator of Equation (6–50) has a pole of second order. The time constant of the exponential source is the same as the time constant of the circuit, so a double-order effect appears.

The function of Equation (6–50) is inverted with the aid of (T–9) of Table 5–1. The function obtained is

$$i(t) = 10te^{-2t} \quad\quad (6\text{--}51)$$

In a sense, the natural and forced responses merge together, since they are of the same form.

FIGURE 6–13

Circuit for Example 6–10 and its *s*-domain model.

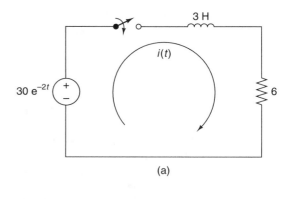

(a)

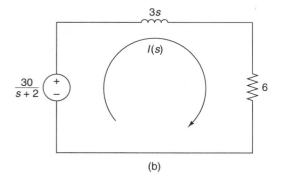

(b)

6–6   SECOND-ORDER CIRCUITS

Second-order circuits are of special interest because they are capable of displaying, on a simpler scale, the types of responses that appear in circuits of arbitrary order. In fact, second-order circuits and systems occur frequently in practical applications, so their behavior is a subject of considerable interest.

We will use one particular second-order circuit as a frame of reference in the development that follows. Many of the general properties that will be deduced apply to all second-order circuits. However, some of the detailed relationships involving circuit parameters for the particular circuit will change when the circuit configuration is modified. Thus, an attempt will be made to distinguish between general properties and those results peculiar to the particular configuration. Through examples at the end of the section and problems at the end of the chapter, other examples of second-order circuits will be considered.

Series *RLC* Circuit

The particular second-order circuit of interest for our development is the series *RLC* circuit of Figure 6–14(a). The particular excitation of interest is a dc (or step function) voltage of magnitude V_i, which is applied at $t = 0$. When the input is a dc voltage switched on at a given time, any resulting response is referred to as the

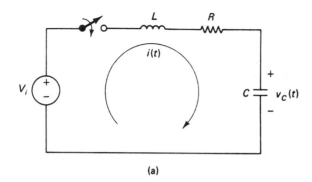

(a)

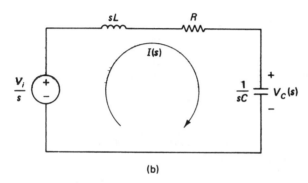

(b)

FIGURE 6–14
Series *RLC* circuit used in second-order development and its *s*-domain model.

step response of the circuit. The two variables of special interest are the loop current $i(t)$ and the capacitor voltage $v_C(t)$. The circuit is assumed to be initially relaxed.

The *s*-domain model is shown in Figure 6–14(b). A loop equation for the circuit reads

$$-\frac{V_i}{s} + sLI(s) + RI(s) + \frac{1}{sC}I(s) = 0 \qquad \textbf{(6–52)}$$

Solution for $I(s)$ and some simplification lead to

$$I(s) = \frac{V_i/L}{s^2 + s\dfrac{R}{L} + \dfrac{1}{LC}} \qquad \textbf{(6–53)}$$

The capacitor voltage $V_C(s)$ is

$$V_C(s) = \frac{1}{sC}I(s) = \frac{V_i/LC}{s\left(s^2 + s\dfrac{R}{L} + \dfrac{1}{LC}\right)} \qquad \textbf{(6–54)}$$

Three Possible Forms

Before attempting to invert either $I(s)$ or $V_C(s)$, some further inspection of the quadratic factor appearing in the denominator of each function is required. Let s_1 and s_2 represent the roots of this factor, which are determined by equating the quadratic to zero. The roots are determined by the quadratic formula to be

$$\begin{matrix} s_1 \\ s_2 \end{matrix} = -\frac{R}{2L} \pm \sqrt{\frac{R^2}{4L^2} - \frac{1}{LC}} \qquad \text{(6–55)}$$

There are three possibilities for the roots s_1 and s_2.

1. *The roots are real and different.* For this condition to be true,

$$\frac{R^2}{4L^2} > \frac{1}{LC} \qquad \text{(6–56)}$$

This will be referred to as the *overdamped* case.

2. *The roots are real and equal.* For this condition to be true,

$$\frac{R^2}{4L^2} = \frac{1}{LC} \qquad \text{(6–57)}$$

This will be referred to as the *critically damped* case.

3. *The roots are complex.* For this condition to be true,

$$\frac{R^2}{4L^2} < \frac{1}{LC} \qquad \text{(6–58)}$$

This will be referred to as the *underdamped* case.

We will next consider the three cases in some detail. While we will focus on the series RLC circuit as a representative example as mentioned earlier, most of the general results to be developed apply to many second-order circuits (and systems). We will leave some functions with unspecified constants where the exact constants would be rather unwieldy in general terms, since the goal is to develop general concepts rather than specific values at this point.

Overdamped Case

The two poles are real and different, so assume that these poles are $s_1 = -\alpha_1$ and $s_2 = -\alpha_2$. The exact forms for s_1 and s_2 in terms of the circuit parameters were given by Equation (6–55).

The forms for $I(s)$ and $V_C(s)$ can be expressed as

$$I(s) = \frac{V_i/L}{(s + \alpha_1)(s + \alpha_2)} \qquad \text{(6–59)}$$

$$V_C(s) = \frac{V_i/LC}{s(s + \alpha_1)(s + \alpha_2)} \qquad \text{(6–60)}$$

The inverse transforms of Equations (6–59) and (6–60) are of the forms

$$i(t) = A_0 e^{-\alpha_1 t} - A_0 e^{-\alpha_2 t} \qquad (6\text{--}61)$$
$$v_C(t) = V_i + A_1 e^{-\alpha_1 t} + A_2 e^{-\alpha_2 t} \qquad (6\text{--}62)$$

where A_0, A_1, and A_2 are determined from the inversion process. When numbers are given, these constants are easily determined, but they will be left in these arbitrary forms at the present time. The final value V_i in Equation (6–62) has been identified by inspecting the steady-state condition for the capacitor.

An important point about the response functions in the overdamped case is that the natural response consists of two exponential terms, each having a different damping factor or time constant. The forced response for the current is zero, and the forced response for the capacitor voltage is the constant final voltage across the capacitor.

Critically Damped Case

In this case, the two poles are real and equal, so assume that $s_1 = s_2 = -\alpha$. For the series RLC circuit, the value of α is readily determined from Equation (6–55) as $\alpha = R/2L$.

The s-domain voltage and current functions can be expressed as

$$I(s) = \frac{V_i/L}{(s + \alpha)^2} \qquad (6\text{--}63)$$

$$V_C(s) = \frac{V_i/LC}{s(s + \alpha)^2} \qquad (6\text{--}64)$$

The current $I(s)$ can be inverted with the aid of transform pair (T–9) of Table 5–1. The voltage $V_C(s)$, however, requires the use of the special procedure given in Section 5–6. The results are of the form

$$i(t) = C_0 t e^{-\alpha t} = \frac{V_i t}{L} e^{-Rt/2L} \qquad (6\text{--}65)$$

$$v_C(t) = V_i + (C_1 t + C_2) e^{-Rt/2L} \qquad (6\text{--}66)$$

The constants $C_0 = V_i/L$ and $\alpha = R/2L$ for the series RLC circuit have been explicitly substituted in the current expression due to their simplicity. The constants C_1 and C_2 in the voltage expression form are retained as arbitrary, since their particular forms are not of interest at this particular time. The final value V_i for the voltage has been recognized and expressed explicitly.

The most significant aspect of the natural response function for the critically damped case is the $te^{-\alpha t}$ form. This type of response is characteristic for a critically damped system, and it is the only type of natural response term for the current. Although the t factor increases with increasing t, the $e^{-\alpha t}$ decreases at a faster rate, so the product eventually approaches zero. The forced response for the current is zero, and the forced response for the capacitor voltage is the final constant value V_i.

Underdamped Response

The two poles are complex, and they will be denoted as

$$s_1 = -\alpha + j\omega_d \quad \text{and} \quad s_2 = -\alpha - j\omega_d$$

From Equation 6–55), with the recognition that the roots are complex, we have for the series RLC circuit:

$$\alpha = \frac{R}{2L} \tag{6–67}$$

$$\omega_d = \sqrt{\frac{1}{LC} - \left(\frac{R}{2L}\right)^2} \tag{6–68}$$

The quantity α is the damping factor, and ω_d is called the damped natural oscillation frequency. The choice of the adjective *damped* and the addition of the subscript d to the frequency will be explained shortly. The forms for $I(s)$ and $V_C(s)$ can be expressed as

$$I(s) = \frac{V_i/L}{(s + \alpha)^2 + \omega_d^2} \tag{6–69}$$

$$V_C(s) = \frac{V_i/LC}{s\left(s^2 + s\dfrac{R}{L} + \dfrac{1}{LC}\right)} \tag{6–70}$$

The current was written in the form for completion of the square, since this transform can by quickly inverted with (T–5) of Table 5–1. The result is

$$i(t) = \frac{V_i}{\omega_d L} e^{-\alpha t} \sin \omega_d t \tag{6–71}$$

The inverse transform for the voltage is somewhat more involved and will be of the form

$$v_C(t) = V_i + Be^{-\alpha t} \sin (\omega_d t + \theta) \tag{6–72}$$

The constants B and θ in the voltage expression are retained as arbitrary because their particular forms are not of interest at this particular time. As in previous cases, the final value V_i for the voltage has been recognized from steady-state conditions.

The significant fact for the underdamped case is that the natural response is oscillatory. Depending on the value of α, this response may damp out very quickly, or it may continue for a reasonable period of time.

The instantaneous forms of the current and voltage step response functions for the three cases will now be compared. Although the exact shapes and absolute levels of the functions will depend on the parameter and source values, as well as possible initial conditions, some general trends are noteworthy.

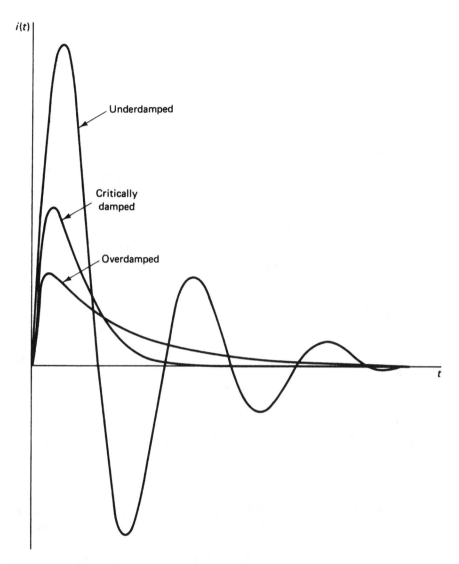

FIGURE 6–15
Typical current response forms for second-order series *RLC* circuit.

Comparison of Response Forms

Typical forms of the current for the three cases are compared in Figure 6–15. A typical overdamped current will increase at a moderately fast rate, but will decrease back toward zero at a sluggish rate. At first glance, the critically damped response appears similar to the overdamped response. However, it is not as sluggish, approaching zero at a much more rapid rate. Neither the overdamped nor critically

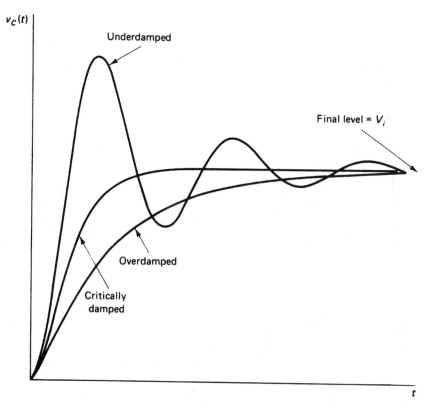

FIGURE 6–16
Typical capacitor voltage response forms for second-order series *RLC* circuit.

damped current responses go negative. Indeed, the critically damped response represents the transition where the current does not quite become negative. The underdamped response does become negative in certain intervals and oscillates as shown.

Typical forms of the capacitor voltage are shown in Figure 6–16. The overdamped response takes a relatively long time to reach the steady-state voltage level. The critically damped response approaches the final level at a faster rate without overshoot. Finally, the underdamped response overshoots and oscillates before settling to the dc steady-state level.

EXAMPLE 6–11
In the series *RLC* circuit of Figure 6–17(a), the switch is closed at $t = 0$, and the circuit is initially relaxed. Determine the current $i(t)$ and the voltage $v_C(t)$.

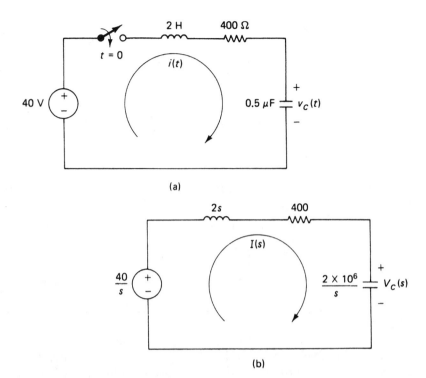

FIGURE 6–17
Circuit for Example 6–11 and its *s*-domain model.

Solution
The *s*-domain model is shown in Figure 6–17(b). The impedance concept will be used in this solution. The transform impedance seen by the source is

$$Z(s) = 2s + 400 + \frac{2 \times 10^6}{s}$$

$$= \frac{2s^2 + 400s + 2 \times 10^6}{s} \qquad (6\text{–}73)$$

The transform current $I(s)$ is

$$I(s) = \frac{40/s}{(2s^2 + 400s + 2 \times 10^6)/s} = \frac{20}{s^2 + 200s + 10^6} \qquad (6\text{–}74)$$

where the coefficient of the highest-degree term was set to unity in the last step.

By factoring the polynomial in the denominator of Equation (6–74), the roots are determined as $s_1, s_2 = -100 \pm j994.987$. The response will thus be *underdamped* and oscillatory. The current transform is most easily inverted by rearranging it in the form of (T–5) of Table 5–1). We have

$$I(s) = \frac{20}{(s + 100)^2 + (994.987)^2} \qquad (6\text{--}75)$$

The time-domain current is

$$i(t) = 0.0201008e^{-100t} \sin 994.987t \qquad (6\text{--}76)$$

The transform of the capacitor voltage $V_C(s)$ is

$$V_C(s) = \frac{2 \times 10^6}{s} \times I(s) = \frac{40 \times 10^6}{s(s^2 + 200s + 10^6)} \qquad (6\text{--}77)$$

From the procedures of Chapter 5, the function $V_C(s)$ can be inverted to determine

$$v_C(t) = 40 + 40.2015e^{-100t} \sin (994.987t - 95.7392°) \qquad (6\text{--}78)$$

The functions $i(t)$ and $v_C(t)$ are shown in Figure 6–18 on the following page.

EXAMPLE 6–12

Consider the series *RLC* circuit of Figure 6–19(a), in which *L* and *C* are fixed, but in which *R* is adjustable. The switch is closed at $t = 0$, and the circuit is initially relaxed. Determine the current $i(t)$ and the voltage $v_C(t)$ for each of the following three possible values of resistance: **(a)** $R = 6\ \Omega$, **(b)** $R = \sqrt{20}\ \Omega$, and **(c)** $R = 2\ \Omega$.

Solution
The general form of the *s*-domain equivalent circuit is shown in Figure 6–19(b) with all parameters except *R* established. We will first obtain expressions for $I(s)$ and $V_C(s)$ with *R* left unspecified. The three values of *R* may then be substituted in the expression.

A loop equation for this circuit reads

$$-\frac{30}{s} + 0.5sI(s) + RI(s) + \frac{10}{s}I(s) = 0 \qquad (6\text{--}79)$$

Solution for $I(s)$ and some simplification yield

$$I(s) = \frac{60}{s^2 + 2Rs + 20} \qquad (6\text{--}80)$$

The capacitor voltage $V_C(s)$ is

$$V_C(s) = \frac{10}{s} \times I(s) = \frac{600}{s(s^2 + 2Rs + 20)} \qquad (6\text{--}81)$$

Each of the three values of resistance will now be considered.

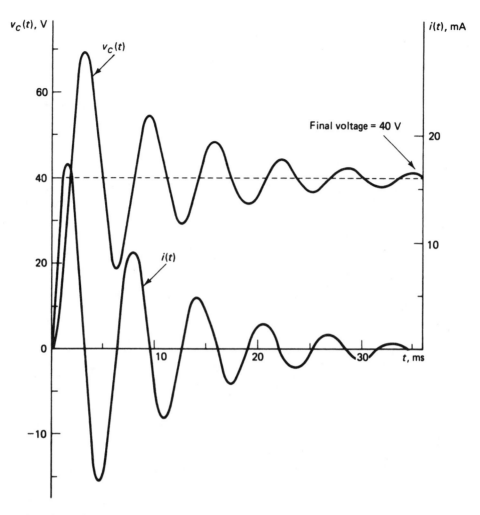

FIGURE 6–18
Current and capacitor voltage for Example 6–11.

(a) $R = 6 \, \Omega$ The current and capacitor voltage are

$$I(s) = \frac{60}{s^2 + 12s + 20} \tag{6–82}$$

$$V_C(s) = \frac{600}{s(s^2 + 12s + 20)} \tag{6–83}$$

The roots of the quadratic in this case are $s_1 = -2$ and $s_2 = -10$ so the quadratic can be expressed as $(s + 2)(s + 10)$. The circuit is thus *overdamped*. By the

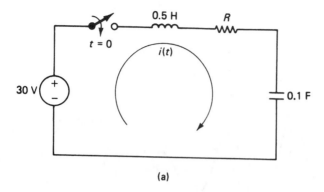

(a)

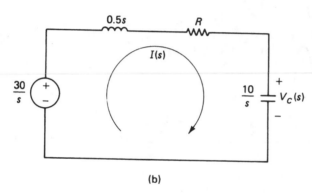

(b)

FIGURE 6–19
Circuit for Example 6–12 and its s-domain model.

methods of Chapter 5, the functions of Equations (6–82) and (6–83) are inverted as

$$i(t) = 7.5e^{-2t} - 7.5e^{-10t} \tag{6-84}$$

$$v_C(t) = 30 - 37.5e^{2t} + 7.5e^{-10t} \tag{6-85}$$

(b) $R = \sqrt{20}\ \Omega$ The current and capacitor voltage are

$$I(s) = \frac{60}{s^2 + 2\sqrt{20}\,s + 20} \tag{6-86}$$

$$V_C(s) = \frac{600}{s(s^2 + 2\sqrt{20}\,s + 20)} \tag{6-87}$$

The roots of the quadratic in this case are $s_1 = s_2 = -\sqrt{20} = 4.47214$, so the quadratic can be expressed as $(s + \sqrt{20})^2 = (s + 4.47214)^2$. The circuit in this case is *critically damped.* Inversion in this case results in

$$i(t) = 60te^{-\sqrt{20}t} \tag{6-88}$$

$$v_C(t) = 30 - 60\sqrt{5}\,te^{-\sqrt{20}t} - 30e^{-\sqrt{20}t} \tag{6-89}$$

(c) $R = 2\,\Omega$ The current and capacitor voltage are

$$I(s) = \frac{60}{s^2 + 4s + 20} \tag{6-90}$$

$$V_C(s) = \frac{600}{s(s^2 + 4s + 20)} \tag{6-91}$$

The roots of this quadratic are $s_1, s_2 = 2 \pm j4$. The circuit is thus *underdamped.* Inversion in this case yields

$$i(t) = 15e^{-2t} \sin 4t \tag{6-92}$$

$$v_C(t) = 30 + 33.54102e^{-2t} \sin(4t - 116.565°) \tag{6-93}$$

EXAMPLE 6–13

The series *RLC* circuit of Figure 6–20(a) is excited at $t = 0$ by the sinusoidal source shown, and the capacitor is initially uncharged. Determine the current $i(t)$ for $t > 0$.

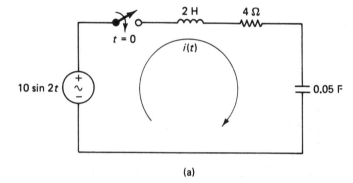

(a)

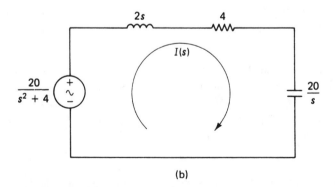

(b)

FIGURE 6–20
Circuit for Example 6–13 and its *s*-domain model.

Solution

The *s*-domain equivalent circuit is shown in Figure 6–20(b). A loop equation for the circuit reads

$$-\frac{20}{s^2 + 4} + 2sI(s) + 4I(s) + \frac{20}{s} I(s) = 0 \qquad \textbf{(6–94)}$$

Solution for $I(s)$ and same rearrangement results in

$$I(s) = \frac{20s}{(2s^2 + 4s + 20)(s^2 + 4)} = \frac{10s}{(s^2 + 2s + 10)(s^2 + 4)} \qquad \textbf{(6–95)}$$

where the factor of 2 for the s^2 term in one of the quadratics was factored out and cancelled in the last step.

The roots of the quadratic $s^2 + 2s + 10$ are $-1 \pm j3$, so the circuit is underdamped. In addition, the roots of the quadratic $s^2 + 4$, which are $\pm j2$, appear in the denominator of Equation (6–95). Complete inversion of Equation (6–95) leads to

$$1.46176e^{-t}\sin(3t - 127.875°) + 1.38675 \sin(2t + 56.3099°) \qquad \textbf{(6–96)}$$

The first term in Equation (6–96) is the natural response, and the second term is the forced response. The natural response is also the transient response, and the forced response is the steady-state response.

6–7 PSPICE EXAMPLES

The PSPICE examples of this chapter will emphasize transient analysis of circuits contained in the text. The application of the SIN transient source will be illustrated, and a second-order transient analysis will be performed.

PSPICE EXAMPLE 6–1

Use PSPICE to analyze the circuit of Example 6–7, (Figure 6–10), and obtain plots of the input and capacitor voltages.

Solution

The circuit adapted to the PSPICE format and the code are shown in Figure 6–21. In this example, the use of the SIN function as a transient souce will be introduced. Following the definition of the voltage *V* connected between nodes 1 and 0, a SIN function is defined. The simplest format for the sine function consists of three entries with the following parameters:

```
SIN  dc offset  amplitude  frequency
```

The "dc offset" represents any dc level added to the sine function. The "amplitude" represents the peak value of the sinusoid, and "frequency" is the frequency in hertz. Other optional parameters will be considered later in the text.

FIGURE 6–21

Circuit and code for PSPICE Example 6–1.

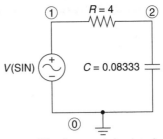

(All values are in basic units.)

```
PSPICE EXAMPLE 6-1
V 1 0 SIN 0 40 0.6366
R 1 2 4
C 2 0 0.08333
.TRAN 1.6 1.6
.PROBE
.END
```

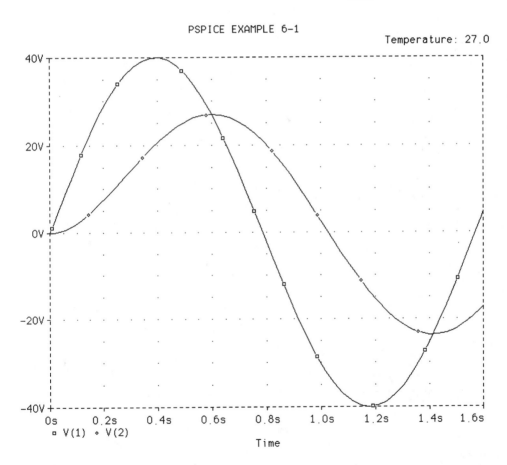

FIGURE 6–22

Waveforms for PSPICE Example 6–1 during early transient phase.

For the case at hand, the SIN part of the code reads

```
SIN 0 40 0.6366
```

The entries 0 and 40 represent, respectively, the dc level and the peak value of the sinusoid. The value 0.6366 is the frequency in hertz or cycles per second. The form of the sine function given is sin 4*t*, and the value 4 represents the radian frequency ω. In general $\omega = 2\pi f$, where f is the cyclic frequency in hertz or cycles per second. Thus, the value required on the code line is $f = \omega/2\pi = 4/2\pi = 0.6366$.

In order to demonstrate more effectively the transient effect, this first run is performed over an internal of 1.6 s, and plots of the input voltage and capacitor voltage are shown in Figure 6–22. In this case, two traces have been placed on the same graph. Whereas the input voltage V(1) is sinusoidal in nature, the capacitor voltage V(2) deviates somewhat from a pure sine function in the first cycle. This

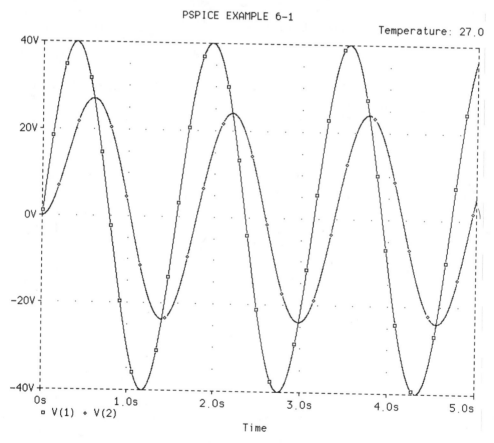

FIGURE 6–23

Waveforms for PSPICE Example 6–1 over longer time interval.

is a result of the natural response's adding to the forced response, which changes the nature somewhat.

A second run over a 5-s interval was taken to better demonstrate the evolution into the forced or steady-state response. After attempting this once with the simplest form of the codes, the plots appeared to be a bit jagged. The transient statement was modified to read

```
.TRAN 5 5 0 1E-3
```

This resulted in a much smoother presentation, and the curves are shown in Figure 6–23. Steady-state conditions are clearly established after a cycle or so, and both voltages continue as sinusoids with different peak values and phase angles.

PSPICE EXAMPLE 6–2

Use PSPICE to analyze the circuit of Example 6–11 (Figure 6–17), and obtain plots of the input and capacitor voltages.

Solution

The circuit adapted to the PSPICE format and the code are shown in Figure 6–24. The input voltage utilizes the PWL function, appropriately defined.

Plots of the input and capacitor voltages obtained with PROBE are shown in Figure 6–25. The capacitor voltage is compatible with the mathematically predicted result of Figure 6–18.

FIGURE 6–24
Circuit and code for PSPICE Example 6–2.

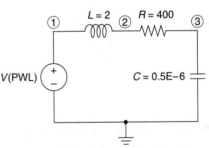

(All values are in basic units.)

```
PSPICE EXAMPLE 6-2
V 1 0 PWL 0 0 10E-6 40
L 1 2 2
R 2 3 400
C 3 0 0.5E-6
.TRAN 50E-3 50E-3 0 10E-6
.PROBE
.END
```

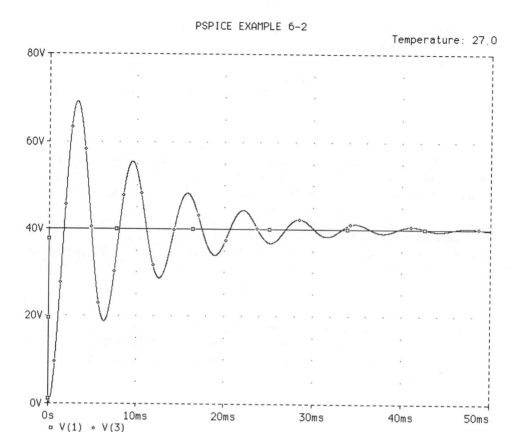

FIGURE 6–25
Waveform for PSPICE Example 6–2.

6–8 MATLAB EXAMPLE

The example that follows is similar to the first example of the preceding chapter, in that it involves the determination of an inverse transform of a Laplace function. However, the function of interest involves the voltage obtained in one of the examples of this chapter.

MATLAB EXAMPLE 6–1
Use MATLAB to determine a plot of the inverse Laplace transform of the capacitor voltage of Example 6–11 (Equation 6–77).

Solution

First, the original equation will be repeated for convenience. It is

$$V_C = \frac{40 \times 10^6}{s(s^2 + 200s + 10^6)} \qquad (6\text{–}97)$$

The denominator is a third-degree polynomial in the variable s. As suggested in MATLAB Example 5–1, it is useful to write (or at least to think of) the function in the form

$$V_C = \frac{0 \cdot s^3 + 0 \cdot s^2 + 0 \cdot s + 40 \times 10^6}{s^3 + 200\, s^2 + 10^6 s + 0} \qquad (6\text{–}98)$$

Refer to the workplace of Figure 6–26 in the discussion that follows. The reader who has followed all MATLAB examples thus far should be able to interpret the various commands without lengthy explanations. First, the numerator and denominator polynomial coefficients, denoted respectively as n and d, are typed and entered. Printing of these row vectors was not suppressed so that they could be easily checked. (They don't occupy much space.) Next, the variable t (for time) was generated at 501 points over an interval from $t = 0$ to $t = 50$ ms, and printing was suppressed. Finally, the inverse Laplace transform, denoted as vc, was generated numerically by the "impulse" command, and printing was suppressed.

The remaining lines were used to plot and label the curves. The "grid" option was employed here, and the results are shown in Figure 6–27. This curve can be compared with the text plot of Figure 6–18 and the PSPICE plot of Figure 6–25.

```
EDU» n=[0 0 0 40e6]

n =

          0          0          0    40000000

EDU» d=[1 200 1e6 0]

d =

          1        200    1000000          0

EDU» t=linspace(0,50e-3,501);
EDU» vc=impulse(n,d,t);
EDU» plot(t,vc)
EDU» xlabel('time, seconds')
EDU» ylabel('voltage, volts')
EDU» title('Capacitor Voltage of Example 6-11')
EDU» grid
EDU»
```

FIGURE 6–26
Workspace for MATLAB Example 6–1.

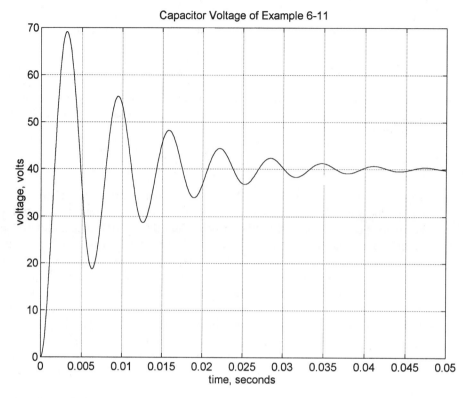

FIGURE 6–27
Plot for MATLAB Example 6–1.

DRILL PROBLEMS

6–1. Draw and label with impedances the s-domain models for the following components: **(a)** a 50-Ω resistor, **(b)** a 0.01-μF uncharged capacitor, and **(c)** a 200-mH unfluxed inductor.

6–2. Draw and label with impedances the s-domain models for the following components: **(a)** a 4.7-kΩ resistor, **(b)** a 5-μF uncharged capacitor, and **(c)** a 500-μH unfluxed inductor.

6–3. Draw and label both the Thevenin and Norton s-domain models for the charged capacitor shown in Figure P6–3.

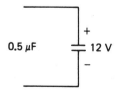

FIGURE P6–3

6-4. Draw and label both the Thevenin and Norton *s*-domain models for the charged capacitor shown in Figure P6–4.

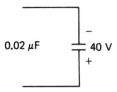

FIGURE P6-4

6-5. Draw and label both the Norton and Thevenin *s*-domain models for the fluxed inductor shown in Figure P6–5.

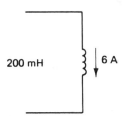

FIGURE P6-5

6-6. Draw and label both the Norton and Thevenin *s*-domain models for the fluxed inductor shown in Figure P6–6.

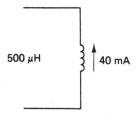

FIGURE P6-6

6-7. The switch in the circuit of Figure P6–7 is closed at $t = 0$. Initial conditions in the circuit are as shown. Determine a complete *s*-domain model for $t > 0$, and write a pair of simultaneous *s*-domain mesh current equations.

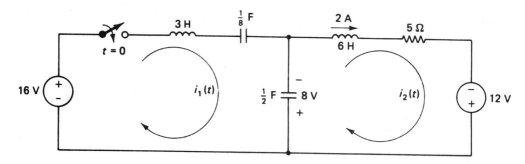

FIGURE P6–7

6–8. The switch in the circuit of Figure P6–8 is closed at $t = 0$. Initial conditions in the circuit are as shown. Determine a complete s-domain model for $t > 0$, and write a pair of simultaneous s-domain mesh current equations.

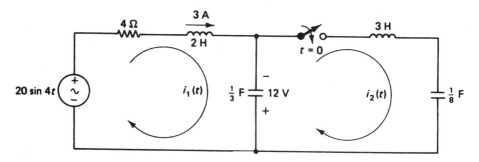

FIGURE P6–8

6–9. The switch in the circuit of Figure P6–9 is closed at $t = 0$. Initial conditions in the circuit are as shown. Determine a complete s-domain model for $t > 0$, and write a pair of simultaneous s-domain node voltage equations.

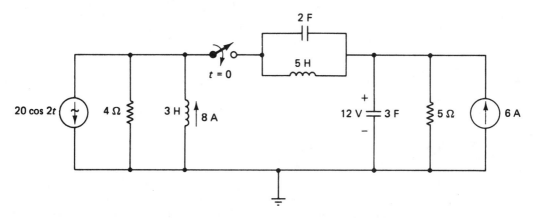

FIGURE P6–9

6–10. The switch in the circuit of Figure P6–10 is closed at $t = 0$. Initial conditions in the circuit are as shown. Determine a complete s-domain model for $t > 0$, and write a pair of simultaneous s-domain node voltage equations.

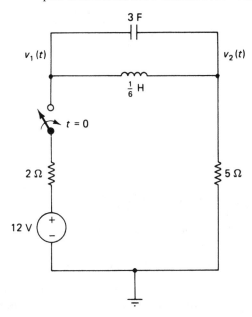

FIGURE P6–10

6–11. Determine the transform impedance $Z(s)$ as viewed from the left-hand terminals of the relaxed circuit of Figure P6–11. Arrange the result as a ratio of polynomials.

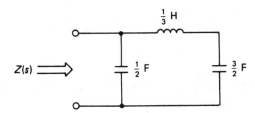

FIGURE P6–11

6–12. Determine the transform impedance $Z(s)$ as viewed from the left-hand terminals of the relaxed circuit of Figure P6–12. Arrange the result as a ratio of polynomials.

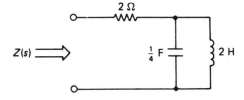

FIGURE P6–12

6–13. The switch in the circuit of Figure P6–13 is closed at $t = 0$, and the capacitor is initially uncharged. Determine a transform Thevenin equivalent circuit as viewed from terminals x–y that would apply for $t > 0$.

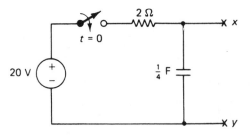

FIGURE P6–13

6–14. Suppose in the circuit of Problem 6–13 that the switch had been closed for a long time, so that steady-state conditions exist. Determine again a transform Thevenin equivalent circuit as viewed from terminals x–y.

6–15. A particular circuit is excited by the single voltage source $v_s(t) = 20 \cos 50t$. The general solution for a particular voltage $v(t)$ of interest is

$$v(t) = 5e^{-20t} \cos(40t + 30°) + 12 \sin(50t + 60°)$$

(a) Determine the natural response and the forced response.

(b) Comment on the possible use of the terms *transient response* and *steady-state response,* and define these terms, if appropriate.

6–16. A particular circuit is excited by the single voltage source $v_s(t) = 20$. The general solution for a particular current $i(t)$ of interest is

$$i(t) = 3e^{-t} + 5e^{-2t} - 8$$

(a) Determine the natural response and the forced response.

(b) Comment on the possible use of the terms *transient response* and *steady-state response,* and define these terms, if appropriate.

6–17. Consider the simple *RC* circuit of Figure P6–17 with a dc input V_s applied at $t = 0$ and an uncharged capacitor. This circuit was analyzed in Chapter 4 using the first-order inspection method. Solve for the current $i(t)$ and capacitor voltage $v_C(t)$ using the Laplace transform methods of this chapter.

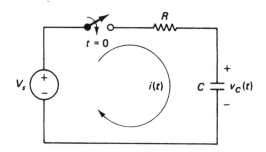

FIGURE P6–17

6-18. Consider the simple *RL* circuit of Figure P6–18 with a dc input V_s applied at $t = 0$ and an unfluxed inductor. This circuit was analyzed in Chapter 4 using the first-order inspection method. Solve for the current $i(t)$ and inductor voltage $v_L(t)$ using the Laplace transform methods of this chapter.

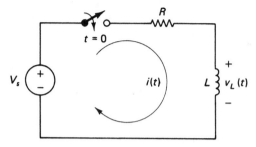

FIGURE P6-18

6-19. The switch in the circuit of Figure P6–19 is closed at $t = 0$, and the capacitor is initially uncharged. **(a)** Using Laplace transform methods, determine the current $i(t)$ and the voltage $v_C(t)$ for $t > 0$, and **(b)** evaluate $i(0)$ and $v_C(0)$, and compare these values with those obtained from the circuit model at $t = 0^+$.

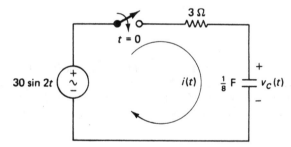

FIGURE P6-19

6-20. The switch in the circuit of Figure P6–20 is closed at $t = 0$, and the capacitor is initially uncharged. **(a)** Using Laplace transform methods, determine the current $i(t)$ and the voltage $v_C(t)$ for $t > 0$, and **(b)** evaluate $i(0)$ and $v_C(0)$, and compare these values with those obtained from the circuit model at $t = 0^+$

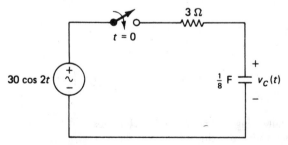

FIGURE P6-20

6–21. Assume in Problem 6–19 that the capacitor is initially charged to a value of 10 V, with the upper terminal positive. Repeat the analysis of Problem 6–19.

6–22. Assume in Problem 6–20 that the capacitor is initially charged to a value of 10 V, with the lower terminal positive. Repeat the analysis of Problem 6–20.

6–23. The switch in the circuit of Figure P6–23 is closed at $t = 0$, and the inductor is initially unfluxed. **(a)** Using Laplace transform methods, determine the current $i(t)$ and the voltage $v_L(t)$ for $t > 0$, and **(b)** evaluate $i(0)$ and $v_L(0)$, and compare these values with those obtained from the circuit model at $t = 0^+$.

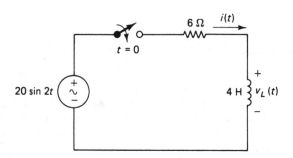

FIGURE P6–23

6–24. The switch in the circuit of Figure P6–24 is closed at $t = 0$, and the inductor is initially unfluxed. **(a)** Using Laplace transform methods, determine the current $i(t)$ and the voltage $v_L(t)$ for $t > 0$, and **(b)** evaluate $i(0)$ and $v_L(0)$, and compare these values with those obtained from the circuit model at $t = 0^+$.

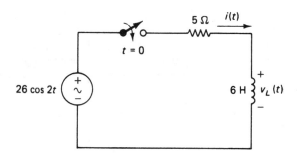

FIGURE P6–24

6–25. In the series RLC circuit of Figure P6–25, the switch is closed at $t = 0$, and the circuit is initially relaxed. Determine the current $i(t)$ and the voltage $v_C(t)$.

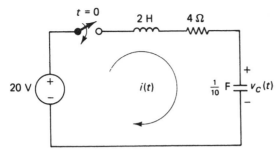

FIGURE P6–25

6–26. In the series RLC circuit of Figure P6–26, the switch is closed at $t = 0$, and the circuit is initially relaxed. Determine the current $i(t)$ and the voltage $v_C(t)$.

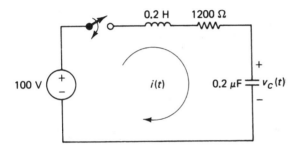

FIGURE P6–26

6–27. In the parallel RLC circuit of Figure P6–27, the dc current source is applied at $t = 0$ (by opening the switch). The circuit is initially relaxed. Determine the voltage $v(t)$.

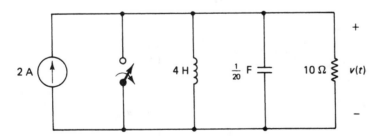

FIGURE P6–27

6–28. In the circuit of Figure P6–28, the switch is closed at $t = 0$, and the circuit is initially relaxed. Determine the voltage $v(t)$. (*Hint:* Use a source transformation to establish a parallel RLC circuit excited by a current source.)

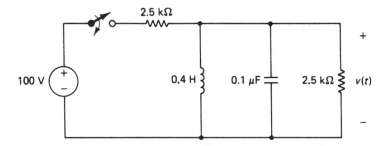

FIGURE P6–28

6–29. For the circuit of Figure P6–29, the switch is closed at $t = 0$, and the circuit is initially relaxed. Determine the current $i(t)$.

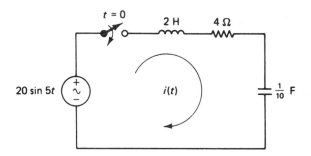

FIGURE P6–29

6–30. For the circuit of Figure P6–30, the switch is closed at $t = 0$, and the circuit is initially relaxed. Determine the current $i(t)$.

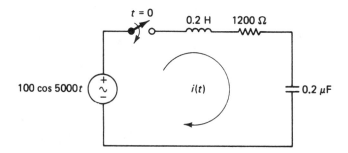

FIGURE P6–30

DERIVATION PROBLEMS

6–31. The derivation of transform capacitive impedance was achieved in the text with operation (O–2) of Table 5–2. Provide an alternate derivation by first starting with the derivative relationship for current in terms of voltage, and then applying operation (O–1) of Table 5–2.

6–32. The derivation of transform inductive impedance was achieved in the text with operation (O–1) of Table 5–1. Provide an alternate derivation by first starting with the integral relationship for current in terms of voltage, and then applying operation (O–2) of Table 5–2.

6–33. Under certain conditions, it is possible for a given initial condition to result in a total cancellation of the natural response. Consider the *RC* circuit of Figure P6–33 with the capacitor initially charged to a voltage V_0 as shown. Assume a sinusoidal excitation

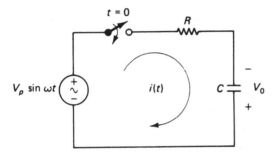

FIGURE P6–33

at $t = 0$. Show that the natural response completely vanishes if the following condition is met:

$$V_0 = \frac{\omega RCV_p}{1 + \omega^2 R^2 C^2}$$

6–34. A parallel *RLC* circuit excited at $t = 0$ by a dc current source of value *I* is shown in Figure P6–34. The circuit is assumed to be initially relaxed. Determine the *s*-domain model and write a single node-pair equation for the circuit. From this equation,

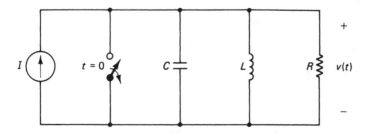

FIGURE P6–34

determine an expression for the transform voltage $V(s)$ across the node-pair. From this information, determine the three conditions for overdamped response, critically damped response, and underdamped response. Finally, write three possible forms for the voltage $v(t)$. This development parallels the development for the series RLC circuit given in the first part of Section 6–6.

6–35. Consider the current response of the critically damped series RLC circuit as given by Equation 6–65. **(a)** Show that the maximum value of the current occurs at a time t_m given by

$$t_m = \frac{2L}{R}$$

(b) Show that the maximum value of the current I_m is given by

$$I_m = 0.7358\frac{V_i}{R}$$

6–36. Consider the current response of the underdamped series RLC circuit as given by Equation 6–71. Show that the first maximum of the current occurs at a time t_m given by

$$t_m = \frac{1}{\omega_d}\tan^{-1}\frac{\omega_d}{\alpha}$$

APPLICATION PROBLEMS

6–37. A series RLC circuit is desired to produce an overdamped response. The two desired time constants are 0.5 ms and 0.25 ms. If the series resistance is 100 Ω, determine the required values of L and C.

6–38. A parallel RLC circuit is desired to produce an underdamped voltage across a 100-kΩ resistor. The oscillatory frequency desired is 1 kHz, and the desired time constant of the envelope is 0.1 s. A dc current source will be used to initiate oscillations in the circuit. If the 100-kΩ resistor is the only resistor used in the circuit, determine the required values of L and C.

6–39. A certain industrial process circuit is modeled by the circuit shown in Figure P6–39. **(a)** Show that the output voltage response is underdamped. **(b)** It is proposed to add

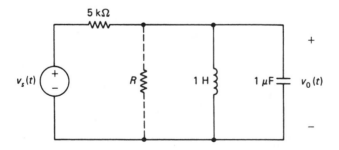

FIGURE P6–39

a shunt resistance R across the output to achieve critical damping. Determine the required value of R.

6–40. Consider the circuit of Figure P6–40 with a dc voltage V_i, and assume that the diode is ideal, i.e., it is assumed to be either a perfect short circuit or a perfect open circuit. It is also assumed that there is no resistance in the circuit and the circuit is initially relaxed. **(a)** If the switch is closed at $t = 0$, determine an expression for $i(t)$ and $v_C(t)$ that will apply while the diode is conducting. **(b)** Sketch the forms of $i(t)$ and $v_C(t)$, noting the point at which the current would become negative. **(c)** Assuming that current stops abruptly as it attempts to reverse direction, what is the final voltage remaining on the capacitor? **(d)** Can you suggest an application for this circuit?

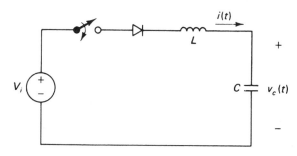

FIGURE P6–40

7

TRANSFER FUNCTIONS

OBJECTIVES

After completing this chapter, the reader should be able to:

- State the conditions required for a system to be linear, lumped, and time invariant.
- Define the transfer function and the input–output relationship for a linear system.
- Determine the transfer function for a given circuit.
- Determine the output of a given system from a knowledge of the transfer function and the input.
- Determine the poles and zeros from a given transfer function.
- Construct an s-plane pole–zero diagram and show its relationship to the transfer function.
- Define a stable system, a marginally stable system, and an unstable system; and discuss the relationship to the poles and zeros of the transfer function.
- Apply block diagram algebra to simplify interconnections of transfer functions.
- Discuss the form of the step response of a second-order system.

7–1   LINEAR SYSTEMS

Throughout the text, numerous circuits have been analyzed by the various methods considered. The structure of the circuit was always evident in the analysis, and the procedure employed was often determined by the form of the circuit. At this point in the text, we should be prepared to make a transition from the circuits level to the systems level. At the systems level, the emphasis is directed toward the input–output point of view for a circuit rather than the actual internal details.

For this purpose, consider a linear system having a *single input* and a *single output* as depicted by Figure 7–1. Although a wide range of physical variables can be adapted to this method of analysis, voltages and currents will be the variables

FIGURE 7–1

Illustration of single input and single output of linear system.

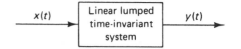

of primary interest in this text. However, to provide some generality in the development that follows, the input variable will be denoted as $x(t)$, and the output variable will be denoted as $y(t)$. The input $x(t)$ is also considered as the *excitation,* and the output $y(t)$ is considered as the *response.*

Definition of Linear System

As a necessary prelude to the development of systems models, the concept of a *linear system* will be explained. In the definition that follows, we will assume an initially *relaxed system.* An initially relaxed system is one with no initial energy storage.

A system is said to be *linear* with respect to an input $x(t)$ and an output $y(t)$ if the following two properties are satisfied:

1. *Amplitude linearity.* If an input $x(t)$ produces an output $y(t)$, then an input $Kx(t)$ should produce an output $Ky(t)$ for any value of K, where K is a *constant.*
2. *Superposition principle.* If an input $x_1(t)$ produces an output $y_1(t)$, and an input $x_2(t)$ produces an output $y_2(t)$, then an input $x_1(t) + x_2(t)$ should produce an output $y_1(t) + y_2(t)$ for any arbitrary functions $x_1(t)$ and $x_2(t)$.

As an illustration of amplitude linearity, assume that an input voltage (in volts) of value 2 to a certain system produces an output voltage (also in volts) given by $4t$. If the system is linear, an input voltage of value 6 should produce an output voltage given by $12t$. As an illustration of superposition, assume that when the input voltage to the system just described is $2 \cos t$, the output is $4 \sin t$. When the system is excited by the sum of the two inputs considered, that is, $2 + 2 \cos t$, the output would be $4t + 4 \sin t$ if the system is linear.

A large body of theory has been developed around the concept of linear systems and circuits. A circuit is linear provided that the circuit parameters (R, L, and C) are either constant or vary only with time. A circuit in which the parameters vary with the levels of voltages and currents will be a nonlinear circuit. Examples of nonlinear elements are a thermistor, in which the resistance varies with the current flow; a varactor diode, in which the capacitance is a function of the voltage across it; and a saturable core reactor, in which the inductance varies with the current.

Time-Invariant System

It was indicated that a linear system has either constant parameter values or parameters that change with time. Circuits with time-variable parameters are rather specialized, and will not be considered in this text. We will restrict our consideration here to *time-invariant* linear systems, the most widely employed types.

Lumped Parameter System

One more restriction of importance to our department is that only lumped parameters will be considered. *Lumped parameter* circuits are those in which actual resistors, capacitors, and inductors as discrete components appear in the circuit models. This is in contrast to *distributed parameter* circuits such as transmission lines. A primary distinction between lumped and distributed circuits is that an ordinary differential equation may be used to describe a lumped circuit, whereas a partial differential equation is required to describe a distributed parameter circuit.

Putting the various modifiers together, our interest is in circuits and systems described as *linear, lumped,* and *time invariant.* To simplify the many references to such circuits and systems, most references will employ the much simpler term *linear system,* since it will usually be obvious that the other two conditions are satisfied.

The use of all these modifiers may lead the reader to believe that we are being overly restrictive. However, most engineering design is based on models employing these restrictions. If any of these modifiers are eliminated, the complexity of the design and analysis process increases markedly. Thus, the restrictions are required to allow a manageable approach to dealing with a large number of real-life problems.

Although much engineering design is based on the theory of linear systems, no real-life systems are truly linear over an arbitrary wide range. For example, a carbon composition resistor fits the linear assumption quite well, provided that its power dissipation rating is not exceeded. A transistor amplifier may be assumed to be linear if operated in a very small signal region. However, if the power or signal level is excessive, these devices can no longer be described as linear. Thus, when we say a system is linear, it is assumed that the linear definitions apply over its normal working range and that this range is not exceeded.

7-2 TRANSFER FUNCTION

Consider a linear system with a single input and a single output as discussed in the preceding section. Except for a few special cases (such as a purely resistive circuit), the relationship between the input and the output in the time domain requires a differential equation, as will be seen later. However, if the entire circuit is converted to an *s*-domain model, all derivative and/or integral relationships become algebraic forms. The output can then be algebraically expressed as a function of the input. It is then possible to formulate a function that provides a fixed relationship between the output and the input such that the output can be determined for any arbitrary input. This is the concept of the transfer function, in which the relationship between the output and the input is established for any arbitrary input.

FIGURE 7–2
Representation of the input-output relationship of a linear system by a transfer function.

Assume, then, an initially relaxed linear system excited at $t = 0$ by an input $x(t)$, and assume that $y(t)$ is the corresponding output. Let

$$X(s) = \mathcal{L}[x(t)] \tag{7–1}$$
$$Y(s) = \mathcal{L}[y(t)] \tag{7–2}$$

Because of the algebraic s-domain relationships for the linear system, it will be possible to express $Y(s)$ as the product of $X(s)$ and an s-domain function $G(s)$. Thus

$$Y(s) = G(s)X(s) \tag{7–3}$$

where $G(s)$ is an algebraic function of s. Solving for $G(s)$, we obtain

$$G(s) = \frac{Y(s)}{X(s)} \tag{7–4}$$

The function $G(s)$ is called the *transfer function* of the circuit or system, and it provides a direct mathematical relationship between the input and the output for any arbitrary input. The concept of replacing the system by an s-domain transfer function is illustrated in Figure 7–2.

Determining the Transfer Function

To determine $G(s)$ for an arbitrary circuit or system, an input $X(s)$ is assumed and the output $Y(s)$ is determined in terms of $X(s)$. By means of Equation (7–4), the transfer function is then determined. If desired, a specific input (e.g., a step function) may be assumed. More often, however, an arbitary input $X(s)$ is assumed, and the output is determined in terms of $X(s)$. In this manner, the direct linear relationship between input and output is clearly delineated.

Application of Transfer Function

Once the transfer function for the circuit or system is known, the output may be determined for any arbitrary input by the use of Equation (7–3). The transform for the input is determined and multiplied by the transfer function, and the result is the transform of the output.

The transfer function is a generalization of the more common gain parameter used to characterize the input–output relationship for linear amplifiers. The gain for an amplifier is the ratio of the output to the input, which in simplest form is a real number. The transfer function is an extension of this concept for the purpose of characterizing the input–output relationship whenever derivative and/or integral relationships are involved. Indeed, a more elaborate model of an amplifier, which includes frequency dependent parameters, can be characterized by a transfer function.

The transfer function is fixed by the nature of the system or circuit and is not dependent on the type of excitation. The poles and zeros of $G(s)$ are due only to the circuit or system.

The transfer function concept has been developed with the assumption that the circuit has been initially relaxed. Although the approach can be modified to incorporate initial energy storage, the major utility is in analyzing circuits under initially relaxed conditions. Unless stated otherwise, initially relaxed circuit conditions will be assumed in all further developments relative to the transfer function. When initial energy is present, a return to the earlier circuit analysis approach is recommended at this point.

EXAMPLE 7–1

Determine the transfer function of the circuit of Figure 7–3(a). The input is $v_1(t)$, and the output is $v_2(t)$.

Solution
The circuit is converted to the s-domain form as shown in Figure 7–3(b). The input is retained in the arbitrary form $V_1(s)$. The output $V_2(s)$ may be determined most easily by means of the voltage divider rule. We have

$$V_2(s) = \frac{40/s}{40/s + 4} \times V_1(s) = \frac{40V_1(s)}{4s + 40} = \frac{10V_1(s)}{s + 10} \qquad \textbf{(7–5)}$$

Some simplification was performed in the last two steps.

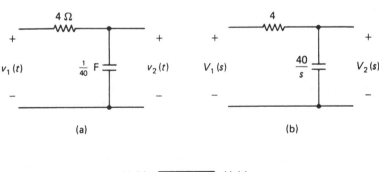

(a)   (b)

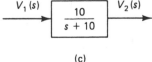

(c)

FIGURE 7–3
Circuit for Example 7–1 and the transfer function.

The result of Equation (7–5) indicates that the output $V_2(s)$ is a function of s times the input $V_1(s)$ as expected. The transfer function $G(s)$ is readily determined by dividing both sides of Equation (7–5) by $V_1(s)$. Thus,

$$G(s) = \frac{V_2(s)}{V_1(s)} = \frac{10}{s + 10} \qquad (7\text{–}6)$$

The transfer function is illustrated in block diagram form in Figure 7–3(c).

EXAMPLE 7–2

Determine the transfer function of the circuit of Figure 7–4(a), and show that it is the same function as obtained for the circuit of Example 7–1. The input is $v_1(t)$, and the output is $v_2(t)$.

Solution

The circuit is converted to the s-domain form as shown in Figure 7–4(b) with $V_1(s)$ retained in arbitrary form. As in Example 7–1, the voltage divider rule may be applied to the circuit. In this case, the desired ratio will be formed directly, and we have

$$G(s) = \frac{V_2(s)}{V_1(s)} = \frac{20}{2s + 20} = \frac{10}{s + 10} \qquad (7\text{–}7)$$

The function is shown in block diagram form in Figure 7–4(c).

Comparing Equation (7–7) of this example with Equation (7–6) of the preceding example, we see that both circuits have identical transfer functions! Yet, one circuit is an *RC* circuit, and the other is an *RL* circuit. In fact, the block diagram

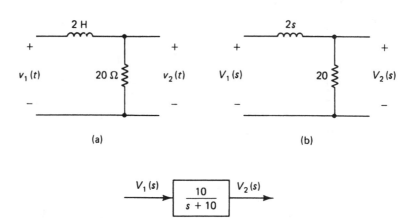

(a)

(b)

(c)

FIGURE 7–4
Circuit for Example 7–2 and the transfer function.

representations depicted by Figures 7–3(c) and 7–4(c) show only the mathematical relationship between input and output and are void of any circuit elements.

These examples illustrate that two or more entirely different circuits may have the same transfer function. If these two circuits were enclosed in boxes, one could not distinguish between them from measurements of the input–output characteristics. (There are other measurements that could be used to identify the separate circuits, but these do not relate to the transfer function under consideration.)

At the systems level of analysis, it is recognized that many different circuit or system configurations could all possess identical transfer functions. Consequently, the input–output relationship is the function of significance rather than the circuit or system details.

EXAMPLE 7–3

For the circuit of either Example 7–1 or Example 7–2 with no initial energy storage, determine (by using the transfer function concept) the output $v_2(t)$ when the input is excited at $t = 0$ by (a) $v_1(t) = 20$ and (b) $v_1(t) = 20 \sin 10t$.

Solution
The transfer function of either circuit is

$$G(s) = \frac{V_2(s)}{V_1(s)} = \frac{10}{s + 10} \tag{7–8}$$

The output $V_2(s)$ can be expressed as

$$V_2(s) = G(s)V_1(s) = \frac{10}{s + 10} \times V_1(s) \tag{7–9}$$

The output can then be determined for an arbitrary input without returning to the circuit configuration. This will now be illustrated with the two input signals given.

(a) The first input is

$$v_1(t) = 20 \quad \text{for } t > 0 \tag{7–10}$$

The input transform is

$$V_1(s) = \frac{20}{s} \tag{7–11}$$

Substitution of Equation (7–11) in Equation (7–9) results in

$$V_2(s) = \frac{10}{s + 10} \times \frac{20}{s} = \frac{200}{s(s + 10)} \tag{7–12}$$

The function can be inverted by the methods of Chapter 5, and the reader is invited to verify that the result is

$$v_2(t) = 20 - 20e^{-10t} \tag{7–13}$$

(b) The next input considered is

$$v_1(t) = 20 \sin 10t \tag{7-14}$$

The transform of the input is

$$V_1(s) = \frac{20(10)}{s^2 + (10)^2} = \frac{200}{s^2 + 100} \tag{7-15}$$

Substitution of Equation (7–15) in Equation (7–9) yields

$$V_2(s) = \frac{2000}{(s + 10)(s^2 + 100)} \tag{7-16}$$

Inverse transformation by the methods of Chapter 5 yields

$$v_2(t) = 10e^{-10t} + 10\sqrt{2} \sin(10t - 45°) \tag{7-17}$$

The results of this example will be used to assist in some explanations in Section 7–3, so any discussion will be deferred to that section.

EXAMPLE 7–4

The circuit of Figure 7–5(a) is one particular passive realization of a third-order Butterworth low-pass filter, and it is normalized to a cutoff frequency of 1 rad/s with a 1-Ω load. This particular structure has been designed for an ideal current source input. With $i_1(t)$ as the input and $v_2(t)$ as the output, determine the transfer

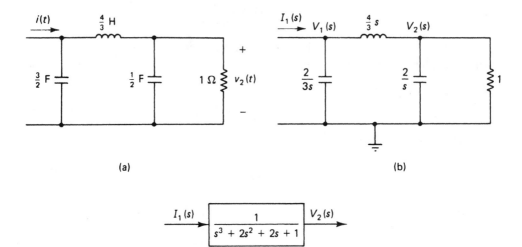

(a) (b)

(c)

FIGURE 7–5
Normalized third-order passive low-pass Butterworth filter, s-domain model, and transfer function as determined in Example 7–4.

function. (An actual realistic circuit is derived from the normalized circuit by scaling the frequency and resistance levels.)

Solution
The *s*-domain form of the circuit is shown in Figure 7–5(b). The input transform current $I_1(s)$ is considered as a source. In view of this current source, node voltage analysis seems quite appropriate. While $V_2(s)$ is the desired output, it is necessary to introduce $V_1(s)$ in order to write the two node equations. The bottom connection is considered as the ground reference.

Summing currents according to conventions established in Chapter 2, the node equation at the V_1 node is

$$-I_1(s) + \frac{3s}{2} V_1(s) + \frac{V_1(s) - V_2(s)}{(4/3)s} = 0 \qquad (7\text{–}18)$$

The equation at the V_2 node is

$$\frac{V_2(s) - V_1(s)}{(4/3)s} + \frac{s}{2} V_2(s) + \frac{V_2(s)}{1} = 0 \qquad (7\text{–}19)$$

Simultaneous solution of these equations for $V_2(s)$ yields

$$V_2(s) = \frac{I_1(s)}{s^3 + 2s^2 + 2s + 1} \qquad (7\text{–}20)$$

The transfer function is then determined as

$$G(s) = \frac{V_2(s)}{I_1(s)} = \frac{1}{s^3 + 2s^2 + 2s + 1} \qquad (7\text{–}21)$$

A block diagram depicting the transfer function relationship is shown in Figure 7–5(c).

EXAMPLE 7–5

The circuit of Figure 7–6(a) is one particular active realization of a second-order Butterworth low-pass filter using an operational amplifier, and it is normalized to a cutoff frequency of 1 rad/s with 1-Ω resistors. With $v_1(t)$ as the input and $v_2(t)$ as the output, determine the transfer function. (An actual realistic circuit is derived from the normalized circuit by scaling the frequency and resistance levels.)

Solution
The operational amplifier is connected in what is known as a "voltage follower" configuration in which the output voltage acts as a VCVS with a voltage gain of 1. The controlling voltage is the voltage from the noninverting input terminal (labeled as +) to ground. No current is assumed to flow into this input terminal.

The *s*-domain circuit is shown in Figure 7–6(b). The input and output node voltages are labeled as $V_1(s)$ and $V_2(s)$, respectively. However, two additional voltages, $V_3(s)$ and $V_4(s)$, are introduced at the two intermediate nodes. Note that the

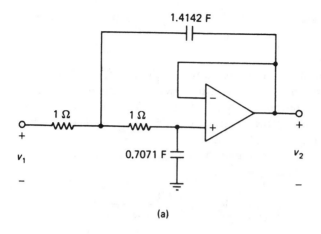

(a)

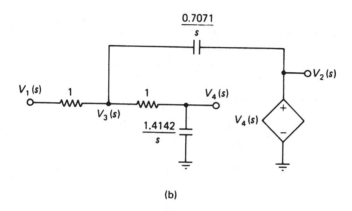

(b)

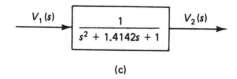

(c)

FIGURE 7–6

Normalized second-order active low-pass Butterworth filter, s-domain model with VCVS, and transfer function as determined in Example 7–5.

dependent source has a voltage value equal to the voltage of the controlling node, since the gain is unity.

If it is desired to eliminate n variables in a set of simultaneous equations, it is necessary to have $n + 1$ independent equations. A relationship between $V_1(s)$ and $V_2(s)$ is desired, so it is necessary to eliminate $V_3(s)$ and $V_4(s)$. Thus, three independent equations are required.

First, a node equation will be written at the $V_3(s)$ node. Applying KCL at this node, we have

$$\frac{V_3(s) - V_1(s)}{1} + \frac{s}{0.7071}[V_3(s) - V_2(s)] + \frac{V_3(s) - V_4(s)}{1} = 0 \qquad \textbf{(7–22)}$$

Next, a node equation will be written at the $V_4(s)$ node. We have

$$\frac{V_4(s) - V_3(s)}{1} + \frac{s}{1.4142}V_4(s) = 0 \qquad \textbf{(7–23)}$$

Finally, the dependent source voltage and the output voltage are exactly the same, namely,

$$V_2(s) = V_4(s) \qquad \textbf{(7–24)}$$

The voltages $V_3(s)$ and $V_4(s)$ are eliminated by the procedure that follows. First, the relationship of Equation (7–24) is used to eliminate $V_4(s)$ in Equations (7–22) and (7–23). This leaves two equations with only $V_3(s)$ remaining as a superfluous variable. Next, Equation (7–23) is solved for $V_3(s)$ in terms of $V_2(s)$ following the earlier substitution. This result is substituted in Equation (7–22). After several manipulations, we obtain

$$V_2(s) = \frac{V_1(s)}{s^2 + 1.4142s + 1} \qquad \textbf{(7–25)}$$

The transfer function is then determined as

$$G(s) = \frac{V_2(s)}{V_1(s)} = \frac{1}{s^2 + 1.4142s + 1} \qquad \textbf{(7–26)}$$

7–3 NATURAL AND FORCED RESPONSES

Consider again the basic input–output relationship expressed in terms of the transfer function, that is,

$$Y(s) = G(s)X(s) \qquad \textbf{(7–27)}$$

In general, both $G(s)$ and $X(s)$ will have poles (as well as zeros). Thus, the poles of $Y(s)$ will consist of the poles of $G(s)$ and the poles of $X(s)$. Recall that each pole (or pair of poles for the complex case) may be considered as producing one of the terms in the associated output function $y(t)$. The poles of $G(s)$ are a function of the system parameters, whereas poles of $X(s)$ are a function of excitation or input.

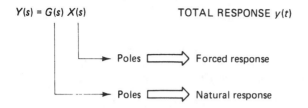

$Y(s) = G(s)\, X(s)$ TOTAL RESPONSE $y(t)$

Poles ⟹ Forced response

Poles ⟹ Natural response

FIGURE 7–7
Relationship between poles and response functions.

Recall that any response $y(s)$ may be considered as the sum of a natural response plus a forced response. *The natural response may be considered as being produced by the poles of the transfer function, and the forced response may be considered as being produced by the poles of the input function.* This is illustrated in Figure 7–7.

To further illustrate these concepts for a relatively simple case, consider the transfer function for the circuits of Examples 1 through 3, which was last given in Equation (7–8). The transfer function has a pole $s = -10$, and the response due to any input will display a natural response function with the exponential term e^{-10t}. Such a term is observed both in Equation (7–13) and in (7–17).

When the input is the constant 20, the output transform has a pole $s = 0$ as observed in Equation (7–12). The output time function corresponding to this pole is a constant term, which is the forced response term as noted in Equation (7–13). However, when the input is a sinusoid with a frequency of 10 rad/s, the output transform has a pair of purely imaginary poles $\pm j10$ as observed in Equation (7–16). The output time function corresponding to this pair of poles is a sinusoidal function of the same frequency, which is the forced response term as noted in Equation (7–17).

The concepts discussed in this section permit the general form of a response to be predicted quite easily in many cases when a full inverse transformation is not required. The poles of the transfer function and the source determine the forms of the terms in the response function, and the general nature of the response may be predictable from a simple inspection process. This concept is illustrated in Example 7–6.

EXAMPLE 7–6
The transfer function of a certain system is given by

$$G(s) = \frac{Y(s)}{X(s)} = \frac{N(s)}{(s^2 + 5s + 6)(s^2 + 2s + 17)} \qquad (7\text{–}28)$$

where $N(s)$ is an unspecified numerator polynomial. It will be assumed, however, that no roots of $N(s)$ will be the same as roots of either the denominator of $G(s)$ or of $X(s)$. Using arbitrary constant magnitudes and phase angles, write the general form of the response $y(t)$ for **(a)** $x(t) = 10$, and **(b)** $x(t) = 10 \cos 6t$. In each case, identify the natural response and the forced response.

Solution
As always, the roots of all quadratics should be checked before any analysis with transforms is attempted. The quadratic $(s^2 + 5s + 6)$ has two real roots, $s = -2$ and $s = -3$. However, the quadratic $s^2 + 2s + 17$ has two complex roots, $s = -1 \pm j4$. The function $G(s)$ is first rewritten with the first quadratic factored as

$$G(s) = \frac{Y(s)}{X(s)} = \frac{N(s)}{(s+2)(s+3)(s^2 + 2s + 17)} \qquad \textbf{(7–29)}$$

The output transform $Y(s)$ is then determined as

$$Y(s) = \frac{N(s)X(s)}{(s+2)(s+3)(s^2 + 2s + 17)} \qquad \textbf{(7–30)}$$

(a) The input in this case is

$$x(t) = 10 \qquad \textbf{(7–31)}$$

The input transform is

$$X(s) = \frac{10}{s} \qquad \textbf{(7–32)}$$

Substitution of Equation (7–32) in Equation (7–30) yields for the output transform

$$Y(s) = \frac{10N(s)}{s(s+2)(s+3)(s^2 + 2s + 17)} \qquad \textbf{(7–33)}$$

The form of the time-domain output is

$$y(t) = A_1 + A_2 e^{-2t} + A_3 e^{-3t} + A_4 e^{-t} \sin(4t + \theta) \qquad \textbf{(7–34)}$$

where $A_1, A_2, A_3, A_4,$ and θ are constants that could be determined by a full inversion if $N(s)$ were given.

All terms in Equation (7–34) except A_1 represent the natural response, which is a transient response in this case. The term A_1 represents the forced response, which is a steady-state response.

(b) The input for this part is

$$x(t) = 10 \cos 6t \qquad \textbf{(7–35)}$$

The input transform is

$$X(s) = \frac{10s}{s^2 + 36} \qquad \textbf{(7–36)}$$

Substitution of Equation (7–36) in Equation (7–30) yields for the output transform

$$Y(s) = \frac{10sN(s)}{(s^2 + 36)(s + 2)(s + 3)(s^2 + 2s + 17)} \quad \text{(7–37)}$$

The form of the time-domain output is

$$y(t) = B_1 \sin(6t + \phi_1) + B_2 e^{-2t} + B_3 e^{-3t} + B_4 e^{-t} \sin(4t + \phi_2) \quad \text{(7–38)}$$

where B_1, B_2, B_3, B_4, ϕ_1, and ϕ_2 are constants that could be determined by a full inversion if $N(s)$ were given.

All terms in Equation (7–38) except the first sinusoidal term represent the natural response, which is a transient response. The term $B_1 \sin(6t + \phi_1)$ represents the forced response, which is steady state.

Once the concept is understood, the general form can be expressed without necessarily showing all the steps that have been performed here. The important point to remember is that the natural response contains terms representing the poles of the transfer function, and the forced response contains terms representing the poles of the input source function. The total response is the sum of these two responses.

7–4 POLES AND ZEROS OF TRANSFER FUNCTIONS

In general, the transfer function of a lumped, linear, time-invariant circuit or system is a ratio of polynomials in the variable s. Let $N(s)$ represent the numerator polynomial, and let $D(s)$ represent the denominator polynomial. Thus, $G(s)$ can be expressed as

$$G(s) = \frac{N(s)}{D(s)} \quad \text{(7–39)}$$

The numerator polynomial $N(s)$ will be assumed to be of order n, and it can be expressed as

$$N(s) = a_n s^n + a_{n-1} s^{n-1} + \cdots + a_1 s + a_0 \quad \text{(7–40)}$$

The denominator polynomial $D(s)$ will be assumed to be of order m, and it can be expressed as

$$D(s) = b_m s^m + b_{m-1} s^{m-1} + \cdots + b_1 s + b_0 \quad \text{(7–41)}$$

Definitions

The definitions of poles and zeros were given in Chapter 5 as pertaining to Laplace transforms. We will now review these definitions as they apply to transfer functions and extend the definitions somewhat. In particular, the adjective *finite* will be included as a modifier in the definitions that follow:

1. *Poles* (*finite*). The m roots of $D(s)$ are called the finite *poles* of the transfer function.
2. *Zeros* (*finite*). The n roots of $N(s)$ are called the finite *zeros* of the transfer function.

The modifier is used here to distinguish between poles and zeros determined from the roots of polynomials and some special definitions called "poles and zeros at infinity," which will be given shortly. When references to poles and zeros without any special designations are made, it will be understood that such poles and zeros are finite and are the roots of appropriate polynomials.

Order of Transfer Function

The order of a transfer function is the value of the larger of the two integers m and n. Thus, if $m > n$, the transfer function is of order m, whereas if $n > m$, the transfer function is of order n. Most transfer functions occurring in circuits are characterized by the condition that $m \geq n$, that is, the denominator has a degree at least as high, and often higher, than that of the numerator. However, a few special functions are encountered in which the numerator polynomial has a higher degree than that of the denominator, so that possibility will be considered here.

Next, the concept of "poles and zeros at infinity" will be defined. The most common case is the "zero at infinity," and that case will be discussed first.

Zeros at Infinity

In order to have one or more zeros at infinity, it is necessary that $m > n$, that is, the degree of the denominator polynomial must be greater than the degree of the numerator polynomial. In this case, the transfer function is said to have n_i zeros at infinity, where

$$n_i = m - n \qquad (7\text{--}42)$$

The logic behind this definition is based on the behavior of $G(s)$ as s increases without limit, that is, approaches infinity. For very large s, the polynomials $N(s)$ and $D(s)$ may each be closely approximated by their highest-degree terms. Thus,

$$N(s) \approx a_n s^n \quad \text{for } |s| \gg 1 \qquad (7\text{--}43)$$
$$D(s) \approx b_m s^m \quad \text{for } |s| \gg 1 \qquad (7\text{--}44)$$

The transfer function $G(s)$ may then be closely approximated by

$$G(s) \approx \frac{a_n s^n}{b_m s^m} = \frac{a_n/b_m}{s^{m-n}} = \frac{a_n/b_m}{s^{n_i}} \quad \text{for } |s| \gg 1 \qquad (7\text{--}45)$$

The result of Equation (7–45) indicates that $G(s)$ can be closely approximated by a constant divided by s^{n_i} for large s. This function approaches 0 as s increases without limit. A zero is a value of s at which the function becomes 0, so it is meaningful then to say that $s = \infty$ is a zero of the function. The rate at which the function approaches 0 is dependent on the integer n_i. As n_i increases, the function approaches 0 more rapidly, so it is meaningful to define n_i as the order of the zero at $s = \infty$. Although this idea may seem like a totally abstract mathematical concept, it turns out that useful information concerning the high-frequency behavior of systems may be deduced from the number of zeros at $s = \infty$.

Poles at Infinity

In order to have one or more poles at infinity, it is necessary that $n > m$, that is, the degree of the numerator polynomial must be greater than the degree of the denominator polynomial. In this case, the transfer function is said to have m_i poles at infinity, where

$$m_i = n - m \qquad (7\text{–}46)$$

To justify this definition, consider again the limiting forms for $N(s)$ and $D(s)$ of Equations (7–43) and (7–44) as s becomes very large. The limiting form for the transfer function can be manipulated as

$$G(s) \approx \frac{a_n s^n}{b_m s^m} = \frac{a_n}{b_m} s^{n-m} = \frac{a_n}{b_m} s^{m_i} \quad \text{for } |s| \gg 1 \qquad (7\text{–}47)$$

The result of Equation (7–47) indicates that $G(s)$ can be closely approximated by a constant times s^{m_i} for large s. This function grows without limit as s increases without limit. A pole is a value of s at which the function grows without limit, so it is meaningful then to say that $s = \infty$ is a pole of the function. As m_i increases, the function increases at a more rapid rate, so it is meaningful to define m_i as the order of the pole at $s = \infty$.

A transfer function will have n finite zeros and m finite poles. When $m > n$, the additional n_i zeros at $s = \infty$ results in $n_i + n = m - n + n = m$ zeros, which is also the number of poles. When $n > m$, the additional m_i poles at $s = \infty$ result in $m_i + m = n - m + m = n$ poles, which is also the number of zeros. The conclusion is that *the number of zeros of a transfer function is equal to the number of poles if zeros or poles at s $= \infty$ are included in the total.*

Examples

As a simple example of these concepts, refer back to the transfer function of Examples 7–1, 7–2, and 7–3. This function, which was last given by Equation (7–8), has a denominator polynomial of degree one and a numerator polynomial of degree zero. (A constant is a limiting form of a polynomial with degree zero.) The transfer function is of first order, and it has one finite pole ($s = -10$). There are no finite zeros, but there is one zero at $s = \infty$.

As a second example, refer back to the transfer function of Example 7–4. The transfer function as given by Equation (7–21) has a denominator polynomial of degree 3 and a numerator polynomial of degree 0. The transfer function is of third order, and there are 3 finite poles. There are no finite zeros, but there are 3 zeros at $s = \infty$.

Factored Form of Transfer Function

Let $z_1, z_2, \ldots, z_n$ represent the n finite zeros of $G(s)$, and let $p_1, p_2, \ldots, p_m$ represent the m finite poles of $G(s)$. If $N(s)$ and $D(s)$ are then expressed in factored forms, $G(s)$ may be expressed as

$$G(s) = \frac{A(s - z_1)(s - z_2) \cdots (s - z_n)}{(s - p_1)(s - p_2) \cdots (s - p_m)} \tag{7–48}$$

where A is a constant multiplier. In terms of the notation of Equations (7–40) and (7–41), this constant is $A = a_n/b_m$, but it is equivalent to a single constant as far as the net transfer function is concerned.

Inspection of Equation (7–48) reveals that *a transfer function may be specified to within a constant multiplier by a knowledge of its poles and zeros.*

S-Plane

Important analysis techniques have been developed around the concept of the *s*-plane. The *s-plane* is a rectangular coordinate system in which certain properties of transfer functions may be visualized. The concept is based on representing s as a complex variable of the form

$$s = \sigma + j\omega \tag{7–49}$$

The quantity σ is the real part of s, and it can be interpreted as a damping factor. The quantity ω is the imaginary part of s, and it can be interpreted as an angular frequency. The quantity s is often denoted as the *complex frequency.*

The form of the *s*-plane is illustrated in Figure 7–8. The horizontal axis is the σ-axis, and the vertical axis is the $j\omega$-axis. The first axis is often called the *real axis,* and the latter axis is called the *imaginary axis.*

Useful results in system theory can be determined from plots of the finite poles and zeros of a transfer function. Finite poles are denoted by x's, and finite zeros are denoted by 0s. A typical pole–zero plot, to be considered in Example 7–7, is illustrated in Figure 7–9. Zeros and poles at $s = \infty$ are not shown in the *s*-plane, but their presence can be inferred by a count of the finite poles and zeros.

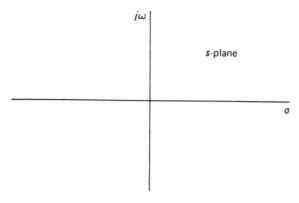

FIGURE 7–8
Form of the s-plane.

Relationship Between Order and Number of Reactive Elements

A final point of interest is the relationship between the type of circuit elements and the order of a circuit. It turns out that *the order of a circuit is the total number of nonredundant inductors and capacitors.* The number of nonredundant elements is recognized after the circuit is reduced to its simplest form. Thus, simple series or parallel combinations of more than one inductance or capacitance are each equivalent to only one inductance or capacitance, since some of the element values are redundant from a mathematical point of view.

As an example of circuit order, the circuit of Figure 7–5 has two capacitors and one inductor, and it is a third-order circuit. The circuit of Figure 7–6 has two capacitors, and it is a second-order circuit.

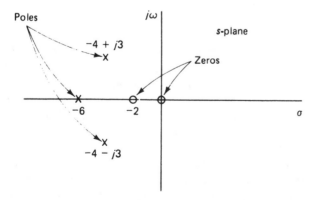

FIGURE 7–9
Pole–zero plot for Example 7–7.

EXAMPLE 7–7
Write the form of the transfer function corresponding to the s-plane plot shown in Figure 7–9.

Solution
The specific values of finite zeros and poles are substituted in the respective polynomials in accordance with the form of Equation (7–48). We have

$$G_s = \frac{A(s - 0)[s - (-2)]}{[s - (-6)][s - (-4 - j3)][s - (-4 + j3)]} \qquad (7\text{–}50)$$

$$= \frac{As(s + 2)}{(s + 6)(s + 4 + j3)(s + 4 - j3)} = \frac{As(s + 2)}{(s + 6)(s^2 + 8s + 25)}$$

The form of the transfer function is now known, but the level of the overall function depends on the value of A. One additional specification would be required to determine A.

EXAMPLE 7–8
A certain transfer function is given by

$$G(s) = \frac{2(s^2 + 6s + 25)}{s^3 + 7s^2 + 10s} \qquad (7\text{–}51)$$

Determine the poles and zeros, and construct an s-plane plot of finite poles and zeros.

Solution
Since the numerator polynomial is a second-degree function, there are two finite zeros. It can be readily determined that the roots of the numerator polynomial are complex, and the two finite zeros z_1 and z_2 are determined to be $z_1, z_2 = -3 \pm j4$. The denominator polynomial is a cubic equation, but it contains the factor s, so

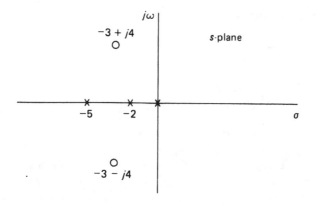

FIGURE 7–10
Pole–zero plot for Example 7–8.

one pole p_1 can be identified as $p_1 = 0$. The remaining two poles are the roots of the polynomial $s^2 + 7s + 10$, and these are determined as $p_2 = -2$ and $p_3 = -5$. Since the denominator polynomial is one degree higher than the numerator polynomial, there is one zero at $s = \infty$. A summary is as follows:

Three finite poles: $0, -2, -5$

Two finite zeros: $-3 \pm j4$

One zero at $s = \infty$

An s-plane plot of the finite poles and zeros is shown in Figure 7–10.

7–5 STABILITY

The concept of stability is very important in the design and analysis of active electronic circuits and closed-loop feedback control systems. A circuit designed for a routine linear application, such as an amplifier, may turn out to oscillate or to move into saturation if circuit conditions are not favorable. In this section, stability will be discussed in terms of the transfer function concept. We will see that stability can be related to the locations of poles in the complex s-plane.

Relationship to Natural Response

In general, the degree of stability of a system is closely related to the natural response of the system. Recall that when a system is excited by an arbitrary input signal, the natural response terms appear in the output. If these terms vanish after a sufficient period of time, the circuit settles into a mode of operation in which the forced response assumes a steady-state nature. The question of stability, then, can be related to whether or not the natural response terms vanish, remain at a fixed level, or possibly even grow without bound.

Definitions

Although there are variations in the definitions of stability in the literature, the following definitions will suffice for our purposes:

1. *Stable system.* A system is stable if all natural response terms vanish or approach zero after a sufficiently long time.
2. *Unstable system.* A system is unstable if at least one term in the natural response grows without bound (i.e., approaches infinity) as time increases.
3. *Marginally stable system.* A system is marginally stable if there are no unstable terms and if at least one term approaches a constant nonzero value or a constant amplitude oscillation as time increases.

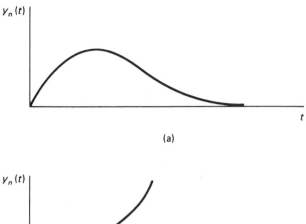

(a)

(b)

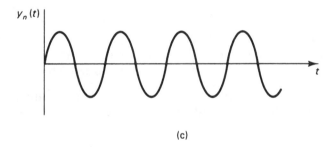

(c)

FIGURE 7–11

Examples of natural response for (a) stable system, (b) unstable system, and (c) marginally stable system.

Examples

Examples of the three possibilities are shown in Figure 7–11. The function shown is the natural response $y_n(t)$ for some arbitrary input, and the forced response is not shown. The function of (a) represents a stable response, the function of (b) represents an unstable response, and the function of (c) represents a marginally stable response.

From the definitions, it can be inferred that the degree of stability of a system is established by the *least* stable term. For example, if any one term in the natural response grows without bound, the system is considered to be unstable, irrespective of all other terms. As an additional example, the response could contain a constant amplitude oscillation, but if a term that grows without bound is present, the system would be classified as unstable. To be stable, *all* terms must be stable. These definitions are logical if one considers the implications of unstable terms.

Practical Perspective

In a practical sense, the concept of an unstable response must be tempered with the reality of the real world. For example, suppose analysis indicates that an amplifier output response will be unstable. As the output signal starts to grow, eventually saturation, clipping, or other nonlinear effects will take over and limit how far the output can go. In the case of closed-loop control systems, the consequences of instability can often be disastrous. The concept of instability, as determined here, is based on the ideal linear model, and one must realize that eventually a nonlinear effect would dominate.

The next point of interest is that of relating the preceding definitions to *s*-plane properties. For this purpose, we will divide the *s*-plane into three regions. Refer to Figure 7–12 for the definitions that follow.

1. *Left-hand half-plane.* All points to the left of the $j\omega$-axis, but *not* including the $j\omega$-axis, will be referred to as the *left-hand half-plane.* This region will be denoted as LHHP for brevity.
2. *Right-hand half-plane.* All points to the right of the $j\omega$-axis, but *not* including the $j\omega$-axis, will be referred to as the *right-hand half-plane.* This region will be denoted as RHHP for brevity.
3. *$j\omega$-axis.* The $j\omega$-axis will be considered as a separate area of the *s*-plane for reasons that will be clear shortly. The $j\omega$-axis includes the point $s = 0$.

In the discussion that follows, the effects of different pole locations will be investigated as they relate to the natural response. In each case, the form of a

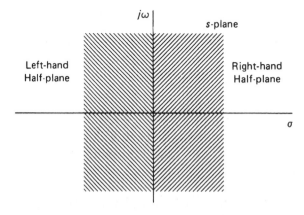

FIGURE 7–12
Division of *s*-plane into left-hand half-plane, right-hand half-plane, and $j\omega$-axis.

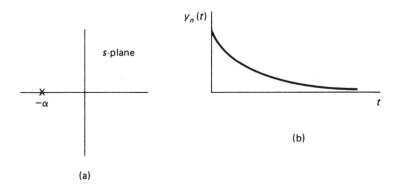

s-plane

$y_n(t)$

$-\alpha$

t

(a)

(b)

FIGURE 7–13
Pole on negative real axis and the corresponding natural response.

natural response term $y_n(t)$ will be shown based on the particular pole or pair of poles assumed.

Pole on the Negative Real Axis

An s-plane plot with a pole on the negative real axis is shown in Figure 7–13(a), and the corresponding form of the natural response is shown in (b). The time response is of the form

$$y_n(t) = Ae^{-\alpha t} \tag{7-52}$$

where A is some arbitrary constant. This term approaches zero as t increases, so it is a *stable* response.

Recall that the time constant $\tau = 1/\alpha$. If α is very close to the origin, τ is very large, and the time that it takes for the natural response to become negligible can be rather long. Conversely, if α is quite far to the left on the negative real axis, τ is very small, and the natural response term will vanish rather quickly. The preceding statements indicate that it is possible to acquire an intuitive relationship between s-domain locations and the form and duration of natural response terms.

If the pole were a multiple-order pole, the natural response terms would be of the form

$$y_n(t) = At^k e^{-\alpha t} \tag{7-53}$$

where k is an integer. Although the factor t^k grows without limit, the exponential approaches zero at a more pronounced rate, and the product approaches zero. Proof of this fact is left as an exercise for the reader (Problem 7–29).

Complex Poles in the Left-Hand Half-Plane

An s-plane plot with a pair of first-order complex poles in the LHHP is shown in Figure 7–14(a), and the corresponding time response is shown in (b). The time

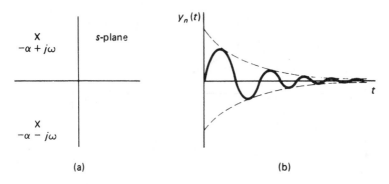

FIGURE 7–14
Complex poles in left-hand half-plane and the corresponding natural response.

response is of the form

$$y_n(t) = Ae^{-\alpha t} \sin(\omega t + \theta) \tag{7–54}$$

where A and θ are arbitrary constants. This term approaches zero as t increases, so it is a *stable* response.

The same argument made earlier for the relative duration of the natural response in terms of the value of α holds for the case of complex poles. The closer α is to the $j\omega$-axis, the longer will be the duration of the natural response. Conversely, the farther α is from the $j\omega$-axis, the shorter will be the duration of the natural response.

A separate, intuitive argument can be given here in terms of the value of ω. If the poles are very close to the negative real axis, the oscillation frequency ω is small, and the period of the oscillation will be long. However, as the poles move away from the negative real axis, the oscillation frequency ω increases, and the period of the oscillation will be shorter.

If the pair of poles were of multiple order, the natural response terms would be of the form

$$y_n(t) = At^k e^{-\alpha t} \sin(\omega t + \theta) \tag{7–55}$$

where k is an integer. As in the case of multiple-order real poles, this function approaches zero, and verification of this fact is left as an exercise (Problem 7–30).

Pole on the Positive Real Axis

An s-plane plot with a pole on the positive real axis is shown in Figure 7–15(a), and the corresponding form of the natural response is shown in (b). The time response is of the form

$$y_n(t) = Ae^{\alpha t} \tag{7–56}$$

This function grows without bound, so it is an *unstable* response.

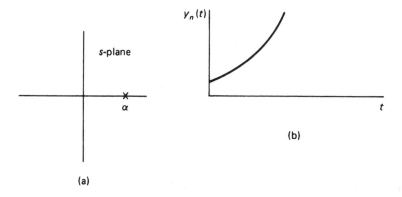

(a)

(b)

FIGURE 7–15
Pole on positive real axis and the corresponding natural response.

Complex Poles in the Right-Hand Half-Plane

An s-plane plot with a pair of first-order complex poles in the RHHP is shown in Figure 7–16(a), and the corresponding natural response is shown in (b). The time response is of the form

$$y_n(t) = Ae^{\alpha t}\sin{(\omega t + \theta)} \qquad (7\text{--}57)$$

This function grows without bound, so it is an *unstable* response.

Pole at the Origin

As one particular point on the $j\omega$-axis, consider a first-order pole at the origin, as shown in Figure 7–17(a). The corresponding time response is shown in (b), and it

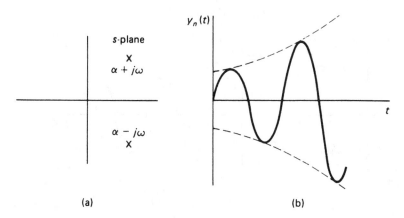

(a)   **(b)**

FIGURE 7–16
Complex poles in right-hand half-plane and the corresponding natural response.

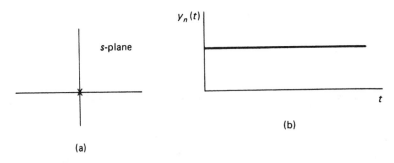

FIGURE 7–17
Pole at $s = 0$ and the corresponding natural response.

is simply

$$y_n(t) = A \qquad (7\text{–}58)$$

This response remains constant as time increases, so it is a *marginally stable* term.

Poles on $j\omega$-Axis

An s-plane plot with a pair of *first-order* poles located on the $j\omega$-axis is shown in Figure 7–18(a), and the corresponding time response is shown in (b). The time function is of the form

$$y_n(t) = A \sin (\omega t + \theta) \qquad (7\text{–}59)$$

This function oscillates with a constant amplitude for all time, so it is a *marginally stable* response.

If the pair of poles were of *multiple order,* the response would be of the form

$$f_n(t) = A t^k \sin (\omega t + \theta) \qquad (7\text{–}60)$$

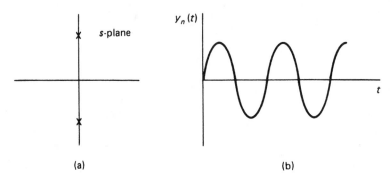

FIGURE 7–18
Poles on imaginary axis (first-order) and the corresponding natural response.

FIGURE 7–19
Response due to second-order poles on imaginary axis.

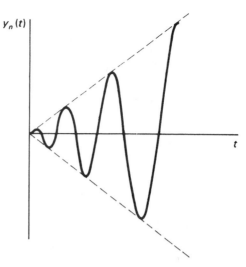

where k is an arbitrary positive integer. A typical response is shown in Figure 7–19 for the case of a pair of second-order poles. (In this case, $k = 1$ in Equation 7–60.) This response is obviously *unstable*.

General Conclusions

We conclude from the preceding discussion that first-order poles on the $j\omega$-axis represent marginally stable terms, whereas multiple-order poles represent unstable terms. The $j\omega$-axis is the only portion of the s-plane where the degree of stability is dependent on the order of the poles.

The possible types of conditions associated with $j\omega$-axis poles deserve some further discussion. The degree of stability desired in a system depends on the desired objectives. In a large number of systems, $j\omega$-axis poles of any order are undesirable. For example, a control system with a pair of first-order $j\omega$-axis poles would constantly "hunt" (oscillate). Indeed, in some applications, first-order $j\omega$-axis poles are classified as "unstable." On the other hand, oscillator circuits are designed specifically to produce a specific oscillatory behavior. In that case, the natural response is the desired output, and there is usually no forced response present.

Consider next the possibility of a system having a pair of first-order poles on the $j\omega$-axis. As it stands, the system is marginally stable. However, suppose the system is excited by a signal having one component with the same natural frequency as that of the system. Since $Y(s) = G(s) X(s)$, the product of the denominator factor $(s^2 + \omega^2)$ of $G(s)$ with a corresponding denominator factor of $X(s)$ results in a denominator factor $(s^2 + \omega^2)^2$ in $Y(s)$. This represents an *unstable* response. Thus, although neither the system nor the excitation was unstable, the combination produced an unstable response. This represents the concept of exciting a system at a natural resonant frequency, which can lead to an unstable response. Some natural disasters (such as bridge collapses) can be attributed to phenomena of this general type.

Returning to electrical circuits, it can be shown that the poles of any passive circuit containing some resistance will always be in the LHHP. Thus, passive circuits with resistance will always be stable. If ideal LC circuits are assumed, passive circuits having poles on the $j\omega$-axis (marginally stable case) would result. However, such an assumption is strictly a "paper exercise" because there will always be some resistance presence in the actual passive circuit, which will shift the poles to the left.

When active circuits (e.g., amplifiers) and negative feedback are considered, the situation changes drastically. First, it is possible for circuits having any combination of passive elements (R, L, or C) plus active elements to be either marginally stable or unstable. With the exception of oscillator circuits, where a controlled, marginally stable situation is desired, instability is usually a by-product of complex circuit interactions and is not desired. Almost everyone working with electronic circuits has encountered the situation of undesired oscillations occurring in what was intended as a stable circuit.

EXAMPLE 7–9

Of all the circuit examples considered at the end of Section 7–2 (Examples 7–1 through 7–5), the only one that possesses the potential for instability is the active filter of Example 7–5, because it contains an active element and feedback. Show that the transfer function obtained for the circuit, i.e., Equation (7–26), represents a stable system based on the assumptions made.

Solution
The poles of the transfer function are the roots of the denominator polynomial $s^2 + 1.4142s + 1$. Setting this polynomial to zero, the poles p_1 and p_2 are determined as

$$p_1, p_2 = -0.7071 \pm j0.7071 \qquad \textbf{(7–61)}$$

The real part (-0.7071) is negative, meaning that the poles are in the LHHP, so the circuit is stable.

Incidentally, the stability of the system in this case could have been predicted by inspection based on the following property: For a second-degree polynomial, if all coefficients are nonzero and have the same sign, all roots of the polynomial will have negative real parts. For higher-degree polynomials, a necessary (but not sufficient) condition for all roots to have negative real parts is that all coefficients are nonzero and have the same sign. For second-degree polynomials, the condition is both necessary and sufficient.

EXAMPLE 7–10

Consider the transfer function of Example 7–7, whose finite pole–zero plot was given in Figure 7–9. Classify the stability of the system according to the three categories.

Solution

Referring to Figure 7–9, the system is *stable* because all poles lie in the LHHP. The presence of the $j\omega$-axis zero (or even a RHHP zero) does not affect the stability of a true transfer function.

EXAMPLE 7–11

Consider the transfer function of Example 7–8, whose finite pole–zero plot was made in Figure 7–10. Classify the stability of the system according to the three categories.

Solution

Referring to Figure 7–10, a first-order pole lies on the $j\omega$-axis (specifically the origin), so the system is *marginally stable*.

7–6   TRANSFER FUNCTION "ALGEBRA"

At the systems level, complete linear circuit blocks are connected together to achieve a composite system function. If the complete system is linear and has a single input and a single output, a composite transfer function may be used to describe the net relationship between the input and the output. This composite transfer function may be expressed as a combination of the individual transfer functions. Certain rules will be developed in this section for combining forms that arise frequently in practice.

At the outset, an important stipulation must be made: *Each transfer function given is either assumed to be unaffected by the interconnection used, or the transfer function is defined under the loaded conditions given.* To elaborate further on this concept, assume a voltage transfer function $V_2(s)/V_1(s) = G(s)$. If the output terminals are connected to the input terminals of one other unit, there may be a loading effect that could result in a different value for $V_2(s)$ than the value determined under open-circuit conditions. If the loading effect does alter $V_2(s)$, a modified transfer function could be defined under such conditions. The point is that one cannot simply "throw together" blocks and assume that transfer functions remain unchanged. The effects of loading will be illustrated with a simple circuit in Example 7–12. Several arrangements will now be considered.

Cascade Connection

A cascade connection of several blocks having individual transfer functions is shown at the top of Figure 7–20. To avoid "clutter" on the figure, only the net input $X(s)$ and the net output $Y(s)$ are shown. However, for the development that follows, assume that the output of $G_1(s)$ is $Y_1(s)$; the output of $G_2(s)$ is $Y_2(s)$; and so on. The input of $G_n(s)$ is $Y_{n-1}(s)$, and the output of $G_n(s)$ is the overall output $Y(s)$.

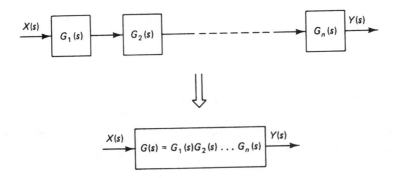

FIGURE 7–20
Cascade connection of transfer functions and an equivalent single transfer function.

The following array of equations can then be written:

$$Y_1(s) = G_1(s)X(s) \tag{7-62a}$$
$$Y_2(s) = G_2(s)Y_1(s) \tag{7-62b}$$

$$\vdots$$

$$Y_{n-1}(s) = G_{n-1}(s)Y_{n-2}(s) \tag{7-62c}$$
$$Y(s) = G_n(s)Y_{n-1}(s) \tag{7-62d}$$

At the risk of redundancy, we stress again that the preceding individual transfer functions are either unaffected by the connections, or the transfer functions are defined under loaded conditions.

The preceding equations must be combined in a manner that will eliminate all variables from $Y_1(s)$ through $Y_{n-1}(s)$. A systematic way of achieving this is as follows: Substitute $Y_{n-1}(s)$ expressed in terms of $Y_{n-2}(s)$ from Equation (7–62c), in (7–62d). Next substitute $Y_{n-2}(s)$ from the assumed next equation (not shown) expressed in terms of $Y_{n-3}(s)$. This pattern continues until $Y_1(s)$ expressed in terms of $X(s)$ is reached. At that point, the resulting equation, appropriately arranged, reads

$$Y(s) = G_1(s)G_2(s) \cdots G_n(s)X(s) \tag{7-63}$$

The net transfer function is thus

$$G(s) = \frac{Y(s)}{X(s)} = G_1(s)G_2(s) \cdots G_n(s) \tag{7-64}$$

Stated in words, *the composite transfer function of a cascade connection of transfer functions is the product of all the individual transfer functions.*

Parallel Connection

A parallel connection of several individual transfer functions is shown at the top of Figure 7–21. Only the net input $X(s)$ and the net output $Y(s)$ are shown. However, define the outputs of the n blocks as $Y_1(s)$ through $Y_n(s)$. Since $X(s)$ is common to all of these, we can write

$$Y_1(s) = G_1(s)X(s) \tag{7-65a}$$
$$Y_2(s) = G_2(s)X(s) \tag{7-65b}$$

$$\begin{array}{c} \cdot \\ \cdot \\ \cdot \end{array}$$

$$Y_n(s) = G_n(s)X(s) \tag{7-65c}$$

These n variables are assumed to be combined together at the output according to the equation

$$Y(s) = Y_1(s) + Y_2(s) + \cdots + Y_n(s) \tag{7-66}$$

When the individual forms are substituted in Equation (7–66), $X(s)$ is a common factor, and we have

$$Y(s) = [G_1(s) + G_2(s) + \cdots + G_n(s)]X(s) \tag{7-67}$$

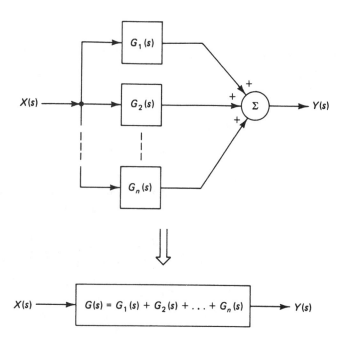

FIGURE 7–21
Parallel connection of transfer functions and an equivalent single transfer function.

The net transfer function is thus

$$G(s) = \frac{Y(s)}{X(s)} = G_1(s) + G_2(s) + \cdots + G_n(s) \tag{7-68}$$

Stated in words, *the composite transfer function of a parallel connection of transfer functions is the sum of all the individual transfer functions.*

Feedback Loop

A feedback loop, consisting of a forward transfer function $G_1(s)$ and a feedback transfer function $G_2(s)$ is shown at the top of Figure 7–22. Let $F(s)$ represent the output of the feedback block, and this quantity is subtracted from the input $X(s)$ to yield a difference variable $D(s)$, defined as

$$D(s) = X(s) - F(s) \tag{7-69}$$

The difference variable is multiplied by the forward transfer function to yield the output $Y(s)$, that is,

$$Y(s) = G_1(s)D(s) \tag{7-70}$$

The output is multiplied by the feedback transfer function to produce the feedback variable.

$$F(s) = G_2(s)Y(s) \tag{7-71}$$

The variables $F(s)$ and $D(s)$ are eliminated by substituting Equation (7–71) in Equation (7–69) and then substituting that result in Equation (7–70). After some manipulation, we obtain

$$Y(s) = \left[\frac{G_1(s)}{1 + G_1(s)G_2(s)} \right] X(s) \tag{7-72}$$

FIGURE 7–22
Transfer function with feedback loop and an equivalent single transfer function.

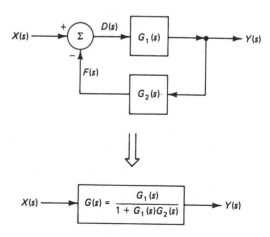

The transfer function is

$$G(s) = \frac{Y(s)}{X(s)} = \frac{G_1(s)}{1 + G_1(s)G_2(s)} \qquad \text{(7–73)}$$

The result of Equation (7–73) is one of the most important relationships of linear system theory. It serves as the basis for much of the design work of stable linear circuits and closed-loop feedback control systems. Our approach here is simply to show this as a simplification of feedback loops that might arise in working with transfer functions.

EXAMPLE 7–12
A simple resistive network will be used in this problem to illustrate how loading effects can affect or alter the simplified relations of this section. Consider the simple voltage divider of Figure 7–23(a). **(a)** Determine the transfer function of this circuit. **(b)** Assume that two of these sections are connected together as shown in Figure 7–23(b). Work out the composite transfer function and compare it with the value that would be obtained by employing the cascade formula of this section.

Solution
(a) Let $G_A(s) = V(s)/V_1(s)$ for the simple network of Figure 7–23(a). This result is immediately determined as

$$G_A(s) = \frac{1000}{1000 + 1000} = \frac{1}{2} \qquad \text{(7–74)}$$

(b) The transfer function of the circuit in Figure 7–23(b) can be determined on a circuit analysis basis by several approaches. This author's favorite method for this circuit is a successive Thevenin transformation approach as illustrated in Figure 7–23(c). Looking back from the output of the first stage, the open-circuit voltage is $V_1(s)/2$, and the equivalent resistance is 500 Ω. A further application of the voltage divider rule reads

$$G(s) = \frac{V_2(s)}{V_1(s)} = \frac{1000}{1000 + 1500} \times \frac{1}{2} = \frac{1}{5} \qquad \text{(7–75)}$$

Suppose, however, that one blindly used the result of the cascade simplification formula of Equation (7–64), as depicted on Figure 7–20. Since each section has an individual transfer function of 1/2, the formula would suggest a net transfer function of $(1/2) \times (1/2) = 1/4$, which is obviously different from the result obtained. What is wrong?

The answer to this difficulty lies in the interaction and loading effects of the two sections. Assume that an input voltage is applied to the single stage of Figure 7–23(a). If connecting the second stage to the output of the first stage alters the output voltage of the first stage, there will be a loading effect. For transfer functions to be multiplicative, there must be no alteration of the output

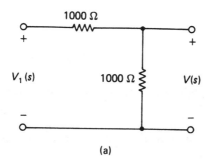

(a)

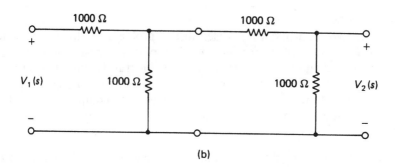

(b)

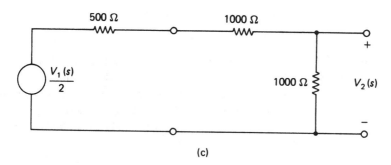

(c)

FIGURE 7–23
Circuits for Example 7–12.

variable of the first stage when the second stage is connected. In this circuit, the voltage across the output stage will no longer be $V_1(s)/2$ when the second stage is connected.

When the output voltage of a stage is the input voltage to the next stage, a necessary condition for minimal loading effect is that the input resistance looking into the second stage must be very large compared with the equivalent output resistance of the first stage (typically 100 times as great or more). In this circuit, the input resistance to the second stage is 2000 Ω and the output

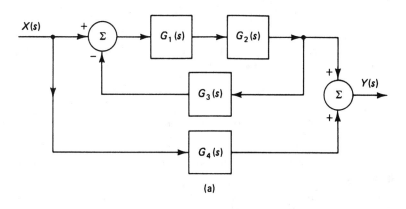

(a)

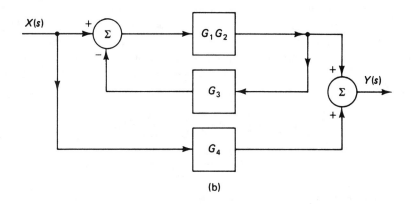

(b)

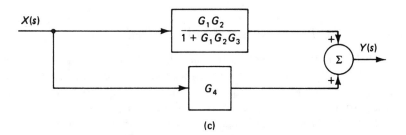

(c)

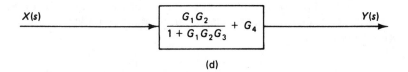

(d)

FIGURE 7–24
System for Example 7–13 and simplification steps.

resistance of the first stage is 500 Ω, so the desired inequality is not satisfied. The preceding discussion explains why the ideal desired output impedance of a voltage amplifier is zero and why the ideal desired input impedance is infinite. For a current amplifier, the desired situation is the opposite.

EXAMPLE 7–13

Determine a single transfer function equivalent to the system of Figure 7–24(a).

Solution

As stated numerous times earlier, it is assumed that the transfer functions shown are applicable under the loaded conditions given. To simplify the notation, the *s*-variable will be eliminated from the transfer functions in parts (b), (c), and (d) of Figure 7–24 and in the discussion that follows.

The first step is to simplify the cascade connection of G_1 and G_2, whose net transfer function is G_1G_2. The resulting system is shown in Figure 7–24(b). Next, the feedback loop consisting of G_1G_2 as the forward transfer function and G_3 as the feedback transfer function is simplified, as shown in Figure 7–24(c). Finally, the resulting parallel structure is simplified as shown in Figure 7–24(d). The overall transfer function is

$$G(s) = \frac{Y(s)}{X(s)} = \frac{G_1G_2}{1 + G_1G_2G_3} + G_4 \qquad \textbf{(7–76a)}$$

$$= \frac{G_1G_2 + G_4 + G_1G_2G_3G_4}{1 + G_1G_2G_3} \qquad \textbf{(7–76b)}$$

where the latter form uses a common denominator.

7–7 SECOND-ORDER STEP RESPONSE FORMS

The series RLC circuit was used in the last chapter to demonstrate the forms of second-order circuit responses. However, there are various other circuits and systems that produce second-order response forms. Any circuit that can be reduced to a form where there are two distinct energy storage elements is a second-order form.

If a circuit contains no feedback, both types of reactive elements (e.g., L and C) are required in order to produce the underdamped oscillatory response. Thus, a passive RC second-order circuit, for example, can exhibit only an overdamped or a critically damped response, but it cannot be underdamped. On the other hand, the presence of active devices and feedback can produce an underdamped response even without inductors.

Terminology

At this point, the terminology for the second-order circuit will be generalized somewhat. Let $P(s)$ represent a quadratic denominator factor with unity coefficient for the s^2 term. There are several forms used in the literature for such factors. Two closely related forms are

$$P(s) = s^2 + 2\alpha s + \omega_n^2 \qquad (7\text{--}77)$$

and

$$P(s) = s^2 + 2\zeta\omega_n s + \omega_n^2 \qquad (7\text{--}78)$$

The quantity ω_n, which appears in both forms, is called the *undamped natural oscillation frequency*. The quantity ζ in Equation (7–78) is called the *damping ratio*, and it is defined as

$$\zeta = \frac{\alpha}{\omega_n} \qquad (7\text{--}79)$$

The damping ratio is a dimensionless quantity, and it is widely used in specifying conditions for second-order circuits and systems.

When the circuit contains no damping, $\alpha = 0$ (and $\zeta = 0$), and the denominator of any pertinent voltage and current functions will contain the quadratic $s^2 + \omega_n^2$. The response will then contain an undamped sinusoidal function of radian frequency ω_n.

Next, the roots of Equations (7–77) and (7–78) will be determined using the notation of those expressions.

Let s_1 and s_2 represent these roots. We have

$$\begin{matrix} s_1 \\ s_2 \end{matrix} = -\alpha \pm \sqrt{\alpha^2 - \omega_n^2} \qquad (7\text{--}80a)$$

$$= -\zeta\omega_n \pm \omega_n\sqrt{\zeta^2 - 1} \qquad (7\text{--}80b)$$

The circuit is overdamped when $\alpha > \omega_n$ or when $\zeta > 1$. In this case, ω_n is not too meaningful because no oscillations actually occur. Instead, the two real roots constitute the two damping factors in the exponential terms of the natural response. The critically damped case corresponds to $\alpha = \omega_n$ or $\zeta = 1$.

The most meaningful situation for which the notation introduced applies is the underdamped case, in which $\alpha < \omega_n$ or $\zeta < 1$. It is convenient in this case to express Equations (7–80a) and (7–80b) as

$$\begin{matrix} s_1 \\ s_2 \end{matrix} = -\alpha \pm j\sqrt{\omega_n^2 - \alpha^2} \qquad (7\text{--}81a)$$

$$= -\zeta\omega_n \pm j\omega_n\sqrt{1 - \zeta^2} \qquad (7\text{--}81b)$$

Let ω_d represent the damped oscillation frequency, which was introduced earlier in the book for the series RLC circuit. The damped frequency is

$$\omega_d = \sqrt{\omega_n^2 - \alpha^2} = \omega_n\sqrt{1 - \zeta^2} \qquad (7\text{--}82)$$

The roots of the polynomial are then denoted as

$$\begin{matrix} s_1 \\ s_2 \end{matrix} = -\alpha \pm j\omega_d \tag{7-83}$$

The frequency ω_d represents the actual oscillation frequency in the natural response for the underdamped case. From Equation (7–82), it is seen that ω_d is smaller than ω_n, although for small values of ζ, the two frequencies are quite close to each other.

We will next consider a common form of a second-order transfer function that arises frequently in practice, and some curves displaying the step response forms will be given. The transfer function to be considered, a *low-pass second-order response,* is given by

$$G(s) = \frac{Y(s)}{X(s)} = \frac{\omega_n^2}{s^2 + 2\zeta\omega_n s + \omega_n^2} \tag{7-84}$$

Let $x(t) = u(t)$ represent a step function excitation of unit magnitude. The transform of $x(t)$ is $X(s) = 1/s$. The output transform is

$$Y(s) = G(s)X(s) = \frac{\omega_n^2}{s(s^2 + 2\zeta\omega_n s + \omega_n^2)} \tag{7-85}$$

When the system is critically damped, the inverse transform of Equation (7–85) can be expressed as

$$y(t) = 1 - (\omega_n t + 1)e^{-\omega_n t} \quad \text{for } \zeta = 1$$

When the system is underdamped, the inverse transform of Equation (7–85) can be expressed as

$$y(t) = 1 - \frac{e^{-\zeta\omega_n t}}{\sqrt{1 - \zeta^2}} \sin\left(\sqrt{1 - \zeta^2}\,\omega_n t + \cos^{-1}\zeta\right) \tag{7-86}$$

Curves depicting the behavior of $y(t)$ for different values of ζ are shown in Figure 7–25. In addition to the critically damped and underdamped cases considered, two curves for the overdamped case are also given ($\zeta = 1.5$ and $\zeta = 2.0$).

Incidentally, the voltage across a capacitor in a series RLC circuit excited by a dc input as considered in Chapter 6 is an example of the type of response just considered.

EXAMPLE 7–14
Consider the series circuit of Example 6–11 [Figure 6–17(a)]. Determine the values of ω_n and ζ.

Solution
Referring back to Equations (6–73) through (6–78), the second-degree polynomial appearing in the response functions was determined to be $s^2 + 200s + 10^6$. When

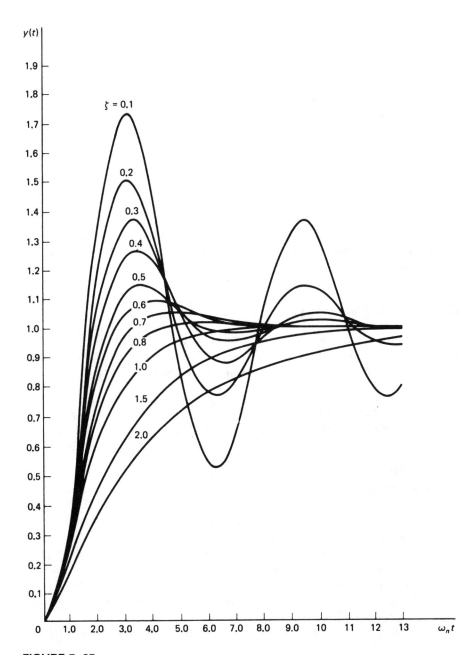

FIGURE 7–25
Normalized step response of second-order low-pass transfer function. (Input step function has unit magnitude.)

this function is compared with the general form $s^2 + 2\zeta\omega_n s + \omega_n^2$ as defined in this section, the following identification can be made:

$$\omega_n^2 = 10^6 \qquad \qquad \textbf{(7–87)}$$

$$2\zeta\omega_n = 200 \qquad \qquad \textbf{(7–88)}$$

This leads to

$$\omega_n = 10^3 \, \text{rad/s} \qquad \qquad \textbf{(7–89)}$$

$$\zeta = \frac{200}{2\omega_n} = 0.1 \qquad \qquad \textbf{(7–90)}$$

The reader may wish to compare the function $v_c(t)$ shown in Figure 6–18 with the step response corresponding to $\zeta = 0.1$ in Figure 7–25. Specific values of magnitude and time were assumed in Figure 6–18, of course, but the forms of the two functions should be identical.

7–8 PSPICE EXAMPLES

In view of the system emphasis in this chapter, the PSPICE examples considered will all use that approach. In particular, the LAPLACE command will be introduced. This operation allows a transfer function to be specified directly in its s-domain form.

PSPICE EXAMPLE 7–1

Use the PSPICE Laplace transform operation and the transfer function to determine the response of Example 7–3(a).

Solution

No circuit diagram is necessary in this example, since the transfer function determined in Examples 7–1 and 7–2 will be employed directly in the analysis. PSPICE allows a direct operation to be performed on a transfer function to determine the response caused by many different types of excitations.

Refer to the code shown in Figure 7–26. First, the form of the input excitation must be defined, which in this case is a constant voltage of 20 V turned on at $t = 0$. This is achieved with a PWL function for V1 starting at a value of 0 at $t = 0$ and reaching a level of 20 in 0.1 ms. (This rise time is very short compared with the time constant of 100 ms.)

The Laplace operation must be entered in the description of either a VCVS

FIGURE 7–26
Code for PSPICE Example 7–1.

```
PSPICE EXAMPLE 7-1
V1 1 0 PWL 0 0 1E-4 20
E 2 0 LAPLACE {V(1)}={10/(S+10)}
.TRAN 0.5 0.5
.PROBE
.END
```

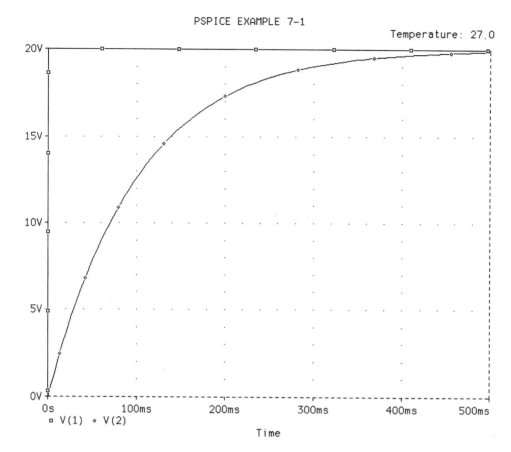

FIGURE 7–27
Waveform for PSPICE Example 7–1.

or a VCIS. Choosing the former in this case, the source E is defined between nodes 2 and 0. The remainder of the line reads

```
LAPLACE {V(1)} = {10/(S+10)}
```

Note the use of the brackets { }, which are required on both sides of the = sign. The left-hand side is the voltage at the node representing the input, and the right-hand side is the transfer function. The "=" is a little misleading, because for this analysis, it is behaving more like a multiplication. The whole expression may be interpreted as "multiply the Laplace transform of the voltage at node 1, by the transfer function $10/(s + 10)$, and perform a numerical inverse transform."

A transient run must be made to provide a printout. The resulting plots for the input and output voltages are shown in Figure 7–27. These results clearly agree with the mathematical form of the solution.

FIGURE 7–28
Code for PSPICE Example 7–2.

```
PSPICE EXAMPLE 7-2
V1 1 0 SIN 0 20 1.59155
E 2 0 LAPLACE {V(1)}={10/(S+10)}
.TRAN 2 2
.PROBE
.END
```

PSPICE EXAMPLE 7–2

Repeat the analysis of PSPICE Example 7–1 as applied to Example 7–3(b).

Solution

The code is shown in Figure 7–28. The form is the same as in PSPICE Example 7–1, except for one difference. The input voltage V1 is now a sinusoidal function with no dc component, an amplitude of 20 V, and a frequency of 1.59155 Hz $(10/2\pi)$. Following the transient run, plots of the input and output voltages are shown in Figure 7–29. The output voltage contains both a natural response and a

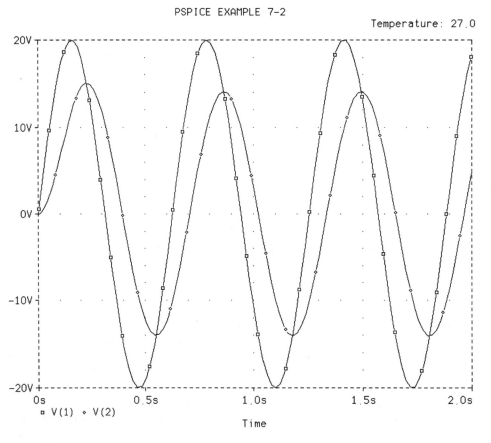

FIGURE 7–29
Waveforms for PSPICE Example 7–2.

```
PSPICE EXAMPLE 7-3
V1 1 0 PWL 0 0 1E-4 1
E 2 0 LAPLACE {V(1)}={1/(S*S+1.4142*S+1)}
.TRAN 10 10
.PROBE
.END
```

FIGURE 7-30
Code for PSPICE Example 7-3.

forced response, but after a short time interval, the output forced response is observed to be a sinusoid with the same frequency as the input.

PSPICE EXAMPLE 7-3

Use the PSPICE LAPLACE function to determine the step response of the Butterworth filter of Example 7-5.

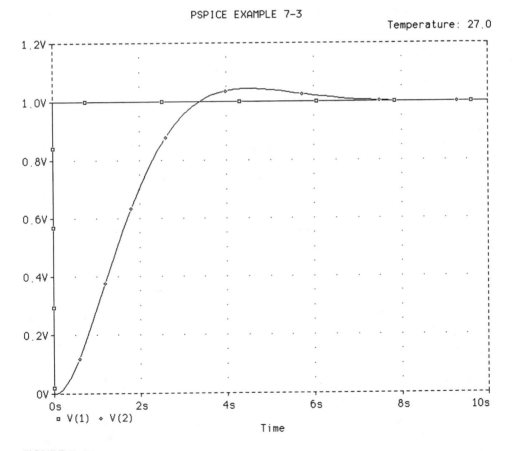

FIGURE 7-31
Waveforms for PSPICE Example 7-3.

Solution

The code is shown in Figure 7–30. The input is a "unit step function," that is, a constant voltage of unity magnitude switched on at $t = 0$. The only further difference between this Example and the previous two is that the transfer function is now a second-order form. When higher powers of S are required, they may be coded by two possible forms. One form is simply to write higher powers of S as successive multiples of S. Thus, in this case S² is written as S*S. An alternate form is PWR(S, N) where N is the required power. Plots of the input and output voltages obtained with PROBE are shown in Figure 7–31.

7–9   MATLAB EXAMPLES

Several different MATLAB operations that are useful in dealing with transfer functions will be illustrated in this section. This will include the impulse response, which was included earlier as a tool for determining the inverse transform, as well as the step response. Other operations relating to the manipulation of zeros and poles in the *s*-plane will also be considered.

MATLAB EXAMPLE 7–1

Use MATLAB to determine numerically and plot the impulse response of the transfer function of Example 7–4 (Equation 7–21).

Solution

We have already used the impulse response command to determine the inverse transforms of two Laplace functions (MATLAB Examples 5–1 and 6–1), so the procedure here should be straightforward. At this point, however, the concept should make a little more sense. Since $\mathscr{L}[\delta(t)] = 1$ and $Y(s) = X(s)G(s) = 1 \cdot G(s) = G(s)$, the inverse Laplace transform of $G(s)$ is the impulse response, which is usually denoted as $g(t)$. For convenience, the transfer function is repeated here as

$$G(s) = \frac{1}{s^3 + 2s^2 + 2s + 1} \tag{7–91}$$

Refer to Figure 7–32 in the discussion that follows. The numerator and denominator polynomials are defined, respectively, as *n* and *d*, and they are written as

```
n = [0 0 0 1]
```

and

```
d = [1 2 2 1]
```

It may take a little trial and error to determine an appropriate time range to fully display the nature of the function. After some experimentation, it was decided

```
EDU» n=[0 0 0 1]

n =

    0    0    0    1

EDU» d=[1 2 2 1]

d =

    1    2    2    1

EDU» t=0:0.05:12;
EDU» g=impulse(n,d,t);
EDU» plot(t,g)
EDU» xlabel('time')
EDU» ylabel('response')
EDU» title('Impulse Response of Filter of Example 7-4')
EDU»
```

FIGURE 7–32
Workspace for MATLAB Example 7–1.

to use a time range from 0 to 12 seconds in steps of 0.05 s. This was achieved through the command.

```
t = 0:0.05:12;
```

The impulse command is then given as

```
g = impulse(n,d,t);
```

The remainder of the workspace is devoted to appropriate labeling, and the final result is displayed in Figure 7–33.

MATLAB EXAMPLE 7–2
Use MATLAB to determine numerically and plot the step response of the transfer function of MATLAB Example 7–1.

Solution
One approach would be to multiply $G(s)$ of Equation (7–21) by $1/s$ and then use the impulse response command to perform an inverse transformation. However, MATLAB provides a direct command for determining the step response from the transfer function. Letting vs represent the step response, and with the use of arrays defined in the preceding example, the command is

```
vs = step(n,d,t)
```

The resulting function after appropriate labeling is shown in Figure 7–34.

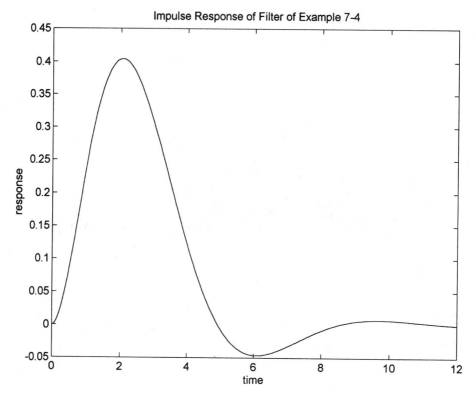

FIGURE 7–33
Waveform for MATLAB Example 7–1.

MATLAB EXAMPLE 7-3

Use MATLAB to construct a pole–zero plot for the transfer function of Example 7–8 (Equation 7–51).

Solution
After eliminating the parentheses in the numerator by multiplying all terms by 2, the function of Equation (7–51) can be rewritten as

$$G(s) = \frac{2s^2 + 12s + 50}{s^3 + 7s^2 + 10s} \tag{7–92}$$

Refer now to Figure 7–35, and mentally visualize both numerator and denominator of Equation (7–92) as cubic polynomials by adding appropriate zeros. The numerator and denominator coefficient row vectors, denoted respectively as n and d, can be expressed as

```
n = [0 2 12 50]
d = [1 7 10 0]
```

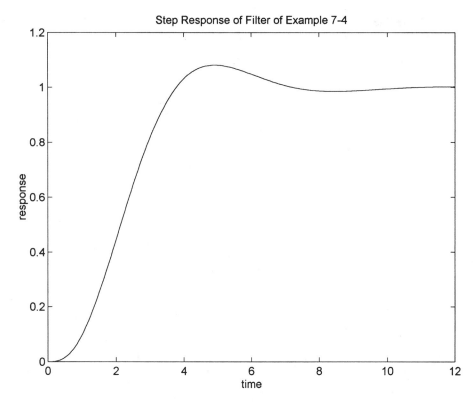

FIGURE 7–34
Waveform for MATLAB Example 7–2.

The command for generating a plot of the poles and zeros is simply

```
zplane(n,d)
```

After the additional "grid" command, the plot is shown in Figure 7–36.

MATLAB EXAMPLE 7–4
Continuing with the function of Example 7–8, demonstrate the manipulation of poles and zeros and the associated transfer functions through three separate operations: **(a)** polynomial factorization, **(b)** transfer function to zero–pole conversion, and **(c)** zero–pole to transfer function conversion.

Solution
(a) The first method is the basic process of factoring the separate polynomials to determine their roots, as illustrated in Figure 7–37. First, the numerator and denominator polynomial coefficients are entered as row vectors in the same manner as in MATLAB Example 7–3. Next, the roots of the numerator polyno-

FIGURE 7–35
Workspace for MATLAB Example 7–3.

```
EDU» n=[0 2 12 50]

n =

        0      2     12     50

EDU» d=[1 7 10 0]

d =

        1      7     10      0

EDU» zplane(n,d)
EDU» grid
EDU»
```

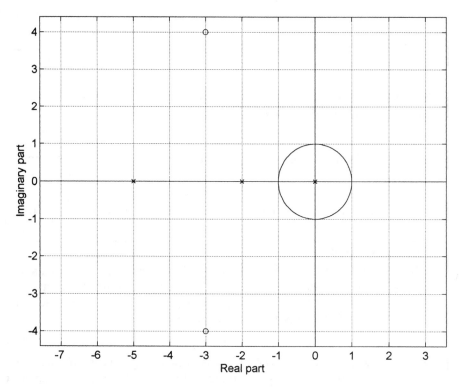

FIGURE 7–36
Pole–zero plot for MATLAB Example 7–3.

FIGURE 7–37
Workspace for MATLAB Example 7–4, part (a).

```
EDU» n=[0 2 12 50]

n =

        0      2     12     50

EDU» d=[1 7 10 0]

d =

        1      7     10      0

EDU» roots(n)

ans =

     -3.0000 + 4.0000i
     -3.0000 - 4.0000i

EDU» roots(d)

ans =

        0
       -5
       -2

EDU»
```

mial n are determined by the simple command

```
roots(n)
```

Finally, the roots of the denominator polynomial are determined by the same command applied to d, which is

```
roots(d)
```

After each command is entered, the roots are automatically displayed. Note that MATLAB uses the more common symbol i to represent $\sqrt{-1}$, and it appears after the imaginary part.

(b) The second method performs the same operation as (a) but accomplishes the result in one step, along with determining the gain constant. Let z represent the column of zeros, p the column of poles, and k the gain constant. Referring to Figure 7–38, the command is

```
[z,p,k] = tf2zp(n,d)
```

The command "tf2zp" can be thought of as "transfer function to zero–pole." The desired variables are then listed as shown.

(c) In this part, we will simply reverse the operation and show how the transfer function can be determined from the zeros, poles, and gain constant. Refer to Figure 7–39. Although MATLAB always prints i for the imaginary number, it

FIGURE 7–38
Workspace for MATLAB Example 7–4, part (b).

```
EDU» n=[0 2 12 50]

n =

     0     2    12    50

EDU» d=[1 7 10 0]

d =

     1     7    10     0

EDU» [z,p,k]=tf2zp(n,d)

z =

  -3.0000 + 4.0000i
  -3.0000 - 4.0000i


p =

     0
    -5
    -2


k =

     2

EDU»
```

will accept either *i* or *j* as input. The *i* or *j* follows the imaginary part. First we enter the zeros as a column vector denoted as *z* by the command

```
z = [-3+4j;-3-4j]
```

Note that the semicolon is required between entries. Next, the poles are entered as a column vector by the command

```
p = [0;-2;-5]
```

Last, the gain constant *k* is entered as

```
k = 2
```

The command for determining the transfer function is

```
[n d] = zp2tf(z,p,k)
```

The command "zp2tf" can be thought of as "zero–pole to transfer function." The results are the numerator and denominator row vectors that we started with, thus completing a full cycle.

FIGURE 7–39
Workspace for MATLAB Example 7–4, part (c).

```
EDU» z=[-3+4j;-3-4j]

z =

   -3.0000 + 4.0000i
   -3.0000 - 4.0000i

EDU» p=[0;-2;-5]

p =

     0
    -2
    -5

EDU» k=2

k =

     2

EDU» [n,d]=zp2tf(z,p,k)

n =

     0     2    12    50


d =

     1     7    10     0

EDU»
```

DRILL PROBLEMS

7–1. Determine the transfer function of the circuit of Figure P7–1. The input is $v_1(t)$ and the output is $v_2(t)$.

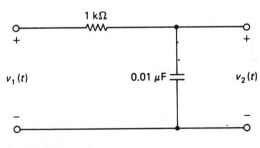

FIGURE P7–1

7-2. Determine the transfer function of the circuit of Figure P7-2. The input is $v_1(t)$ and the output is $v_2(t)$.

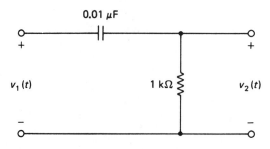

FIGURE P7-2

7-3. Determine the transfer function of the circuit of Figure P7-3, and show that it is the same function as obtained for the circuit of Problem 7-1. The input is $v_1(t)$ and the output is $v_2(t)$.

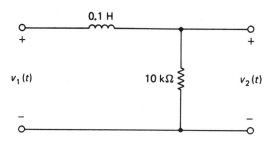

FIGURE P7-3

7-4. Determine the transfer function of the circuit of Figure P7-4, and show that it is the same function as obtained for the circuit of Problem 7-2. The input is $v_1(t)$ and the output is $v_2(t)$.

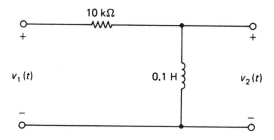

FIGURE P7-4

7–5. For the circuit of either Problem 7–1 or 7–3, determine (by using the transfer function concept) the output $v_2(t)$ when the input is excited at $t = 0$ by **(a)** $v_1(t) = 20$ V and **(b)** $v_1(t) = 20 \cos 10^4 t$. For each solution, determine $v_2(0^+)$, and comment on its significance.

7–6. For the circuit of either Problem 7–2 or 7–4, determine (by using the transfer function concept) the output $v_2(t)$ when the input is excited at $t = 0$ by **(a)** $v_1(t) = 20$ V and **(b)** $v_1(t) = 20 \cos 10^4 t$. For each solution, determine $v_2(0^+)$, and comment on its significance.

7–7. The transfer function of a normalized third-order passive low-pass Butterworth filter was determined in Example 7–4. That particular filter was designed for a current source input. A corresponding filter designed for a voltage source input is shown in Figure P7–7. Determine the transfer function $G(s) = V_2(s)/V_1(s)$ and show that it is the same as obtained in Example 7–4, namely, Equation (7–21).

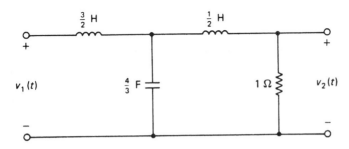

FIGURE P7–7

7–8. Determine the transfer function of the circuit of Figure P7–8. The input is $v_1(t)$ and the output is $v_2(t)$.

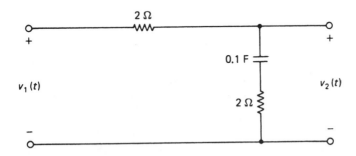

FIGURE P7–8

7–9. Determine the transfer function of the circuit of Figure P7–9. The input is $v_1(t)$ and the output is $v_2(t)$.

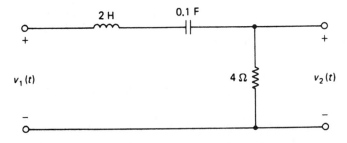

FIGURE P7-9

7-10. Determine the transfer function of the circuit of Figure P7-10. The input is $i_1(t)$ and output is $v_2(t)$.

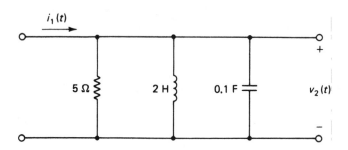

FIGURE P7-10

7-11. The transfer function of a normalized second-order active low-pass Butterworth filter was determined in Example 7-5. That particular filter used a unity gain amplifier. A different realization in which the two capacitors have the same values is shown in Figure P7-11. The VCVS now has a gain of 1.5858. Determine the transfer function

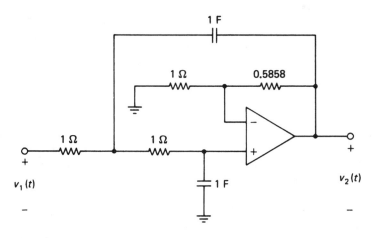

FIGURE P7-11

$G(s) = V_2(s)/V_1(s)$ and show that it has the same *form* as obtained in Example 7–5. (The denominator polynomial is the same, but the numerator constant is different.)

7–12. The circuit of Figure P7–12 is one particular active realization of a second-order Butterworth high-pass filter using an operational amplifier, and it is normalized to a cutoff frequency of 1 rad/s with 1-Ω resistors. With $v_1(t)$ as the input and $v_2(t)$ as the output, determine the transfer function. The operational amplifier is connected as a voltage follower and it is modeled in the same manner as in Example 7–5.

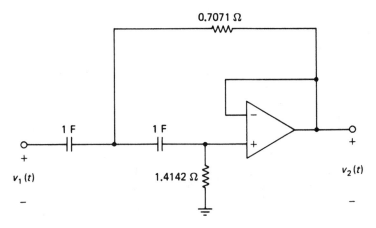

FIGURE P7–12

7–13. The transfer function of a certain system is given by

$$G(s) = \frac{Y(s)}{X(s)} = \frac{N(s)}{(s^2 + 7s + 12)(s^2 + 4s + 29)}$$

where $N(s)$ is an unspecified numerator polynomial. Using arbitrary constant magnitudes and phase angles, write the general form of the response $y(t)$ for **(a)** $x(t) = 20$, and **(b)** $x(t) = 20 \sin 4t$. In each case, identify the natural response and the forced response. Assume that no roots of $N(s)$ are the same as poles of either $G(s)$ or $X(s)$.

7–14. The transfer function of a certain circuit is given by

$$G(s) = \frac{V_2(s)}{V_1(s)} = \frac{N(s)}{(s^2 + 2s + 5)(s^2 + 4s + 13)}$$

where $N(s)$ is an unspecified numerator polynomial. Using arbitrary constant magnitudes and phase angles, write the general form of the response $y(t)$ for **(a)** $x(t) = 10 \sin 2t$ and **(b)** $x(t) = 10e^{-2t}$. In each case, identify the natural response and the forced response. Comment on the term *steady-state response* as it relates to case (b). Assume that no roots of $N(s)$ are the same as poles of either $G(s)$ or $X(s)$.

7–15. Write the form of the transfer function corresponding to the s-plane plot shown in Figure P7–15. An arbitrary constant multiplier A may be assumed.

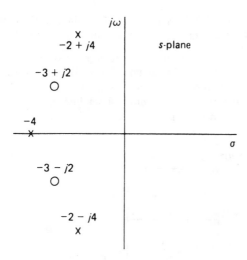

FIGURE P7–15

7–16. Write the form of the transfer function corresponding to the *s*-plane plot shown in Figure P7–16. An arbitrary constant multiplier *A* may be assumed.

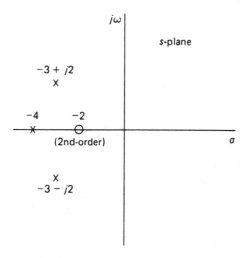

FIGURE P7–16

7–17. A certain transfer function is given by

$$G(s) = \frac{s(s+2)}{(s+4)(s^2 + 6s + 34)}$$

Determine the poles and zeros, and construct an *s*-plane plot of the finite poles and zeros.

7–18. A certain transfer function is given by

$$G(s) = \frac{s + 3}{(s + 2)(s^2 + 9)}$$

Determine the poles and zeros, and construct an s-plane plot of the finite poles and zeros.

7–19. For the pole–zero plot of Problem 7–15, classify the stability of the system according to the three categories defined in the text.

7–20. For the pole–zero plot of Problem 7–16, classify the stability of the system according to the three categories defined in the text.

7–21. For the transfer function of Problem 7–17, classify the stability of the system according to the three categories defined in the text.

7–22. For the transfer function of Problem 7–18, classify the stability of the system according to the three categories defined in the text.

7–23. Determine a single transfer function equivalent to the system of Figure P7–23.

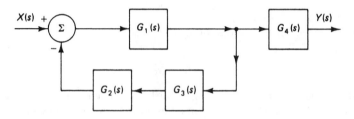

FIGURE P7–23

7–24. Determine a single transfer function equivalent to the system of Figure P7–24.

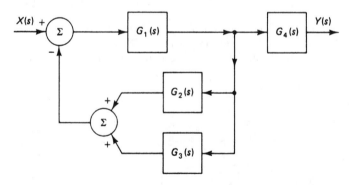

FIGURE P7–24

7–25. In the series RLC circuit of Problem 6–25 (Figure P6–25), determine the values of ω_n and ζ.

7–26. In the series RLC circuit of Problem 6–26 (Figure P6–26), determine the values of ω_n and ζ.

7–27. In the parallel RLC circuit of Problem 6–27 (Figure P6–27), determine the values of ω_n and ζ.

7–28. In the circuit of Problem 6–28 (Figure P6–28), determine the values of ω_n and ζ.

DERIVATION PROBLEMS

7–29. Prove that the response of Equation (7–53) is a stable response by showing that

$$\lim_{t \to \infty} y_n(t) = 0$$

(*Hint:* Consider L'Hôpital's rule.)

7–30. Prove that the response of Equation (7–55) is a stable response by showing that

$$\lim_{t \to \infty} y_n(t) = 0$$

(See hint in Problem 7–29.)

7–31. Consider a transfer function of the form

$$G(s) = \frac{N(s)}{s^2 + bs + c}$$

where $N(s)$ is the numerator polynomial, which is assumed to be of degree two or less. (Note that the factor of s^2 has been adjusted to be unity.) Show that necessary and sufficient conditions for $G(s)$ to be stable are $b > 0$ and $c > 0$.

7–32. Consider the normalized active filter circuit shown in Figure P7–32.

(a) Show that the transfer function is

$$G(s) = \frac{K}{s^2 + s(3 - K) + 1}$$

(b) Show that the circuit is stable for $K < 3$. (*Hint:* Refer to Problem 7–31.)

(c) Show that the circuit becomes marginally stable for $K = 3$.

(d) Determine the radian frequency of the natural oscillation under the condition of (c).

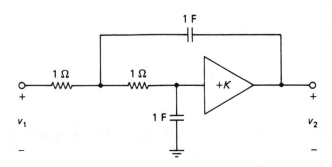

FIGURE P7–32

7-33. A somewhat general circuit form from which a variety of active filter types, such as considered in this chapter, can be derived is shown in Figure P7–33. The amplifier is assumed to be an ideal VCVS with voltage gain K. Each of the blocks is a passive admittance. Show that the transfer function is

$$G(s) = \frac{V_2(s)}{V_1(s)} = \frac{KY_1Y_2}{Y_1Y_2 + Y_1Y_4 + Y_2Y_4 + Y_3Y_4 + Y_2Y_3(1 - K)}$$

(For brevity, the s arguments have been deleted from the admittances in this expression.)

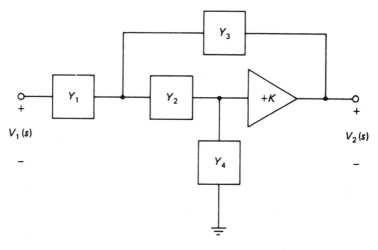

FIGURE P7–33

7-34. Consider the net transfer function for a feedback loop as given by Equation (7–73). Show that the system is marginally stable if

$$G_1(j\omega_0)G_2(j\omega_0) = -1$$

where the $j\omega$-axis poles associated with the marginally stable condition are $\pm j\omega_0$.

APPLICATION PROBLEMS

7-35. In network *synthesis,* a desired transfer function is stated, and a network is determined that will achieve the given transfer function. Although the subject is primarily a graduate-level topic, certain limited forms can be considered here. Assume that it is desired to realize the transfer function

$$G(s) = \frac{V_2(s)}{V_1(s)} = \frac{5}{s^2 + 2s + 5}$$

From "experience," the circuit of Figure P7–35 will work. By matching the transfer function for the circuit in general to the particular form desired, determine the values

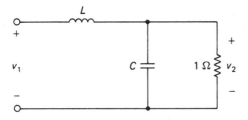

FIGURE P7–35

of L and C required if the resistance is selected as $R = 1\ \Omega$ in this *normalized* design. (The reason one element value is specified is that there is one more parameter value to be determined than the number of constraints. In such a case, one element value is arbitrarily selected.)

7–36. Assume that it is desired to realize the transfer function

$$G(s) = \frac{V_2(s)}{V_1(s)} = \frac{2s}{s^2 + 2s + 5}$$

The circuit of Figure P7–36 can be used. Select $R = 1\ \Omega$, and determine L and C. (See discussion with Problem 7–35.)

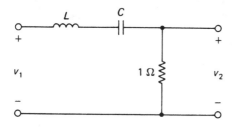

FIGURE P7–36

7–37. Assume that it is desired to realize the transfer function

$$G(s) = \frac{s^2}{s^2 + 2s + 5}$$

The circuit of Figure P7–37 can be used. Select $R = 1\ \Omega$, and determine L and C. (See discussion with Problem 7–35.)

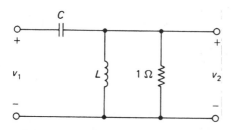

FIGURE P7–37

7–38. Several *normalized* circuits were considered in the chapter examples and in earlier problems. Such normalized circuits are developed with simple frequency values and resistance levels. To convert these circuits to practical levels, we define a frequency scaling constant K_f and a resistance scaling constant K_r as follows:

$$K_f = \frac{\text{desired cutoff frequency}}{\text{normalized cutoff frequency}}$$

$$K_r = \frac{\text{desired resistance level}}{\text{normalized resistance level}}$$

Element values are modified as follows:

1. Multiply all L's by $\dfrac{K_r}{K_f}$.

2. Multiply all C's by $\dfrac{1}{K_f K_r}$.

3. Multiply all R's by K_r.

Consider the Butterworth filter of Example 7–5 [Figure 7–6(a)], which used 1-Ω resistors and had a cutoff frequency of 1 rad/s. Design a practical circuit having a cutoff frequency of 1 kHz and 10-kΩ resistors, that is, determine the resistance and the capacitance values. (*Note:* the normalized design frequency is expressed in rad/s, and the desired design frequency is expressed in Hz. The frequency scales must have the same units in defining K_f.)

8

SINUSOIDAL STEADY-STATE ANALYSIS

OBJECTIVES

After completing this chapter, the reader should be able to:

- Represent a sinusoidal voltage or current as a complex phasor.
- Obtain a single sinusoid equivalent to the sum of several sinusoids of the same frequency using phasor analysis.
- Define the following terms based on the sinusoidal steady-state: *impedance, resistance, reactance, admittance, conductance,* and *susceptance.*
- Determine inductive reactance and susceptance for a given inductance.
- Determine capacitive reactance and susceptance for a given capacitance.
- Transform a complete circuit to the steady-state phasor form.
- Apply basic circuit analysis methods to phasor circuit models to determine complete solutions.
- Compute average power in a sinusoidal steady-state circuit.

8-1 GENERAL DISCUSSION

The term *sinusoidal steady state* refers to a class of circuits and circuit analysis procedures in which the following two conditions are generally met:

1. All sources are sinusoidal in form, and all have the same frequency. (The analysis can actually be extended, with some care, to more than one frequency by the use of Fourier series, and that concept will be developed in Chapter 11.)

2. The circuit is stable, that is, all natural response terms vanish after a sufficient time has passed.

Recall that the forced response has the same general form as the source or sources exciting the circuit. Specifically, when all sources are sinusoidal functions of a given frequency, any forced response is a sinusoidal function of the same frequency. Since the natural response is a transient response, the forced response represents a steady-state response, which remains after the transient disappears.

The conclusion of the preceding paragraph must be strongly emphasized: *Any steady-state voltage or current response of a linear, stable, constant parameter circuit excited by a sinusoid is a sinusoid of the same frequency.*

Since any steady-state circuit response will have the same frequency as the source, the form of the response can be readily specified. However, two parameters of any desired response must be determined: (a) amplitude or peak value of the response and (b) phase shift of the response with respect to that of the source or sources.

Phasor Analysis

Sinusoidal steady-state circuit analysis is a technique for determining the magnitude and angle of any response within a circuit under the steady-state conditions previously discussed. Other terms referring to this method are *ac circuit analysis* and *phasor analysis.* The complex number system is used extensively in sinusoidal steady-state analysis.

Historically, steady-state sinusoidal methods were widely used in circuit analysis before transform methods became popular. In fact, many readers will likely have studied ac circuit analysis in basic courses prior to studying transform methods. The decision to postpone steady-state sinusoidal methods in this book was based on the premise that a better understanding of the capabilities and limitations of this method would result after studying the more general Laplace transform methods.

Why is the sinusoidal steady state such an important part of circuit analysis? First, sinusoidal functions are among the most common of all voltage and current waveforms encountered in practice. Examples range from commercial power generation to the waveforms produced by laboratory function generators. Second, in many circuits actually designed for nonsinusoidal waveforms, sinusoidal techniques are used to measure and characterize the circuits. For example, the frequency response of an audio amplifier is specified in terms of its response to a sinusoidal input at different frequencies. The logic in doing this is based on the property that nonsinusoidal waveforms can be represented as a combination of sinusoidal functions. (This concept will be developed in Chapter 11.)

The material to be presented in the remainder of this chapter and in Chapter 9 requires some facility in performing arithmetic operations with complex numbers. Readers not having a background in this subject may refer to Appendix B for assistance.

8-2 PHASOR REPRESENTATION OF VOLTAGES AND CURRENTS

Steady-state ac circuit analysis is based on representing all voltages and currents by *complex phasors*. To develop the phasor concept, we start with *Euler's formula*, which reads

$$e^{j\theta} = \cos\theta + j\sin\theta \tag{8-1}$$

where $j = \sqrt{-1}$. The quantity $e^{j\theta}$ is a complex value, and it consists of both a real part and an imaginary part. The real part is $\cos\theta$, and the imaginary part is $\sin\theta$. Let $Re[\ ]$ represent a short-hand notation for the "real part of," with reference to the quantity in brackets, and let $Im[\ ]$ represent a corresponding notation for the "imaginary part of." We can then state that

$$\cos\theta = Re[e^{j\theta}] \tag{8-2}$$

$$\sin\theta = Im[e^{j\theta}] \tag{8-3}$$

Thus, the cosine and sine functions can be expressed as the real and imaginary parts, respectively, of a complex exponential function. Note that the imaginary part of a function is actually a real quantity. The term *imaginary part* refers to that real quantity which is "tagged" by the j factor.

Next, the argument of the exponential will be changed from $j\theta$ to $j(\omega t + \theta)$, and we can now write

$$e^{j(\omega t+\theta)} = \cos(\omega t + \theta) + j\sin(\omega t + \theta) \tag{8-4}$$

for which we can say

$$\cos(\omega t + \theta = Re[e^{j(\omega t+\theta)}] \tag{8-5}$$

$$\sin(\omega t + \theta) = Im[e^{j(\omega t+\theta)}] \tag{8-6}$$

The argument of the sine and cosine functions is now in the form in which sinusoidal sources in circuit analysis are best represented.

The concept of phasor analysis is based on representing all sinusoidal functions as the real or the imaginary parts of complex exponentials. The reason is that all differentiation and integration operations become simple algebraic operations with the exponential function, which is not the case with sine and/or cosine functions.

From Equation (8-5) or (8-6), we note that neither the sine nor the cosine function is *equal* to the complex exponential, but each is a *part of* the exponential. By a process called *analytic continuation* in complex variable theory, we extend or represent a sine or cosine function by the complex exponential. Since the complex exponential function can always be separated into its real and imaginary parts, any eventual results arising from linear analysis can be separated by the real and imaginary designations. Thus, to represent a sine or a cosine by a complex exponential, we have to "carry along extra baggage," but it can be readily sorted out and eliminated later.

Either the cosine or the sine function can be used as the basis for the development. If the cosine function is used, the real part concept is utilized, and if the sine function is used, the imaginary part concept is utilized. To correlate with the earlier transform analysis, the imaginary part concept will be used in this book.

Representation of Sinusoid by a Phasor

Assume a reference sinusoidal current $i(t)$ of the form

$$i(t) = I_p \sin(\omega t + \theta) \tag{8–7}$$

where I_p is the *peak* value or *amplitude*. From Equation (8–6), $i(t)$ can be expressed exactly as

$$i(t) = Im[I_p e^{j(\omega t + \theta)}] \tag{8–8}$$

The imaginary part designation is awkward, and it has become accepted practice simply to write

$$i(t) = I_p e^{j(\omega t + \theta)} \tag{8–9}$$

with the imaginary part concept understood. As discussed in the preceding paragraph, we are actually representing the real current by the complex exponential, but in the final analysis, it is the imaginary part to which our interest is directed.

Next, Equation (8–9) can be rewritten slightly as

$$i(t) = I_p e^{j\theta} e^{j\omega t} \tag{8–10}$$

We will define a complex quantity $\bar{I}$ as

$$\bar{I} = I_p e^{j\theta} \overset{\text{def}}{=} I_p \underline{/\theta} \tag{8–11}$$

where the last form in Equation (8–11) is a popular shorthand form used in electrical circuit analysis.

Before commenting on this result, consider next a voltage waveform $v(t)$ of the form

$$v(t) = V_p \sin(\omega t + \phi) \tag{8–12}$$

In accordance with the preceding development, $v(t)$ can be expressed as

$$v(t) = V_p e^{j(\omega t + \phi)} = V_p e^{j\phi} e^{j\omega t} = \bar{V} e^{j\omega t} \tag{8–13}$$

where

$$\bar{V} = V_p e^{j\phi} \overset{\text{def}}{=} V_p \underline{/\phi} \tag{8–14}$$

The quantities $\bar{I}$ in Equation (8–11) and $\bar{V}$ in Equation (8–14) are complex *phasors*. All voltages and currents in steady-state ac circuit analysis are represented by phasors of this type. Note that the $e^{j\omega t}$ factor does not appear in the phasor definition. Rather, a phasor is multiplied by the $e^{j\omega t}$ factor before the sinusoidal time function is reconstructed.

FIGURE 8–1
Phasor representations for typical voltage and current.

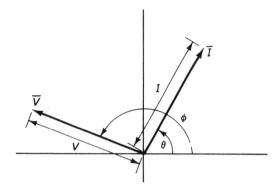

A phasor is characterized by a *magnitude* and a *phase angle*. The magnitude of $\bar{I}$ in Equation (8–11) is I_p, and its phase angle is θ. The magnitude of $\bar{V}$ in Equation (8–14) is V_p, and its phase angle is ϕ. The graphical representations for these phasors are shown in Figure 8–1 with arbitrary values of magnitudes and phase angles assumed. The graphical interpretation is a most useful property of the phasor concept, as we will see shortly.

EXAMPLE 8–1
Represent the voltage $v(t) = 50 \sin (10^3t + 60°)$ as a phasor.

Solution
The phasor form is $\bar{V} = 50\underline{/60°}$.

8–3 PHASOR COMBINATIONS OF SINUSOIDS

Before studying the specific details of ac circuit analysis, we will investigate how a sum of several sinusoids of the same frequency can be combined to produce a single sinusoid of the given frequency by the use of phasor analysis. The method works with voltages, currents, or any physical variables represented by sinusoids of a given frequency. However, to establish a notational form for reference in the development that follows, we will assume voltage functions.

Sine Function

First, consider the instantaneous voltage

$$v(t) = V_p \sin \omega t \overset{\text{def}}{=} \bar{V} e^{j\omega t} \tag{8–15}$$

in which $\bar{V} = V_p\underline{/0°}$ in accordance with the phasor concept of the last section. The phasor $\bar{V}$ appears along the positive real axis as shown in Figure 8–2(a). If the phasor were allowed to rotate in a counterclockwise direction with an angular speed ω, its projection along the vertical (imaginary) axis would be the function $V_p \sin \omega t$

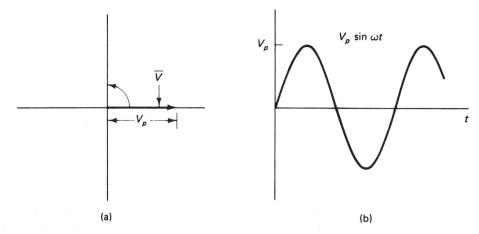

FIGURE 8–2
Phasor representation of $V_p \sin \omega t$ and the vertical projection if the phasor rotates.

as shown in Figure 8–2(b). The rotation would result from multiplying the phasor by the $e^{j\omega t}$ factor. This graphical interpretation corresponds with the more rigorous mathematical interpretation given earlier.

Cosine Function

Next, consider

$$v(t) = V_p \cos \omega t = V_p \sin \left(\omega t + \frac{\pi}{2} \right) \stackrel{\text{def}}{=} V_p e^{j(\omega t + \pi/2)} = \overline{V} e^{j\omega t} \qquad \textbf{(8–16)}$$

where in this case

$$\overline{V} = V_p e^{j\pi/2} = V_p \underline{/90^\circ} \qquad \textbf{(8–17)}$$

This phasor is aligned with the positive imaginary axis as shown in Figure 8–3(a). If this phasor rotates, the function $V_p \cos \omega t$ is generated as shown in Figure 8–3(b).

Negative Sine Function

The third form of the sinusoidal function to be considered is

$$v(t) = -V_p \sin \omega t = V_p \sin (\omega t + \pi)$$
$$\stackrel{\text{def}}{=} V_p e^{j(\omega t + \pi)} = \overline{V} e^{j\omega t} \qquad \textbf{(8–18)}$$

with $\overline{V}$ defined as

$$\overline{V} = V_p e^{j\pi} = V_p \underline{/180^\circ} \qquad \textbf{(8–19)}$$

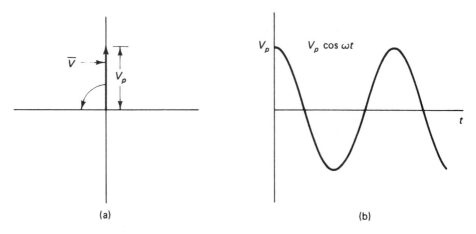

(a)   (b)

FIGURE 8–3
Phasor representation of $V_p \cos \omega t$ and the vertical projection if the phasor rotates.

This phasor is aligned with the negative real axis as shown in Figure 8–4(a). If rotated, the function $-V_p \sin \omega t$ is generated as shown in Figure 8–4(b).

Negative Cosine Function

The last form of the sinusoidal function to be considered is

$$v(t) = -V_p \cos \omega t = V_p \sin \left(\omega t - \frac{\pi}{2} \right)$$

(8–20)

$$\stackrel{\text{def}}{=} V_p e^{j(\omega t - \pi/2)} = \overline{V} e^{j\omega t}$$

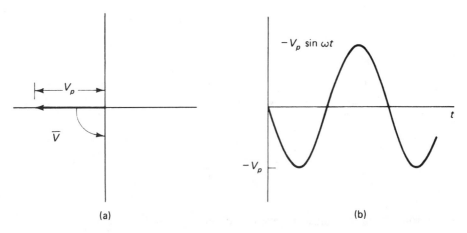

(a)   (b)

FIGURE 8–4
Phasor representation of $-V_p \sin \omega t$ and the vertical projection if the phasor rotates.

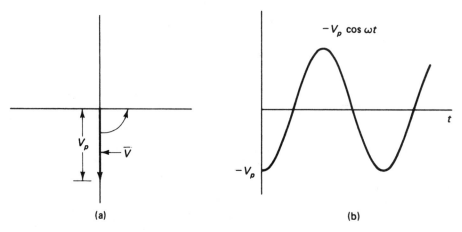

FIGURE 8-5
Phasor representation of $-V_p \cos \omega t$ and the vertical projection if the phasor rotates.

with $\overline{V}$ defined as

$$\overline{V} = V_p e^{-j\pi/2} = V_p \underline{/-90°} \tag{8-21}$$

This phasor is aligned with the negative imaginary axis as shown in Figure 8–5(a). If rotated, the function $-V_p \cos \omega t$ is generated as shown in Figure 8–5(b).

Relative Phase Diagram

From the preceding development, each of the functions—sine, cosine, −sine, and −cosine—can be represented as one of the four axes in the complex plane. A summary of these results is shown by the diagram in Figure 8–6.

Examples

Sinusoidal functions having a phase angle that is not an integer multiple of 90° are represented as phasors located between appropriate axes. For example, consider the two functions

$$v_1(t) = 4 \sin (\omega t + 20°) \tag{8-22}$$

and

$$v_2(t) = 4 \sin (\omega t + 135°) \tag{8-23}$$

From the preceding work, we can define $v_1(t)$ and $v_2(t)$ as

$$v_1(t) \overset{\text{def}}{=} \overline{V}_1 e^{j\omega t} \tag{8-24}$$

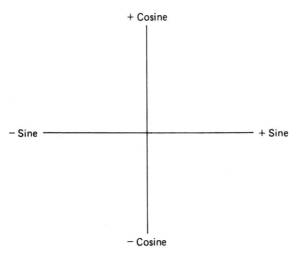

FIGURE 8–6
Correspondence of the four axes to the four forms of sinusoidal functions in the phasor representation of this text.

and

$$v_2(t) \overset{\text{def}}{=} \overline{V}_2 e^{j\omega t} \tag{8–25}$$

with

$$\overline{V}_1 = 4\underline{/20^\circ} \tag{8–26}$$

and

$$\overline{V}_2 = 4\underline{/135^\circ} \tag{8–27}$$

The preceding phasors are shown in Figure 8–7(a), and the corresponding sinusoidal functions are shown in (b). Note that $v_2(t)$ leads $v_1(t)$ by 115°. This can be observed from the functions by noting that $v_2(t)$ crosses the ωt-axis 115° ahead of $v_1(t)$. These curves illustrate how the functions would be observed on an oscilloscopic display. However, on the phasor display, the phasor $\overline{V}_2$ is 115° more positive (more counterclockwise), and this provides the same information.

The relative phase sequence also allows conversion between sines and cosines. For example, suppose it is desirable to express $v_1(t)$ and $v_2(t)$ of Equations (8–22) and 8–23) as cosine functions. The function $v_2(t)$ leads the +cosine axis by 135° − 90° = 45°, so it can be expressed as

$$v_2(t) = 4 \cos (\omega t + 45^\circ) \tag{8–28}$$

However, $v_1(t)$ *lags* the +cosine axis by 90° − 20° = 70°, so it can be written as

$$v_1(t) = 4 \cos (\omega t - 70^\circ) \tag{8–29}$$

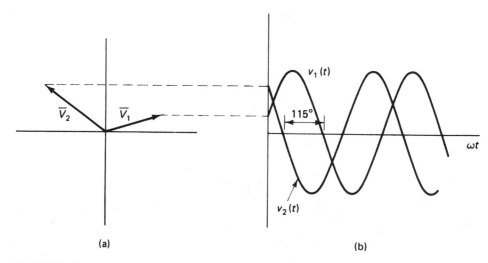

(a)   (b)

FIGURE 8–7
Phasor and time functions for sinusoids given by Equations (8–22) and (8–23).

If the two voltages had initially been given in the forms of Equations (8–28) and (8–29), the phasor locations could have been deduced directly from their angles relative to the cosine axes. However, since the sine function is the basis for the overall representation in this text, the angles with respect to that axis were determined in arriving at the complex phasor forms as given by Equations (8–26) and (8–27).

Addition of Sinusoids

A very useful application of the phasor concept is that of combining a sum of sinusoids of the same frequency. Using voltage functions again for discussion, consider the function

$$v(t) = V_{p1} \sin(\omega t + \theta_1) + V_{p2} \sin(\omega t + \theta_2)$$
$$+ \cdots + V_{pn} \sin(\omega t + \theta_n) \tag{8-30}$$

A basic property of sinusoidal functions is that *the sum of an arbitrary number of sinusoids of the same frequency is equivalent to a single sinusoid of the given frequency.* It must be emphasized that *all sinusoids must be of the same frequency.* Thus, the sum given by Equation (8–30) may be expressed in the form

$$v(t) = V_p \sin(\omega t + \theta) \tag{8-31}$$

where V_p and θ are to be determined.

The sum can be readily determined by means of the phasor concept. For convenience, let

$$v_1(t) = V_{p1} \sin(\omega t + \theta_1)$$
$$v_2(t) = V_{p2} \sin(\omega t + \theta_2)$$
$$\vdots \qquad \vdots \tag{8-32}$$
$$v_n(t) = V_{pn} \sin(\omega t + \theta_n)$$

Each of the sinusoidal components is represented by its phasor form. The resulting phasors are

$$\overline{V}_1 = V_{p1} \underline{/\theta_1}$$
$$\overline{V}_2 = V_{p2} \underline{/\theta_2}$$
$$\vdots \tag{8-33}$$
$$\overline{V}_n = V_{pn} \underline{/\theta_n}$$

The addition of the preceding phasors produces a phasor that represents the composite sum of the sinusoids. Thus, let $\overline{V} = V_p \underline{/\theta}$, and we have

$$\overline{V} = \overline{V}_1 + \overline{V}_2 + \cdots + \overline{V}_n \tag{8-34}$$

Thus, the peak value and phase angle of the sinusoid representing the sum are determined directly from the magnitude and angle of the phasor $\overline{V}$.

EXAMPLE 8–2
In solving a certain steady-state ac circuit problem, the following two voltages are obtained:

$$v_1(t) = 20 \cos(100t - 120°) \tag{8-35}$$
$$v_2(t) = -15 \sin(100t + 60°) \tag{8-36}$$

Determine the phase angle between the two voltages and express both in terms of the positive sine function.

Solution
For the purpose of representing sine and cosine functions and their relative phase sequence, the diagram of Figure 8–6 is employed. The voltage $v_1(t)$ is given in terms of the positive cosine function, and its phase angle is $-120°$ with respect to that axis; that is, it *lags* the +cosine function by 120°. The phasor form is shown as $\overline{V}_1$ in Figure 8–8. The angle of $\overline{V}_1$ is $-30°$ with respect to the +sine axis. The voltage $v_2(t)$ is given in terms of the negative sine function, and its phase angle is $+60°$ with respect to that axis; that is, it *leads* the −sine function by 60°. The phasor form is shown as $\overline{V}_2$ in Figure 8–8. The angle of $\overline{V}_2$ is $-120°$ with respect to the +sine axis.

FIGURE 8–8
Phasor representations of sinusoids in Example 8–2.

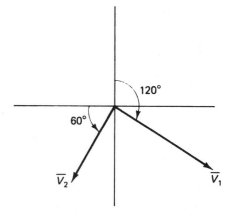

The phasor $\overline{V}_1$ is observed to *lead* $\overline{V}_2$ by 90° since it is 90° more positive (counterclockwise). An alternate way to convey the same meaning is to say that $\overline{V}_2$ *lags* $\overline{V}_1$ by 90°. Once sinusoids are converted to phasors, the phase difference can be readily determined, but in the original forms of Equations (8–35) and (8–36), the phase difference is not readily apparent.

Actually, the concepts of lead and lag are arbitrary if angles with magnitudes greater than 180° are used. We could, for example, say that $\overline{V}_2$ leads $\overline{V}_1$ by 270° or that $\overline{V}_1$ lags $\overline{V}_2$ by 270°. However, this interpretation is awkward and not as meaningful as the earlier deduction. Unless otherwise stated, the angle whose magnitude is *smaller* will be used in specifying phase differences throughout the book. Of course, when the angle is 180°, it is ambiguous as to which is leading.

To express the functions $v_1(t)$ and $v_2(t)$ as sine functions, we observe again $\overline{V}_1$ and $\overline{V}_2$ in Figure 8–8 and note the angles with respect to the positive sine axis. The functions are readily expressed as

$$v_1(t) = 20 \sin (100t - 30°) \qquad\qquad \textbf{(8–37)}$$
$$v_2(t) = 15 \sin (100t - 120°) \qquad\qquad \textbf{(8–38)}$$

Alternately, these functions could be written in terms of positive angles measured from the positive sine axis. In this case, the results are

$$v_1(t) = 20 \sin (100t + 330°) \qquad\qquad \textbf{(8–39)}$$
$$v_2(t) = 15 \sin (100t + 240°) \qquad\qquad \textbf{(8–40)}$$

This author prefers the form in which the magnitudes of the angles are smaller, namely, Equations (8–37) and (8–38).

For the requirements of this problem, it has not been necessary to express the phasor representations of $v_1(t)$ and $v_2(t)$ explicitly in their complex forms. Instead, the phasors were located with respect to the respective sine and cosine axes, and various angles were determined as required. An effort has been made to demonstrate that relative phase angles and expressions for sine functions in terms of different forms can be obtained quite easily without assuming the more formal phasor forms. However, the next example involves the addition of two sinusoids, and the complex phasor forms will be desirable for that purpose.

EXAMPLE 8–3

A certain circuit performs the operation of summing two voltages $v_1(t)$ and $v_2(t)$ to produce a third voltage $v_0(t)$ according to

$$v_0(t) = v_1(t) + v_2(t) \tag{8–41}$$

Determine an expression for $v_0(t)$ if $v_1(t)$ and $v_2(t)$ are the functions of Example 8–2, that is, Equations (8–35) and (8–36).

Solution
Referring back to Figure 8–8 and the associated discussion, the phasors $\overline{V}_1$ and $\overline{V}_2$ can be expressed as

$$\overline{V}_1 = 20\underline{/-30°} = 17.3205 - j10 \tag{8–42}$$

$$\overline{V}_2 = 15\underline{/-120°} = -7.5 - j12.9904 \tag{8–43}$$

The phasor $\overline{V}_0$ representing the output voltage is

$$\overline{V}_0 = \overline{V}_1 + \overline{V}_2 = 17.3205 - j10 - 7.5 - j12.9904$$
$$= 9.8205 - j22.9904 = 25\underline{/-66.87°} \tag{8–44}$$

A graphical interpretation of this addition is illustrated in Figure 8–9. The phasor $\overline{V}_2$ can be translated over to where it starts at the end of $\overline{V}_1$, and a phasor drawn from the origin to the end of $\overline{V}_2$ represents the sum. The corresponding time function for the sum can be expressed as

$$v_0(t) = 25 \sin (100t - 66.87°) \tag{8–45}$$

FIGURE 8–9
Phasor diagram for Example 8–3 illustrating $\overline{V}_0 = \overline{V}_1 + \overline{V}_2$.

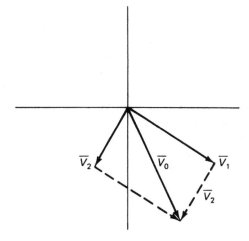

8–4 STEADY-STATE IMPEDANCE

Consider a passive *RLC* circuit with a time-varying voltage $v(t)$ and time-varying current $i(t)$ as indicated in block form in Figure 8–10(a). Assume that the time-varying voltage and current are restricted to steady-state sinusoidal functions. These functions can then be represented as phasors $\overline{V}$ and $\overline{I}$ as shown in Figure 8–10(b). The *steady-state impedance* $\overline{Z}$ is then defined as

$$\overline{Z} = \frac{\overline{V}}{\overline{I}} \qquad (8\text{–}46)$$

The steady-state impedance has the same relationship to the sinusoidal steady-state circuit as the transform impedance has to the *s*-domain circuit. It is an operational function that allows an algebraic basis for voltage and current relationships under the assumed conditions.

Since $\overline{Z}$ is a complex function, it can be represented as the complex sum of a real and an imaginary part. We define

$$\overline{Z} = R + jX \qquad (8\text{–}47)$$

where R is the resistance and X is defined as the *reactance*. The reactance can be either positive or negative. As we will see later, a net *inductive reactance* will be *positive,* and a net *capacitive reactance* will be *negative.*

In a purely resistive circuit, $X = 0$, and in a circuit containing only inductance and/or capacitance, $R = 0$. Capacitance and inductance are referred to as *reactive* elements, and a circuit containing only capacitance and/or inductance is called a *reactive circuit.*

FIGURE 8–10
Passive *RLC* circuit and its phasor or steady-state impedance.

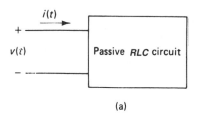

(a)

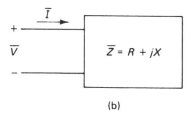

(b)

Admittance

The reciprocal of steady-state impedance $\overline{Z}$ is called *steady-state admittance* $\overline{Y}$. Thus,

$$\overline{Y} = \frac{1}{\overline{Z}} \tag{8-48}$$

The admittance can also be represented as a combination of real and imaginary parts in the form

$$\overline{Y} = G + jB \tag{8-49}$$

where G is the conductance and B is defined as the *susceptance*. The susceptance can be either positive or negative. As we will see later, a net inductive *susceptance* will be *negative*, and a net capacitive *susceptance* will be *positive*.

In a purely resistive circuit, $B = 0$, and in a purely reactive circuit, $G = 0$.

Whereas $\overline{Y}$ and $\overline{Z}$ are reciprocal functions, R and G are not simply reciprocal quantities (except in a purely resistive circuit), and X and B are not reciprocal quantities (except in a purely reactive circuit). To convert from $\overline{Z}$ to $\overline{Y}$ (or vice versa), it is necessary to form the reciprocal of the complete function as given by Equation (8-48) (or its inverse) and separate the new function into its real and imaginary parts.

EXAMPLE 8-4

At the terminals of a certain passive circuit, the net steady-state impedance (in ohms) at a specific frequency is

$$\overline{Z} = 30 + j40 \tag{8-50}$$

(a) Identify the resistance and the reactance.
(b) Construct an equivalent circuit for the impedance at the given frequency.
(c) Determine the admittance $\overline{Y}$, and identify the conductance and the susceptance.
(d) Determine an alternate equivalent circuit at the given frequency.

Solution
(a) Comparing Equation (8-50) and the definition of $\overline{Z}$ in Equation (8-47), we have

$$R = 30 \ \Omega \tag{8-51}$$

$$X = 40 \ \Omega \tag{8-52}$$

In accordance with the sign pattern given in this section, an inductive reactance is positive, so the given reactance is inductive. This property will be established mathematically in the next section.

(b) An impedance expressed as a sum describes a series connection, so an equivalent circuit can be readily constructed as shown in Figure 8–11(a). Note that an inductor symbol is used to characterize the inductive reactance. It should be emphasized that *this equivalent circuit, in general, is valid only at one specific*

FIGURE 8–11
Circuits for Example 8–4.

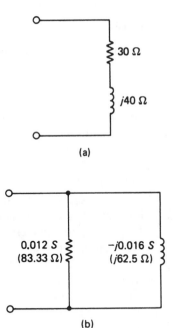

(a)

(b)

frequency. Although the actual circuit may be a simple series connection of a resistor and an inductor, it could actually be a complex combination of resistors, inductors, and capacitors. The point is that the impedance of any circuit is equivalent, at a single frequency, to a single resistance in series with a single reactance. The value of the impedance given is as much as we can deduce about this circuit from the available information.

(c) The admittance $\overline{Y}$ is determined as

$$\overline{Y} = \frac{1}{\overline{Z}} = \frac{1}{30 + j40} \tag{8–53}$$

The quantity $\overline{Y}$ is best simplified by multiplying the denominator and numerator by the complex conjugate of the denominator, which is $30 - j40$. We have

$$\overline{Y} = \frac{1}{(30 + j40)} \times \frac{(30 - j40)}{(30 - j40)} = \frac{30 - j40}{2500} = 0.012 - j0.016 \tag{8–54}$$

Comparing this result with the general form of $\overline{Y}$ given in Equation (8–49), we identify

$$G = 0.012 \, \text{S} \tag{8–55}$$

$$B = -0.016 \, \text{S} \tag{8–56}$$

Note that the susceptance turns out to be negative, which is a characteristic of an inductive susceptance, as mentioned in the text. Thus, as impedance is converted to admittance, the sign pattern between reactance and susceptance changes as required for a given type.

(d) An admittance expressed as a sum describes a parallel connection, so an equivalent circuit can be constructed as shown in Figure 8–11(b). Since it is more common to label impedance values in ohms rather than admittance values in siemens, the individual impedance values are listed in parentheses. The discussion about the single-frequency validity of the circuit discussed in part (b) applies here as well.

Reviewing (a) and (b) of Figure 8–11, we deduce that it is possible to represent the impedance of a circuit at a specific frequency by either a series or a parallel circuit. The series form is obtained directly from $\overline{Z}$, and the parallel form is obtained directly from $\overline{Y}$.

8–5 DERIVATION OF PHASOR VOLTAGE–CURRENT RELATIONSHIPS

The phasor forms for the voltage–current relationships for each of the three basic circuit parameters will now be derived. We will refer to Figure 8–12 in the developments that follow. The time-domain circuit form is shown on the left for each parameter, and the corresponding phasor circuit form is shown in the middle. The right-hand figure in each case is a representative phasor diagram for the given parameter depicting the phasor forms and their relationships.

For each of the circuits, we will start with an assumed current $i(t)$ of the form

$$i(t) \stackrel{\text{def}}{=} \overline{I}e^{j\omega t} \tag{8–57}$$

where

$$\overline{I} = I_p\underline{/\theta} \tag{8–58}$$

The corresponding voltage $v(t)$ will be of the form

$$v(t) = \overline{V}e^{j\omega t} \tag{8–59}$$

where

$$\overline{V} = V_p\underline{/\phi} \tag{8–60}$$

In each case, $\overline{V}$ will be determined in terms of $\overline{I}$.

Resistance

The time-domain voltage–current relationship for a resistance is

$$v(t) = Ri(t) \tag{8–61}$$

Time domain

Phasor domain

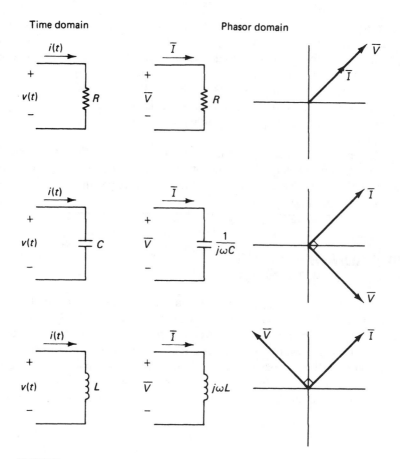

FIGURE 8–12
Conversion to phasor-domain models for circuit parameters.

Substituting phasor forms for voltage and current from Equations (8–57) and (8–59), we have

$$\overline{V}e^{j\omega t} = R\overline{I}e^{j\omega t} \tag{8–62}$$

Cancellation of the exponential factor results in

$$\overline{V} = R\overline{I} \tag{8–63}$$

The steady-state impedance $\overline{Z}$ for the resistance is simply

$$\overline{Z} = R \tag{8–64}$$

The corresponding admittance is

$$\overline{Y} = \frac{1}{\overline{Z}} = \frac{1}{R} = G \tag{8–65}$$

The various forms for the resistance are shown along the top row of Figure 8–12. These results show that the phasor representation of a resistance is simply the value of the resistance. Further, the voltage and current phasors are aligned in the same direction, that is, there is no phase angle between them. Thus, *the steady-state voltage and current for a resistor always have the same phase angle.*

Capacitance

The time-domain voltage–current relationship for a capacitance is

$$v(t) = \frac{1}{C} \int_0^t i(t)\, dt \tag{8–66}$$

Next, substitute the phasor forms for voltage and current from Equations (8–57) and (8–59) on both sides of Equation (8–66). We have

$$\overline{V} e^{j\omega t} = \frac{1}{C} \int_0^t \overline{I} e^{j\omega t}\, dt \tag{8–67}$$

When the integration on the right of Equation (8–67) is performed, both the upper and lower limits would normally be evaluated. However, the lower limit, which would be a constant, will be ignored because our interest is in steady-state sinusoidal conditions for a stable circuit, and the effects of a constant value will have disappeared when those conditions are reached.

The result of Equation (8–67) is thus written as

$$\overline{V} e^{j\omega t} = \frac{1}{j\omega C} \overline{I} e^{j\omega t} \tag{8–68}$$

Cancellation of the exponential factor results in

$$\overline{V} = \frac{1}{j\omega C} \overline{I} = \frac{-j}{\omega C} \overline{I} \tag{8–69}$$

where $1/j = -j$ was employed in the last form.

The steady-state impedance for the capacitor is

$$\overline{Z} = \frac{\overline{V}}{\overline{I}} = \frac{1}{j\omega C} = \frac{-j}{\omega C} \tag{8–70}$$

The corresponding admittance is

$$\overline{Y} = j\omega C \tag{8–71}$$

The results for the capacitance are shown on the middle row of Figure 8–12. The factor $-j = 1\underline{/-90°}$ means that the voltage phasor is 90° behind the current phasor. Thus, *the steady-state voltage across a capacitor lags the current through it by 90°.*

Inductance

The time-domain voltage–current relationship for an inductor is

$$v(t) = L\frac{di(t)}{dt} \tag{8-72}$$

Substituting the phasor forms for voltage and current of Equations (8–57) and (8–59), we have

$$\overline{V}e^{j\omega t} = L\frac{d}{dt}(\overline{I}e^{j\omega t}) \tag{8-73}$$

When the differentiation on the right-hand side of Equation (8–73) is performed, there results

$$\overline{V}e^{j\omega t} = j\omega L\overline{I}e^{j\omega t} \tag{8-74}$$

Cancellation of the rotational factor results in

$$\overline{V} = j\omega L\overline{I} \tag{8-75}$$

The steady-state impedance of the inductor is

$$\overline{Z} = \frac{\overline{V}}{\overline{I}} = j\omega L \tag{8-76}$$

The admittance is

$$\overline{Y} = \frac{1}{j\omega L} = \frac{-j}{\omega L} \tag{8-77}$$

The results for the inductor are shown in the bottom row of Figure 8–12. The factor $j = 1\underline{/90°}$ means that the voltage phasor is 90° ahead of the current phasor. Thus, *the steady-state voltage across an inductor leads the current through it by* 90°.

Reactance Functions

We have seen that the impedance or admittance of either an ideal capacitor or an ideal inductor is a purely imaginary number. For either case, the impedance is simply $\overline{Z} = jX$. When jX is compared with Equation (8–70) for a capacitor and Equation (8–76) for an inductor, expressions for the reactance functions may be readily determined. Let X_C represent the capacitive reactance, and let X_L represent the inductive reactance. It is readily determined that

$$X_C = \frac{-1}{\omega C} = \frac{-1}{2\pi fC} \tag{8-78}$$

and

$$X_L = \omega L = 2\pi fL \tag{8-79}$$

Similar definitions for capacitive susceptance B_C and inductive susceptance B_L may be made, and they can be determined as

$$B_C = \omega C = 2\pi f C \tag{8-80}$$

$$B_L = \frac{-1}{\omega L} = \frac{-1}{2\pi f L} \tag{8-81}$$

If X represents either reactance and B represents the corresponding susceptance, we have

$$B = -\frac{1}{X} \tag{8-82}$$

Thus, when the given reactance or susceptance is inverted, a negation of the sign results in the pattern previously defined. When the complex form with $\overline{Z}$ or $\overline{Y}$ is used, the negation is automatically achieved by the relationship $1/j = -j$.

Incidentally, some books define all of the reactive forms of Equations (8–78) through (8–81) with positive signs, and the negative sign results only when the complex form is used. However, this author advocates the sign pattern given here because it is more natural and less prone to certain types of sign difficulties.

From Equation (8–78), observe that capacitive reactance varies inversely with frequency, whereas Equation (8–79) indicates that inductive reactance varies directly with frequency. Curves displaying these properties are shown in Figure 8–13. (To simplify the presentation, the magnitude of capacitive reactance, namely, $|X_C|$, is shown.)

EXAMPLE 8–5

Calculate the reactance and susceptance of a 0.01-μF capacitor at **(a)** $f = 50$ Hz and **(b)** $f = 1$ kHz.

Solution
The capacitive reactance is given by

$$X_C = \frac{-1}{\omega C} = \frac{-1}{2\pi \times f \times 1 \times 10^{-8}} \tag{8-83}$$

(a) At $f = 50$ Hz,

$$X_C = \frac{-1}{2\pi \times 50 \times 1 \times 10^{-8}} = -318.3 \text{ k}\Omega \tag{8-84}$$

$$B_C = \frac{-1}{X_C} = 3.142 \ \mu\text{S} \tag{8-85}$$

(b) At $f = 1$ kHz,

$$X_C = \frac{-1}{2\pi \times 10^3 \times 1 \times 10^{-8}} = -15.92 \text{ k}\Omega \tag{8-86}$$

$$B_C = \frac{-1}{X_C} = 62.83 \ \mu\text{S} \tag{8-87}$$

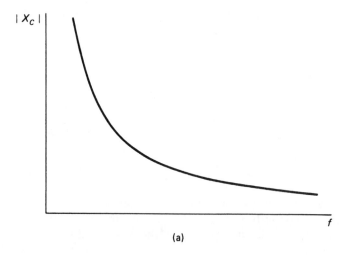

(a)

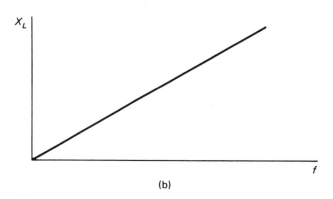

(b)

FIGURE 8-13
Forms of capacitive and inductive reactance as functions of frequency.

Note that the magnitude of the capacitive reactance decreases as the frequency increases, whereas the susceptance increases as the frequency increases.

EXAMPLE 8-6
Calculate the reactance and susceptance of a 100-mH inductor at **(a)** $f = 50$ Hz and **(b)** $f = 1$ kHz.

Solution
The inductive reactance is given by

$$X_L = \omega L = 2\pi \times f \times 0.1 \qquad \text{(8-88)}$$

(a) At $f = 50$ Hz,

$$X_L = 2\pi \times 50 \times 0.1 = 31.42 \ \Omega \qquad \text{(8–89)}$$

$$B_L = \frac{-1}{X_L} = -0.03183 \ \text{S} \qquad \text{(8–90)}$$

(b) At $f = 1$ kHz,

$$X_L = 2\pi \times 10^3 \times 0.1 = 628.3 \ \Omega \qquad \text{(8–91)}$$

$$B_L = \frac{-1}{X_L} = -0.001592 \ \text{S} \qquad \text{(8–92)}$$

Note that the inductive reactance increases as the frequency increases, whereas the magnitude of the susceptance decreases as the frequency increases.

8–6 COMPLETE SINUSOIDAL STEADY-STATE CIRCUIT SOLUTIONS

A general procedure for solving circuits in the sinusoidal steady state can now be formulated. Assume that all sources are sinusoidal and have the same frequency. Assume also that the circuit is a stable form. The procedure is given by the steps that follow.

1. Transform the complete circuit from the time domain to the phasor domain. All quantities in the time domain are replaced by phasor-domain quantities as follows:
   a. Any voltage variable $v(t)$ is replaced by a phasor variable $\overline{V}$.
   b. Any current variable $i(t)$ is replaced by a phasor variable $\overline{I}$.
   c. Assume that some representative voltage source $v_s(t)$ in the circuit is given by

   $$v_s(t) = V_p \sin(\omega t + \phi) \qquad \text{(8–93)}$$

   This source is replaced by a complex phasor $\overline{V}_s$ given by

   $$\overline{V}_s = V_p\underline{/\phi} \qquad \text{(8–94)}$$

   This procedure is performed on all voltage and current sources within the circuit. At the risk of redundancy, we stress again that all sources must have the same frequency.
   d. Passive parameters are replaced by their phasor impedances as discussed in Section 8–5.
2. Solve for any desired voltage or current variables using the phasor-domain circuit model. In general, standard circuit analysis methods may be employed, and all mathematical manipulations are algebraic in form.

3. Assume, for example, that the phasor form of some representative current response of interest is

$$\bar{I} = I_p \underline{/\phi} \tag{8–95}$$

The corresponding time-domain steady-state form of the current is

$$i(t) = I_p \sin(\omega t + \theta) \tag{8–96}$$

This procedure is repeated for all response functions of interest.

The procedure delineated in the preceding steps provides only the steady-state sinusoidal response functions for the circuit. It provides no information concerning the natural or transient response functions. If a circuit is first excited with sinusoidal sources, natural or transient terms will initially appear, and the length of time it will take for the steady-state response to dominate is a function of the circuit time constants. The fact that we are restricting our consideration to stable circuits ensures that the natural response terms will eventually vanish.

The procedure as given is based on the assumption that we start with the basic parameter values in the time domain, convert to the phasor domain, and eventually return to the time domain. In many cases, it may not be necessary to perform all three operations. In some areas of application where phasor analysis is routinely employed, virtually all the analysis is performed directly in the phasor domain. Obviously, it is necessary to maintain a perspective on what the various qualities, such as phasor magnitudes and angles, actually mean when converted back to the time-domain forms, but the whole framework of analysis can be performed directly in the phasor domain. An inductance parameter might then be specified directly as a reactance in ohms rather than in the basic units of henries.

Peak Value vs. Effective Value

One further point concerning the phasor representation of sinusoidal functions deserves discussion. Early in this chapter, it was established that the magnitude of a phasor is equal to the peak value of a sinusoid. Thus, the sinusoid current $i(t) = I_p \sin(\omega t + \theta)$ is represented by the phasor $I_p \underline{/\theta}$. This is the most natural form from a mathematical point of view, because the phasor is part of the complex exponential representation. However, a widely employed convention in some segments of the electrical industry is to express the phasor magnitude in terms of the effective (or *root-mean-square*) value. The concept of the effective value will be partially developed in Section 8–7 and further explained in general terms in Chapter 10. For the moment, however, the relationship for a sinusoidal function will be stated and used.

Let I_{rms} represent the root-mean-square (rms) or effective value of a sinusoidal current $i(t) = I_p \sin(\omega t + \theta)$. The effective value for a sinusoid is

$$I_{\text{rms}} = \frac{I_p}{\sqrt{2}} = 0.7071 I_p \tag{8–97}$$

A similar relationship applies to a voltage function as well. The effective value for a sinusoid is thus the peak value divided by $\sqrt{2}$.

The significance of the effective value is that it is the value used in computing average power. In those segments of the electrical industry where power calculations are frequently made, it is much more convenient to employ effective values of voltage and current. For example, commercial power system voltages and currents are generally expressed in terms of effective values.

When the effective value is to be employed directly in phasor analysis, a current $i(t) = I_p \sin (\omega t + \theta)$ is converted to a phasor $I_{rms}\underline{/\theta}$, where $I_{rms} = I_p/\sqrt{2}$. Similarly, a voltage $V_p \sin (\omega t + \phi)$ is converted to a phasor $V_{rms}\underline{/\phi}$, where $V_{rms} = V_p/\sqrt{2}$. For example, the instantaneous voltage $141.4 \sin \omega t$ is converted to a phasor $100\underline{/0°}$.

EXAMPLE 8–7

The circuit of Figure 8–14(a) is assumed to be operating under sinusoidal steady-state conditions. Convert the circuit to the sinusoidal steady-state phasor form, and solve for the phasor current and the phasor voltages across the three passive

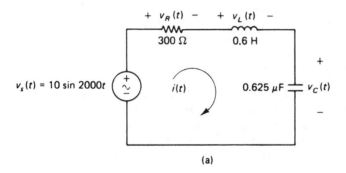

(a)

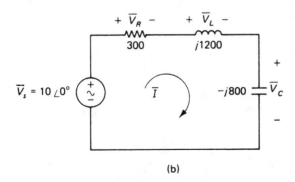

(b)

FIGURE 8–14
Circuit for Example 8–7 and the phasor model.

parameters. Determine the corresponding time-domain forms for the current and the voltages.

Solution

Inspection of the sinusoidal function reveals that the radian frequency is $\omega = 2000$ rad/s, which corresponds to 318.3 Hz. In many practical problems, the frequency is specified in hertz directly, and the form of the sinusoidal function is not given.

Refer to the phasor domain circuit model of Figure 8–14(b) in the steps that follow. The sinusoidal source $v_s(t) = 10 \sin 2000t$ is converted to a phasor $\overline{V}_s = 10\underline{/0°}$. The current $i(t)$ is replaced by a phasor $\overline{I}$, and the voltages $v_R(t)$, $v_L(t)$, and $v_C(t)$ are replaced by $\overline{V}_R$, $\overline{V}_L$, and $\overline{V}_C$, respectively.

The 300-Ω resistor remains simply as a resistive impedance of 300 Ω. The 0.6-H inductor is changed to a purely imaginary impedance $j\omega L = j(2000 \times 0.6) = j1200\ \Omega$. The capacitor is changed to a purely imaginary impedance $1/(j\omega C) = -j/(\omega C) = -j/(2000 \times 0.625 \times 10^{-6}) = -j800$.

To minimize the number of symbols in steady-state phasor circuit diagrams, units will often be omitted in this text. However, when not shown, it will be understood that impedances are expressed in ohms, and voltages and currents are expressed in their basic units.

Although a loop equation could be readily written for the phasor circuit of Figure 8–14(b), the impedance concept will be used instead because of certain insights that it provides. The complex impedance $\overline{Z}$ "seen" by the source is determined by adding the three series impedances. We have

$$\overline{Z} = 300 + j1200 - j800 = 300 + j400 \qquad \textbf{(8–98)}$$

Observe that the inductive and capacitive impedances partially cancel and that the net imaginary component is smaller in magnitude for this example than either of the separate components. In fact, it is possible to completely cancel the reactive effects, and this is the basis of *resonance*. (See Problems 8–33 and 8–34.) Observe that the net impedance is *inductive* in this example, since the net reactance has a positive imaginary sign.

The current phasor $\overline{I}$ is determined by applying Ohm's law in phasor form to the phasor source voltage $\overline{V}_s$ and the impedance $\overline{Z}$. We have

$$\overline{I} = \frac{\overline{V}_s}{\overline{Z}} = \frac{10\underline{/0°}}{300 + j400} = \frac{10\underline{/0°}}{500\underline{/53.13°}} = 0.02\underline{/-53.13°} \qquad \textbf{(8–99)}$$

where the impedance was converted to polar form to simplify the phasor division. This result indicates that the peak value of the current is 0.02 A, and the angle is $-53.13°$ with respect to the reference.

The voltages for the different elements are determined by multiplying $\overline{I}$ by the different impedances. In the case of $\overline{V}_L$, $j = 1\underline{/90°}$ is used, and in the case of $\overline{V}_C$, $-j = 1\underline{/-90°}$ is used. The steps involved follow.

$$\overline{V}_R = 300\overline{I} = 300(0.02\underline{/-53.13°}) = 6\underline{/-53.13°} \qquad \textbf{(8–100)}$$

$$\overline{V}_L = j1200\overline{I} = (1200\underline{/90°})(0.02\underline{/-53.13°}) = 24\underline{/36.87°} \qquad \textbf{(8-101)}$$

$$\overline{V}_C = -j800\overline{I} = (800\underline{/-90°})(0.02\underline{/-53.13°}) = 16\underline{/-143.13°} \qquad \textbf{(8-102)}$$

Readers with only minimal background in ac circuits may be disturbed by some of these results. Whereas the source has a peak voltage of only 10 V, the peak values of the inductive and capacitive voltages are 24 V and 16 V, respectively. This situation is common in ac circuits. In fact, with some resonant circuits, it is possible for voltages hundreds of times greater than the source voltage to appear in the circuit.

The manner in which the preceding phenomenon occurs will be illustrated with the phasor diagram of Figure 8-15. The current $\overline{I}$ lags the reference sine axis by 53.13°, and the voltage $\overline{V}_R$ is in the same direction. (Note that since voltage and current have different units, the relative magnitude scale used is arbitrary, so current has been made larger on the diagram so that it doesn't "get in the way" with subsequent voltage comparisons.)

The inductive voltage leads the current by 90°, so $\overline{V}_L$ is 36.87° ahead of the sine axis. However, the capacitive voltage lags the current by 90°, so $\overline{V}_C$ is 143.13° behind the positive sine axis. Note that the inductive and capacitive voltages are 180° out of phase with each other.

By KVL as applied to phasors, we have

$$\overline{V}_s = \overline{V}_R + \overline{V}_L + \overline{V}_C \qquad \textbf{(8-103)}$$

The last two terms in Equation (8-103), namely, $\overline{V}_L + \overline{V}_C$, represent the sum of the inductive and capacitive voltages, and it will be in the direction of either $\overline{V}_L$ or $\overline{V}_C$, depending on which is larger. In this case, it is in the direction of $\overline{V}_L$, as noted in Figure 8-15. The phasor $\overline{V}_R$ can then be added to the sum $\overline{V}_L + \overline{V}_C$, and the result is $\overline{V}_s$ as shown.

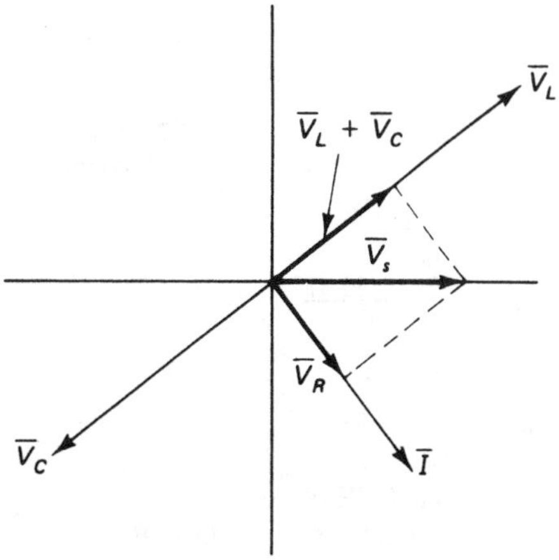

FIGURE 8-15
Phasor diagram for Example 8-7.

The end result seems much more reasonable when viewed on a phasor diagram than from the basic mathematical forms. Phasor diagrams are often helpful in explaining and interpreting results. In the present problem, the large inductive and capacitive voltages are 180° out of phase with each other, and these pronounced effects partially cancel as far as the circuit external terminals are concerned. Much of the stored energy in these reactances is being circulated back and forth between the components.

The instantaneous forms of the steady-state sinusoidal current and voltages for the circuit can be expressed as

$$i(t) = 0.02 \sin (2000t - 53.13°) \tag{8-104}$$
$$v_R(t) = 6 \sin (2000t - 53.13°) \tag{8-105}$$
$$v_L(t) = 24 \sin (2000t + 36.87°) \tag{8-106}$$
$$v_C(t) = 16 \sin (2000t - 143.13°) \tag{8-107}$$

EXAMPLE 8-8

The circuit of Figure 8-16(a) is in steady-state phasor form. **(a)** Determine the Thevenin equivalent circuit looking back from x–y with the load resistor not connected. **(b)** Using the result of (a), determine the phasor load current $\bar{I}$ with the resistor connected.

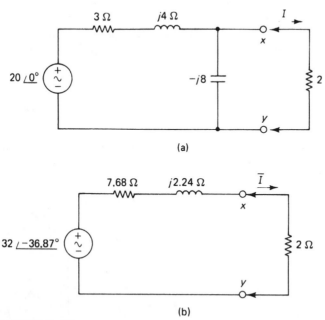

FIGURE 8-16
Circuit for Example 8-8.

Solution

(a) The open-circuit voltage $\overline{V}_{oc}$ across the capacitor can be determined by the voltage divider rule. We have

$$\overline{V}_{oc} = \frac{-j8}{3+j4-j8} \times 20\underline{/0°} = \frac{-j160}{3-j4}$$

$$= \frac{160\underline{/-90°}}{5\underline{/-53.13°}} = 32\underline{/-36.87°} \qquad \textbf{(8–108)}$$

With the source denergized, the equivalent impedance $\overline{Z}_{eq}$ is

$$\overline{Z}_{eq} = \frac{(3+j4)(-j8)}{3+j4-j8} = \frac{32-j24}{3-j4} \qquad \textbf{(8–109)}$$

A rectangular form for this impedance is most easily determined by rationalizing the denominator. We have

$$\overline{Z}_{eq} = \frac{(32-j24)(3+j4)}{(3-j4)(3+j4)} = \frac{96+96-j72+j128}{25} = 7.68 + j2.24 \qquad \textbf{(8–110)}$$

The Thevenin equivalent circuit is shown in Figure 8–16(b). Note that the net impedance is inductive.

(b) When the 2-Ω resistor is connected, the equivalent circuit shown in Figure 8–16(b) is a single-loop form. The current $\overline{I}$ is determined as

$$\overline{I} = \frac{32\underline{/-36.87°}}{7.68+j2.24+2} = \frac{32\underline{/-36.87°}}{9.936\underline{/13.03°}} \qquad \textbf{(8–111)}$$

$$= 3.221\underline{/-49.90°}$$

8–7 POWER IN ac CIRCUITS

Consider a portion of a circuit operating in steady-state ac with a current $i(t)$ and a voltage $v(t)$ as shown in Figure 8–17. The assumed forms for the current and voltage are

$$i(t) = I_p \sin(\omega t + \phi) \qquad \textbf{(8–112)}$$

FIGURE 8–17
Steady-state ac circuit form.

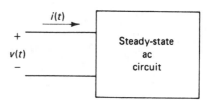

FIGURE 8–18
Phasor forms for power development.

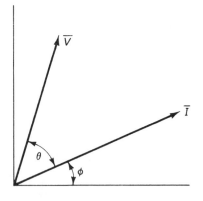

and

$$v(t) = V_p \sin(\omega t + \phi + \theta) \tag{8–113}$$

In the forms given, the voltage is assumed to lead the current if $\theta > 0$. The phasor forms are illustrated in Figure 8–18.

Instantaneous and Average Power

The instantaneous power $p(t)$ absorbed by this circuit is

$$p(t) = v(t)i(t) = V_p I_p \sin(\omega t + \phi + \theta) \sin(\omega t + \phi) \tag{8–114}$$

The average power P can then be determined by integrating $p(t)$ over a period and dividing by the period, that is,

$$P = \frac{1}{T} \int_0^T p(t)\, dt = \frac{1}{T} \int_0^T v(t)i(t)\, dt \tag{8–115a}$$

$$= \frac{1}{T} \int_0^T V_p I_p \sin(\omega t + \phi + \theta) \sin(\omega t + \phi)\, dt \tag{8–115b}$$

The integrand can be expanded with the assistance of the trigonometric identity

$$\sin A \sin B = \tfrac{1}{2}\cos(A - B) - \tfrac{1}{2}\cos(A + B) \tag{8–116}$$

Application of this identity to Equation (8–115b) leads to

$$P = \frac{V_p I_p}{2T}\left[\int_0^T \cos\theta\, dt - \int_0^T \cos(2\omega t + 2\phi + \theta)\, dt \right] \tag{8–117}$$

The second integral in Equation (8–117) involves an integration over two cycles of a sinusoidal component having a frequency twice that of the reference frequency, so the net value is zero. However, the first integral gives

$$P = \frac{V_p I_p}{2T}(\cos\theta)t \Big]_0^T = \frac{V_p I_p}{2T}(\cos\theta)T = \frac{V_p I_p \cos\theta}{2} \tag{8–118}$$

Definition of rms Value

We will now define the rms voltage V_{rms} and rms current I_{rms} as

$$V_{rms} = \frac{V_p}{\sqrt{2}} \tag{8–119}$$

$$I_{rms} = \frac{I_p}{\sqrt{2}} \tag{8–120}$$

Substitution of the rms values in Equation (8–118) leads to

$$P = V_{rms} I_{rms} \cos \theta \tag{8–121}$$

The rms value is also called the *effective value*. A full discussion of this concept will be developed in Section 10–1 as it applies to different waveforms. However, for the moment, accept the rms value as the quantity used in ac power calculations. For the sinusoid, the rms value is the peak value divided by $\sqrt{2}$ as noted in Equations (8–119) and (8–120). In general, however, the ratio of the peak value to the rms value is a different constant.

Apparent Power

Observe from Equation (8–121) that for ac, the actual power is not simply the product of voltage and current as it is for dc. For example, if an ac load draws 2 A at 120 V, the power is not necessarily 240 W. The product $V_{rms}I_{rms}$ is referred to as the *apparent power*. For the example just considered, the apparent power is 240 W,* but the actual power is determined by multiplying the apparent power by $\cos \theta$.

Power Factor

The quantity $\cos \theta$ in Equation (8–121) is defined as the *power factor*. Although it was assumed that the current was leading the voltage in the development, the result for average power applies for either case, since $\cos (-\theta) = \cos \theta$.

The magnitude of the power factor is bounded by $0 \leq |\cos \theta| \leq 1$. For $\theta = 90°$, $\cos 90° = 0$, and $P = 0$. This corresponds to a purely reactive network. In this case, energy alternately is accepted from and released back to the external circuitry by the reactive circuit.

When $\theta = 0$, $\cos 0° = 1$, and the power is maximum for given voltage and current values. This corresponds to a resistive equivalent circuit. The circuit may actually contain reactance, but the effects cancel at the given frequency, and the net impedance at the terminals is resistive.

---

*In industrial practice, the units of apparent power are stated as VA (volt-amperes) rather than watts.

Power formulas for Resistor

For a resistance R, the power may be expressed in either of the forms

$$P = V_{rms}I_{rms} = I_{rms}^2 R = \frac{V_{rms}^2}{R} \qquad (8\text{--}122)$$

These equations for a resistor have the same forms as those for dc, but with rms values of voltage and current used for ac. The form of Equation (8–121), however, is more general for ac, and it applies in all cases for average power calculations.

EXAMPLE 8–9

For the circuit of Example 8–7, determine the average power dissipated two ways: **(a)** by computing the power dissipated directly in the resistor and **(b)** by determining the power delivered by the source.

Solution

(a) Since there is only one resistor in the circuit, it must necessarily dissipate all the power. The solution in Example 8–7 was developed on a peak basis, so the magnitude of the phasor current (0.02 A) determined in Equation (8–99) is a peak value. The corresponding rms value I_{rms} is

$$I_{rms} = \frac{0.02}{\sqrt{2}} = 0.01414 \text{ A} \qquad (8\text{--}123)$$

Using one form of Equation (8–122), the average power is

$$P = (0.01414)^2 \times 300 = 0.06 \text{ W} \qquad (8\text{--}124)$$

(b) The rms value V_{rms} of the source voltage is

$$V_{rms} = \frac{10}{\sqrt{2}} = 7.071 \text{ volts} \qquad (8\text{--}125)$$

The angle θ between the source voltage and current is 53.13°. The power factor is

$$\cos \theta = \cos 53.13° = 0.6 \qquad (8\text{--}126)$$

The power delivered by the source is

$$P = V_{rms}I_{rms} \cos \theta = (7.071)(0.01414)(0.6) = 0.06 \text{ W} \qquad (8\text{--}127)$$

This result obviously agrees with the result of Equation (8–124).

8–8 PSPICE EXAMPLES

The major emphasis for PSPICE examples in this chapter is on sinusoidal analysis at a single frequency. The first example will consider the instantaneous process of adding two sinusoids using transient analysis. This will be followed by examples illustrating phasor analysis using the .AC command.

PSPICE EXAMPLE 8–1

Use PSPICE to add together the two sinusoids of Example 8–3 and use PROBE to plot the various waveforms.

Solution

The code is shown in Figure 8–19. First, the two sinusoidal functions are defined on the lines representing V1 and V2. The full code for a sinusoid has been employed in each of the waveforms; the numbers represent the following format:

`SIN `V_{dc}` `V_p` `f` `T_d` `α` `θ

The first three entries have already been utilized, but will be reviewed here for completeness. The six entries are as follows:

$$V_{dc} = \text{dc level added to sine function}$$
$$V_p = \text{peak value of sinusoid}$$
$$f = \text{frequency of sinusoid in hertz} = \omega/2\pi$$
$$T_d = \text{time delay before sinusoid begins}$$
$$\alpha = \text{damping factor (associated with a possible } e^{-\alpha t} \text{ factor)}$$
$$\theta = \text{phase shift in degrees}$$

The first sinusoid is described as

`SIN 0 20 15.915 0 0 -30`

Thus, the function has no dc component, a peak value of 20 V, and a frequency of 15.915 Hz (100 rad/s). There is no time delay and the damping factor is 0 (meaning that the multiplicative factor for the function is $e^0 = 1$). Finally, the phase shift is $-30°$. All the parameters produce the equivalent of Equation (8–37), namely,

$$v_1(t) = 20 \sin (100t - 30°) \tag{8–128}$$

The second sinusoid has the same form as the first, except that the peak value is 15 V, and the phase shift is $-120°$.

FIGURE 8–19
Code for PSPICE Example 8–1.

```
PSPICE EXAMPLE 8-1
V1 1 0 SIN 0 20 15.915 0 0 -30
V2 2 0 SIN 0 15 15.915 0 0 -120
E 3 0 VALUE={V(1)+V(2)}
.TRAN 62.83E-3 62.83E-3
.PROBE
.END
```

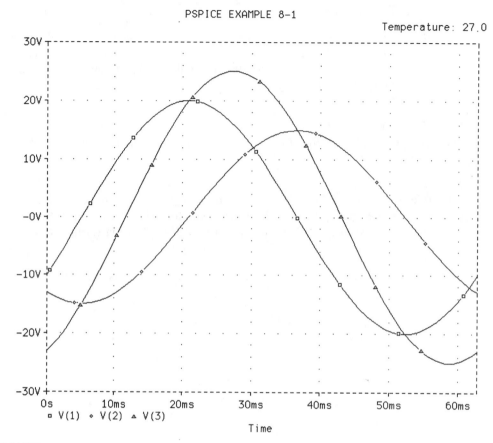

FIGURE 8–20
Waveforms for PSPICE Example 8–1.

The two voltages could be added together by either connecting them in series or by a suitable adding circuit. However, we will employ one of the PSPICE VALUE function operations. This operation utilizes a dependent source, which is denoted here as E and is connected between nodes 3 and 0. The value of this voltage is defined through a VALUE statement, which reads as follows:

```
VALUE = {V(1)+V(2)}
```

This operation performs exactly what it says. The value of the voltage E is the sum of the values of the voltage at node 1 and the voltage at node 2.

To display the voltage, a transient run is performed over an interval of 62.83 ms, which corresponds to one cycle. Note that since there are no reactive elements, steady-state conditions should be established immediately. The two separate voltages and their sum are displayed in Figure 8–20.

PSPICE EXAMPLE 8–2

Repeat the analysis of PSPICE Example 8–1 using a phasor domain approach, and determine the magnitude and phase angle of the sum voltage.

Solution

For a sinusoid of the form $V_p \sin(\omega t + \theta)$, the PSPICE phasor representation is

$$V_p \underline{/\theta}$$

Refer to the code of Figure 8–21 for the discussion that follows. Following the options line, V1 is defined as a phasor by the code line

```
V1 1 0 AC 20 -30
```

Similarly, V2 is defined as

```
V2 2 0 AC 15 -120
```

Once again, a VALUE function is used to sum the two voltages, except in this case, it is two phasors that are being added.

When phasor analysis is performed, it is necessary to employ an .AC control statement. One or more frequencies must be specified for the analysis. This statement requires that three frequency values must be specified. When only a single frequency is desired, as is the case here, it is necessary to "fool" the control.

In general, there are three types of ac sweeps: linear (LIN), decade (DEC), and octave (OCT). Any of them could be used, but it is probably easier to use the linear sweep for a single point. The command reads

```
.AC LIN 1 15.915 15.915
```

Following LIN, the number 1 indicates the number of frequencies to be employed. The next value represents the beginning frequency (15.915 Hz), and the last value represents the ending frequency (also 15.915 Hz).

To produce numerical output a .PRINT statement must be employed, and it reads as follows:

```
.PRINT AC VM(3) VP(3)
```

The type of analysis desired in the printout (AC) must follow the .PRINT command. Desired variables are then listed. For magnitude, M can follow the first symbol. (Actually, PSPICE defaults to magnitude when M is omitted, but it has

FIGURE 8–21

Code for PSPICE Example 8–2.

```
PSPICE EXAMPLE 8-2
.OPTIONS NOECHO NOPAGE NOBIAS
V1 1 0 AC 20 -30
V2 2 0 AC 15 -120
E 3 0 VALUE={V(1)+V(2)}
.AC LIN 1 15.915 15.915
.PRINT AC VM(3) VP(3)
.END
```

```
****                     ********** Evaluation PSpice (July 1996) *************

PSPICE EXAMPLE 8-2


****      CIRCUIT DESCRIPTION


*************************************************************************


.OPTIONS NOECHO NOPAGE NOBIAS


****      AC ANALYSIS                        TEMPERATURE =    27.000 DEG C


 FREQ        VM(3)        VP(3)

  1.592E+01    2.500E+01   -6.687E+01


          JOB CONCLUDED

          TOTAL JOB TIME              .38
```

FIGURE 8–22
Output data for PSPICE Example 8–2.

been included for clarity.) For phase, P must follow the first letter of the variable for which phase is desired. Thus, VM(3) and VP(3) represent, respectively, the magnitude and phase of the voltage at node 3.

The printout is shown in Figure 8–22. The phasor form of the voltage at node 3 can be read as $25.00\underline{/-66.87°}$. This result is in perfect agreement with that of Example 8–3, as given by Equation (8–44).

PSPICE EXAMPLE 8–3

Use PSPICE to perform a steady-state ac analysis of the circuit of Example 8–7 (Figure 8–14).

Solution
The circuit adapted to the PSPICE format and the code are shown in Figure 8–23. Following the .OPTIONS line, the input voltage V1 is defined for AC analysis as a phasor with a peak value of 10 V and an angle of 0. (When the angle is zero, it may be omitted, but it is included here for clarity.) The three circuit parameters R, L, and C are defined in terms of their basic units.

As in the previous example, a single frequency analysis is desired, and the .AC command line utilizes the LIN form with a frequency of 318.3 Hz ($2000/2\pi$)

FIGURE 8–23
Circuit and code for PSPICE Example 8–3.

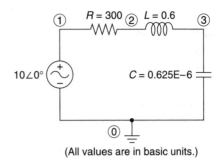

(All values are in basic units.)

```
PSPICE EXAMPLE 8-3
.OPTIONS NOECHO NOPAGE NOBIAS
V1 1 0 AC 10 0
R 1 2 300
L 2 3 0.6
C 3 0 0.625E-6
.AC LIN 1 318.31 318.31
.PRINT AC IM(R) VM(R) VM(L) VM(C)
.PRINT AC IP(R) VP(R) VP(L) VP(C)
.END
```

specified. For convenience in the printout, two .PRINT statements are used with the phase values directly underneath the corresponding magnitude values.

The output data file is shown in Figure 8–24. Phasor magnitude and angle values can be readily paired. For example, the resistive current, which is also the loop current, is readily seen to be of the form $2 \times 10^{-2} \underline{/-53.13°}$. All values are identical with the results of Example 8–7.

DRILL PROBLEMS

8–1. Represent the voltage $v(t) = 200 \sin (500t + 30°)$ as a phasor.

8–2. Represent the current $i(t) = 0.04 \sin (2000t - 120°)$ as a phasor.

8–3. Express each of the following sinusoidal functions in terms of the positive sine function, and write the phasor form in each case.

    a. $v(t) = 20 \sin (100t + 60°)$

    b. $i(t) = 12 \cos 200t$

    c. $v(t) = 20 \sin (150t + 150°)$

    d. $f(t) = -15 \cos 75t$

    e. $v(t) = -170 \sin (377t - 15°)$

    f. $i(t) = -4 \sin (50t + 25°)$

    g. $f(t) = -25 \cos (200t - 40°)$

    h. $v(t) = -10 \cos (200t + 20°)$

```
****                  ********** Evaluation PSpice (July 1996) *************

  PSPICE EXAMPLE 8-3


  ****       CIRCUIT DESCRIPTION


  ************************************************************************


.OPTIONS NOECHO NOPAGE NOBIAS


  ****       AC ANALYSIS                        TEMPERATURE =    27.000 DEG C


  FREQ         IM(R)        VM(R)        VM(L)        VM(C)

  3.183E+02    2.000E-02    6.000E+00    2.400E+01    1.600E+01


  ****       AC ANALYSIS                        TEMPERATURE =    27.000 DEG C


  FREQ         IP(R)        VP(R)        VP(L)        VP(C)

  3.183E+02   -5.313E+01   -5.313E+01    3.687E+01   -1.431E+02


             JOB CONCLUDED

             TOTAL JOB TIME            .44
```

FIGURE 8–24
Output data for PSPICE Example 8–3.

8–4. Express each of the sinusoidal functions in Problem 8–3 in terms of the positive cosine function.

8–5. Express each of the sinusoidal functions in Problem 8–3 in terms of the negative sine function.

8–6. Express each of the sinusoidal functions in Problem 8–3 in terms of the negative cosine function.

8–7. A certain circuit performs the operation

$$v_0(t) = v_1(t) + v_2(t)$$

Determine $v_0(t)$ in the form of a single sinusoidal function if

$$v_1(t) = 5 \sin (1000t + 30°)$$
$$v_2(t) = 12 \cos (1000t + 30°)$$

8–8. Repeat the analysis of Problem 8–7 if

$$v_1(t) = 5 \cos (500t + 60°)$$
$$v_2(t) = 12 \sin (500t - 120°)$$

8–9. Repeat the analysis of Problem 8–7 if

$$v_1(t) = 20 \sin (100t + 30°)$$
$$v_2(t) = -12 \sin 100t$$

8–10. Repeat the analysis of Problem 8–7 if

$$v_1(t) = 12 \cos (200t - 30°)$$
$$v_2(t) = 10 \sin (200t - 60°)$$

8–11. Determine a single sinusoidal function equivalent to

$$v(t) = V_p \sin 377t - V_p \sin (377t + 120°)$$

8–12. Determine a single sinusoidal function equivalent to

$$i(t) = 10 \sin (377t) + 10 \sin (377t + 120°) + 10 \sin (377t - 120°)$$

8–13. Determine a single sinusoidal function equivalent to

$$v(t) = 240\sqrt{2} \sin (377t + 120°) + 120\sqrt{2} \sin 377t$$

8–14. Determine a single sinusoidal function equivalent to

$$v(t) = 100 \sin (\omega t + 1°) - 100 \sin \omega t$$

8–15. At the terminals of a certain circuit, the net steady-state impedance (in ohms) at a specific frequency is

$$\overline{Z} = 120 + j50$$

    a. Identify the resistance and the reactance.
    b. Construct an equivalent circuit for the impedance at the given frequency.
    c. Determine the admittance $\overline{Y}$, and identify the conductance and the susceptance.
    d. Determine an alternate equivalent circuit at the given frequency.

8–16. At the terminals of a certain circuit, the net steady-state admittance (in siemens) at a specific frequency is

$$\overline{Y} = 70 + j240$$

    a. Identify the conductance and the susceptance.
    b. Construct an equivalent circuit for the admittance at the given frequency.
    c. Determine the impedance $\overline{Z}$, and identify the resistance and the reactance.
    d. Determine an alternate equivalent circuit at the given frequency.

8–17. Calculate the reactance and susceptance of a 25-mH inductor at **(a)** $f = 200$ Hz and **(b)** $f = 5$ kHz.

8–18. Calculate the reactance and susceptance of a 50-μH inductor at **(a)** $f = 2$ kHz and **(b)** $f = 1$ MHz.

8–19. Calculate the reactance and susceptance of a 0.05-μF capacitor at **(a)** $f = 200$ Hz and **(b)** $f = 5$ kHz.

8–20. Calculate the reactance and susceptance of a 40-pF capacitor at **(a)** $f = 5$ kHz and **(b)** $f = 20$ MHz.

8–21. Convert the circuit of Figure P8–21 to the sinusoidal steady-state phasor form, and:
   a. Solve for the phasor current and the phasor voltages across the three passive parameters.
   b. Draw a phasor diagram showing all voltages and the current.
   c. Determine the corresponding time-domain form for the current and the voltages.

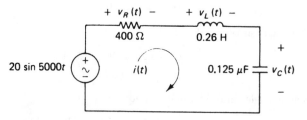

FIGURE P8–21

8–22. The sinusoidal voltage source in the circuit of Figure P8–22 has a frequency $f = 500$ Hz.
   a. Convert the circuit to the sinusoidal steady-state phasor form, and solve for the phasor current and the phasor voltages across the three passive parameters.
   b. Draw a phasor diagram showing all voltages and the currents.
   c. Determine the corresponding time-domain forms.

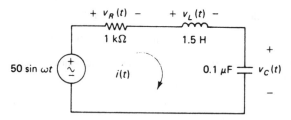

FIGURE P8–22

8–23. The sinusoidal voltage source in the circuit of Figure P8–23 has a frequency $f = 20$ kHz. Determine the steady-state functions $i_R(t)$, $i_C(t)$, $i_L(t)$, and $i_0(t)$.

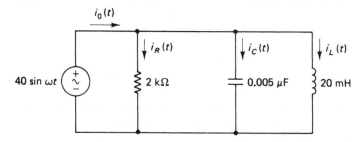

FIGURE P8–23

8–24. The sinusoidal current source in the circuit of Figure P8–24 has a frequency $f = 1$ MHz. Determine the steady-state functions $v(t)$, $i_R(t)$, $i_C(t)$, and $i_L(t)$.

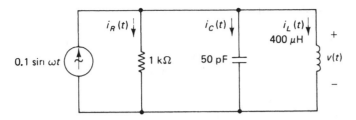

FIGURE P8–24

8–25. The steady-state phasor form for a certain circuit is shown in Figure P8–25. Determine the phasor currents $\bar{I}_1$ and $\bar{I}_2$ using *mesh current* analysis.

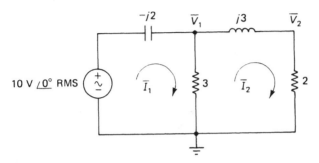

FIGURE P8–25

8–26. For the circuit of Problem 8–25, determine the phasor voltages $\bar{V}_1$ and $\bar{V}_2$ using *node voltage* analysis.

8–27. For the circuit of Problem 8–25, start on the right and successively combine impedances to obtain the net impedance $\bar{Z}$ seen by the source. Using this impedance, solve for the current $\bar{I}_1$ using Ohm's law in phasor form.

8–28. For the circuit of Problem 8–25, determine the Thevenin equivalent circuit looking back from the 2-Ω resistor. Use this circuit in conjunction with the 2-Ω resistor to determine the phasor current $\bar{I}_2$.

8–29. Determine the average power in the circuit of Problem 8–21 two ways: **(a)** by determining the power delivered by the source and **(b)** by considering the power dissipated in the resistor.

8–30. Determine the average power dissipated in the circuit of Problem 8–25 two ways: **(a)** by determining the power delivered by the source and **(b)** by considering the power dissipated in the two resistors.

DERIVATION PROBLEMS

8–31. In deriving the steady-state impedance for capacitance in the text, the integral form expressing voltage in terms of current as given by Equation (8–66) was used as the starting point. Provide an alternate derivation starting with the derivative relationship expressing current in terms of voltage.

8–32. In deriving the steady-state impedance for inductance in the text, the derivative form expressing voltage in terms of current as given by Equation (8–72) was used as the starting point. Provide an alternate derivation starting with the integral relationship expressing current in terms of voltage. Ignore the lower limit of the integral because it would be a constant and would disappear in the steady state.

8–33. Steady-state resonance is a process in which the net capacitive and inductive reactances cancel at a given frequency, thus providing a purely resistive impedance. Consider the series RLC circuit of Figure P8–33.

a. Determine the impedance $\overline{Z}$.

b. Impose the condition that the net reactance be zero at a given frequency f_0 and show that

$$f_0 = \frac{1}{2\pi\sqrt{LC}}$$

c. By a simple argument, show that the magnitude of the impedance $|\overline{Z}|$ is *minimum* at the resonant frequency f_0.

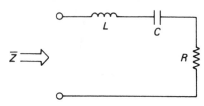

FIGURE P8–33

8–34. In Problem 8–33, the series resonant circuit was discussed. Consider now the parallel RLC circuit of Figure P8–34.

a. Determine the admittance $\overline{Y}$.

b. Impose the condition that the net susceptance be zero at a given frequency f_0 and show that

$$f_0 = \frac{1}{2\pi\sqrt{LC}}$$

c. By a simple argument, show that the magnitude of the impedance is *maximum* at the resonant frequency f_0.

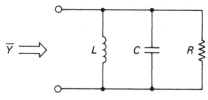

FIGURE P8–34

8–35. In Problem 8–34, the basic parallel resonant circuit was discussed. Many practical LC combinations used in parallel resonant circuits are often best modeled by the circuit of Figure P8–35, in which the inductor losses are represented by a series resistance R. Show that the net impedance measured at the terminals is resistive at a frequency

f_0 given by

$$f_0 = \frac{1}{2\pi} \sqrt{\frac{1}{LC} - \frac{R^2}{L^2}}$$

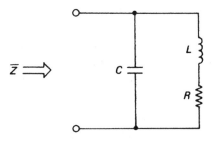

FIGURE P8–35

8–36. The circuit of Figure P8–36 is one form of a *reactive impedance matching* network. It is possible with this particular configuration to achieve an input $\overline{Z}_{in} = R_{in}$ at a specific frequency ω, where $R_{in} < R_L$. Show that this condition requires the following values of L and C:

$$C = \frac{1}{\omega R_L} \sqrt{\frac{R_L}{R_{in}} - 1}$$

$$L = \frac{R_{in}}{\omega} \sqrt{\frac{R_L}{R_{in}} - 1}$$

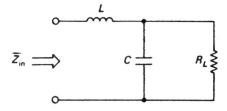

FIGURE P8–36

8–37. Consider the combination of a series resistance R_s and a series reactance X_s shown on the left in Figure P8–37. Assume that a parallel equivalent circuit is to be determined

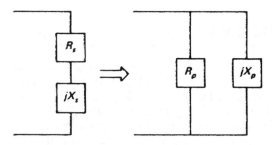

FIGURE P8–37

as shown on the right. Define $Q_s = X_s/R_s$. Derive the following relationships for the parallel values:

$$R_p = (1 + Q_s^2)R_s$$

$$X_p = \left(1 + \frac{1}{Q_s^2}\right)X_s$$

8–38. Consider the combination of a parallel resistance R_p and a parallel reactance X_p shown on the left in Figure P8–38. Assume that a series equivalent circuit is to be determined as shown on the right. Define $Q_p = R_p/X_p$. Derive the following relationships for the series values:

$$R_s = \frac{R_p}{1 + Q_p^2}$$

$$X_s = \frac{X_p}{1 + \dfrac{1}{Q_p^2}}$$

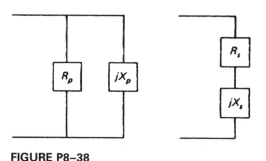

FIGURE P8–38

APPLICATION PROBLEMS

8–39. It is desired to establish series resonance at a frequency of 10 kHz with an inductance of 100 mH. Determine the required value of capacitance. (See Problem 8–33.)

8–40. It is desired to establish parallel resonance at a frequency of 5 MHz with a capacitance of 200 pF. Determine the required value of inductance. (See Problem 8–34.)

8–41. A certain inductance is in series with a resistance as shown in Figure P8–41. It is desired to connect a capacitance C across the combination so that the net impedance is resistive at a frequency of 1 MHz. Determine the value of C.

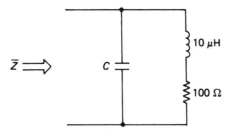

FIGURE P8-41

8-42. A certain capacitance is in parallel with a resistance as shown in Figure P8-42. It is desired to connect an inductance L in series with the combination so that the net impedance is resistive at a frequency of 2 MHz. Determine the value of L.

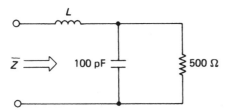

FIGURE P8-42

8-43. Design a reactive impedance matching network of the type considered in Problem 8-36 that will transform a 1000-Ω resistive load to a 50-Ω resistive input impedance at a frequency of 1 MHz.

8-44. The equivalent circuit of a certain signal source at a *particular* frequency of interest is shown on the left in Figure P8-44. It is desired to extract *maximum* power at the given frequency from this source into a 40-Ω load resistor shown on the right. Design a circuit to accomplish the purpose. Specify the turns ratio of any transformer used, and specify any reactance value used. (You will not be able to specify an actual component value, since the frequency is not given.)

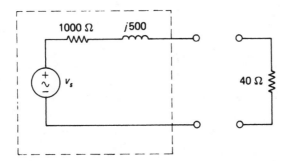

FIGURE P8-44

9

FREQUENCY RESPONSE ANALYSIS AND BODE PLOTS

OBJECTIVES

After completing this chapter, the reader should be able to:

- Discuss and show the mathematical relationship between the s-domain and the phasor domain.
- Define the steady-state transfer function, and show how it relates to the s-domain transfer function.
- Determine the steady-state transfer function for a given circuit.
- Define the linear amplitude response, the decibel amplitude response, and the phase response, and determine these functions from the steady-state transfer function.
- Discuss the form of a Bode plot, and explain its significance.
- Construct a Bode plot decibel amplitude response from a given transfer function or circuit.
- Construct a Bode plot phase response from a given transfer function or circuit.

9–1 RELATIONSHIP BETWEEN LAPLACE DOMAIN AND PHASOR DOMAIN

In Chapter 6, the concept of the Laplace or s-domain was established; and in Chapter 8, the concept of the phasor or steady-state domain was established. In the s-domain, passive parameters are represented by transform impedances; and in the phasor domain, passive parameters are represented by phasor impedances. A comparison of the circuit elements, the s-domain impedance forms, and the phasor-domain impedance forms is given in Figure 9–1. Recall that for the time-domain forms, the concept of impedance is meaningful only for resistance, since derivative and/or integral relationships are required for capacitance and inductance.

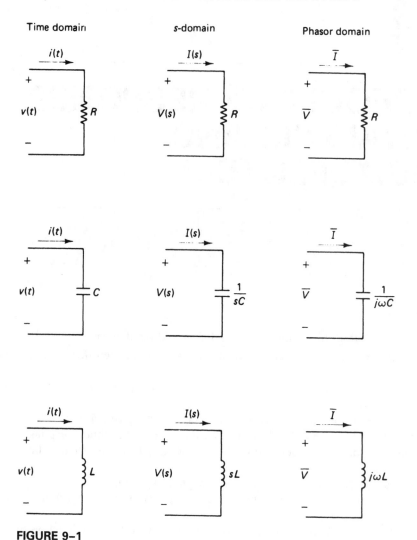

FIGURE 9–1
Passive circuit elements and the s-domain and phasor-domain impedance forms.

A resistance is always represented by the real value R. For a capacitance of value C, the s-domain impedance is $1/sC$, and the phasor domain impedance is $1/j\omega C$. For an inductance of value L, the s-domain impedance is sL, and the phasor-domain impedance is $j\omega L$.

Allowing ω to Vary

In virtually all the work of Chapter 8, the radian frequency ω was considered as a single frequency, and the quantities $1/j\omega C$ and $j\omega L$ were always evaluated at that specific frequency. The results are fixed impedances at the particular frequency. In the present chapter, however, many developments will be made in which we consider ω as unspecified. Strictly speaking, we are still considering the circuit from a single frequency point of view, so the basic phasor assumption has not been violated. However, by leaving ω unspecified, we are able to observe the manner in which various functions, such as impedances, depend on the actual circuit operating frequency. Thus, ω is considered a variable in the sense that it will be unspecified, but we must interpret all results as applicable when the source or sources are sinusoids with a frequency ω.

Relationship Between *s* and *jω*

The link between the s-domain and the phasor domain is the relationship between s and $j\omega$. To convert from an s-domain impedance to a phasor impedance, we replace s in all impedances by $j\omega$ and vice versa. Since this operation applies for each individual element, it must apply to any combination of elements. Let $Z(s)$ represent any passive RLC impedance as represented by s-domain functions. The corresponding phasor impedance $\overline{Z}$ is

$$\overline{Z} = Z(s)]_{s=j\omega} = Z(j\omega) \qquad (9\text{--}1)$$

When the focus of analysis is on a single frequency, the notation $\overline{Z}$, as defined in Chapter 8 for impedance, is most appropriate. However, where the frequency dependency of the impedance is to be delineated, the notation $Z(j\omega)$ is more desirable.

Resistance and Reactance Functions

The steady-state frequency-dependent impedance function $Z(j\omega)$ can be expressed as

$$Z(j\omega) = R(\omega) + jX(\omega) \qquad (9\text{--}2)$$

where $R(\omega)$ is the *resistance function* and $X(\omega)$ is the *reactance function*. The quantity $j\omega$ is used as the argument for the complete impedance function to emphasize the direct interchange between s and $j\omega$. However, both $R(\omega)$ and $X(\omega)$ are real functions of ω, so it is customary to use the simpler quantity ω as the argument in those functions.

Conductance and Susceptance Functions

The preceding development for $Z(j\omega)$ applies equally well to admittance. Thus, a phasor admittance $\overline{Y}$ at a given frequency may be determined from an s-domain admittance $Y(s)$ as

$$\overline{Y} = Y(s)]_{s=j\omega} = Y(j\omega) \qquad (9\text{--}3)$$

When ω is interpreted as a variable, $Y(j\omega)$ can be expressed as

$$Y(j\omega) = G(\omega) + jB(\omega) \qquad (9\text{--}4)$$

where $G(\omega)$ is the *conductance function*, and $B(\omega)$ is the *susceptance function*. The earlier discussion of $Z(j\omega)$ (concerning the use of $j\omega$ as the argument for the complete function) and ω for the two parts applies for admittance as well.

Although replacing s by $j\omega$ converts impedance functions from the s-domain form to the steady-state phasor form, the same process does *not* apply for sources. The reason is that sources in the s-domain are represented by their Laplace transforms. However, phasor-domain sources are represented as phasors having magnitudes and angles, and these forms are not determined by simply replacing s by $j\omega$.

EXAMPLE 9–1

Consider the *RC* circuit at the top of Figure 9–2, in which the excitation is a sinusoidal function. **(a)** Assuming that the circuit is initially relaxed and that the switch is closed at $t = 0$, determine the s-domain model and use it to determine the complete time-domain current $i(t)$. **(b)** Determine the phasor-domain model and use it to determine the steady-state current.

Solution
This problem represents a rehash of topics already covered, but the intent is to bring together s-domain and phasor analysis in the same problem so that the forms of the models and the solutions can be carefully compared.

(a) The s-domain model for the circuit is shown in Figure 9–2(a). The transform current $I(s)$ can be readily determined as

$$I(s) = \frac{\dfrac{40}{s^2 + 4}}{4 + \dfrac{6}{s}} = \frac{\dfrac{40}{s^2 + 4}}{\dfrac{4s + 6}{s}} = \frac{10s}{(s + 1.5)(s^2 + 4)} \qquad (9\text{--}5)$$

The reader is invited to verify that the inverse transform is

$$i(t) = -2.4e^{-1.5t} + 4\sin(2t + 36.87°) \qquad (9\text{--}6)$$

(b) Consider next the phasor-domain model given in Figure 9–2(b). The s-domain impedance $6/s$ is replaced by a phasor-domain impedance $6/j\omega$. In this problem, the circuit is excited by a single frequency sinusoid (at 2 rad/s), so the impedance is evaluated at the specific frequency $j\omega = j2$, so $6/j\omega = 6/j2 = -j3$. Observe, however, that the source $20\sin 2t$ is replaced by a phasor $20\underline{/0°}$, and this function is not derived from the s-domain model. The 4-Ω resistor remains unchanged.

The phasor current $\bar{I}$ is

$$\bar{I} = \frac{20\underline{/0°}}{4 - j3} = \frac{20\underline{/0°}}{5\underline{/-36.87°}} = 4\underline{/36.87°} \qquad (9\text{--}7)$$

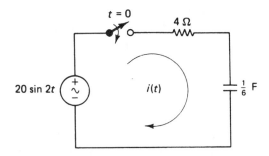

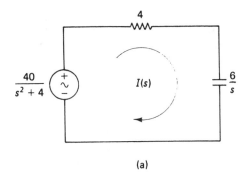

(a)

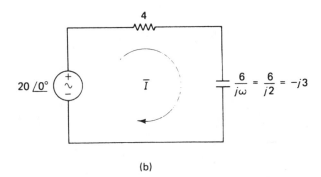

(b)

FIGURE 9–2
Circuit of Example 9–1 with s-domain and phasor-domain models.

The result indicates that the peak value of the steady-state current is 4, and the angle is leading the reference sine function by 36.87°. To distinguish the time-domain form of this current from the complete current of Equation (9–6), the steady-state current will be denoted as $i_{ss}(t)$. We have

$$i_{ss}(t) = 4 \sin(2t + 36.87°) \tag{9–8}$$

Comparing Equation (9–8) with Equation (9–6), we note that the s-domain approach provides the entire solution, while the phasor approach yields only the sinusoidal steady-state portion of the solution.

EXAMPLE 9–2

The equivalent circuit of many carbon-composition resistors can be represented over a reasonable frequency range by the given resistance in parallel with a shunt capacitance as shown in Figure 9–3(a). However, many impedance bridges measure a *series* combination of a resistance and a reactance, so it is necessary to relate the two forms. Considering ω as a variable, determine the composite impedance function $Z(j\omega)$, the resistance function $R(\omega)$, and the reactance function $X(\omega)$.

Solution

The phasor-domain form for the equivalent circuit is shown in Figure 9–3(b). Note that ω is unspecified, so it is treated as a variable. Since the two impedances are

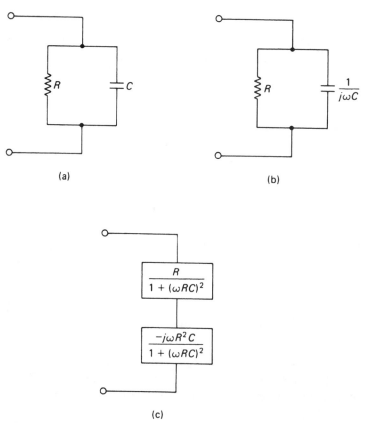

(a)

(b)

(c)

FIGURE 9–3

Circuit for Example 9–2 and equivalent series form.

in parallel, it is convenient first to determine the admittance $Y(j\omega)$, since admittances in parallel add. We have

$$Y(j\omega) = \frac{1}{R} + j\omega C = \frac{1 + j\omega RC}{R} \tag{9-9}$$

The impedance $Z(j\omega)$ is

$$Z(j\omega) = \frac{1}{Y(j\omega)} = \frac{R}{1 + j\omega RC} \tag{9-10}$$

To determine the resistance and reactance, it is necessary to represent $Z(j\omega)$ in terms of its real and imaginary parts. This is best achieved by rationalizing the denominator of Equation (9–10). We have

$$
\begin{aligned}
Z(j\omega) &= \frac{R}{(1 + j\omega RC)} \frac{(1 - j\omega RC)}{(1 - j\omega RC)} = \frac{R - j\omega R^2 C}{1 + (\omega RC)^2} \\
&= \frac{R}{1 + (\omega RC)^2} - j\frac{\omega R^2 C}{1 + (\omega RC)^2}
\end{aligned}
\tag{9-11}
$$

Identifying real and imaginary parts of Equation (9–11), we have

$$R(\omega) = \frac{R}{1 + (\omega RC)^2} \tag{9-12}$$

$$X(\omega) = \frac{-\omega R^2 C}{1 + (\omega RC)^2} \tag{9-13}$$

An equivalent series circuit based on the mathematical results of Equations (9–12) and (9–13) is shown in Figure 9–3(c). As should be expected, the reactive part of the impedance is negative, since the reactance in the circuit is capacitive. This series reactance varies with frequency in a more complex manner than for a simple capacitive reactance. An interesting fact is that the equivalent series resistance also varies with the frequency, even though the actual resistance is assumed to be constant. This is a result, of course, of the series representation of the basic parallel structure. This property is often puzzling to someone using an impedance bridge who does not understand this conversion process. For example, at the frequency where $\omega RC = 1$, the equivalent series resistance of a 1000-Ω resistor measured on an impedance bridge would be only 500 Ω. (This circuit is considered further in Problems 9–17 and 9–25.)

9–2 STEADY-STATE TRANSFER FUNCTION

In Chapter 7, the concept of the s-domain transfer function $G(s)$ for a linear system was established. The relationship between an output transform $Y(s)$ and an input transform $X(s)$ was determined to be

$$\frac{Y(s)}{X(s)} = G(s) \tag{9-14}$$

The function $G(s)$ can be expressed completely in terms of the parameter values R, L, and C converted to appropriate s-domain impedances.

In the preceding section, it was determined that a steady-state phasor impedance $Z(j\omega)$ can be determined from an s-domain function by simply replacing s with $j\omega$. Since $G(s)$ is a function of s-domain impedances, it seems reasonable that if s in $G(s)$ is replaced by $j\omega$, a corresponding phasor transfer function is determined.

We have seen that sources in the phasor domain are obtained by substituting phasor sources for s-domain sources rather than by replacing s with $j\omega$. This implies that $Y(s)$ and $X(s)$ on the left-hand side of Equation (9–14) are replaced with phasors $\overline{Y}$ and $\overline{X}$ when $s = j\omega$ is substituted on the right.

To formalize the preceding discussion, we define a *steady-state transfer function* $G(j\omega)$ in the following manner:

$$\frac{\overline{Y}}{\overline{X}} = G(j\omega) \tag{9–15}$$

where $G(j\omega)$ is obtained from the Laplace transfer function as

$$G(j\omega) = G(s)]_{s=j\omega} \tag{9–16}$$

The phasors $\overline{Y}$ and $\overline{X}$ are interpreted as phasors at a single frequency. However, ω may be interpreted as a variable in the sense that it could represent *any* particular frequency, so the steady-state transfer function is considered as a frequency-dependent function.

Amplitude and Phase Response Functions

The steady-state transfer function $G(j\omega)$ is a complex function of frequency and can be expressed in polar form as a magnitude and angle as

$$G(j\omega) = A(\omega)\underline{/\beta(\omega)} \tag{9–17}$$

The function $A(\omega) = |G(j\omega)|$ is called the *amplitude response* or *magnitude response*, and $\beta(\omega)$ is called the *phase response*. Observe that $j\omega$ is retained as the argument for the composite transfer function due to the interchange between s and $j\omega$, but the amplitude and phase functions are expressed in terms of ω, since these latter functions are both real. One could also express the transfer function as a real and an imaginary part, but this is seldom done because there is no special significance to such a form.

Assume next that $x(t)$ and $y(t)$ are both steady-state sinusoids of the forms

$$x(t) = X \sin (\omega t + \theta_x) \tag{9–18}$$

$$y(t) = Y \sin (\omega t + \theta_y) \tag{9–19}$$

The phasor forms of $x(t)$ and $y(t)$ are

$$\overline{X} = X\underline{/\theta_x} \tag{9-20}$$

$$\overline{Y} = Y\underline{/\theta_y} \tag{9-21}$$

From Equation (9–15), we note that

$$\overline{Y} = G(j\omega)\overline{X} \tag{9-22}$$

Substitution of Equations (9–20), (9–21), and (9–17) in Equation (9–22) yields

$$Y\underline{/\theta_y} = (A(\omega)\underline{/\beta(\omega)})(X\underline{/\theta_x}) \tag{9-23a}$$

$$= XA(\omega)\underline{/\theta_x + \beta(\omega)} \tag{9-23b}$$

The form of Equation (9–23b) can be separated into two different results as follows:

$$Y = XA(\omega) \tag{9-24}$$

$$\theta_y = \theta_x + \beta(\omega) \tag{9-25}$$

These results are very important and should be strongly emphasized. Stated in words, Equation (9–24) indicates that *the magnitude of an output phasor (or sinusoid) is the magnitude of the input phasor (or sinusoid) times the amplitude response evaluated at the frequency of the input.* The result of Equation (9–25) indicates that *the phase angle of an output phasor (or sinusoid) is the phase angle of the input phasor (or sinusoid) plus the phase response of the system evaluated at the frequency of the input.*

Frequency Response Analysis

The preceding concepts form the basis of frequency response analysis, which is one of the most important concepts in the theory of linear systems. Complex signals can be considered to be composed of a sum of sinusoidal functions through the process of Fourier analysis. (The subject of Fourier analysis will be considered in Chapter 11.) A system that is designed to transmit or process signals can then be described by its frequency response.

The composite frequency response can be considered as the combination of an amplitude response and a phase response. The response functions can be either calculated or measured, depending on the system. The manner in which they can be measured is illustrated in Figure 9–4. A constant amplitude sinusoidal oscillator, whose frequency can be varied, is applied to the input. At a number of different frequencies, the output amplitude and the phase shift of the system are measured. The resulting data provide the frequency response of the system. The extent to which the system can faithfully reproduce complex waveforms can then be judged on the basis of the frequency response over the band of frequencies comprising the complex signal.

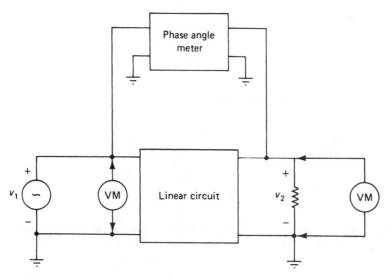

FIGURE 9–4
Measurement of amplitude and phase response functions for a circuit. (The input is v_1 and the output is v_2.)

Decibel Response

Whereas the amplitude response $A(\omega)$ is useful for many purposes, decibel forms of amplitude response are widely used in many applications. Decibel functions are derived from logarithmic forms of the amplitude response. Such forms have been in use since the early days of the electrical industry. Although a newcomer might logically question the value of such forms, their use has become embedded within the industry, and it is virtually impossible to read and understand the specifications for many systems without understanding such forms. We will see that there are significant advantages in using decibel forms once the basic concept is understood.

Although there are a number of variations of the basic decibel form, all have similar patterns. We will start with a decibel function $A_{dB}(\omega)$ defined as

$$A_{dB}(\omega) = 20 \log_{10} \left[\frac{A(\omega)}{A_0} \right] \qquad (9\text{--}26)$$

where A_0 is any reference level desired, and the response is measured in decibels (abbreviated dB). The number of decibels calculated from this formula represents the level with respect to A_0. If $A(\omega) > A_0$, the result is a positive number of decibels; and if $A(\omega) < A_0$, the result is a negative number of decibels.

Having established the somewhat general form, we will now choose to set $A_0 = 1$ for subsequent developments. Thus, from this point on in the book, $A_{dB}(\omega)$ will be defined as

$$A_{dB}(\omega) = 20 \log_{10} A(\omega) \qquad (9\text{--}27)$$

where the reference level of unity is understood. Thus, if the amplitude response is greater than one, the dB level is positive, and if it is less than one, the dB level is negative.

Where it is necessary to distinguish between the functions $A(\omega)$ and $A_{dB}(\omega)$ in subsequent discussions, the former will be referred to as the *linear amplitude response* and the latter as the *decibel amplitude response*.

EXAMPLE 9–3

For the circuit of Figure 9–5, determine mathematical expressions for the steady-state transfer function, the linear amplitude response, the decibel amplitude response, and the phase response.

Solution

If desired, we could start with the *s*-domain model, determine the *s*-domain transfer function, and then set $s = j\omega$. However, a more direct route is to establish the steady-state phasor model directly with $j\omega$ considered as a variable, as shown in Figure 9–5(b). The input and output variables are assumed to be phasors.

Using the voltage divider rule, the output phasor $\overline{V}_2$ is expressed in terms of the input phasor $\overline{V}_1$ as

$$\overline{V}_2 = \frac{(10^8/j\omega)\overline{V}_1}{10^4 + 10^8/j\omega} = \frac{10^8\overline{V}_1}{10^8 + 10^4 j\omega} \tag{9–28}$$

The steady-state transfer function is

$$G(j\omega) = \frac{\overline{V}_2}{\overline{V}_1} = \frac{10^8}{10^8 + 10^4 j\omega} = \frac{10^4}{10^4 + j\omega} \tag{9–29}$$

The linear amplitude response $A(\omega)$ is the magnitude of $G(j\omega)$. The magnitude of the ratio of two complex numbers is the ratio of the two magnitudes. The numerator magnitude is simply 10^4. Thus,

$$A(\omega) = \frac{10^4}{\sqrt{10^8 + \omega^2}} \tag{9–30}$$

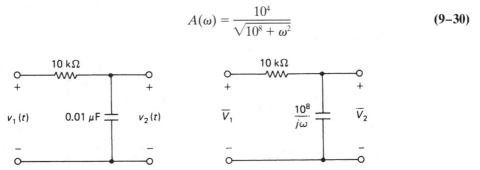

(a)   (b)

FIGURE 9–5
Circuit for Example 9–3.

The corresponding decibel amplitude response is

$$A_{dB}(\omega) = 20 \log_{10} A(\omega) = 20 \log_{10} \left(\frac{10^4}{\sqrt{10^8 + \omega^2}} \right) \qquad (9\text{--}31)$$

The phase response is the numerator angle minus the denominator angle. Since the numerator angle is zero, we have

$$\beta(\omega) = -\tan^{-1} \frac{\omega}{10^4} \qquad (9\text{--}32)$$

This problem has illustrated with a simple circuit the process of calculating amplitude and phase functions. No interpretations of these particular results will be made at this time because the basic *RC* circuit form will be considered after Bode plots are introduced. With that approach, it will be easier to deduce actual properties of the circuit.

9–3   BODE PLOT CONCEPTS

In previous sections, the concept of the amplitude and phase response functions for a linear circuit was developed. For a relatively simple circuit, the forms of these functions may be readily plotted or inspected. However, the corresponding functions for elaborate network structures may be more difficult to visualize. Programs for computers or programmable calculators may be written as an aid in this endeavor.

Although the computer has eliminated, or at least reduced, the necessity for many graphical techniques, one graphical method for determining amplitude and phase functions still plays a major role in circuit analysis. This technique is the *Bode plot method*, named after the late Henrik W. Bode of Bell Telephone Laboratories, who introduced the concept. This method is straightforward and easy to apply for many networks and transfer functions, and it provides an intuitive approach that is often lacking with the "brute-force" computational and analytical approaches.

Logarithmic Scales

The Bode plot method is most often used in conjunction with the amplitude response, and we will devote a major part of our focus to that function. The amplitude response is either plotted on *semilogarithmic* graph paper (referred to as *semilog* paper) or on *logarithmic-logarithmic* graph paper (referred to as *log-log* paper). For readers not familiar with such paper, examples are shown in Figures 9–6 and 9–7. The semilog paper of Figure 9–6 is characterized by one linear scale and one logarithmic horizontal scale. The log-log paper of Figure 9–7 is characterized by logarithmic scales for both vertical and horizontal scales.

A logarithmic scale is further classified by the number of cycles, which is not the same as frequency in a steady-state sense. Strictly speaking, the term should be "number of decades," but the former term is used by paper manufacturers. The

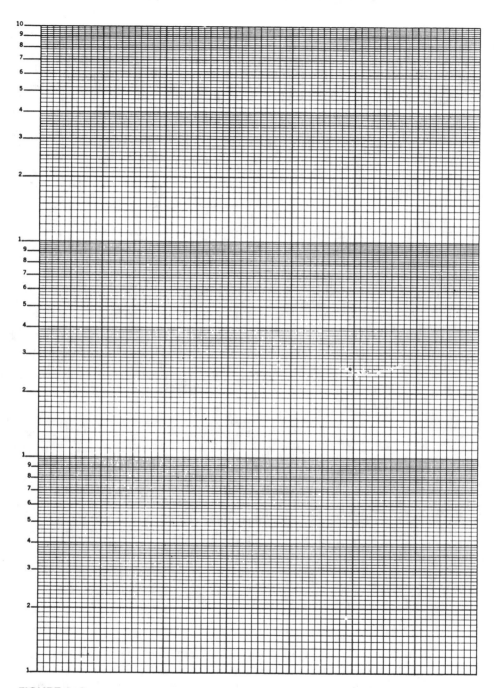

FIGURE 9–6
Example of semilog graph paper with three "cycles" on log scale.

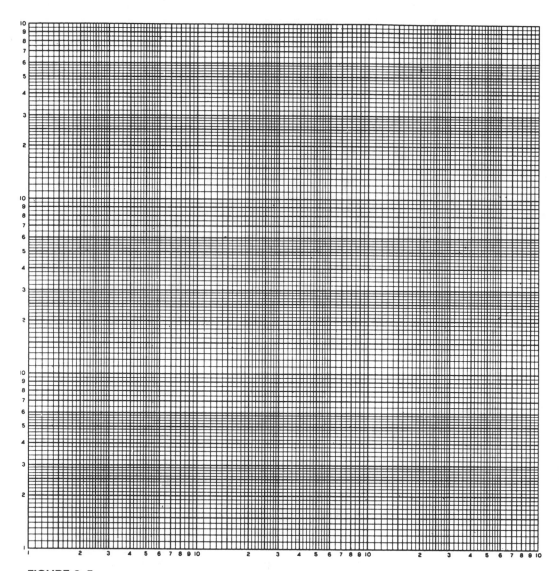

FIGURE 9-7

Example of log-log graph paper with three "cycles" on each scale.

logarithmic scales in Figures 9–6 and 9–7 are all of the 3-cycle type. The number of cycles varies from one to six or more on a standard-size sheet.

Frequency, expressed either as ω in radians/second or as f in hertz, is represented on a logarithmic scale. The "number of cycles" desired on the paper is a function of the frequency range of interest. For example, the 3-cycle forms of Figures 9–6 and 9–7 would be appropriate for a frequency range from 1 rad/s to

1000 rad/s. The actual three decades involved would be 1–10 rad/s, 10–100 rad/s, and 100–1000 rad/s. Likewise, three cycles would be appropriate for a frequency range from 10 kHz to 10 MHz or for any frequency limits with a 1000 to 1 ratio. In general, the number of "paper cycles" n required for a frequency range from ω_1 to ω_2 must satisfy

$$n \geq \log_{10} \frac{\omega_2}{\omega_1} \tag{9–33}$$

where n must be rounded to the next highest integer value. In view of the labeling of most log scales, it is customary to select the frequency limits ω_2 and ω_1 (or f_2 and f_1, when cyclic frequency is used) as integer powers of 10. The actual frequencies on the scales will then simply be integer powers of 10 times the values provided on the scales.

When log-log paper is used, the linear amplitude response $A(\omega)$ is plotted on the vertical log scale. When semilog paper is used, however, the decibel amplitude response $A_{dB}(\omega)$ is plotted on the vertical scale. It turns out that the curve of $A(\omega)$ on log-log paper *has the same shape as* $A_{dB}(\omega)$ on semi-log paper. Both points of view are used in the literature, but the decibel form is somewhat more widely used. Therefore, that approach will be used here, and thus, all further analysis will be based on a semilog presentation with $A_{dB}(\omega)$ on the vertical linear scale, and frequency on the horizontal logarithmic scale.

We have noted that either ω or f could be used as the frequency variable. Although virtually all practical frequency measurements and specifications relate to cyclic frequency f, it is easier (at least in the beginning) to work with ω on Bode plots. We will start with that approach and, after the reader has acquired sufficient skill, we will consider the means for using cyclic frequency f.

The types of transfer functions that are most amenable to Bode plot analysis are those containing real poles and zeros, including poles or zeros at the origin. Whereas limited extension to complex poles and/or zeros can be made, the simplicity and beauty of the method tends to become obscured in the process. The philosophy promoted here is to employ Bode plots primarily for real poles and/or zeros and to use computer (or calculator) analysis for most cases involving complex poles and/or zeros. Thus, the major focus of the effort will be aimed at functions with real poles and/or zeros. Some consideration will be devoted in Section 9–5, however, to certain common forms of complex poles.

Development of Bode Plot Approach

Consider a transfer function having only real poles and zeros (including the origin), and assume that the degree of the numerator is n and the degree of the denominator is m. This function may be expressed as

$$G(s) = \frac{A_0 G_{N1}(s) G_{N2}(s) G_{N3}(s) \cdots G_{Nn}(s)}{G_{D1}(s) G_{D2}(s) G_{D3}(s) \cdots G_{Dm}(s)} \tag{9–34}$$

The quantities $G_{N1}(s)$, $G_{N2}(s)$, . . . , $G_{Nn}(s)$ represent the n first-order numerator factors, and the quantities $G_{D1}(s)$, $G_{D2}(s)$, . . . , $G_{Dm}(s)$ represent the m first-order denominator factors. The exact forms of these factors most suitable for Bode plot analysis will be developed later. The quantity A_0 represents a constant multiplier.

To convert to the steady-state transfer function, the substitution $s = j\omega$ is made.

$$G(j\omega) = \frac{A_0 G_{N1}(j\omega) G_{N2}(j\omega) G_{N3}(j\omega) \cdots G_{Nn}(j\omega)}{G_{D1}(j\omega) G_{D2}(j\omega) G_{D3}(j\omega) \cdots G_{Dm}(j\omega)} \qquad \textbf{(9–35)}$$

Next, the magnitude of $G(j\omega)$ will be determined. For Equation (9–35), this involves determining the magnitude of a product and the magnitude of a quotient. The magnitude of a product is the product of the magnitudes, and the magnitude of a quotient is the quotient of the magnitudes. Thus, the magnitude $|G(j\omega)|$ of the transfer function may be expressed as

$$|G(j\omega)| = \frac{|A_0|\,|G_{N1}(j\omega)|\,|G_{N2}(j\omega)|\,|G_{N3}(j\omega)| \cdots |G_{Nn}(j\omega)|}{|G_{D1}(j\omega)|\,|G_{D2}(j\omega)|\,|G_{D3}(j\omega)| \cdots |G_{Dm}(j\omega)|} \qquad \textbf{(9–36)}$$

(The only reason for placing magnitude bars with A_0 is that it could be a negative value.)

For any numerator factor $|G_{Nk}(j\omega)|$ and for any denominator factor $|G_{Dk}(j\omega)|$, we will define respective amplitude factors $A_{Nk}(\omega)$ and $A_{Dk}(\omega)$ as

$$A_{Nk}(\omega) = |G_{Nk}(j\omega)| \quad \text{for any } k \qquad \textbf{(9–37)}$$

$$A_{Dk}(\omega) = |G_{Dk}(j\omega)| \quad \text{for any } k \qquad \textbf{(9–38)}$$

The net amplitude factor $A(\omega)$ is, of course, defined as

$$A(\omega) = |G(j\omega)| \qquad \textbf{(9–39)}$$

Performing substitutions of the preceding quantities in Equation (9–36), we have

$$A(\omega) = \frac{|A_0| A_{N1}(\omega) A_{N2}(\omega) A_{N3}(\omega) \cdots A_{Nn}(\omega)}{A_{D1}(\omega) A_{D2}(\omega) A_{D3}(\omega) \cdots A_{Dm}(\omega)} \qquad \textbf{(9–40)}$$

Next, the logarithm to the base 10 will be taken on both sides of Equation (9–40). The following two properties of logarithms will be used in this operation:

$$\log_{10} xy = \log_{10} x + \log_{10} y \qquad \textbf{(9–41)}$$

$$\log_{10} \frac{x}{y} = \log_{10} x - \log_{10} y \qquad \textbf{(9–42)}$$

When the preceding results are applied to both sides of Equation (9–40), the logarithms of the numerator factors add, whereas the logarithms of the denominator factors subtract. Finally, both sides are multiplied by 20, and the resulting expansion has the form

$$20 \log_{10} A(\omega) = 20 \log_{10}|A_0|$$
$$+ 20 \log_{10} A_{N1}(\omega) + 20 \log_{10} A_{N2}(\omega)$$
$$+ \cdots + 20 \log_{10} A_{Nn}(\omega) \qquad \textbf{(9–43)}$$
$$- 20 \log_{10} A_{D1}(\omega) - 20 \log_{10} A_{D2}(\omega)$$
$$- \cdots - 20 \log_{10} A_{Dm}(\omega)$$

The left-hand side of Equation (9–43) is the net decibel response $A_{dB}(\omega)$ of the composite function as defined in Equation (9–27). Let us define a series of decibel terms for the right-hand side as follows:

$$A_{Nk}^{(dB)}(\omega) = 20 \log_{10} A_{Nk}(\omega) \quad \text{for any } k \qquad \textbf{(9–44a)}$$

$$A_{Dk}^{(dB)}(\omega) = 20 \log_{10} A_{Dk}(\omega) \quad \text{for any } k \qquad \textbf{(9–44b)}$$

$$A_0^{(dB)} = 20 \log_{10}|A_0| \qquad \textbf{(9–44c)}$$

Substituting the forms of Equation (9–44) in (9–43), there results

$$A_{dB}(\omega) = A_0^{(dB)} + A_{N1}^{(dB)}(\omega) + A_{N2}^{(dB)}(\omega) + \cdots + A_{Nn}^{(dB)}(\omega)$$
$$- A_{D1}^{(dB)}(\omega) - A_{D2}^{(dB)}(\omega) - \cdots - A_{Dm}^{(dB)}(\omega) \qquad \textbf{(9–45)}$$

Don't be disturbed by the unwieldy notation we have created, since this is the only time we'll need to write a "general" equation of this type. The concept established by Equation (9–45) is the important point for consideration.

According to Equation (9–45), the net decibel response of a composite transfer function is the sum of the numerator decibel response terms minus the denominator decibel response terms. The process of multiplication and division of response terms is reduced to addition and subtraction of simpler decibel response terms.

Forms of the Factors

Now that the concept has been established, the next step is to investigate the forms of the individual factors. Since we are considering only real poles and zeros at this point, all numerator and denominator polynomial factors will initially be of the form $s + \alpha$ or, in the case of a zero or pole at the origin, s. For this purpose, multiple-order zero or pole factors can be thought of as products of equal first-order factors. Recall that the factored form $s + \alpha$, in which the coefficient of s is unity, is the ideal form for inverse Laplace transformation. However, for Bode plot analysis, this form is not the best.

The best form for a real pole or zero factor in Bode plot analysis is to force the constant term to be unity. The resulting desired form for each polynomial factor is $1 + s/\alpha$, in which the unity term has been written first. Any multiplied or divided constants are lumped together and collectively form the A_0 factor. For example, suppose one of the polynomial factors is initially $s + 4$. By dividing internally by 4 and multiplying externally by 4 to compensate, this factor can be expressed as $4(1 + s/4)$. The desired form of the factor is $1 + s/4$, and the surplus 4 is combined with other constant factors in the composite function.

Phase Response

Thus far, we have considered only the amplitude response corresponding to Equation (9–35). For the individual phase contributions, let $\beta_{Nk}(\omega)$ = phase response corresponding to $G_{Nk}(j\omega)$, and let $\beta_{Dk}(\omega)$ = phase responding corresponding to $G_{Dk}(j\omega)$. Momentarily assume that $A_0 > 0$. The composite phase response corresponding to Equation (9–35) is determined from the concept that the net angle is the sum of the numerator angles minus the sum of the denominator angles. Letting $\beta(\omega)$ represent the net phase response, we have

$$\begin{aligned}\beta(\omega) = \beta_{N1}(\omega) + \beta_{N2}(\omega) + \cdots + \beta_{Nn}(\omega) \\ - \beta_{D1}(\omega) - \beta_{D2}(\omega) - \cdots - \beta_{Dm}(\omega)\end{aligned} \tag{9–46}$$

If $A_0 < 0$, an additional phase shift of 180° should be added to Equation (9–46).

EXAMPLE 9–4

A certain transfer function is given by

$$G(s) = \frac{2000(s + 4)}{s^2 + 116s + 1600} \tag{9–47}$$

Arrange the function in a form suitable for Bode plot analysis.

Solution
It can be readily determined that the two poles are real and are −16 and −100. The factored form of the function is

$$G(s) = \frac{2000(s + 4)}{(s + 16)(s + 100)} \tag{9–48}$$

This would be the desired form if inverse Laplace transformation were to be performed.

For Bode plot analysis, the constant 4 should be removed from the $s + 4$ factor, the constant 16 should be removed from the $s + 16$ factor, and the constant 100 should be removed from the $s + 100$ factor. This is achieved by dividing and multiplying each factor by the appropriate constant as follows:

$$G(s) = \frac{2000 \times 4\left(\dfrac{s}{4} + \dfrac{4}{4}\right)}{16\left(\dfrac{s}{16} + \dfrac{16}{16}\right)100\left(\dfrac{s}{100} + \dfrac{100}{100}\right)} \tag{9–49}$$

Cancelling, combining constants, and rearranging lead to

$$G(s) = \frac{5\left(1 + \dfrac{s}{4}\right)}{\left(1 + \dfrac{s}{16}\right)\left(1 + \dfrac{s}{100}\right)} \tag{9-50}$$

The α quantities have been arranged in a left-to-right ascending order for convenience, as will be noted later.

This particular function will be considered again in Example 9–5 at the end of the next section.

9–4 BODE PLOT FORMS

The concept of the Bode plot was established in the preceding section, in which the frequency response of a composite function is expressed as the sum and difference of simpler functions. In this section, the forms of these simpler funtions will be developed, and certain simplified approximations will be introduced. The results can then be tied together to allow full use for composite functions. Each of several basic forms will now be considered. The same basic notation, namely, $G(s)$, $A(\omega)$, $\beta(\omega)$, will be applied to each case as it is considered.

Zero on Negative Real Axis

Consider the s-domain transfer function

$$G(s) = 1 + \frac{s}{\alpha} \quad \text{with } \alpha > 0 \tag{9-51}$$

The steady-state transfer function is

$$G(j\omega) = 1 + \frac{j\omega}{\alpha} \tag{9-52}$$

The amplitude response is

$$A(\omega) = \sqrt{1 + \left(\frac{\omega}{\alpha}\right)^2} \tag{9-53}$$

The corresponding decibel response is

$$A_{dB}(\omega) = 20 \log_{10} \sqrt{1 + \left(\frac{\omega}{\alpha}\right)^2} \tag{9-54}$$

The phase response is

$$\beta(\omega) = \tan^{-1}\frac{\omega}{\alpha} \qquad (9\text{--}55)$$

An exact plot of $A_{dB}(\omega)$ is shown as curve E of Figure 9–8. Since α is arbitrary, the horizontal variable used is ω/α, which makes the result applicable for any value of α.

For $\omega/\alpha = 1$ (or $\omega = \alpha$), the value of A_{dB} is $20\log_{10}\sqrt{1 + (1)^2} = 3.01$ dB. This value is usually rounded to 3 dB. If the independent variable were ω (instead of ω/α), this level would occur at the radian frequency $\omega = \alpha$.

Next, consider the behavior of Equation (9–54) first at frequencies well below and then at frequencies well above $\omega = \alpha$. For $\omega \ll \alpha$, the expression of Equation (9–54) can be approximated by

$$A_{dB}(\omega) \simeq 20\log_{10}\sqrt{1} = 0 \text{ dB} \quad \text{for } \omega \ll \alpha. \qquad (9\text{--}56)$$

For $\omega \gg \alpha$, the expression of Equation (9–54) can be approximated as

$$A_{dB}(\omega) \simeq 20\log_{10}\sqrt{\left(\frac{\omega}{\alpha}\right)^2} = 20\log_{10}\frac{\omega}{\alpha} \quad \text{for } \omega \gg \alpha \qquad (9\text{--}57)$$

The result of Equation (9–56) indicates that at low frequencies relative to $\omega = \alpha$, the response is nearly a flat level of 0 dB. The result of Equation (9–57), which is applicable at high frequencies compared to $\omega = \alpha$, turns out to be a straight line on a semilog decibel versus frequency plot. It can be readily verified that this function increases 6.02 dB (usually rounded to 6 dB) for each doubling of the frequency, and it increases by 20 dB when the frequency is multiplied by 10.

These high and low frequency approximations lead to the Bode plot approximation, which is shown as curve BP of Figure 9–8. This approximation assumes that the response has a flat level of 0 dB up to $\omega = \alpha$, at which point the response changes slope abruptly and turns upward. This function is called a *breakpoint approximation.* For somewhat obvious reasons, the radian frequency $\omega = \alpha$ is called the *break frequency,* or the *corner frequency.* The first term will be used in this text.

The Bode plot approximation to $A_{dB}(\omega)$ is thus a piecewise linear function consisting of a constant level of 0 dB up to the break frequency and then a straight line with a positive slope beyond the break frequency. The slope of this function above the break frequency is +6 dB/octave or + 20 dB/decade. A change of one octave is the process of doubling the frequency, and a change of one decade is the process of increasing the frequency by a factor of ten.

From Figure 9–8, note that the exact curve and the Bode plot approximation are very close together at frequencies well below and well above the break frequency. The worst error occurs at the break frequency, and this error is 3 dB. Thus, whereas the Bode plot technique in its simplest form is an approximate method, corrections may be readily applied to achieve very high accuracy if desired. Actually, many users show only the breakpoint approximations, and it is understood by all parties that more nearly exact results can be obtained if desired.

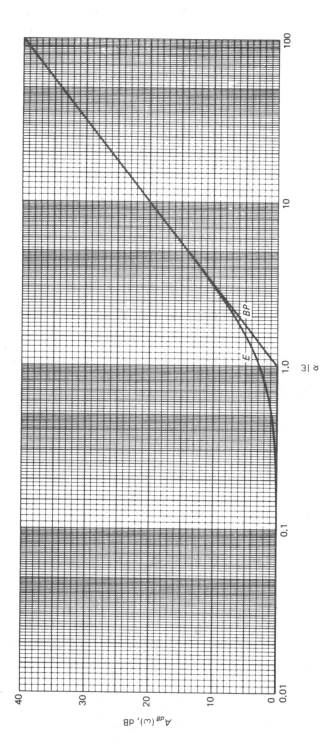

FIGURE 9-8
Exact (*E*) and breakpoint (*BP*) approximation of amplitude response for first-order zero on negative real axis.

Consider next the phase associated with the given transfer function. The exact phase shift as given by Equation (9–55) is shown as curve E of Figure 9–9. Over the frequency interval $0 \le \omega < \infty$, the range of phase shift is $0 \le \beta < 90°$. At a radian frequency $\omega = 0.1\alpha$, the phase shift is 5.71°, and at a radian frequency $\omega = 10\alpha$, the phase shift is 84.29°. At a frequency $\omega = \alpha$, the phase shift is 45°.

The breakpoint approximation to the phase shift consists of three straight-line segments as shown by curve BP of Figure 9–9. In the frequency range $\omega < 0.1\alpha$, the assumed phase shift is 0°. For $\omega > 10\alpha$, the assumed phase shift is $+90°$. For $0.1\alpha < \omega < 10\alpha$, the assumed phase shift is a straight line on a semi-log scale. This straight line predicts the exact phase shift of 45° at $\omega = \alpha$. The worst errors are at $\omega = 0.1\alpha$ and $\omega = 10\alpha$, where the magnitude of the error is 5.71° at both points.

Pole on Negative Real Axis

Consider the transfer function

$$G(s) = \frac{1}{1 + \dfrac{s}{\alpha}} \quad \text{with } \alpha > 0 \qquad \text{(9–58)}$$

From the decibel amplitude properties developed in the last section, it can be readily inferred that the decibel amplitude response of this function is simply the negative of the case of a real axis zero, as developed earlier. A similar result applies to the phase shift as well. However, because of its importance, this case will be developed separately.

The steady-state transfer function is

$$G(j\omega) = \frac{1}{1 + j\dfrac{\omega}{\alpha}} \qquad \text{(9–59)}$$

The amplitude response is

$$A(\omega) = \frac{1}{\sqrt{1 + \left(\dfrac{\omega}{\alpha}\right)^2}} \qquad \text{(9–60)}$$

The corresponding decibel response is

$$A_{\text{dB}}(\omega) = 20 \log_{10} \left[\frac{1}{\sqrt{1 + \left(\dfrac{\omega}{\alpha}\right)^2}} \right] \qquad \text{(9–61a)}$$

$$= -20 \log_{10} \sqrt{1 + \left(\dfrac{\omega}{\alpha}\right)^2} \qquad \text{(9–61b)}$$

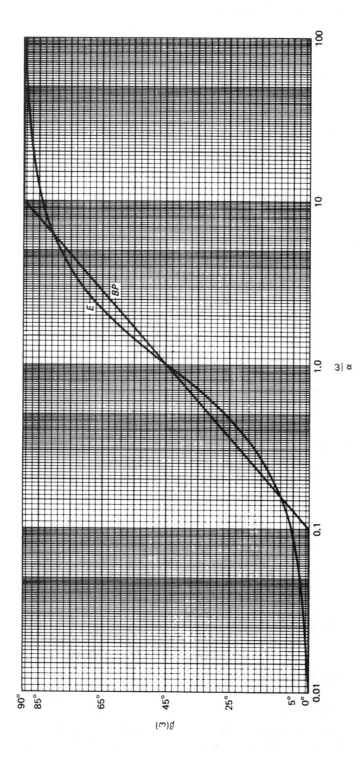

FIGURE 9–9

Exact (E) and breakpoint (BP) approximation of phase response for first-order zero on negative real axis.

The phase response is

$$\beta(\omega) = -\tan^{-1}\frac{\omega}{\alpha} \qquad (9\text{–}62)$$

Comparing Equation (9–61b) with Equation (9–54), one can readily observe that the decibel response for the real axis pole is the negative of the response for the real axis zero.

An exact plot of $A_{dB}(\omega)$ for the real pole case is given by curve E of Figure 9–10. For $\omega = \alpha$ (or $\omega/\alpha = 1$), the value of A_{dB} is now -3 dB. If the independent variable were ω (instead of ω/α), this level would occur at the frequency $\omega = \alpha$.

The breakpoint approximation is shown as curve BP of Figure 9–10. For $\omega < \alpha$, the breakpoint approximation is a flat 0 dB level. The appropriate curve is assumed to change slope abruptly at $\omega = \alpha$; and for $\omega > \alpha$, the breakpoint approximation is a straight line with a slope of -6 dB/octave or -20 dB/decade.

The exact phase response of the real axis pole is shown as curve E of Figure 9–11. At a radian frequency $\omega = 0.1\alpha$, the phase shift is $\beta = -5.71°$, and at a radian frequency $\omega = 10\alpha$, the phase shift is $\beta = -84.29°$. At a frequency $\omega = \alpha$, the phase shift is $\beta = -45°$.

The breakpoint approximation to phase shift consists of three straight-line segments as shown by curve BP of Figure 9–11. In the frequency range $\omega < 0.1\alpha$, the assumed phase shift is $0°$; and for $\omega > 10\alpha$, the assumed phase shift is $-90°$. for $0.1\alpha < \omega < 10\alpha$, the assumed phase shift is a straight line on a semilog scale of phase shift versus frequency.

Zero at Origin

Consider the transfer function

$$G(s) = s \qquad (9\text{–}63)$$

The steady-state transfer function is

$$G(j\omega) = j\omega \qquad (9\text{–}64)$$

The amplitude response is

$$A(\omega) = \omega \qquad (9\text{–}65)$$

The corresponding decibel response is

$$A_{dB}(\omega) = 20 \log \omega \qquad (9\text{–}66)$$

The phase response is simply

$$\beta(\omega) = 90° \qquad (9\text{–}67)$$

The function $A_{dB}(\omega)$ is shown on a semilog plot in Figure 9–12. In this case, there is no "break," so the exact curve can be used directly with other Bode plot

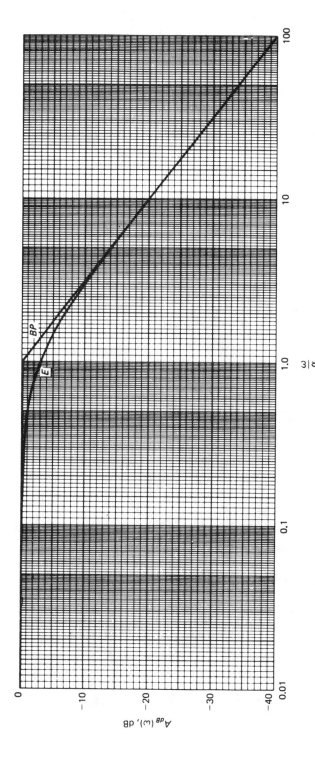

FIGURE 9–10

Exact (*E*) and breakpoint (*BP*) approximation of amplitude response for first-order pole on negative real axis.

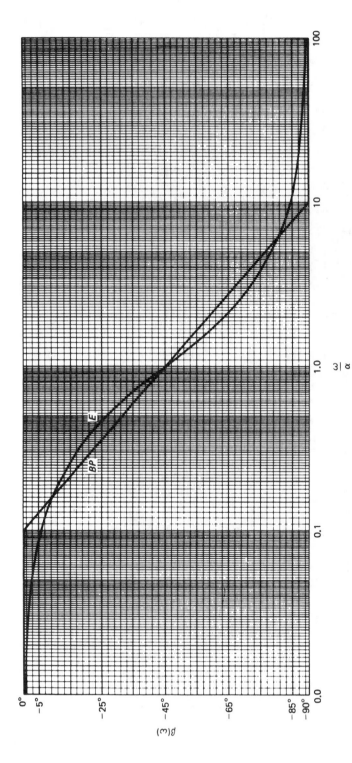

FIGURE 9–11

Exact (*E*) and breakpoint (*BP*) approximation of phase response for first-order pole on negative real axis.

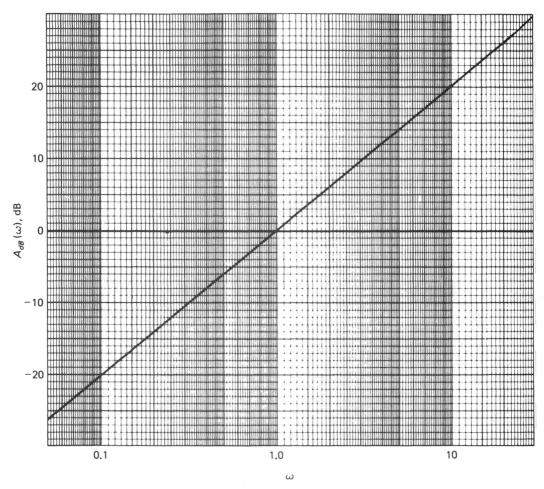

FIGURE 9–12
Exact amplitude response for first-order zero at origin.

factors. The slope of the curve is $+6$ dB/octave or $+20$ dB/decade at all points. The abscissa in this case is simply ω.

A key point of significance is the level at $\omega = 1$ rad/s, which can be readily determined from Equation (9–66) as $A_{\mathrm{dB}}(1) = 0$ dB. This easily remembered fact leads to a quick determination of one point on the curve; and with the slope known, the complete response can be readily determined. Note that we cannot get to the frequency $\omega = 0$ on a logarithmic scale no matter how far to the left we go. Thus, the traditional use of the origin is of little value in a Bode plot.

The phase response for a zero at the origin is shown in Figure 9–13. This function is a simple constant value of $+90°$ at all frequencies and, like the amplitude response, can be applied directly without further refinement.

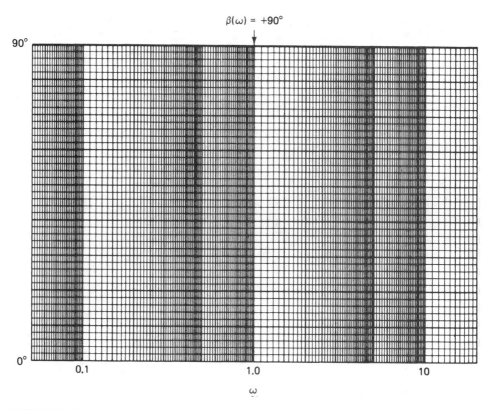

FIGURE 9–13
Exact phase response for first-order zero at origin.

Pole at Origin

Consider the transfer function

$$G(s) = \frac{1}{s} \tag{9-68}$$

The steady-state transfer function is

$$G(j\omega) = \frac{1}{j\omega} \tag{9-69}$$

The amplitude response is

$$A(\omega) = \frac{1}{\omega} \tag{9-70}$$

The corresponding decibel response is

$$A_{dB}(\omega) = 20 \log_{10} \frac{1}{\omega} \tag{9-71a}$$

$$= -20 \log_{10} \omega \tag{9-71b}$$

Comparing Equation (9–71b) with Equation (9–66), observe that the response for the pole at the origin is the negative of the response for the zero at the origin.

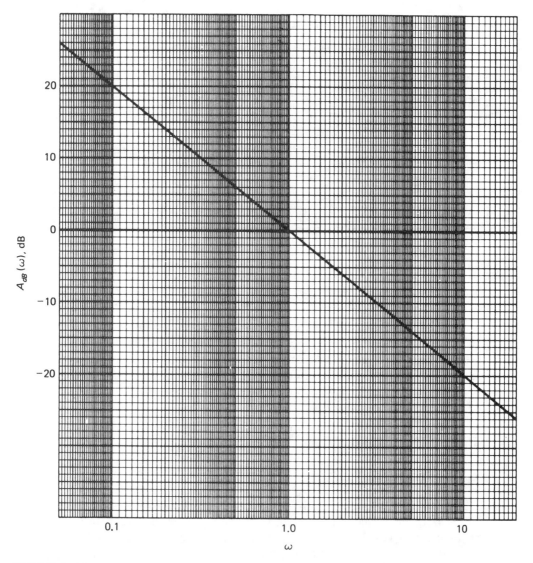

FIGURE 9–14

Exact amplitude response for first-order pole at origin.

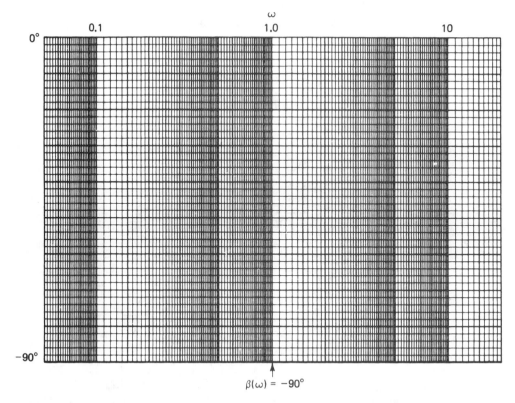

FIGURE 9–15
Exact phase response for first-order pole at origin.

The function $A_{dB}(\omega)$ is shown on a semilog plot in Figure 9–14. As in the case of the zero at the origin, the exact curve is a straight line. The slope in this case is -6 dB/octave or -20 dB/decade. The abscissa is simply ω. Once again, the print $\omega = 1$ rad/s is a convenient reference, since $A_{dB}(1) = 0$ dB.

The phase response in this case is

$$\beta(\omega) = -90° \tag{9–72}$$

This function is shown in Figure 9–15.

Frequency Scale

Before getting into some details about combining various factors, some discussion of the frequency scale is warranted. In converting from the s-domain form to the steady-state phasor form, the radian frequency ω occurs naturally, and all analysis up to this point has retained that quantity. However, virtually all real-life tests are made, and specifications are given, as a function of the cyclic frequency f. Since $\omega = 2\pi f$, there is a simple scale factor between these two quantities. Yet this difference in frequency variables is a constant source of annoyance, and even "experts" sometimes get the 2π factor in the wrong place or accidentally omit it, which can lead to gross errors.

The basic dilemma is that, on one hand, we can stay with ω as the mathematically "clean" variable in Bode plots but it is harder to relate to practical results. On the other hand, we can modify the functions appropriately so that cyclic frequency f is the variable of interest, but the mathematical forms tend to be more clumsy.

Although separate approaches for making plots in terms of either f or ω can be formulated, the approach taken here will be as follows: Basically, all plots will be made as a function of radian frequency ω because of simplicity and direct correlation with s-domain forms. In addition to the horizontal scale of radian frequency ω, however, a separate horizontal scale of $f = \omega/2\pi$ will be shown underneath some of the curves. Any cyclic frequency on this scale is determined by dividing the corresponding point on the ω scale by 2π. Conversely, if it is desired to check the response at some value of f, the corresponding radian frequency ω is determined by multiplying f by 2π. For example, $\omega = 1$ rad/s corresponds to $f \simeq 0.159$ Hz, and $f = 5$ Hz corresponds to $\omega = 31.42$ rad/s.

Strategies for Making Plots

Some strategies for making Bode amplitude plots will now be discussed. The basic concept is based on the addition of decibel amplitude response forms as discussed in the preceding section. Consider first the case where there are no poles or zeros at the origin. Since all $(1 + s/\alpha)$ factors are assumed to have 0 dB levels below the break frequencies, the resulting function must have a flat level equal to the dB level of the constant A_0 at frequencies lower than any of the break frequencies.

At a frequency equal to the lowest break frequency, one of the factors is assumed to change slope abruptly. If the factor is a numerator factor, the composite function turns upward from the initial level at a slope of +6 dB/octave. If the factor is a denominator factor, the composite function turns downward with a slope of −6 dB/octave, of course. When the next break frequency is encountered, the result may either reinforce the first slope, in which case the slope is doubled, or it may cancel it out, in which case the net slope returns to zero. The process is accumulative, and the slope at any frequency is the algebraic sum of all slopes that have been initiated up to that point.

When there is a pole or zero at the origin, the slope going all the way back toward dc will correspond to the pole or zero form. In this case, it is sometimes convenient to use $\omega = 1$ as a convenient point of reference, since the dB level corresponding to either a pole or a zero at the origin is 0 dB at that point. If all break frequencies are larger than $\omega = 1$, the constant sets the total level at $\omega = 1$.

If the given frequency response has a pole or zero at the origin but has a "flat" region, that is, an interval with a slope of zero at higher frequencies, it is often convenient to use that property as a basis for determining the level. To employ this concept, perform approximations on all $1 + s/\alpha$ factors as follows. (a) If $\omega > \alpha$ in the flat region, assume $1 + s/\alpha \simeq s/\alpha$. (b) If $\omega < \alpha$ in the flat region, assume

$1 + s/\alpha \simeq 1$. The resulting approximation must necessarily reduce to a constant value after these assumptions have been made. (Otherwise, the region in question is not flat.) When this value is converted to decibels, it is the level of the constant region on a Bode plot. This concept will be illustrated in Example 9–9.

EXAMPLE 9–5

Consider the transfer function of Example 9–4, which was arranged in a form suitable for Bode plot analysis in Equation (9–50). Plot the breakpoint approximation to the decibel amplitude response.

Solution
For convenience, the transfer function is repeated here.

$$G(s) = \frac{5\left(1 + \dfrac{s}{4}\right)}{\left(1 + \dfrac{s}{16}\right)\left(1 + \dfrac{s}{100}\right)} \qquad (9\text{–}73)$$

There are no poles or zeros at $s = 0$, so the low-frequency amplitude response will be constant at a level $20 \log_{10} 5 = 14$ dB. The three radian break frequencies are 4 rad/s, 16 rad/s, and 100 rad/s. The first break frequency is associated with a numerator factor, and the second two are associated with denominator factors. The two denominator factors are arranged in a left to right ascending order of break frequencies.

The composite form of the Bode plot approximation is shown in Figure 9–16. A reasonable choice of scale is 3 decades, with $\omega = 1$ rad/s as the lowest frequency and $\omega = 1000$ rad/s as the highest frequency. In the lowest frequency range ($\omega < 4$ rad/s), the response has a flat level of 14 dB. The first break is at $\omega = 4$ rad/s, and since this occurs in a numerator factor, the curve breaks upward at +6 dB/octave or +20 dB/decade. This slope is determined by first locating a frequency of either two times or ten times the break frequency. A point at this frequency of either 6 dB above or 20 dB above the level at the break is located, and a straightedge is used to draw a line from the breakpoint in the direction of the new point. The use of the frequency at ten times the break is usually a little more accurate.

In this example, the numbers are "rigged" so that the results can be readily predicted. Two octaves occur between $\omega = 4$ rad/s and $\omega = 16$ rad/s, that is, the frequency is doubled twice. Thus, a change of $2 \times 6 = 12$ dB is expected, so the level at $\omega = 16$ rad/s is 14 dB + 12 dB = 26 dB.

The second break, which occurs at $\omega = 16$ rad/s, is a denominator break. If this factor stood alone, it would result in a downward break with a negative slope. However, since the numerator term is continuing to increase, the net slope becomes 0 above $\omega = 16$ rad/s. In effect, the −6 dB/octave slope of the denominator cancels the +6 dB/octave slope of the numerator.

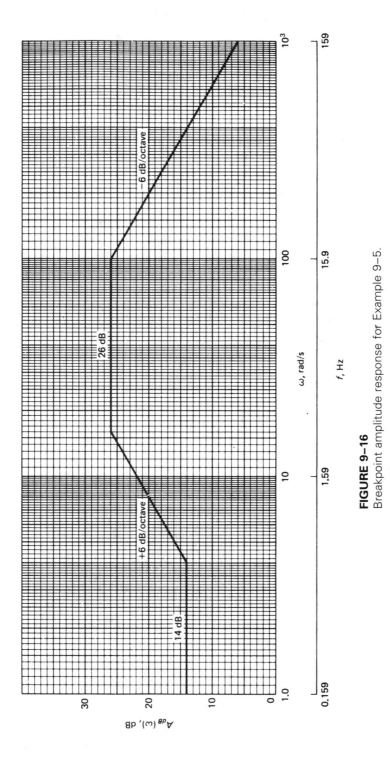

FIGURE 9–16
Breakpoint amplitude response for Example 9–5.

The third break occurs at $\omega = 100$ rad/s, and it is a denominator break. Since the net slope to the left of 100 rad/s is 0, the net slope to the right of 100 rad/s changes to -6 dB/octave. Since this is the last break, the function continues at this slope for higher frequencies. This property is related to the fact that the transfer function has one zero at $s = \infty$, and this condition results in a high-frequency rolloff of -6 dB/octave.

The development thus far has been rather intuitive in nature and, with some practice, can be done almost by inspection. Herein lies the beauty of the Bode plot in permitting a quick, simple inspection of the general form of the frequency response, and this may be perfectly adequate in many cases. Suppose, however, there is a need to actually compute the exact amplitude (or the phase) at certain frequencies. Even then, construction of the Bode plot may be worthwhile, since it can guide us in selecting certain points at which the response should be calculated. The rule about the -3 dB difference at a breakpoint is correct only for a single factor, since the effects of all factors must be considered in a composite function. However, when the various break frequencies are widely separated, the 3-dB difference at break frequencies is a good approximation.

In the event that we need to know the exact amplitude and phase response of this system, it will now be calculated. First, the steady-state transfer function $G(j\omega)$ is determined from Equation (9–73) by substituting $s = j\omega$:

$$G(j\omega) = \frac{5\left(1 + \dfrac{j\omega}{4}\right)}{\left(1 + \dfrac{j\omega}{16}\right)\left(1 + \dfrac{j\omega}{100}\right)} \tag{9–74}$$

The amplitude response is determined by taking the magnitudes of the individual factors and forming the composite product and quotient.

$$A(\omega) = \frac{5\sqrt{1 + \left(\dfrac{\omega}{4}\right)^2}}{\left(\sqrt{1 + \left(\dfrac{\omega}{16}\right)^2}\right)\left(\sqrt{1 + \left(\dfrac{\omega}{100}\right)^2}\right)} \tag{9–75}$$

The decibel amplitude response is

$$A_{dB}(\omega) = 20 \log_{10} A(\omega) \tag{9–76a}$$

$$\begin{aligned} = 14 &+ 20 \log_{10}\sqrt{1 + \left(\frac{\omega}{4}\right)^2} \\ &- 20 \log_{10}\sqrt{1 + \left(\frac{\omega}{16}\right)^2} - 20 \log_{10}\sqrt{1 + \left(\frac{\omega}{100}\right)^2} \end{aligned} \tag{9–76b}$$

The preceding form results from the additive and subtractive property of the dB factors developed earlier. The phase response is

$$\beta(\omega) = \tan^{-1}\frac{\omega}{4} - \tan^{-1}\frac{\omega}{16} - \tan^{-1}\frac{\omega}{100} \qquad (9\text{-}77)$$

EXAMPLE 9–6

Plot the breakpoint decibel amplitude response approximation for the transfer function

$$G(s) = \frac{40s}{\left(1 + \dfrac{s}{2}\right)\left(1 + \dfrac{s}{10}\right)\left(1 + \dfrac{s}{100}\right)} \qquad (7\text{-}78)$$

Solution
The function is already arranged in a form suitable for plotting, and the three break frequencies are readily identified as $\omega = 2$ rad/s, 10 rad/s, and 100 rad/s. All three are denominator breaks. There are no numerator breaks, but the presence of the numerator s-factor (zero at origin) results in a slope of +6 dB/octave at very low frequencies. Since all break frequencies are greater than 1 rad/s, their contributions are all 0 dB at $\omega = 1$, and since the dB level of s at $\omega = 1$ is 0 dB, the level at $\omega = 1$ is simply $20 \log_{10}40 = 32$ dB.

The breakpoint approximation is shown in Figure 9–17. A line with slope +6 dB/octave is drawn through a level of 32 dB at $\omega = 1$. At $\omega = 2$, the level reaches $32 + 6 = 38$ dB, and the first break occurs. Since this is a denominator break, the net slope changes to 0. At $\omega = 10$ rad/s, a second denominator break occurs, and the net slope now changes to -6 dB/octave. At $\omega = 100$ rad/s, the level is down to $38 - 20 = 18$ dB, and the final break occurs. The net slope is now -12 dB/octave, and the function continues at this slope for all higher frequencies. This property is related to the fact that the transfer function has two zeros at $s = \infty$, and this condition results in a high-frequency rolloff of -12 dB/octave.

In the event that any computations of the exact amplitude and phase are to be computed, the reader is invited to verify that the dB amplitude and phase functions are

$$A_{dB}(\omega) = 32 + 20 \log_{10}\omega - 20 \log_{10}\sqrt{1 + \left(\frac{\omega}{2}\right)^2}$$
$$-20 \log_{10}\sqrt{1 + \left(\frac{\omega}{10}\right)^2} - 20 \log_{10}\sqrt{1 + \left(\frac{\omega}{100}\right)^2} \qquad (9\text{-}79)$$

and

$$\beta(\omega) = 90° - \tan^{-1}\frac{\omega}{2} - \tan^{-1}\frac{\omega}{10} - \tan^{-1}\frac{\omega}{100} \qquad (9\text{-}80)$$

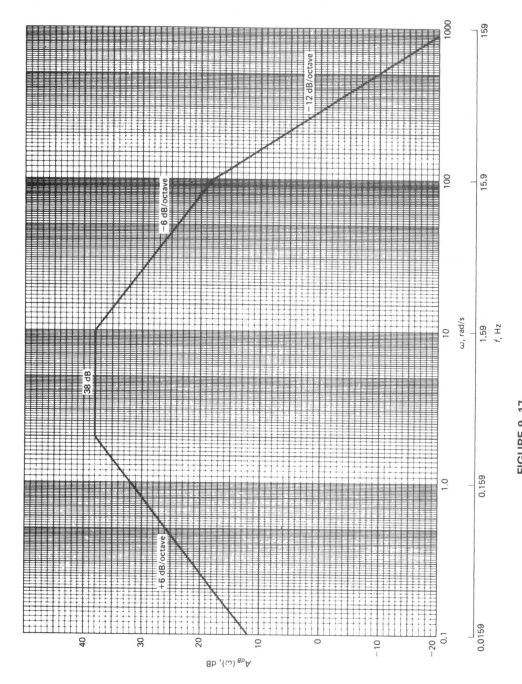

FIGURE 9-17
Breakpoint amplitude response for Example 9-6.

514

EXAMPLE 9–7

Determine the forms of the breakpoint amplitude and phase response approximations of the simple RC low-pass circuit of Figure 9–18(a). The input is v_1 and the output is v_2.

Solution

The reader is invited to verify that the transfer function is

$$G(s) = \frac{V_2(s)}{V_1(s)} = \frac{1}{1 + sRC} \tag{9–81}$$

The denominator has the correct constant "1," but the s term should be divided by a quantity rather than multiplied by a quantity. This situation is simply a matter of interpretation, since we can define a radian break frequency ω_b as

$$\omega_b = \frac{1}{RC} \tag{9–82}$$

The transfer function is then expressed as

$$G(s) = \frac{1}{1 + \dfrac{s}{\omega_b}} \tag{9–83}$$

Whenever a function has the multiplicative form of Equation (9–81), it is a simple matter to define the radian break frequency as the reciprocal of the multiplicative factor. For an example with numbers: if the factor is $1 + 20s$, the radian break frequency is $\omega_b = 1/20 = 0.05$ rad/s, and the factor could be rewritten as $1 + s/0.05$, with the break frequency clearly evident.

Returning to the form of Equation (9–83), the transfer function has one finite denominator break frequency. Referring to Figure 9–18(b), for $\omega < \omega_b$, the response is assumed to be at a flat level corresponding to $20 \log_{10} 1 = 0$ dB. At $\omega = \omega_b$, the function breaks downward at a slope of -6 dB/octave. The exact curve is, of course, 3 dB below the breakpoint approximation at $\omega = \omega_b$. The corresponding cyclic frequency f_b is

$$f_b = \frac{1}{2\pi RC} \tag{9–84}$$

The breakpoint phase approximation is shown in Figure 9–18(c). For convenience, both ω and f frequency scales are given.

EXAMPLE 9–8

Determine the forms of the breakpoint amplitude and phase response approximations of the simple RC high-pass circuit of Figure 9–19(a). The input is v_1 and the output is v_2.

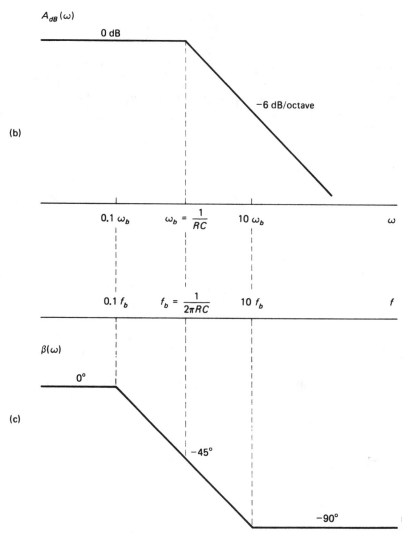

FIGURE 9–18
RC low-pass circuit for Example 9–7 and breakpoint approximations of amplitude and phase.

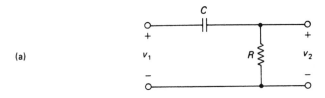

(a)

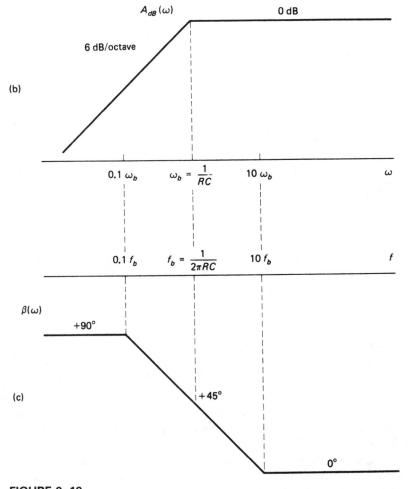

(b)

(c)

FIGURE 9–19

RC high-pass circuit for Example 9–8 and breakpoint approximations of amplitude and phase.

Solution

The reader is invited to verify that the transfer function is

$$G(s) = \frac{V_2(s)}{V_1(s)} = \frac{sRC}{1 + sRC}$$

(9–85)

As in Example 9–7, we define ω_b as

$$\omega_b = \frac{1}{RC}$$

(9–86)

The transfer function can then be expressed as

$$G(s) = \frac{\dfrac{s}{\omega_b}}{1 + \dfrac{s}{\omega_b}}$$

(9–87)

The breakpoint amplitude approximation is shown in Figure 9–19(b). At low frequencies, the slope of $A(\omega)$ is +6 dB/octave. The best way to obtain the level of this function is based on the fact that for $\omega \gg \omega_b$, $A(\omega) \simeq 1$ or $A_{dB}(\omega) \simeq 0$ dB. The line with a slope of 6 dB/octave for $\omega < \omega_b$ must meet the flat line at $\omega = \omega_b$. As in the previous case, the cyclic break frequency is

$$f_b = \frac{1}{2\pi RC}$$

(9–88)

The breakpoint phase approximation is shown in Figure 9–19(c). The *s*-factor in the numerator provides a phase shift of $+90°$ at very low frequencies, and the denominator phase lag eventually brings the phase back to $0°$. As in Example 9–7, both ω and f scales are given.

EXAMPLE 9–9

This problem illustrates how low- and high-frequency degradation in the frequency response affect the transient response. The transfer function used for this purpose is of a form similar to that of a single-stage *RC* coupled amplifier in which the effect of a coupling capacitor causes a drop in response at low frequencies, and the effect of the input capacitance of an active device causes a drop in response at high frequencies.

Assume that the measured amplitude response of a certain *RC* coupled amplifier is of the form shown in Figure 9–20, and assume that the amplifier is an inverting type. The magnitude of the gain is essentially flat at a level of 40 dB over a broad frequency range, but there are both low-frequency and high-frequency "rolloff" rregions. The two 3-dB frequencies are $f_1 = 31.83$ Hz and $f_2 = 159.2$ kHz.

(a) Determine the transfer function $G(s) = V_2(s)/V_1(s)$.

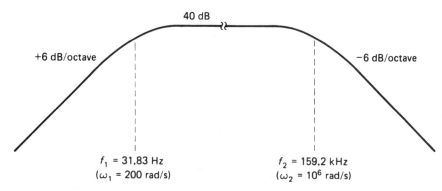

FIGURE 9–20
Amplitude response for Example 9–9.

(b) Assume that the circuit is excited at $t = 0$ by a dc voltage of value -0.1 V, that is, $v_1(t) = -0.1u(t)$; determine the output voltage $v_2(t)$.

(c) Determine an approximation for the output voltage displaying the effect of the high-frequency rolloff on the transient behavior.

(d) Repeat (c) with respect to the low-frequency rolloff.

Solution

(a) The measured 3-dB frequencies are given in Hz, as would normally be the case for experimental data, but they should be converted to rad/s for Bode plot analysis. Let $\omega_1 = 2\pi f_1 = 2\pi \times 31.83 = 200$ rad/s and $\omega_2 = 2\pi f_2 = 2\pi \times 159.2 \times 10^3 = 10^6$ rad/s. These numbers were obviously "loaded" by the author so that we can see the principle of the problem without getting bogged down in computational details. However, these values are in the range of typical amplifier values. The 40-dB gain level corresponds to a linear gain magnitude level of 100. Since the amplifier is an inverting type, the actual linear gain is -100. The transfer function must have an s-factor in the numerator to provide the $+6$ dB/octave slope at low frequencies. A denominator first-order pole factor must provide the first break to flatten the response out, and a second denominator first-order pole factor must eventually supply a second break to provide the proper high-frequency rolloff. Since the two break frequencies will be very far apart, the two 3-dB frequencies may be assumed to be equal to the respective break frequencies. The form of the transfer function is thus deduced as

$$G(s) = \frac{A_0 s}{\left(1 + \dfrac{s}{200}\right)\left(1 + \dfrac{s}{10^6}\right)} \tag{9–89}$$

where A_0 is a constant to be determined.

The best way to determine A_0 is to use the flat region as a basis. In this region, corresponding to $200 < \omega < 10^6$, we assume $1 + s/200 \simeq s/200$ and $1 + s/10^6 \simeq 1$. Simultaneously, we know that $A(\omega) = -100$ in this region, so

$$\frac{A_0 s}{(s/200)(1)} = 200 A_0 = -100 \tag{9-90}$$

This leads to $A_0 = -0.5$, and thus

$$G(s) = \frac{-0.5s}{\left(1 + \dfrac{s}{200}\right)\left(1 + \dfrac{s}{10^6}\right)} \tag{9-91}$$

Note that the gain constant A_0 in the transfer function is not the same value as the amplifier gain in the mid-frequency range.

(b) Our goal in this part is to determine a time-domain response, so the first step is to rearrange $G(s)$ from Equation (9–91) in the form most suitable for inverse transformation. This is done by multiplying numerator and denominator of Equation (9–91) by 200×10^6. After regrouping, the function becomes

$$G(s) = \frac{-10^8 s}{(s + 200)(s + 10^6)} \tag{9-92}$$

The input excitation is $v_1(t) = -0.1u(t)$, so $V_1(s) = -0.1/s$. The output transform is then

$$V_2(s) = G(s)V_1(s)$$

$$= \frac{-10^8 s}{(s + 200)(s + 10^6)} \times \left(\frac{-0.1}{s}\right) = \frac{10^7}{(s + 200)(s + 10^6)} \tag{9-93}$$

The inverse transform will have the form:

$$v_2(t) = A_1 e^{-200t} + A_2 e^{-10^6 t} \tag{9-94}$$

The reader is invited to show that the result (with slight rounding) is

$$v_2(t) = 10e^{-200t} - 10e^{-10^6 t} \tag{9-95}$$

(c) The high-frequency rolloff manifests itself in the form of the second exponential term in Equation (9–95). This exponential has a very short time constant ($1 \ \mu s$) compared with the first term, which has a time constant of $1/200 \ s = 5 \ ms$. Consider, then, the interval from $t = 0$ to about $5 \ \mu s$. During this interval, $e^{-200t} \simeq 1$ and we may approximate Equation (9–95) as

$$v_2(t) \simeq 10 - 10e^{-10^6 t} \tag{9-96}$$

This function is shown in Figure 9–21(a) on a "fine" time scale and represents a simple rising exponential function. After about $5 \ \mu s$ for most practical purposes, the output voltage will have reached the level of 10 V, which corresponds

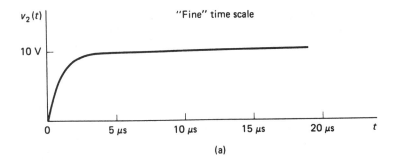

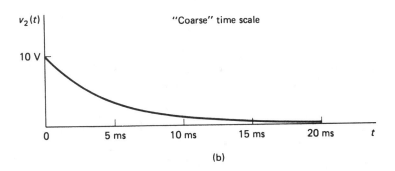

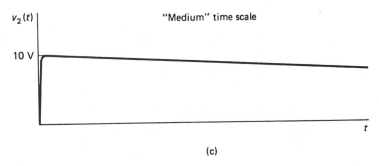

FIGURE 9–21
Step response of amplifier for Example 9–9 on three different time scales.

to the gain of 100 times the input step voltage of 0.1 V. It can be concluded that the effect of the high-frequency rolloff is *to cause the output to take some time (rise time) to reach the final value.*

(d) The low-frequency rolloff manifests itself in the form of the first exponential term in Equation (9–95). Consider, then, the time interval after the second exponential has, for all practical purposes, reached a level of zero. During this interval $e^{-10^6 t} \simeq 0$, and we may approximate Equation (9–95) as

$$v_2(t) \simeq 10e^{-200t} \qquad \qquad \textbf{(9–97)}$$

This function is shown on a "coarse" time scale in Figure 9–21(b) and represents a simple decaying exponential function. Note that the time scale for this part of the figure is different from that of Figure 9–21(a). It can be concluded that the effect of the low-frequency rolloff is *to cause the output to sag (or droop) toward 0.* This phenomenon can be explained by the fact that there is presumably a coupling capacitor in the amplifier, and it cannot sustain a dc source indefinitely.

The combined effects are illustrated on a "medium" time scale in Figure 9–21(c). To repeat, the high-frequency degradation results in nonzero rise time, and the low-frequency degradation results in a sag effect of the dc level. Although illustrated here for a particular case, these effects are general in application.

EXAMPLE 9–10

For the transfer function of the *RC* coupled amplifier in Example 9–9, plot the breakpoint approximation of the phase response.

Solution

For convenience, the transfer function is repeated here, and it is

$$G(s) = \frac{-0.5s}{\left(1 + \dfrac{s}{200}\right)\left(1 + \dfrac{s}{10^6}\right)} \tag{9–98}$$

The negative sign results in a phase shift of 180° at all frequencies. (Either +180° or −180° could be used.) The *s*-factor produces a phase shift of +90° at all frequencies. The two first-order denominator factors follow the forms established earlier.

The construction of a breakpoint phase approximation of a composite function is somewhat more involved than for the amplitude approximation. It is helpful to construct a table of the form given by Table 9–1. The first column provides a tabulation of certain critical frequencies, to be discussed shortly. Each subsequent

TABLE 9–1
Tabulated values used in constructing breakpoint phase approximation in Example 9–10

ω, rad/s	−sign	s	$\dfrac{1}{1 + s/200}$	$\dfrac{1}{1 + s/10^6}$	$\beta(\omega)$
<20	180°	+90°	0°	0°	270°
20	180°	+90°	0°	0°	270°
200	180°	+90°	−45°	0°	225°
2000	180°	+90°	−90°	0°	180°
10^5	180°	+90°	−90°	0°	180°
10^6	180°	+90°	−90°	−45°	135°
10^7	180°	+90°	−90°	−90°	90°
>10^7	180°	+90°	−90°	−90°	90°

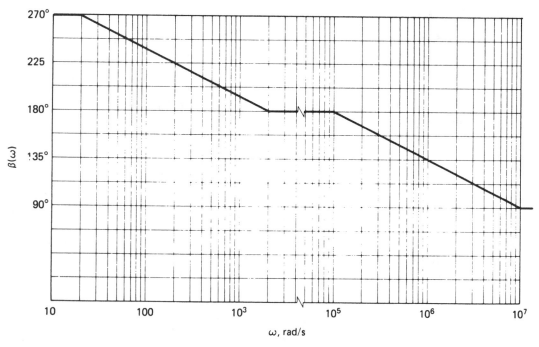

FIGURE 9–22
Phase response for Example 9–10.

column provides the phase contribution from the respective numerator and denominator factor. The column headed "−sign" signifies the fact that the negative gain constant produces a phase shift of 180°. Obviously, this column would be unnecessary if the gain constant were positive. The next column indicates that the contribution of the s-factor is +90 at all frequencies.

For each of the additional denominator factors, three particular frequencies are significant: $0.1\omega_b$, ω_b, and $10\omega_b$. The contribution to the phase at each of these frequencies is determined in accordance with the rules discussed earlier. Since these are denominator factors, the phase contributions at the respective frequencies are assumed to be 0°, −45°, and −90°. The net phase approximation is shown in the last column.

A plot of the composite breakpoint phase approximation is shown in Figure 9–22. The breaks are associated with the values of $0.1\omega_b$ and $10\omega_b$ for each of the two denominator factors. In this example, the effect of one factor has already reached an assumed upper limit when the second factor produces a break. When the break frequencies are closer together, the situation is a bit more complicated and requires careful preparation of the tabulated data.

There is a rather broad mid-frequency range in which the amplifier phase shift is about 180°. To simplify the presentation, the frequency scale is broken in this mid-range.

9-5 SECOND-ORDER LOW-PASS FUNCTION

As explained earlier, the breakpoint approximations are most useful for systems with poles and zeros on the real axis. However, because of its practical importance, we will extend our analysis to the case of complex poles for a second-order low-pass function. The intent is not to provide a simplified plotting approach as was done for real poles, but to provide graphical and mathematical forms of the frequency response for direct use when required.

In accordance with the second-order polynomial notation established in Section 7–7, assume a transfer function of the form

$$G(s) = \frac{\omega_n^2}{s^2 + 2\zeta\omega_n s + \omega_n^2} \tag{9-99}$$

where $\zeta = \alpha/\omega_n$ is the damping ratio and ω_n is the undamped radian oscillatory frequency. For frequency response plots, it is more convenient to divide numerator and denominator of Equation (9–99) by ω_n^2 and rewrite it in the form

$$G(s) = \frac{1}{1 + 2\zeta\dfrac{s}{\omega_n} + \left(\dfrac{s}{\omega_n}\right)^2} \tag{9-100}$$

The steady-state transfer function is

$$G(j\omega) = \frac{1}{1 - \left(\dfrac{\omega}{\omega_n}\right)^2 + j2\zeta\dfrac{\omega}{\omega_n}} \tag{9-101}$$

The linear amplitude response is

$$
A(\omega) = \frac{1}{\sqrt{\left[1 - \left(\dfrac{\omega}{\omega_n}\right)^2\right]^2 + 4\zeta^2\left(\dfrac{\omega}{\omega_n}\right)^2}}
$$

$$
= \frac{1}{\sqrt{1 + (4\zeta^2 - 2)\left(\dfrac{\omega}{\omega_n}\right)^2 + \left(\dfrac{\omega}{\omega_n}\right)^4}} \tag{9-102}
$$

The corresponding decibel amplitude response is

$$A_{\mathrm{dB}}(\omega) = -20\log_{10}\left[\sqrt{1 + (4\zeta^2 - 2)\left(\dfrac{\omega}{\omega_n}\right)^2 + \left(\dfrac{\omega}{\omega_n}\right)^4}\right] \tag{9-103}$$

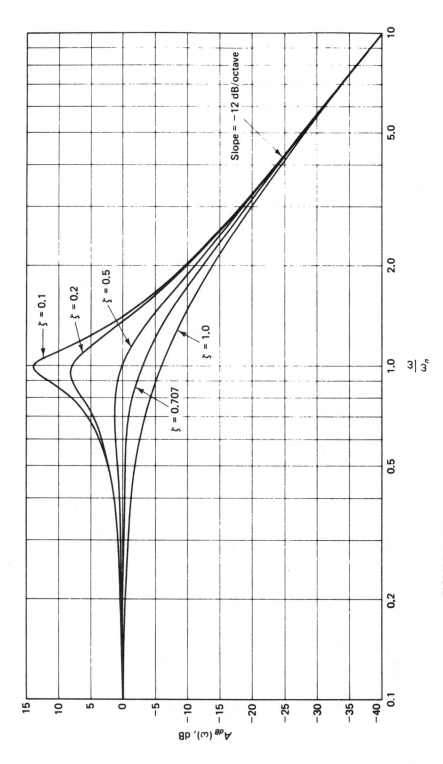

FIGURE 9–23
Decibel amplitude response curves for second-order low-pass transfer function.

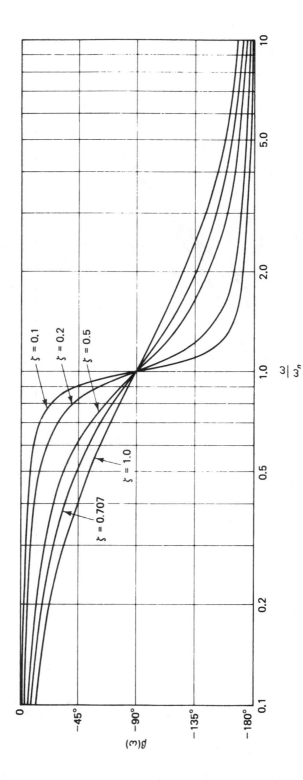

FIGURE 9-24
Phase response curves for second-order low-pass transfer function.

The phase response is

$$\beta(\omega) = -\tan^{-1} \frac{2\zeta \dfrac{\omega}{\omega_n}}{1 - \left(\dfrac{\omega}{\omega_n}\right)^2} \qquad (9\text{--}104)$$

Amplitude and Phase Curves

Curves displaying the decibel amplitude response with ζ as a parameter are shown in Figure 9–23. The relative decibel level at low frequencies for all curves is 0 dB, which corresponds to a linear amplitude response of unity. This means that components of a signal having frequencies low compared with ω_n will undergo no change in passing through such a system. However, at frequencies much higher than ω_n, the decibel response is much lower, indicating a significant reduction in the amplitudes of such components. Thus, the amplitude response has a low-pass form. It should be noted that small values of ζ result in a profound peaking effect in the neighborhood of ω_n. This could be an advantage in some applications, and it could be troublesome in others.

All the curves of Figure 9–23 eventually become asymptotic to a straight line having a slope of -12 dB/octave on a semilog scale. Observe from Equation (9–99) that the transfer function has two zeros at $s = \infty$. To generalize this concept, it can be shown (Problem 9–21) that *the amplitude response of a transfer function having n_i zeros at s $= \infty$ will be asymptotic at high frequencies to a straight line with a slope of $-6n_i$ dB/octave.*

Curves displaying the phase response for the second-order function with ζ as a parameter are shown in Figure 9–24.

EXAMPLE 9–11

Consider the second-order low-pass circuit in Figure 9–25(a).
(a) Determine the transfer function $G(s) = V_2(s)/V_1(s)$.
(b) Determine the values of ζ and ω_n and determine also $f_n = \omega_n/2\pi$.
(c) Determine which amplitude curve of Figure 9–23 and which phase curve of Figure 9–24 are closest to or equal to the actual response.

Solution
(a) The *s*-domain model is shown in Figure 9–25(b). The transfer function may be determined by applying either the voltage divider rule or by node voltage analysis. Choosing the first method, the parallel impedance Z_p of $10^7/s$ and 5000 is

$$Z_p = \frac{\dfrac{10^7}{s} \times 5000}{5000 + \dfrac{10^7}{s}} = \frac{5000 \times 10^7}{5000s + 10^7} = \frac{10^7}{s + 2000} \qquad (9\text{--}105)$$

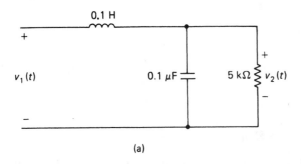

(a)

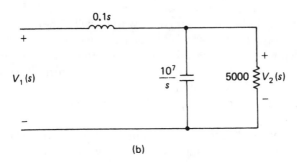

(b)

FIGURE 9–25
Circuuit for Example 9–11 and s-domain model.

The transfer function is then determined as

$$G(s) = \frac{V_2(s)}{V_1(s)} = \frac{Z_p}{Z_p + 0.1s}$$

(9–106)

$$= \frac{\dfrac{10^7}{s + 2000}}{\dfrac{10^7}{s + 2000} + 0.1s} = \frac{10^7}{0.1s^2 + 200s + 10^7}$$

We next arrange $G(s)$ in the form of Equation (9–100) for convenience.

$$G(s) = \frac{1}{1 + 2 \times 10^{-5}s + 1 \times 10^{-8}s^2}$$

(9–107)

(b) To determine ω_n and ζ, we match the coefficients of Equation (9–107) with those of Equation (9–100). The last coefficient in the denominator leads to

$$\omega_n^2 = 10^8$$

(9–108)

or

$$\omega_n = 10^4 \text{ rad/s}$$

(9–109)

and

$$f_n = 1592 \text{ Hz} \qquad \text{(9–110)}$$

We next have

$$\frac{2\zeta}{\omega_n} = 2 \times 10^{-5} \qquad \text{(9–111)}$$

or

$$\zeta = \frac{2 \times 10^{-5} \times 10^4}{2} = 0.1 \qquad \text{(9–112)}$$

(c) Inasmuch as $\zeta = 0.1$ is one of the curves given in Figure 9–23, the form of the response may be readily predicted. (The problem was "rigged!") This particular response is very underdamped, so the amplitude response displays close to a 14-dB peak in the vicinity of 10^4 rad/s (1592 Hz).

9–6 PSPICE EXAMPLES

The examples presented in this section will focus on the concept of steady-state ac analysis with the frequency swept over a specific range. PSPICE thus generates the frequency response for the circuit under consideration. Plots may be generated for the linear amplitude response, the decibel amplitude response, or the phase response.

PSPICE EXAMPLE 9–1

Use PSPICE to plot decibel amplitude and phase response curves for the circuit of Example 9–3.

Solution

The circuit adapted for PSPICE analysis and the code are shown in Figure 9–26. In contrast to the examples of Chapter 8, which were performed at a single frequency, the analysis of this example is to be performed over a frequency range. The 3-dB break frequency for this circuit is 10^4 rad/s or about 1592 Hz. As a reasonable choice, it was decided to sweep from 100 Hz to 10 kHz. The command is

```
.AC DEC 50 100 1E4
```

A decade sweep was chosen, which is indicated by DEC. The first number (50) indicates the number of points to be used in each decade. The number 100 represents the lowest frequency in hertz, and the number 1E4 represents the highest frequency in hertz. When a decade sweep is employed, it is best to select the ends of the frequency range as integer powers of 10.

The desired variables to be plotted are entered directly on the PROBE menu. The entry for the decibel response for V(2) is VDB(2). The entry for the phase

FIGURE 9–26

Circuit and code for PSPICE Example 9–1.

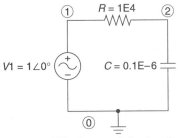

(All values are in basic units.)

PSPICE EXAMPLE 9–1
V1 1 0 AC 1 0
R 1 2 1E4
C 2 0 0.01E–6
.AC DEC 50 100 1E4
.PROBE
.END

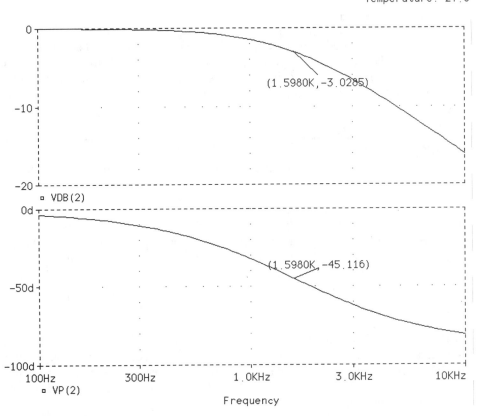

FIGURE 9–27

Decibel amplitude and phase response curves for PSPICE Example 9–1.

response for V(2) is VP(2). Plots of these two variables are shown in Figure 9–27. The cursors were used to mark the response values near the break frequency. Because of the manner in which PROBE extrapolates the data, the values are not exact, but are very close to those expected.

PSPICE EXAMPLE 9–2

Use the PSPICE Laplace function to plot the decibel amplitude response for the transfer function of Examples 9–4 and 9–5.

Solution

We used the Laplace function in Chapter 8 to determine the transient response for several input waveforms in their instantaneous forms. Similarly, we may use the Laplace function to determine steady-state amplitude and/or phase functions if the input is described in terms of its ac phasor form.

We will employ the form of the transfer function as given by Equation (9–47), which is repeated here for convenience:

$$G(s) = \frac{2000(s + 4)}{s^2 + 116s + 1600} \tag{9-113}$$

Refer now to the code of Figure 9–28. The input voltage V is connected between nodes 1 and 0 and is an ac phasor of the form $1\underline{/0°}$. The line defining the transfer function is a VCVS connected between nodes 2 and 0, and the operation portion of this line reads

```
LAPLACE {V(1)} = {2E3*(S+4)/(S*S+116*S+1600)}
```

Remember that the "=" in the Laplace operation acts more like a multiplication than what it implies. A decade sweep is chosen from 0.1 Hz to 1000 Hz, with 50 points per decade.

The decibel amplitude response as obtained from PROBE is shown in Figure 9–29, which can be compared with the breakpoint approach of Figure 9–16. In comparing these figures, bear in mind that the plot obtained with PSPICE should be exact, whereas the breakpoint approximation is generally not an exact representation. Thus, there are no "sharp corners," as they appear in the approximation. The true maximum level of the curve is about 24.8 dB, which is less than the value of 26 dB in the breakpoint approximation. Another difference is that the exact plot obtained with PSPICE uses cyclic frequency in hertz, whereas the breakpoint ap-

```
PSPICE EXAMPLE 9-2
V 1 0 AC 1 0
E 2 0 LAPLACE {V(1)}={2E3*(S+4)/(S*S+116*S+1600)}
.AC DEC 50 0.1 1E3
.PROBE
.END
```

FIGURE 9–28
Code for PSPICE Example 9–2.

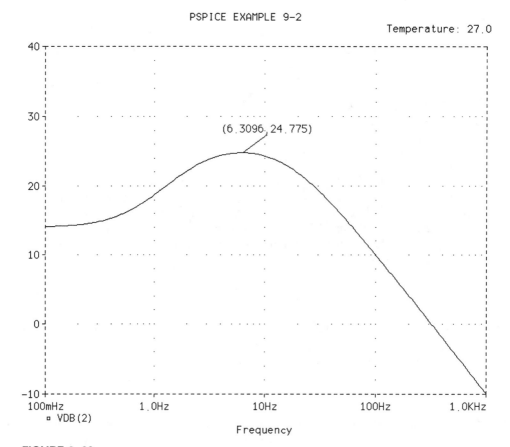

FIGURE 9–29
Decibel amplitude response for PSPICE Example 9–2.

proximation used radian frequency (although an extra horizontal scale was provided in Figure 9–16 to show cyclic frequency).

PSPICE EXAMPLE 9–3
Use the PSPICE Laplace function to plot the decibel amplitude response for the transfer function of Example 9–6.

Solution
This example appears to be of the same type as the last one, except that we will use a factored form of the transfer function in this case. We begin with the form of the transfer function given in Equation (9–78), which will be repeated here in a slightly different form for convenience:

$$G(s) = \frac{40s}{(1 + s/2)(1 + s/10)(1 + s/100)} \tag{9–114}$$

```
PSPICE EXAMPLE 9-3
V 1 0 AC 1 0
E 2 0 LAPLACE {V(1)}={40*S/((1+S/2)*(1+S/10)*(1+S/100))}
.AC DEC 50 0.01 100
.PROBE
.END
```

FIGURE 9–30
Code for PSPICE Example 9–3.

Refer now to Figure 9–30 for the code. The pertinent Laplace function is now expressed as

```
{V(1)} = {40*S/((1+S/2)*(1+S/10)*(1+S/100))}
```

An ac decade sweep is performed from 0.01 Hz to 100 Hz, with 50 points per decade. After appropriate labeling, the decibel amplitude response is shown in Figure 9–31. This curve can be compared with the breakpoint approximation of Figure 9–17.

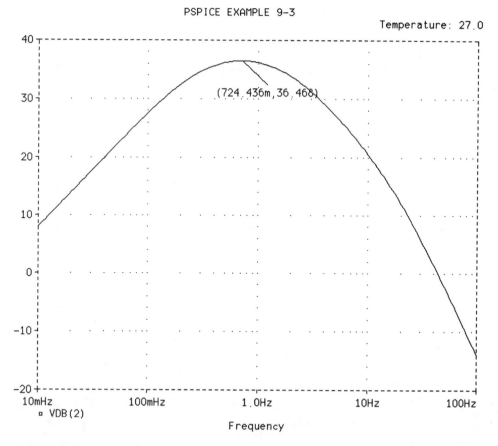

FIGURE 9–31
Decibel amplitude response for PSPICE Example 9–3.

```
PSPICE EXAMPLE 9-4
V 1 0 AC 1 0
E 2 0 LAPLACE {V(1)}={-0.5*S/((1+S/200)*(1+S/1E6))}
.AC DEC 50 10 1E6
.PROBE
.END
```

FIGURE 9–32
Code for PSPICE Example 9–4.

PSPICE EXAMPLE 9-4

Use PSPICE to plot the decibel amplitude response of Example 9–9.

Solution
The transfer function was determined within the example and was given by Equation (9–91). With a slight alteration, this function is given by

$$G(s) = \frac{-0.5s}{(1 + s/200)(1 + s/10^6)} \qquad (9\text{–}115)$$

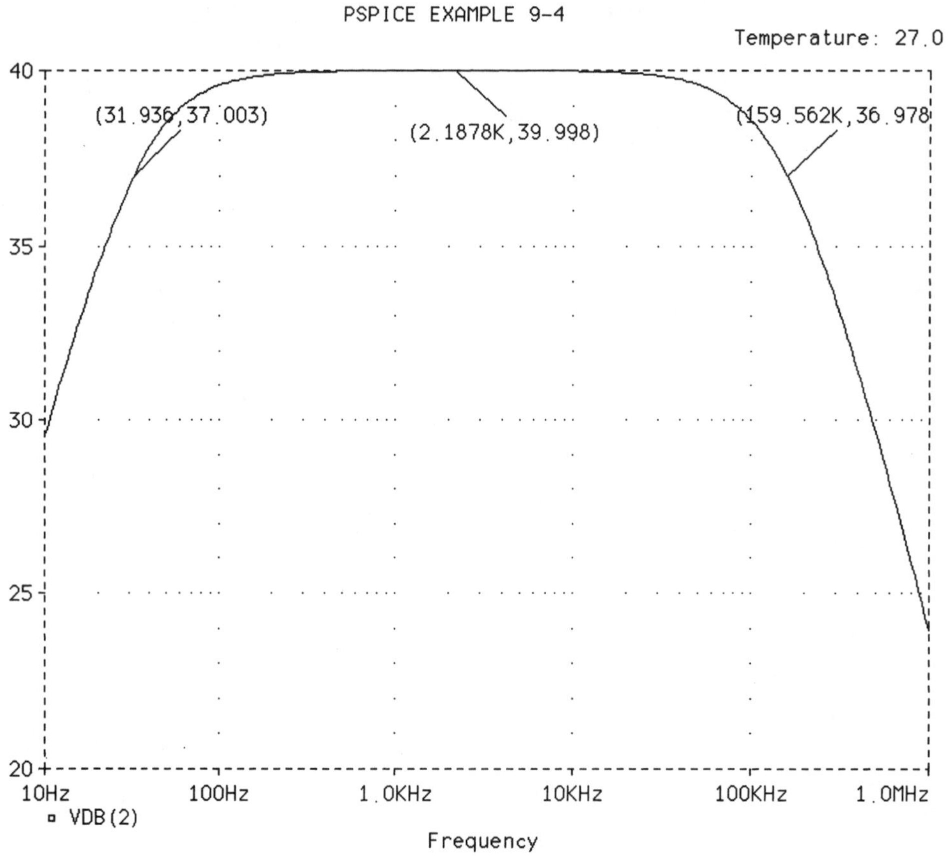

PSPICE EXAMPLE 9-4

Temperature: 27.0

□ VDB(2)

Frequency

FIGURE 9–33
Decibel amplitude response for PSPICE Example 9–4.

The code is shown in Figure 9–32. The Laplace function is written as

```
LAPLACE {V(1)} = {-0.5*S/((1+S/200)*(1+S/1E6))}
```

A decade sweep running from 10 Hz to 1 MHZ, with 50 points per decade, is employed, and the decibel amplitude response is shown in Figure 9–33.

Comparing this response with that of Figure 9–20, note that the curves compare much more closely with each other than for previous examples. The primary reason is that the break frequencies are much farther apart than in previous examples.

9–7 MATLAB EXAMPLES

MATLAB offers some powerful features for determining the amplitude and phase response functions from a transfer function. These features permit a straightforward approach to the computation and graphical presentation of a Bode plot. Some of these features will be explored in this section.

MATLAB EXAMPLE 9–1

Use MATLAB to plot the decibel amplitude and phase response curves for the transfer function of Example 9–5 using the Bode plot command.

Solution
The best format for entering the data is that of the unfactored transfer function form as given in Example 9–4 (Equation 9–47). The factor 2000 is multiplied by the terms in parentheses, and the result is

$$G(s) = \frac{2000s + 8000}{s^2 + 116s + 1600} \qquad (9\text{–}116)$$

Refer to the workspace shown in Figure 9–34. First, the numerator array is entered as

```
n = [0 2e3 8e3]
```

FIGURE 9–34
Workspace for MATLAB Example 9–1.

```
EDU» n=[0 2e3 8e3]

n =

         0        2000        8000

EDU» d=[1 116 1600]

d =

         1         116        1600

EDU» bode(n,d)
EDU»
```

with the first zero representing the coefficient of the nonexistent cubic term. Note that the e (for exponent) can be entered as either lowercase as given or as capital E.

Next, the denominator array is entered as

```
d = [1 116 1600]
```

There are a variety of options for the Bode plot operations, including the possibility of customizing the frequency range. However, MATLAB will choose an appropriate frequency range if desired, and the simplest choice is the command

```
bode(n,d)
```

The decibel amplitude and phase functions are immediately plotted and labeled, and the results are shown in Figure 9–35. Note that the frequency is labeled in radians/second, which is compatible with the text. The MATLAB label "GAIN dB" is the amplitude response $A_{dB}(\omega)$, and "Phase deg" is the phase response $\beta(\omega)$.

In the next example, an alternate method for generating amplitude and phase response curves will be demonstrated.

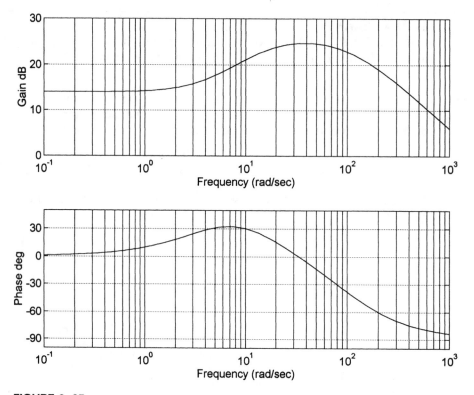

FIGURE 9–35
Decibel amplitude and phase response curves obtained from Bode command in MATLAB Example 9–1.

MATLAB EXAMPLE 9–2

Use MATLAB to plot the decibel amplitude and phase response curves for the transfer function of Example 9–6 using the "frequency response" command.

Solution

The approach taken in this example produces the same type of result as the Bode plot approach of the previous example, but is a little easier to customize to the user's specific needs. First, the transfer function of Example 9–6 as given by Equation (9–78) will be repeated for convenience:

$$G(s) = \frac{40s}{\left(1 + \frac{s}{2}\right)\left(1 + \frac{s}{10}\right)\left(1 + \frac{s}{100}\right)} \tag{9–117}$$

Although this is the best form for manually generating a Bode plot, it is not the best form for entering MATLAB. As a first step, we need to rearrange all factors in the form $(s + \alpha)$, and this can be done by multiplying numerator and denominator by $2 \times 10 \times 100 = 2000$. This results in

$$G(s) = \frac{8 \times 10^4 s}{(s + 2)(s + 10)(s + 100)} \tag{9–118}$$

Next, we need to generate the denominator polynomial coefficients. We could, of course, manually multiply out the three denominator factors. However, an easier approach is to use the "zp2tf" operation used in MATLAB Example 7–4. The one finite zero is $s = 0$, and the three finite poles are $s = -2, -10$, and -100. Referring to Figure 9–36, we define column vectors z and p and a constant k as

```
z = [0]
p = [-2;-10;-100]
k = 8e4
```

The numerator and denominator polynomial arrays n and d are determined as

```
[n d] = zp2tf(z,p,k)
```

The result provides the required numerator and denominator arrays.

All of the preceding steps were preliminary to the main goal of this problem, and could have been eliminated if the transfer function had been given directly in the proper form. However, they illustrate different ways of manipulating the function forms directly on the computer.

To perform a frequency response analysis, it is desirable first to generate a set of frequency values at which the response is to be evaluated. Although linear spacing could be used, a logarithmic scale is preferred, as we have seen in the text. Letting w represent radian frequency, the following format is used:

```
w = logspace(a,b,n)
```

```
EDU» z=[0]

z =

     0

EDU» p=[-2;-10;-100]

p =

    -2
   -10
  -100

EDU» k=8e4

k =

     80000

EDU» [n d]=zp2tf(z,p,k)

n =

         0            0       80000            0


d =

         1          112        1220         2000
EDU» w=logspace(-1,3,400);
EDU» G=freqs(n,d,w);
EDU» ADB=20*log10(abs(G));
EDU» semilogx(w,ADB);
EDU» xlabel('Frequency, rad/s')
EDU» ylabel('Amplitude Response, dB')
EDU» title('Decibel Amplitude Response versus Radian Frequency')
EDU» grid
EDU»
```

FIGURE 9–36
Workspace for MATLAB Example 9–2.

The coding is a little awkward, but meaningful in that a and b are the powers of 10 defining the beginning and ending radian frequencies. Thus, 10^a is the beginning radian frequency, and 10^b is the ending radian frequency. The integer n is the number of points in the range based on a logarithmic spacing. For our example, we will sweep from 0.1 rad/s to 1000 rad/s, with 400 points, which corresponds to 10^{-1} to 10^3, respectively. Thus, the code is

```
w = logspace(-1,3,400);
```

where the semicolon suppresses the printing.

The basic command for performing a frequency response analysis is

```
freqs(n,d,w)
```

where the s in "freqs" can be thought of as "s-plane." In this basic form, the amplitude generated will be the linear amplitude, and the phase generated will be the phase in radians. We prefer to have the amplitude response in decibels, so some alteration of the data is required. To that end, we first generate the linear amplitude and phase by defining G as follows:

```
G = freqs(n,d,w);
```

G can be thought of as $G(j\omega)$, since all the values are complex.

To determine decibel amplitude response, we need to use three separate operations:

1. Determine the magnitude of G, which can be performed in MATLAB by the command "abs(G)." (The code "abs" represents "absolute value.")

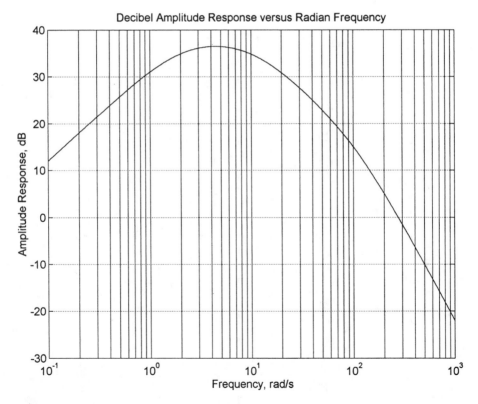

FIGURE 9–37
Decibel amplitude response for MATLAB Example 9–2.

2. Determine the logarithm to the base 10 of the resulting quantity, for which the command in MATLAB is log(10( )).
3. Multiply the result by 20, in accordance with the basic definition of decibels.

All three of the preceding operations can be combined in one statement by the following:

```
ADB = 20*log10(abs(G));
```

where ADB represents the decibel amplitude response, and printing has been suppressed. The plot command used to create a logarithmic horizontal scale and a linear vertical scale is

```
semilogx(w,ADB)
```

Additional labeling is then provided, and the result is shown in Figure 9–37.

Consider now that we desire to relabel the frequency scale in terms of cyclic frequency *f*. This can be accomplished easily through a scale conversion. MATLAB uses "pi" to represent π, and an appropriate operation to change the scale is simply

```
f = w/(2*pi);
```

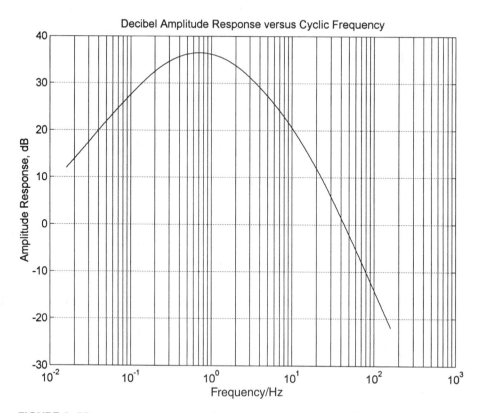

FIGURE 9–38
Decibel amplitude response for MATLAB Example 9–2 with cyclic frequency.

The values of ADB have already been determined, so all we need to do is plot ADB versus *f*, which is accomplished with

```
semilogx(f,ADB)
```

The resulting graph after labeling is shown in Figure 9–38.

MATLAB EXAMPLE 9–3

Repeat the analysis of MATLAB Example 9–2 with the cyclic frequency range specified at the beginning of the analysis. In addition, plot the phase response in degrees.

Solution

This example is presented to show a different perspective for the analysis previously performed. In most practical problems, it is the cyclic frequency range which is initially specified, and that is the approach which will be taken here.

Refer to the workspace of Figure 9–39. The coefficients of the previous problem are first entered as *n* and *d*, respectively. Assume that we desire to sweep the frequency range from 0.01 Hz to 100 Hz. We will first generate a logarithmic array of cyclic frequency values *f* by the command

```
f = logspace(-2,2,400);
```

```
EDU» n=[0 0 8e4 0]

n =

          0           0       80000           0

EDU» d=[1 112 1220 2e3]

d =

          1         112        1220        2000

EDU» f=logspace(-2,2,400);
EDU» G=freqs(n,d,2*pi*f);
EDU» ADB=20*log10(abs(G));
EDU» semilogx(f,ADB)
EDU» grid
EDU» xlabel('Frequency, Hz')
EDU» ylabel('Amplitude Response, dB')
EDU» title('Decibel Amplitude Response versus Cyclic Frequency')
EDU» beta=(180/pi)*angle(G);
EDU» semilogx(f,beta)
EDU» xlabel('Frequency, Hz')
EDU» ylabel('Phase Response, degrees')
EDU» title('Phase Response versus Cyclic Frequency')
EDU» grid
EDU»
```

FIGURE 9–39
Workspace for MATLAB Example 9–3.

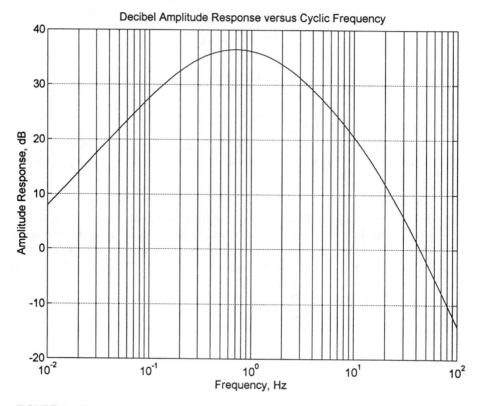

FIGURE 9–40
Decibel amplitude response for MATLAB Example 9–3.

This is the desired independent variable for plotting, but the frequency response command requires radian frequency ω. A suitable way to accomplish this is the following command:

```
G = freqs(n,d,2*pi*f);
```

Alternately, we could have first defined w = 2*pi*f and entered *w* as the last entry in the command.

The remainder of the amplitude response analysis follows essentially the format of the previous example. The plot is controlled directly by *f* and ADB. After appropriate labeling, the decibel amplitude response is shown in Figure 9–40.

The basic command for determining the phase is "angle(G)." However, the result will be given in radians. Since we desire the angle to be expressed in degrees, it is necessary to multiply each value by $180/\pi$. Letting beta represent the required angle, we invoke the following command:

```
beta = (180/pi)*angle(G)
```

A plot is then made of beta versus *f*, and after additional labeling is provided, the result is shown in Figure 9–41.

Phase Response versus Cyclic Frequency

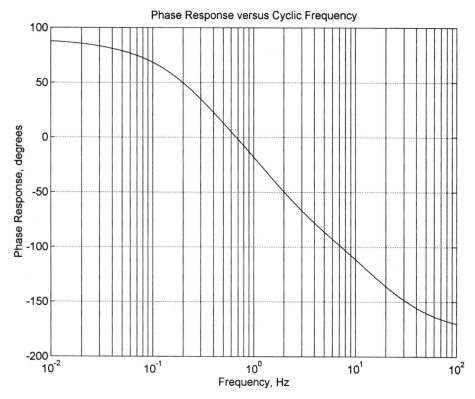

FIGURE 9–41
Phase response for MATLAB Example 9–3.

DRILL PROBLEMS

9–1. Consider the *RC* circuit of Figure P9–1, in which the excitation is a sinusoidal function. **(a)** Assuming that the circuit is initially relaxed, determine the *s*-domain model, and use it to determine the complete time-domain current $i(t)$ and voltage $v(t)$. **(b)** Determine the phasor-domain model, and use it to determine the steady-state current $i_{ss}(t)$ and voltage $v_{ss}(t)$.

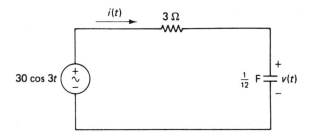

FIGURE P9–1

9–2. Consider the *RL* circuit of Figure P9–2 in which the excitation is a sinusoidal function. **(a)** Assuming that the circuit is initially relaxed, determine the *s*-domain model, and use it to determine the complete time-domain current $i(t)$ and voltage $v(t)$. **(b)** Determine the phasor-domain model, and use it to determine the steady-state current $i_{ss}(t)$ and voltage $v_{ss}(t)$.

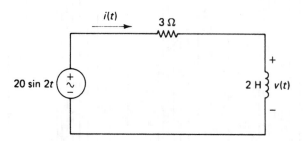

FIGURE P9–2

9–3. For the circuit of Figure P9–3, determine mathematical expressions for the steady-state transfer function, the linear amplitude response, the decibel amplitude response, and the phase response. The input is v_1 and the output is v_2.

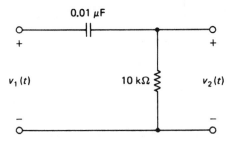

FIGURE P9–3

9–4. For the circuit of Figure P9–4, determine mathematical expressions for the steady-state transfer function, the linear amplitude response, the decibel amplitude response, and the phase response. The input is v_1 and the output is v_2.

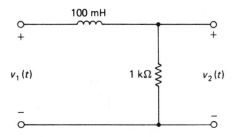

FIGURE P9–4

9–5. Consider the transfer function of the normalized active Butterworth filter of Example 7–5, which was determined to be

$$G(s) = \frac{1}{s^2 + 1.4142s + 1}$$

Determine mathematical expressions for the steady-state transfer function, the linear amplitude response, the decibel amplitude response, and the phase response.

9–6. Consider the transfer function of the normalized passive Butterworth filter of Example 7–4, which was determined to be

$$G(s) = \frac{1}{s^3 + 2s^2 + 2s + 1}$$

Determine mathematical expressions for the steady-state transfer function, the linear amplitude response, the decibel amplitude response, and the phase response.

9–7. Using semilog paper, plot the breakpoint decibel amplitude response approximation for the transfer function

$$G(s) = \frac{20}{\left(1 + \dfrac{s}{10}\right)\left(1 + \dfrac{s}{100}\right)}$$

9–8. Using semilog paper, plot the breakpoint decibel amplitude response approximation for the transfer function

$$G(s) = \frac{5\left(1 + \dfrac{s}{100}\right)}{\left(1 + \dfrac{s}{10}\right)\left(1 + \dfrac{s}{1000}\right)}$$

9–9. Using semilog paper, plot the breakpoint decibel amplitude response approximation for the transfer function

$$G(s) = \frac{500}{s^2 + 120s + 2000}$$

9–10. Using semilog paper, plot the breakpoint decibel amplitude response approximation for the transfer function

$$G(s) = \frac{200(s + 10)}{s(s + 100)}$$

9–11. Using semilog paper, plot the breakpoint phase response approximation for the transfer function of Problem 9–7.

9–12. Using semilog paper, plot the breakpoint phase response approximation for the transfer function of Problem 9–8.

9–13. For the simple RC low-pass circuit of Example 9–7 (Figure 9–18), calculate the break frequency f_b for $R = 10\ \text{k}\Omega$ and $C = 0.1\ \mu\text{F}$.

9–14. For the simple RC high-pass circuit of Example 9–8 (Figure 9–19), calculate the break frequency f_b for $R = 1\ \text{k}\Omega$ and $C = 0.001\ \mu\text{F}$.

9–15. Consider the second-order low-pass circuit shown in Figure P9–15. **(a)** Determine the transfer function $G(s) = V_2(s)/V_1(s)$. **(b)** Determine the values of ζ and ω_n. Determine also $f_n = \omega_n/2\pi$. **(c)** Determine which amplitude curve of Figure 9–23 and which phase curve of Figure 9–24 are closest to or equal to the actual response.

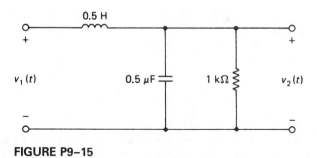

FIGURE P9–15

9–16. Consider the second-order low-pass circuit shown in Figure P9–16. **(a)** Determine the transfer function $G(s) = I_2(s)/I_1(s)$. **(b)** Determine the values of ζ and ω_n. Determine also $f_n = \omega_n/2\pi$. **(c)** Determine which amplitude curve of Figure 9–23 and which phase curve of Figure 9–24 are closest to or equal to the actual response.

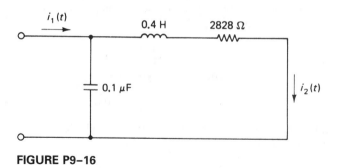

FIGURE P9–16

DERIVATION PROBLEMS

9–17. Consider the model of the carbon composition resistor given in Example 9–2 (Figure 9–3). Assume that the series impedance measured on a bridge is given by $\overline{Z} = R_s + jX_s$. Show that the values of R and C for the resistor are

$$R = \frac{X_s^2 + R_s^2}{R_s}$$

$$C = \frac{-X_s}{\omega(X_s^2 + R_s^2)}$$

9–18. Some bridges provide an *admittance* value at balance based on a parallel equivalent circuit. Referring to Figure P9–18(a), let G_p represent the equivalent parallel conduc-

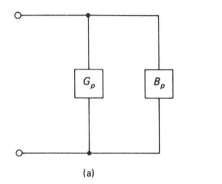 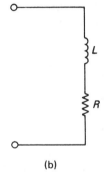

(a)   (b)

FIGURE P9–18

tance, and let B_p represent the equivalent parallel susceptance. The measured admittance is thus

$$\overline{Y} = G_p + jB_p$$

Many practical inductors can be modeled over a reasonable frequency range as a series inductance in series with a resistance R as shown in Figure P9–18(b). Assume that the admittance of such an inductor is measured on an admittance bridge. Show that the values of L and R are given by

$$R = \frac{G_p}{G_p^2 + B_p^2}$$

$$L = \frac{-B_p}{\omega(G_p^2 + B_p^2)}$$

9–19. The circuit of Figure P9–19 can be used as a high-frequency *preemphasis* network. It has the property that the amplitude response has a larger magnitude for higher-frequency components of a given signal. When proper compensation is used, the signal-to-noise ratio in certain processes such as frequency modulation can be improved.
   a. For the circuit as given, derive an expression for the transfer function $G(s) = V_2(s)/V_1(s)$ and arrange in the form most suitable for a Bode plot.
   b. Show that the amplitude response has a numerator break frequency f_{bn} and a denominator break frequency f_{bd}. Obtain expressions for these quantities, and show that $f_{bn} < f_{bd}$.

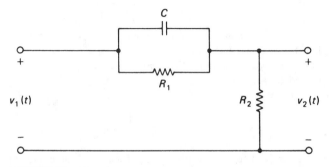

FIGURE P9–19

   c. Sketch the form of the breakpoint decibel amplitude response approximation, and label the levels of the flat portions of the response.

9–20. The circuit of Figure P9–20 can be used as a high-frequency *deemphasis* network. It has the property that the amplitude response has a larger magnitude for lower-frequency components of a given signal. It can be used to compensate for the preemphasis characteristics of a circuit such as that of Problem 9–19.

   a. For the circuit as given, derive an expression for the transfer function $G(s) = V_2(s)/V_1(s)$ and arrange in the form most suitable for a Bode plot.

   b. Show that the amplitude response has a numerator break frequency f_{bn} and a denominator break frequency f_{bd}. Obtain expressions for these quantities and show that $f_{bd} < f_{bn}$.

   c. Sketch the form of the breakpoint decibel amplitude response approximation, and label the levels of the flat portions of the response.

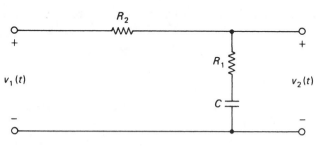

FIGURE P9–20

9–21. Consider a transfer function having n_i zeros at $s = \infty$. It was shown in Chapter 7 by the work leading up to Equation (7–45) that $G(s)$ could then be approximated as

$$G(s) \simeq \frac{K}{s^{n_i}} \quad \text{for } |s| \gg 1$$

and $K = a_n/b_m$ in the notation of Equation (7–45). Show in the range where this result is valid that the slope of the decibel amplitude response is very nearly $-6n_i$ dB/octave.

9–22. Consider the second-order low-pass amplitude response $A(\omega)$ given in general form by Equation (9–102).

   a. Applying differential calculus, show that a maximum in the amplitude response occurs at a frequency ω_{max} given by

$$\omega_{max} = \omega_n\sqrt{1 - 2\zeta^2}$$

   (*Hint:* It is easier to define a new function $y = 1/A^2(\omega)$ and determine the frequency at which this response has a minimum level.)

   b. Show that there is a maximum in the vicinity of ω_n when $\zeta < 0.707$. (This corresponds to a peaked response.)

   c. Show that the value of the peak A_{max} is

$$A_{max} = \frac{1}{2\zeta\sqrt{1 - \zeta^2}}$$

   d. Check the result of (c) by computing A_{max} for one of the curves of Figure 9–23.

APPLICATION PROBLEMS

9–23. A simple RC low-pass circuit of the type considered in Example 9–7 (Figure 9–18) is to be designed to reduce some high-frequency noise in a given appliction. If the desired break frequency is 100 kHz, determine the value of R if C is selected as 0.001 μF.

9–24. A simple RC high-pass circuit of the type considered in Example 9–8 (Figure 9–19) is to be designed to block the dc component in an amplifier while passing the time-varying component. If the desired break frequency is 20 Hz, determine the value of R if C is selected as 1 μF.

9–25. Consider the model of the carbon composition resistor given in Example 9–2 (Figure 9–3) and developed further in Problem 9–17. The impedance of a certain unknown resistance is measured on an RF impedance bridge at 10 MHz as $\overline{Z} = 3600 - j2000\ \Omega$. Determine the values of R and C.

9–26. Assume that a lossy inductor is measured with an *admittance* bridge, as discussed in Problem 9–18. The admittance, as measured at 1 MHz, is

$$\overline{Y} = 1 \times 10^{-3} - j25 \times 10^{-3}\ \text{S}.$$

Determine the values of L and R.

9–27. Referring to Problem 9–19, design a preemphasis network having a value $f_{bn} = 2122$ Hz. (This corresponds to a time constant of 75 μs, and is the value used in commercial FM transmitter preemphasis.) Select a low-frequency value for A_{dB} of -20 dB, and select $C = 0.01\ \mu$F.

9–28. Referring to Problem 9–20, design a deemphasis network having a value $f_{dn} = 2122$ Hz. (This corresponds to a time constant of 75 μs, and is the value used in commercial FM receiver deemphasis.) Select a high-frequency value for A_{dB} of -20 dB, and select $C = 0.01\ \mu$F.

9–29. A certain cable connecting two amplifier stages is modeled by the simple RC low-pass circuit of Example 9–7 (Figure 9–18). The shunt capacitance represents the combination of the cable capacitance plus the second-stage input capacitance. If the imput resistance of the second stage is assumed to be infinite, the resistance of the model represents the source resistance. Assume that the net capacitance of the given system is 800 pF. Determine the maximum source resistance if the lowest acceptable 3-dB frequency is 200 kHz.

9–30. Consider the second-order low-pass circuit of Figure P9–30. An application is required in which the response is to show a +6-dB peak above the low-frequency level in the vicinity of 2 kHz. If the load termination is required to be 1 kΩ as shown, determine the values of L and C. (*Hint:* Refer to the results of Problem 9–22.)

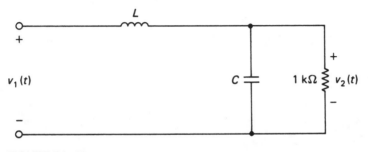

FIGURE P9–30

10

WAVEFORM ANALYSIS

OBJECTIVES

After completing this chapter, the reader should be able to:

- Determine the dc value of a periodic waveform and explain its significance.
- Determine the rms value of a periodic waveform and explain its significance.
- Determine the average power dissipated in a resistor by a periodic waveform of voltage or current.
- Define and sketch the step and ramp functions.
- Express the mathematical equation for and sketch the form of a delayed function.
- Express various piecewise linear waveforms in terms of step and ramp functions starting at appropriate times.
- Apply superposition to determine the response of a circuit excited by a waveform composed of various step and ramp components.
- Define the unit impulse function and explain its relationship to the unit step function.

10–1   dc AND rms VALUES OF PERIODIC SIGNALS

The development in this section will be restricted to periodic functions. A function $f(t)$ is said to be *periodic* with a period T if

$$f(t + T) = f(t) \tag{10–1}$$

for all t. An example of a periodic function is shown in Figure 10–1. The given function repeats the basic pattern during each interval of T seconds. Periodic signals appear extensively in various electrical and electronic circuits. The most common example of a periodic waveform is the sinusoidal function.

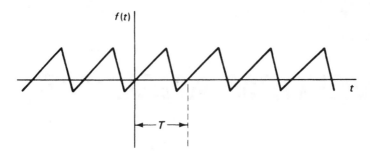

FIGURE 10–1
Example of a periodic function.

Two important properties of periodic signals are the *dc value* and the *rms value*. Actually, these properties may be applied to functions that are not periodic by appropriate mathematical techniques, and this is done in advanced signal analysis theory. However, the common situations for which these terms are used in most electrical circuit analysis are with periodic functions, and the treatment here will be limited to that most important case. Each of the terms will now be defined and explained.

dc Value

The *direct current* (or dc) value F_{dc} of a function $f(t)$ is defined as

$$F_{dc} = \frac{1}{T} \int_0^T f(t)\, dt \qquad \text{(10–2a)}$$

$$= \frac{\text{area under curve of } f(t) \text{ in one cycle}}{\text{period}} \qquad \text{(10–2b)}$$

The form of Equation (10–2b) indicates that the dc value is calculated by determining the area under the curve of $f(t)$ in one cycle and dividing by the period. For piecewise linear functions, that is, those composed of straight-line segments, this area is usually computed most easily by simple geometric considerations. However, functions having more complex forms will require analytical or, in some cases, numerical integration techniques. Area is, of course, weighted both positively and negatively, according to the sign of $f(t)$ in a given interval.

The dc value of a periodic function is the same as the *mean* or *average* value and, in many references, is defined in that manner. The choice of the term *dc value* in this text is based on an important circuit property: *The dc (or average) value of a periodic waveform is the value that will be read by a true dc instrument.* This property is based on the fact that most dc instruments operate on an averaging principle, and thus produce a response proportional to the true average value of a waveform. For this to be true, the frequency must be high enough (often a few hertz) so that all time-varying effects have been suppressed by the instrument.

rms Value

The *root-mean-square* (rms) or *effective* value of a function $f(t)$ is defined as

$$F_{\text{rms}} = \sqrt{\frac{1}{T} \int_0^T f^2(t)\, dt} \tag{10-3a}$$

$$= \sqrt{\frac{\text{area under curve of } f^2(t) \text{ in one cycle}}{\text{period}}} \tag{10-3b}$$

Stated in words, the function $f(t)$ is first squared, and the area under the curve of the squared function is determined. When this quantity is divided by T, the result is the average (or mean) value of the squared function. The square root of this result is then determined. Thus, a simplified algorithm for computing the desired quantity is to take the *root* of the *mean* of the *square*. Therein is the logic for the description *rms,* since it conveys a simplified approach to the computation involved. However, the term *effective value* is also widely used.

Basis of rms Value

The significance of the rms or effective value is best illustrated by actually investigating the basis of the definition. For convenience in terminology, we will momentarily change notation and assume a periodic current waveform $i(t)$ flowing in a resistor R. Let $p(t)$ represent the instantaneous power. This quantity is readily expressed as

$$p(t) = Ri^2(t) \tag{10-4}$$

In specifying the power dissipated in a system, it is usually inconvenient to deal with instantaneous power, since it is a time-varying quantity. More commonly, the average power P is used, and it is

$$P = \frac{1}{T} \int_0^T p(t)\, dt = \frac{1}{T} \int_0^T Ri^2(t)\, dt \tag{10-5}$$

The effective value of the current is then defined as a fixed value of the current such that the power in the resistor is calculated *as if* the current were a dc current. Using I_{rms} in accordance with the earlier definition, if the current is a pure dc current, the power P would be

$$P = RI_{\text{rms}}^2 \tag{10-6}$$

Since Equation (10–6) must produce the same result as Equation (10–5), the expressions can be equated. The common R factor cancels, and we have

$$I_{\text{rms}}^2 = \frac{1}{T} \int_0^T i^2(t)\, dt \tag{10-7}$$

or

$$I_{\text{rms}} = \sqrt{\frac{1}{T} \int_0^T i^2(t)\, dt}$$ (10–8)

Once the rms value is determined for a given current or voltage waveform, it is used exactly like a true dc voltage or current in calculating average power. Thus, Equation (10–6) can be used for computing average power in a resistance when I_{rms} is known. If the rms value V_{rms} of a voltage $v(t)$ is known, the average power P in a resistance R is

$$P = \frac{V_{\text{rms}}^2}{R}$$ (10–9)

In applying the definition of F_{rms} in Equation (10–3), the function is squared before determining an area. Since the square of either positive or negative values is positive, all area under the curve of $f^2(t)$ is necessarily positive. This is in sharp contrast to the computation of the dc value in Equation (10–2), where area may be either positive or negative. Because of these differences, as well as the related magnitude weighting of a squared function compared with the basic function, it can be established that

$$F_{\text{rms}} \geq F_{\text{dc}}$$ (10–10)

for all periodic signals.

Form Factor

The result of Equation (10–10) tells us that the rms value of a periodic signal is never smaller than the dc value. This result is very important when specifying the power ratings of resistors used with periodic waveforms. A term called the *form factor* is sometimes used to relate the two quantities as follows:

$$\text{Form factor} = \frac{F_{\text{rms}}}{F_{\text{dc}}}$$ (10–11)

The form factor is useful in dealing with the properties of certain dc and ac instruments.

The reader may be confused at this point about several earlier points, so some discussion is warranted. Note that *average power* is computed with *rms voltage or current* and *not* with dc or average voltage or current. The most significant circuit interpretation of average voltage or current is that it is the quantity read by a dc instrument. In certain power computations where it is necessary to compute the portion of the total power produced by the dc value, a limited type of power computation may be made with the dc value. However, as a general rule, *the rms value is used to compute the average power*.

A final comment concerns the process for computing the rms value. The function is first squared and the area of the squared-function is determined, as previously noted. However, *the area of the squared function is not the same as the square of the area.* The latter erroneous interpretation would lead to a squaring of both magnitude and time, and this result is meaningless.

EXAMPLE 10–1

Consider the sinusoidal voltage $v(t)$ shown in Figure 10–2 and given by

$$v(t) = V_p \sin \omega t \qquad (10\text{–}12)$$

where $\omega = 2\pi f = 2\pi/T$. Determine **(a)** the dc value and **(b)** the rms value.

Solution

(a) The average or dc value is obtained from the equation

$$V_{dc} = \frac{1}{T} \int_0^T v(t)\, dt = \frac{1}{T} \int_0^T V_p \sin \omega t\, dt \qquad (10\text{–}13)$$

The integral in Equation (10–13) yields

$$V_{dc} = \frac{-V_p}{\omega T} \cos \omega t \Big]_0^T = \frac{-V_p}{\omega T} (\cos \omega T - 1) \qquad (10\text{–}14)$$

The quantity ωT can be simplified to $\omega T = 2\pi f T = (2\pi/T)T = 2\pi$. Substitution of this quantity in Equation (10–14) results in

$$V_{dc} = \frac{-V_p}{2\pi} (\cos 2\pi - 1) = \frac{-V_p}{2\pi} (1 - 1) = 0 \qquad (10\text{–}15)$$

The preceding steps represented a "long way" to get to a result that could have been obvious from the beginning. In one complete cycle of any sinusoidal function, the magnitude of the positive area is exactly equal to the magnitude

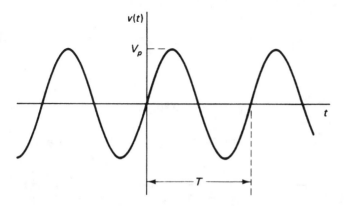

FIGURE 10–2
Sinusoidal voltage for Example 10–1.

of the negative area. Thus, the definite integral of a sinusoidal function over an *integer* number of cycles is exactly zero. This property will be used as a shortcut in appropriate places in the remainder of the book.

The result of Equation (10–15) tells us that the average or dc value of a sine wave is exactly zero. Thus, an ideal dc instrument will not respond to the sinusoidal function. (An exception is when the frequency is so low that the averaging process is not performed, so the meter attempts to "follow" the sine wave.)

The preceding result should not be confused with a half-wave or full-wave rectified sine wave, in which the dc value is *not* zero. (See Problems 10–23 and 10–22.)

(b) In computing rms values, this author prefers to work first with the expression for V_{rms}^2 in order to simplify the notation until the last step. We have

$$V_{rms}^2 = \frac{1}{T} \int_0^T v^2(t)\, dt \tag{10–16}$$

Squaring Equation (10–12) and substituting in Equation (10–16) leads to

$$V_{rms}^2 = \frac{1}{T} \int_0^T V_p^2 \sin^2 \omega t\, dt \tag{10–17}$$

A standard trigonometric identity is

$$\sin^2 \omega t = \tfrac{1}{2} - \tfrac{1}{2} \cos 2\omega t \tag{10–18}$$

When Equation (10–18) is substituted in Equation (10–17), the result can be expanded as

$$V_{rms}^2 = \frac{V_p^2}{2T} \int_0^T dt - \frac{V_p^2}{2T} \int_0^T \cos 2\omega t\, dt \tag{10–19}$$

The second term in Equation (10–19) represents the integral of a cosine function over two cycles of the "double frequency" 2ω, so the net value is zero. Thus,

$$V_{rms}^2 = \frac{V_p^2}{2T} \int_0^T dt = \frac{V_p^2}{2T} t \Big]_0^T = \frac{V_p^2}{2T} \times T = \frac{V_p^2}{2} \tag{10–20}$$

The rms value is thus

$$V_{rms} = \frac{V_p}{\sqrt{2}} = 0.7071 V_p \tag{10–21}$$

This result was stated without proof in Chapter 8, but we have now formally derived it.

EXAMPLE 10–2

Assume the periodic voltage waveform shown in Figure 10–3(a) has a period of 8 ms. (Only one cycle is shown.) Determine **(a)** the dc value and **(b)** the rms value. **(c)** If this voltage appears across a 50-Ω resistance, determine the average power.

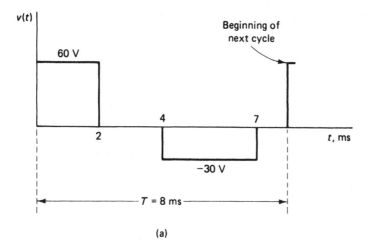

(a)

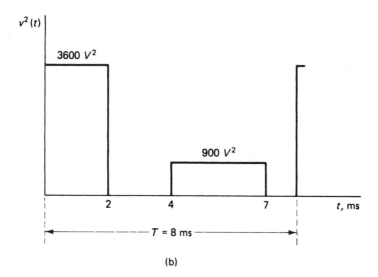

(b)

FIGURE 10–3
Waveform for Example 10–2 and the squared function.

Solution

(a) To determine the dc value, the area under the curve in one cycle is determined. For a "flat-top" function such as this, the area is best determined by simply summing the areas of the individual rectangles. We have

$$\int_0^{8\times10^{-3}} v(t)\, dt = (60\text{ V}) \times (2 \times 10^{-3}\text{ s}) - (30\text{ V}) \times (3 \times 10^{-3}\text{ s}) \qquad \textbf{(10–22a)}$$

$$= 120 \times 10^{-3}\text{ V}\cdot\text{s} - (90 \times 10^{-3})\text{V}\cdot\text{s}$$

$$= 30 \times 10^{-3}\text{ V}\cdot\text{s} \qquad \textbf{(10–22b)}$$

Note that the second rectangle is weighted negatively. The dc value is

$$V_{dc} = \frac{1}{8 \times 10^{-3}} \int_0^{8 \times 10^{-3}} v(t) \, dt = \frac{30 \times 10^{-3}}{8 \times 10^{-3}} = 3.75 \text{ V} \qquad \textbf{(10–23)}$$

Thus, an ideal dc voltmeter would read 3.75 V when connected across this voltage.

(b) To determine the rms voltage, the function $v^2(t)$ must first be determined. This function is shown in Figure 10–3(b) for the given waveform. Each point on the $v^2(t)$ curve is determined by squaring the corresponding point on the $v(t)$ curve. Since there are only two values of $v(t)$, there are only two values of $v^2(t)$. Observe that $(-30 \text{ V})^2 = 900 \text{ V}^2$, so that all area segments of the $v^2(t)$ function are positive.

The area of the squared function in one cycle is

$$\int_0^{8 \times 10^{-3}} v^2(t) \, dt = (3600 \text{ V}^2) \times (2 \times 10^{-3} \text{ s})$$

$$+ (900 \text{ V}^2) \times (3 \times 10^{-3} \text{ s}) \qquad \textbf{(10–24a)}$$

$$= 7.2 \text{ V}^2 \cdot \text{s} + 2.7 \text{ V}^2 \cdot \text{s} = 9.9 \text{ V}^2 \cdot \text{s} \qquad \textbf{(10–24b)}$$

As pointed out earlier, the area of the squared function is *not* the square of the area. The latter calculation would yield $(30 \times 10^{-3} \text{ V} \cdot \text{s})^2 = 900 \times 10^{-6} \text{ V}^2 \cdot \text{s}^2$, which is a useless result.

Proceeding with the computation, we have

$$V_{rms}^2 = \frac{1}{8 \times 10^{-3}} \int_0^{8 \times 10^{-3}} v^2(t) \, dt = \frac{9.9}{8 \times 10^{-3}} = 1237.5 \text{ V}^2 \qquad \textbf{(10–25)}$$

The rms value is thus

$$V_{rms} = \sqrt{1237.5} = 35.18 \text{ V} \qquad \textbf{(10–26)}$$

(c) The average power in a 50-Ω resistance is

$$P = \frac{V_{rms}^2}{50} = \frac{1237.5}{50} = 24.75 \text{ W} \qquad \textbf{(10–27)}$$

One could, of course, take V_{rms} from Equation (10–26) and square it, but V_{rms}^2 was given directly in Equation (10–25).

To illustrate possible pitfalls from misuse of the dc value, suppose one were determining the required power rating of a 50-Ω resistor across which this voltage appears. Based on Equation (10–27), the actual power dissipated is 24.75 W, so a power rating exceeding that value would be required to provide some leeway. If one erroneously used the dc value of 3.75 V, the value predicted would be $(3.75)^2/50 = 0.281$ W. Selecting a power rating based on this incorrect value would lead to some disastrous results!

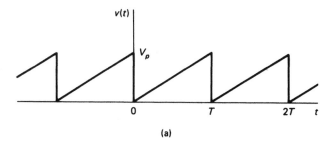

(a)

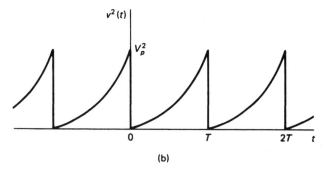

(b)

FIGURE 10–4
Waveform for Example 10–3 and the squared function.

EXAMPLE 10–3
Consider the periodic sawtooth waveform shown in Figure 10–4(a). Determine **(a)** the dc value, **(b)** the rms value, and **(c)** the form factor.

Solution
(a) The area under the curve in one cycle is the area of a triangle with base T and height V_p. The result is determined by inspection as

$$\int_0^T v(t)\, dt = \frac{V_p T}{2} \tag{10–28}$$

The dc value is then

$$V_{dc} = \frac{1}{T}\int_0^T v(t)\, dt = \frac{1}{T} \times \frac{V_p T}{2} = \frac{V_p}{2} \tag{10–29}$$

This result makes sense in view of the fact that the voltage changes linearly from a level of zero to a level of V_p, so the mean or average level is $V_p/2$.
(b) The form of $v^2(t)$ is illustrated in Figure 10–4(b). For a "curved" function such as this, the geometric approach is not so obvious, so a more analytical approach will be used. First, the voltage in one cycle can be expressed in terms of the

straight-line equation

$$v(t) = \frac{V_p}{T}t \quad \text{for } 0 < t < T \tag{10–30}$$

Computation of V^2_{rms} follows.

$$V^2_{rms} = \frac{1}{T}\int_0^T v^2(t)\,dt = \frac{1}{T}\int_0^T \frac{V_p^2}{T^2}t^2\,dt = \frac{V_p^2}{3T^3}t^3\Big]_0^T$$
$$= \frac{V_p^2}{3T^3}(T^3 - 0) = \frac{V_p^2}{3} \tag{10–31}$$

The rms value is

$$V_{rms} = \frac{V_p}{\sqrt{3}} \simeq 0.577V_p \tag{10–32}$$

(c) The form factor is

$$\text{form factor} = \frac{0.577V_p}{0.5V_p} \simeq 1.155 \tag{10–33}$$

10–2   STEP AND RAMP FUNCTIONS

In this section, we begin the study of ways for describing switching operations in electrical circuits in a systematic manner. Certain waveforms that can be used for that purpose will be defined.

Step Function

The first waveform of interest, which was introduced in Chapter 5, is the *unit step function u(t)*. This function is defined as

$$u(t) = 0 \quad \text{for } t < 0$$
$$= 1 \quad \text{for } t > 0 \tag{10–34}$$

The unit step function is shown in Figure 10–5. The unit step function starting at $t = 0$ is nothing more than a dc level of unity "switched on" at $t = 0$.

The process of turning on any arbitrary function $f(t)$ at $t = 0$ can be expressed mathematically by multiplying $f(t)$ by $u(t)$. Thus,

$$f(t)u(t) = 0 \quad \text{for } t < 0$$
$$= f(t) \quad \text{for } t > 0 \tag{10–35}$$

This process is illustrated in Figure 10–6. A function having variation for negative and positive time is shown in (a). When the product of this function and the unit

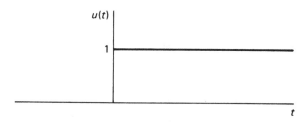

FIGURE 10–5
Basic form of the unit step function.

step is formed, the portion for $t < 0$ is eliminated, as shown in (b).

A *delayed unit step function* is denoted as $u(t - T)$, and it is defined as

$$u(t - T) = 0 \quad \text{for } t < T$$
$$= 1 \quad \text{for } t > T \quad \textbf{(10–36)}$$

This function is shown in Figure 10–7. The significance of the delayed step function will be discussed in Section 10–3.

Ramp Function

The next waveform of interest is the *unit ramp function* $r(t)$, which is defined as

$$r(t) = tu(t) \quad \textbf{(10–37)}$$

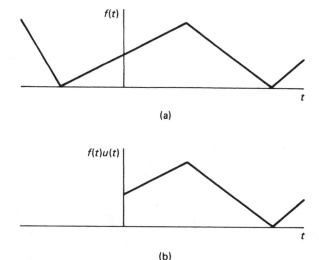

(a)

(b)

FIGURE 10–6
Turning on a function at $t = 0$ with the unit step function.

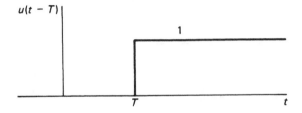

FIGURE 10-7
Delayed unit step function.

This function is shown in Figure 10–8. The unit ramp function is simply a straight line starting at the origin and increasing linearly with a slope of unity. From Equation (10–35), the unit ramp is equivalent to

$$r(t) = 0 \quad \text{for } t < 0$$
$$\quad\quad = t \quad \text{for } t > 0$$

(10–38)

A straight line starting at the origin with a slope K can be expressed as $Kr(t)$.

EXAMPLE 10–4
A certain linearly increasing voltage $v(t)$ is turned on at $t = 0$, as shown in Figure 10–9. Write an equation for the waveform.

Solution
Since the voltage reaches a level of 6 V in 2 s, the slope is $K = 6 \text{ V}/2 \text{ s} = 3 \text{ V/s}$. The voltage can then be expressed as

$$v(t) = 3tu(t)$$

(10–39)

or

$$v(t) = 3r(t)$$

(10–40)

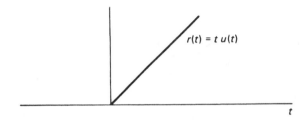

$r(t) = t\, u(t)$

FIGURE 10-8
Unit ramp function.

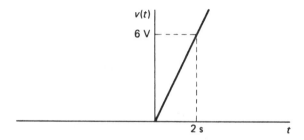

FIGURE 10–9
Voltage for Example 10–4.

10–3 SHIFTED FUNCTIONS

In the analysis of electric circuits considered thus far in the book, virtually all waveforms have assumed $t = 0$ as a starting point. Obviously, this assumption is an artificial point of reference chosen for convenience, and it is the standard approach when only one waveform is of interest. However, in complex situations where more than one waveform is applied at different times, or when a sequence of switching operations is required, a more elaborate scheme is required.

The concept of a *shifted function* is useful for dealing with delayed signals and switching operations. To develop this concept, consider first a function starting at $t = 0$ as shown in Figure 10–10(a). To emphasize the fact that this function starts at 0 and to avoid ambiguity with other functions starting at different times, the function will be denoted as $f(t)u(t)$, as discussed in Section 10–2.

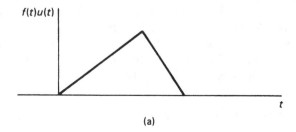

(a)

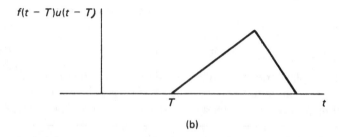

(b)

FIGURE 10–10
Development of the concept of a shifted function.

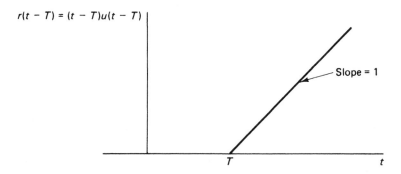

$r(t - T) = (t - T)u(t - T)$

Slope = 1

T

t

FIGURE 10–11
Delayed unit ramp function.

Assume next that this signal is delayed in time by T seconds.* For example, the signal could pass through an ideal delay line in which the function is delayed in time. The process of delaying a signal in time by T seconds is denoted mathematically by replacing t everywhere it appears in the function expression by $t - T$. The reason this works is that for a given argument t_1, the new value of t required to produce the same argument is $t = t_1 + T$, since $t - T = t_1 + T - T = t_1$, indicating that t must be T seconds greater to achieve the same value of the function as before. The function has thus been *delayed* or shifted to the right by T seconds.

The form of a shifted function is thus denoted mathematically as $f(t - T)$ $u(t - T)$, and it is shown in Figure 10–10(b). The exact function is preserved, but all points are delayed by T seconds. The $u(t - T)$ factor results in the function being 0 for $t < T$. It is also possible to shift to the left by replacing t by $t + T$, but this represents an advance in time and is of relatively little interest in circuit analysis. The focus, therefore, will be on delayed shifted functions.

If the unit ramp function defined by Equation (10–37) is delayed by T seconds, the result can be represented by either of the forms $r(t - T)$ or $(t - T)u(t - T)$. A unit delayed ramp is illustrated in Figure 10–11.

Laplace Transform of Shifted Function

As a final point, the Laplace transform form of the shifting operation will be noted. Operation (O–4) of Table 5–2 provides the following Laplace transform operation pair:

$$\mathscr{L}[f(t - T)u(t - T)] = e^{-sT}F(s) \tag{10–41}$$

Stated in words, the Laplace transform of a shifted function is the unshifted function multiplied by e^{-sT}. In much circuit analysis, the e^{-sT} factor serves simply as a "tag" to denote the starting points for various signals. In certain types of design situations,

---

*The reader may be bothered by the use of the same symbol here for delay as was used for period in Section 10–1. However, T is widely used for both purposes. Obviously, if a waveform is both periodic and delayed, separate symbols or subscripts are required.

however, steady-state properties of the e^{-sT} factor may be used to advantage to create desired filter characteristics. In fact, the e^{-sT} form is a direct link to a mathematical transform called the z-transform, from which the theory of discrete-time signals can be modeled.

Based on Equation (10–41), the Laplace transforms of the delayed unit step and unit ramp can be expressed as

$$\mathcal{L}[u(t - T)] = \frac{e^{-sT}}{s} \tag{10–42}$$

and

$$\mathcal{L}[r(t - T)] = \frac{e^{-sT}}{s^2} \tag{10–43}$$

EXAMPLE 10–5
The voltage of Example 10–4 is delayed by 3 seconds. Using the original time scale, write an expression for the delayed voltage $v_d(t)$ and sketch it.

Solution
The original $v(t)$ was shown in Figure 10–9, and the delayed voltage $v_d(t)$ is shown in Figure 10–12. The two voltages are identical in form, but $v_d(t)$ is shifted to the right by 3 s. Two ways of expressing $v(t)$ were given in Equations (10–39) and (10–40). The delayed function $v_d(t)$ is determined by replacing t in either expression by $t - 3$. Thus,

$$v_d(t) = 3(t - 3)u(t - 3) \tag{10–44}$$

or

$$v_d(t) = 3r(t - 3) \tag{10–45}$$

EXAMPLE 10–6
An ideal delay line has the following input–output relationship:

$$v_2 = v_1(t - T) \tag{10–46}$$

FIGURE 10–12
Voltage for Example 10–5.

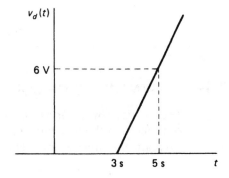

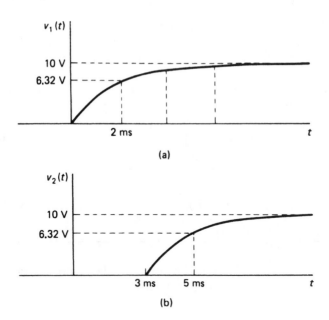

FIGURE 10–13
Waveforms for Example 10–6.

where $v_1(t)$ is the input voltage and $v_2(t)$ is the output voltage. Assume that the input to a certain ideal delay line having a delay of 3 ms is the voltage

$$v_1(t) = 10(1 - e^{-500t})u(t) \qquad (10\text{–}47)$$

Sketch the input and output signals and write an expression for the output voltage.

Solution
The input voltage starts at zero and increases exponentially toward 10 V as shown in Figure 10–13(a). The time constant of the exponential is $\tau = 1/500 = 2$ ms, so the voltage has reached a level of about 6.32 V at $t = 2$ ms. For most practical purposes, the input voltage will have reached steady state in about 10 ms.

The output voltage $v_2(t)$ is shown in Figure 10–13(b). No output is sensed prior to $t = 3$ ms, at which time the signal starts to appear. The output voltage can be expressed as

$$v_2(t) = 10[1 - e^{-500(t-3\times10^{-3})}]u(t - 3 \times 10^{-3}) \qquad (10\text{–}48)$$

10–4 SYNTHESIS OF COMPLEX WAVEFORMS

Several basic waveforms and the means for shifting these functions have been considered in the preceding sections. In this section, these waveforms will be used to create or synthesize more elaborate waveforms such as piecewise linear functions.

To that end, we will first consider the forms of the sums of certain combinations of the basic waveforms.

Sum of Step Functions

First, consider the function $f(t)$, formed from the sum of a step function starting at $t = 0$ and a delayed step function starting at $t = T$. Let A and B represent the respective levels of the steps and, for illustration, assume that both A and B are positive. We have

$$f(t) = Au(t) + Bu(t - T) \qquad \qquad \textbf{(10-49)}$$

The process of generating this function is shown in Figure 10–14. For $t < T$, only the first step function contributes to the response, so $f(t) = A$ in that interval. However, for $t > T$, both functions contribute, so $f(t) = A + B$ in that interval. In this illustration, A and B are both positive, so the net function is larger after the second step is initiated. Depending on the signs of constants or, equivalently,

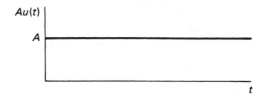

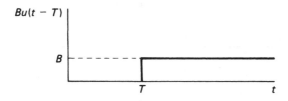

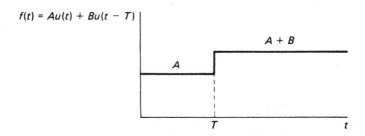

FIGURE 10–14
Addition of two positive step functions.

whether a subsequent step is added or subtracted, a number of possible forms can be generated. This process is shown at the end of this section.

Sum of Ramp Functions

Next, consider the process of adding ramp functions starting at different times. Important points to remember are:

1. In each interval between points at which ramp functions are generated, the resulting function is a straight line.
2. The slope at any point is the algebraic sum of the slopes of all previously generated ramps.
3. The starting point in a given interval is the value reached by the function at the end of the preceding interval. This means that ramp functions alone cannot produce any sudden jumps. A step function is required to produce a discontinuity.

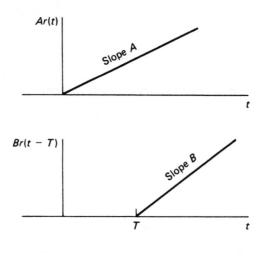

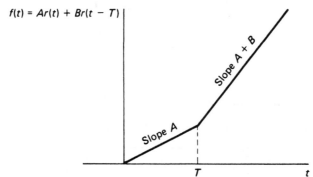

FIGURE 10–15
Addition of two positive ramps.

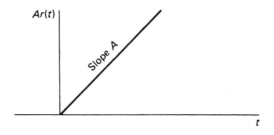

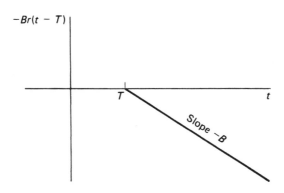

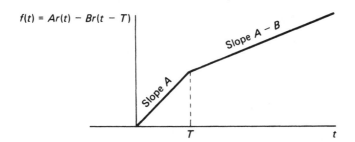

FIGURE 10–16
Addition of positive and negative ramps.

Several representative cases of additive ramps are shown in Figures 10–15, 10–16, and 10–17. These cases will be briefly discussed.

The waveform $f(t)$ of Figure 10–15 corresponds to

$$f(t) = Ar(t) + Br(t - T) \qquad (10\text{–}50)$$

where A and B are both positive. For $t < T$, the net function is a straight line with slope A. For $t > T$, the slope is $A + B$.

The waveform $f(t)$ of Figure 10–16 corresponds to

$$f(t) = Ar(t) - Br(t - T) \qquad (10\text{–}51)$$

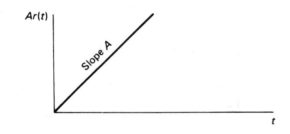

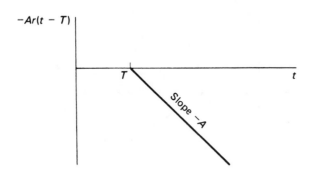

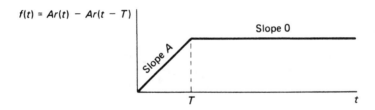

FIGURE 10–17
Addition of ramps with equal and opposite slopes.

where A and B are both positive, but where $A > B$. In this case, the slope for $t >$ T is $A - B$. Since $A > B$, the net slope is still positive for $t > T$. If B were greater than A, the net slope for $t > T$ would be negative.

An interesting case is that of Figure 10–17, in which the function is given by

$$f(t) = Ar(t) - Ar(t - T) \tag{10–52}$$

The net slope for $t > T$ is zero, so the level remains constant for all later time. Understand that each of the two ramps is assumed to continue, but for each increase in the first ramp, there is an equal and opposite decrease in the second ramp, so the net level remains constant. Thus, an "imperfect step function," that is, one with a nonzero rise time, could be expressed in this manner.

FIGURE 10–18
Waveforms for Example 10–7.

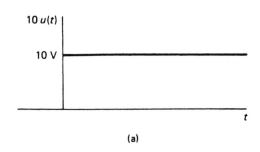

(a)

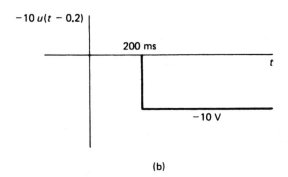

(b)

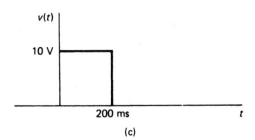

(c)

EXAMPLE 10–7
Express the pulse waveform of Figure 10–18(c) as a combination of the basic waveforms of this section.

Solution
The finite width pulse is a common waveform, and it can be represented as the sum of a positive step function starting at $t = 0$ and a negative step function starting at a time equal to the pulse width. These two functions are shown in (a) and (b) of Figure 10–18 for the particular pulse given. The function is expressed as

$$v(t) = 10u(t) - 10u(t - 0.2) \tag{10–53}$$

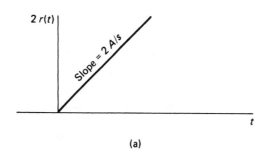

(a)

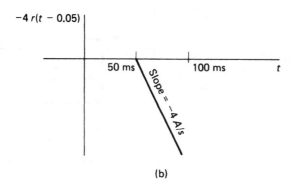

(b)

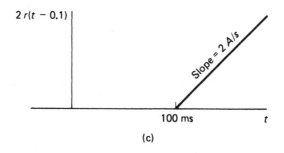

(c)

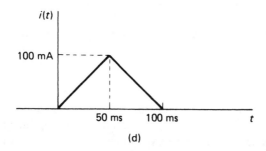

(d)

FIGURE 10–19
Waveforms for Example 10–8.

EXAMPLE 10–8

Express the triangular pulse of Figure 10–19(d) as a combination of the basic waveforms of this section.

Solution

The signal is first initiated at $t = 0$ by the ramp function shown in Figure 10–19(a). The slope of this function is 0.1 A/0.05 s = 2 A/s, so this function can be expressed as $2r(t)$. In the interval from 50 ms to 100 ms, the slope is negative and is −0.1 A/0.05 s = −2 A/s. The function added at $t = 50$ ms must have a sufficiently large negative slope to bring the net slope to a level of −2 A/s. This is achieved by adding a ramp with slope −4 A/s starting at $t = 50$ ms, so this second function is $-4r(t - 0.05)$ as shown in Figure 10–19(b). Finally, at $t = 100$ ms, the net slope must return to zero, and this can be achieved by adding a ramp with slope 2 A/s. This latter function is $2r(t - 0.1)$, and it is shown in Figure 10–19(c). The composite function can thus be expressed as

$$i(t) = 2r(t) - 4r(t - 0.05) + 2r(t - 0.1) \qquad \text{(10–54)}$$

Alternately, this function could be expressed as

$$i(t) = 2tu(t) - 4(t - 0.05)u(t - 0.05) + 2(t - 0.1)u(t - 0.1) \qquad \text{(10–55)}$$

where basic units (seconds and amperes) are assumed for the variables.

It may provide some additional insight to the reader to evaluate the preceding equation at a few points to show that it does produce the correct value. There is little need to do so in the interval $0 < t < 50$ ms, since the function is simply $2t$, and it is clear that the correct pattern is produced. However, consider the interval 50 ms $< t < 100$ ms. In that interval, the first two step functions in Equation (10–55) are "on," and the third is "off," so the current is

$$i(t) = 2t - 4(t - 0.05) = -2t + 0.2 \quad \text{for 50 ms} < t < 100 \text{ ms} \qquad \text{(10–56)}$$

The reader can show that if the straight line is extended, it will intersect the vertical axis at a current 200 mA = 0.2 A, which is the vertical intercept in Equation (10–56). Of course, the slope is −2 A/s in that interval, as already noted.

Next, consider the interval for $t > 100$ ms, in which all step functions are "on." The current is

$$\begin{aligned} i(t) &= 2t - 4(t - 0.05) + 2(t - 0.1) \\ &= 2t - 4t + 0.2 + 2t - 0.2 \\ &= 0 \quad \text{for } t > 100 \text{ ms} \end{aligned} \qquad \text{(10–57)}$$

as expected.

10–5 SUPERPOSITION REVISITED

In the past few sections, we have seen how certain complex waveforms can be expressed in terms of simpler functions. This has consisted of representing such waveforms in terms of steps and ramps starting at $t = 0$, as well as delayed versions of the same signals. We will now investigate the motive for such a process.

Assume that a circuit is excited by some elaborate input function $x(t)$ and that it is desired to determine the response $y(t)$. Although $x(t)$ is assumed to be a rather formidable input with which to analyze the circuit, assume that it can be decomposed into a sum of N "elementary functions" as follows:

$$x(t) = \sum_{n=1}^{N} x_n(t) \tag{10–58}$$

where each $x_n(t)$ is assumed to be simpler in form. More important, the response of the circuit to each $x_n(t)$ can be more easily determined than the response to $x(t)$.

If the circuit were excited by $x_1(t)$, the response is assumed to be $y_1(t)$; if the circuit were excited by $x_2(t)$, the response is assumed to be $y_2(t)$; and so on. On the assumption that the circuit is a linear time-invariant form, the net response can then be expressed from the superposition principle as

$$y(t) = \sum_{n=1}^{N} y_n(t) \tag{10–59}$$

Stated in words, *the response to the complex input is the sum of the responses that would arise from the individual inputs.* This concept is at the heart of the motivation for representing complex waveforms in terms of simpler waveforms.

EXAMPLE 10–9

The purpose of this example is to illustrate with a relatively simple circuit how the concept of superposition, as discussed in this section, can be used to determine the response to a pulse-type waveform. The particular situation was analyzed in Example 4–11 of Chapter 4 using a different approach. For convenience, the circuit and pulse input are shown again in (a) and (b) of Figure 10–20. Determine the output voltage $v_2(t)$, making use of the principles of the preceding few sections.

Solution

From the similarity of this waveform to that of Example 10–7, it is evident that the pulse can be expressed as the sum of a positive step occurring at $t = 0$ and a negative step occurring at $t = T$. We have

$$v_1(t) = V_p u(t) - V_p u(t - T) \tag{10–60}$$

From the principle of superposition, if we can determine the separate responses to the two functions on the right-hand side of Equation (10–60), the total response will be the sum of the two individual responses.

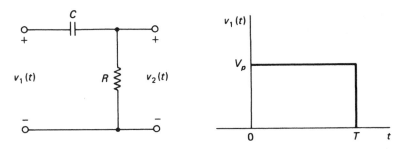

FIGURE 10–20
Circuit and pulse excitation for Example 10–9.

Let $v_1'(t)$ represent the first term of Equation (10–60), and let $v_1''(t)$ represent the second term. Let $v_2'(t)$ and $v_2''(t)$ represent the respective response terms. The respective Laplace transforms of $v_1'(t)$ and $v_1''(t)$ are

$$V_1'(s) = \frac{V_p}{s} \qquad (10\text{–}61)$$

and

$$V_1''(s) = \frac{-V_p}{s} e^{-sT} \qquad (10\text{–}62)$$

The s-domain forms for the circuit and the two source components are illustrated in Figure 10–21. Without showing all the steps, it can be readily verified that

$$V_2'(s) = \frac{V_p}{s + \dfrac{1}{RC}} \qquad (10\text{–}63)$$

and

$$V_2''(s) = \frac{-V_p}{s + \dfrac{1}{RC}} e^{-sT} \qquad (10\text{–}64)$$

Observe that the s-domain form for $V_2''(s)$ differs from that of $V_2'(s)$ only by the negative sign and the exponential factor. Although these functions do not require partial fraction expansion, it should be emphasized that any e^{-sT} factors should be separately associated with all terms arising from a particular expansion. Said differently, each e^{-sT} multiplier remains as an identification factor with all terms with which it is associated.

The inverse transform of Equation (10–63) is readily expressed as

$$v_2'(t) = V_p e^{-t/RC} u(t) \qquad (10\text{–}65)$$

FIGURE 10–21

Forms of the s-domain models in Example 10–9.

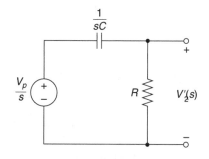

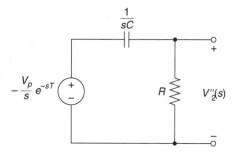

The step function factor has been included to identify clearly the starting time for this term.

The inverse transform of Equation (10–64) is obtained by momentarily ignoring the e^{-sT} factor, taking the inverse transform, and then replacing t by $t - T$. We have

$$v_2''(t) = -V_p e^{-(t-T)/RC} u(t - T) \qquad (10\text{–}66)$$

With appropriate insight, it is unnecessary to belabor a problem such as this to the extent that we have done. In fact, once the response to the first step is determined, the response to the second step is obtained by shifting the time scale (i.e., replacing t by $t - T$), gating the function on at $t = T$, and employing a negative sign, since the second step function is negative.

Another point to emphasize is that we did not consider any initial voltage on the capacitor in determining the second response, since superposition was being applied, and the response to each component is determined *as if* that input were the only source.

The approach used back in Example 4–11 was to determine the voltage on the capacitor at $t = T$ and then simply treat the problem for $t > T$ as the problem of discharging a capacitor through a resistance. With the approach of this chapter, the assumption of the separate input terms in the superposition process takes care of the boundary conditions at transition points.

The net response can now be written as

$$v_2(t) = v_2'(t) + v_2''(t) \tag{10-67a}$$

$$= V_p e^{-t/RC} u(t) - V_p e^{-(t-T)/RC} u(t-T) \tag{10-67b}$$

If it is desired to determine separate equations for the two major intervals, we can write for the case of $0 < t < T$

$$v_2(t) = V_p e^{-t/RC} \tag{10-68}$$

and for $t > T$

$$v_2(t) = V_p e^{-t/RC} - V_p e^{-(t-T)/RC} \tag{10-69}$$

By some algebraic manipulations, Equation (10–69) can be expressed as

$$v_2(t) = -V_p (1 - e^{-T/RC}) e^{-(t-T)/RC} \tag{10-70}$$

The reader is invited to verify that these results are the same as those obtained in Example 4–11, after slight changes in notation are made.

EXAMPLE 10–10

The current $i(t)$ in Figure 10–22(a) is the triangular current pulse of Example 10–8. Determine an expression for the voltage $v(t)$ and sketch it. The inductor is assumed to be initially relaxed at $t = 0$.

Solution
Expressions for the current pulse were given in Equations (10–54) and (10–55). The latter form will be repeated here for convenience:

$$i(t) = 2tu(t) - 4(t - 0.05)u(t - 0.05) + 2(t - 0.1)u(t - 0.1) \tag{10-71}$$

The s-domain form of the circuit is shown in Figure 10–22(b). The transform current $I(s)$ is

$$I(s) = \frac{2}{s^2} - \frac{4}{s^2} e^{-0.05s} + \frac{2}{s^2} e^{-0.1s} \tag{10-72}$$

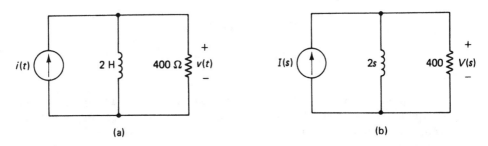

(a)   (b)

FIGURE 10–22
Circuit for Example 10–10 and s-domain model.

The transform voltage $V(s)$ may be expressed as

$$V(s) = Z(s)I(s) \tag{10-73}$$

where $Z(s)$ is the net impedance of the two parallel elements, which is

$$Z(s) = \frac{2s \times 400}{2s + 400} = \frac{400s}{s + 200} \tag{10-74}$$

Substituting Equations (10–72) and (10–74) in Equation (10–73), there results

$$V(s) = \frac{400s}{s + 200}\left(\frac{2}{s^2} - \frac{4}{s^2}e^{-0.05s} + \frac{2}{s^2}e^{-0.1s}\right) \tag{10-75}$$

This result is equivalent to three terms as follows:

$$V(s) = \frac{800}{s(s + 200)} - \frac{1600e^{-0.05s}}{s(s + 200)} + \frac{800e^{-0.1s}}{s(s + 200)} \tag{10-76}$$

Each of the terms in Equation (10–76) can be expanded in a partial fraction expansion. All of the forms are virtually the same except for sign and magnitude differences. The exponential factors are not altered during the expansion. The resulting function is

$$\begin{aligned}
V(s) = &\frac{4}{s} - \frac{4}{s + 200} \\
&- \left(\frac{8}{s} - \frac{8}{s + 200}\right)e^{-0.05s} \\
&+ \left(\frac{4}{s} - \frac{4}{s + 200}\right)e^{-0.1s}
\end{aligned} \tag{10-77}$$

The inverse transform is

$$\begin{aligned}
v(t) = &[4 - 4e^{-200t}]u(t) \\
&- [8 - 8e^{-200(t-0.05)}]u(t - 0.05) \\
&+ [4 - 4e^{-200(t-0.1)}]u(t - 0.1)
\end{aligned} \tag{10-78}$$

The form of the voltage is shown in Figure 10–23. The time constant of each exponential is 0.005 s = 5 ms, and the second component begins at 50 ms, so the first component will have reached a steady-state level of 4 V for most practical purposes. A similar conclusion holds for the level of −4 V at the point where the third component is initiated.

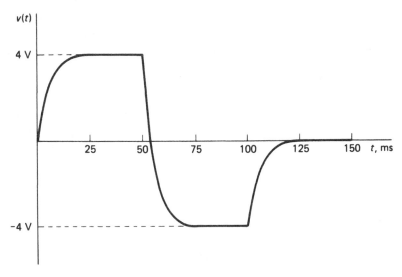

FIGURE 10–23
Voltage waveform for Example 10–10.

10–6 IMPULSE FUNCTION

In various portions of the text, vague references have been made to impulsive conditions. The process of explaining such conditions begins with a consideration of the unit impulse function.

The concept of the impulse function will be developed in this text through a somewhat intuitive process, illustrated in Figure 10–24. Consider two functions $f_1(t)$ and $f_2(t)$ with the following relationship assumed between the two functions:

$$f_2(t) = \frac{df_1(t)}{dt} \tag{10–79}$$

The function $f_1(t)$ of Figure 10–24(a) rises rather slowly to its final value of unity, so its derivative is a pulse of moderate height $1/t_1$. The area of the pulse is $t_1 \times (1/t_1) = 1$. The function $f_1(t)$ of (b) rises more quickly than the function of (a), so its slope is larger. However, the area of the resulting pulse is again $t_1 \times (1/t_1) = 1$, the same as before.

Assume now that the rise time of the input approaches zero as shown in (c). The height $1/t_1$ of the differentiated pulse now increases without limit. However, the area of the pulse has been shown to be independent of the pulse width, so it is assumed to be still unity! Stretching our imaginations, we might be tempted to say that "the differentiated pulse has zero width, infinite height, and unit area!"

The limiting pulse case as developed in Figure 10–24(c) is the basis of the unit impulse function, which is denoted as $\delta(t)$. The function $f_1(t)$ in this case is equivalent to a unit step function.

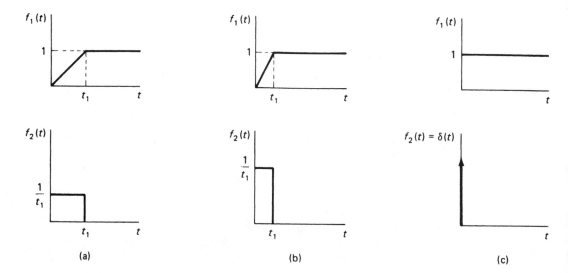

FIGURE 10–24
Development of the impulse function as the derivative of the step function.

Impulse as Derivative of Step

Summarizing the outcome of the preceding development, we will define the unit impulse function $\delta(t)$ as

$$\delta(t) = \frac{du(t)}{dt} \qquad (10\text{–}80)$$

If the step function is delayed in time, a delayed impulse function $\delta(t - T)$ can be defined as

$$\delta(t - T) = \frac{du(t - T)}{dt} \qquad (10\text{–}81)$$

The preceding results are summarized in Figures 10–25 and 10–26. Note the symbol for the impulse function. Stated in words, *the derivative of a unit step function is an impulse function occurring at the same point at which the step function begins.*

   It should be emphasized that we are not violating the student's basic rules of calculus, in which the derivative of a constant is zero. The "pure constant" is assumed to be constant everywhere, so its derivative is zero. The step function is a "switched-on" constant, and the derivative is not zero at the point of switching.

Integral of Impulse

The inverse relationships may be inferred by integrating both sides of Equations (10–80) and (10–81). Since most of our integrals are initiated at $t = 0$, that reference time will be assumed here. Further, we will assume that all the area of the impulse

FIGURE 10–25
Unshifted unit step function and its derivative.

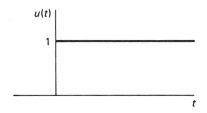

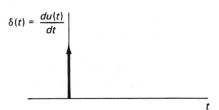

at $t = 0$ will be covered, and we have

$$\int_0^t \delta(t)\, dt = u(t) \tag{10–82}$$

and

$$\int_0^t \delta(t - T)\, dt = u(t - T) \tag{10–83}$$

Simply stated, the preceding results mean that when an impulse is passed during an integration process, the result is an instantaneous change in the area, which assumes the form of a step function in the integral. This is a result of the fact that an impulse is assumed to contain a nonzero area even though its width is infinitesimal.

FIGURE 10–26
Shifted or delayed unit step function and its derivative.

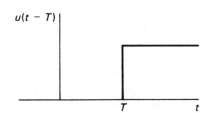

Applications of Impulse Function

No doubt the practically oriented, skeptical reader may be wondering whether there is any real value to such a hypothetical concept. The answer is a definite yes—if we will first temper the results carefully. First, the ideal impulse function (zero width, infinite height, etc.) certainly doesn't exist. However, sudden switching processes that create conflict with initial values give rise to sudden changes whose mathematical limits would be ideal impulse functions. Such situations as the sudden termination of current in an inductor or the sudden application of voltage across a capacitor result in impulse types of situations. In the real world, these situations correspond to the generation of very large, sudden voltages or surges of currents, whose actual values are usually limited by controlling physical parameters. The results predicted by impulsive conditions may be totally undesirable, and can often indicate disastrous circuit conditions!

A second use of the impulse function is that of approximating a real nonzero width pulse. It turns out that if the width of a real pulse is short compared with the time constant of a circuit, the response to that real pulse may be closely approximated by the predicted response to a fictitious impulse. This phenomenon will be illustrated in Example 10–11.

A third use of the impulse function is in conjunction with initial conditions, and this concept was introduced indirectly in Chapter 6. When the Norton equivalent circuit of a charged capacitor or the Thevenin equivalent circuit of a fluxed inductor is used, an impulse source is required. This is strictly a "bookkeeping" approach, and need not concern us when s-domain models are directly employed, as discussed in Chapter 6.

Whereas the impulse function may be rather strange, its Laplace transform has the simplest form of all functions. It can be shown that

$$\mathcal{L}[\delta(t)] = 1 \qquad\qquad (10\text{–}84)$$

This was given as (T–10) in Table 5–1. The result of Equation (10–84) explains why the Norton equivalent circuit of a charged capacitor and the Thevenin equivalent circuit of a fluxed inductor each contain constant sources in the s-domain. These functions correspond to impulse functions in the time domain.

EXAMPLE 10–11

The circuit of Figure 10–27 is excited by the voltage $v_1(t)$, which is a narrow pulse, as shown in Figure 10–28(a). The capacitor is initially uncharged. **(a)** Determine the exact form of the output voltage $v_2(t)$ and sketch it. **(b)** Justify the use of an impulse approximation for the source voltage, and determine the corresponding output voltage. Compare the two results.

Solution

(a) The input pulse can be represented as the sum of a positive step of 500 V starting at $t = 0$, followed by a negative step of -500 V starting at $t = 10$ ms.

FIGURE 10–27
Circuit for Example 10–11.

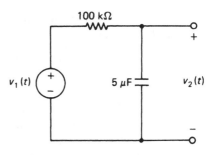

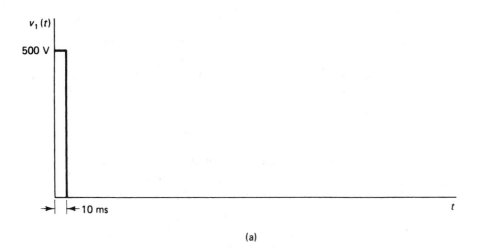

(a)

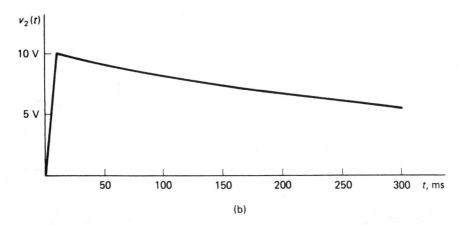

(b)

FIGURE 10–28
Input voltage pulse and response of circuit for Example 10–11.

The reader is invited to show that an application of the concepts of Section 10–5 yields

$$v_2(t) = 500[1 - e^{-2t}]u(t)$$
$$-500[1 - e^{-2(t-0.01)}]u(t - 0.01) \tag{10-85}$$

Note that the time constant is $\tau = 10^5 \ \Omega \times 5 \times 10^{-6} \ F = 0.5$ s.

Although $v_2(t)$ starts charging toward 500 V, it reaches only a small fraction of that value because of the narrow pulse width. When $t = 10$ ms, the output voltage is $v_2(0.01) = 500(1 - e^{-0.02}) \approx 9.9$ V.

For $t > 10$ ms, the expression of Equation (10–85) reduces to

$$v_2(t) = 500 - 500e^{-2t} - 500 + 500e^{-2(t-0.01)}$$
$$= (500e^{0.02} - 500)e^{-2t} \approx 10.1e^{-2t} \quad \text{for } t > 10 \text{ ms} \tag{10-86}$$

The form of the exact output voltage is shown in Figure 10–28(b).

(b) The time constant of 0.5 s is 50 times the pulse width of 10 ms. An impulse approximation for the source voltage may then be reasonable. The area of the pulse is 500 V × 0.01 s = 5 V · s. Let $\hat{v}_1(t)$ represent the impulse approximation, which is given by

$$\hat{v}_1(t) = 5\delta(t) \tag{10-87}$$

Let $\hat{v}_2(t)$ represent the corresponding output voltage. Time-domain and s-domain circuits based on the impulse approximation are shown in (a) and (b) of Figure 10–29, respectively. The output voltage transform is

$$\hat{V}_2(s) = \frac{\dfrac{2 \times 10^5}{s} \times 5}{\dfrac{2 \times 10^5}{s} + 10^5} = \frac{10}{s + 2} \tag{10-88}$$

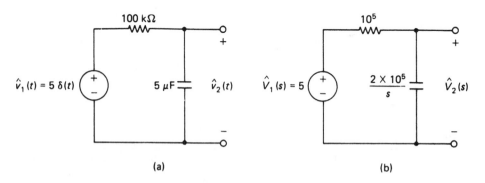

(a)   (b)

FIGURE 10–29
Time-domain and s-domain models of circuit for Example 10–11 with impulse source approximation.

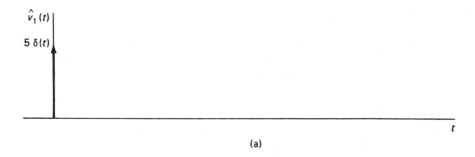

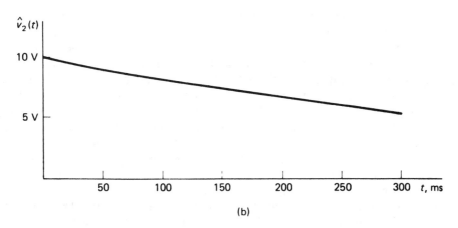

FIGURE 10–30
Impulse source approximation and response of circuit for Example 10–11.

The corresponding time-domain voltage is

$$\hat{v}_2(t) = 10e^{-2t} \tag{10–89}$$

The impulse approximation and the corresponding impulse response are shown in (a) and (b) of Figure 10–30, respectively. The response of Figure 10–30(b) can be compared with the exact response of Figure 10–28(b). The impulse response indicates that the output voltage reaches a level of 10 V instantaneously, whereas the exact response rises from 0 to nearly 10 V in 10 ms. If the response in this first, short time interval is important for the particular application, the impulse approximation may not yield satisfactory results. On the other hand, for $t > 10$ ms, the two functions differ by less than 1%. If this larger time scale is the region of primary interest, the response predicted by the impulse approximation is very close to the exact response.

10-7 PSPICE EXAMPLES

In this section, PSPICE will be used to perform waveform analysis. The major focus will be on the applications of the mathematical operations of PROBE to determine various properties of waveforms.

PSPICE EXAMPLE 10-1

Use PROBE post-processor functions to determine the average and rms values of a sine wave of unit amplitude.

Solution

The intent is to illustrate the power of the PROBE functions, which include differentiation, integration, averaging, and many others. Refer to the code of Figure 10–31. First a simple circuit form is created, consisting of an ideal voltage source in parallel with a 1-Ω resistor. The sine function has a peak value of 1 V and a frequency of 1 Hz, corresponding to a period of 1 s. A transient run is made for 1 s, and since there are no reactive elements, this provides an immediate steady-state form. The voltage is V(1) of Figure 10–32, and so far, nothing exciting has happened.

We now turn to the post-processor functions. First, an averaging operation is performed on the waveform. The code for this is AVG(V(1)), and the result is shown on the figure. The averaging process is a running average, which could have meaning at any time for some applications. However, for determining the dc value, it must be performed over one cycle (or over an integer number of cycles). Thus, the only point of significance for us is the value at $t = 1$ s. The cursor C1 was employed, and the value is 1.3572×10^{-6}, which is about as close to zero as one could expect for the numerical approximation involved. Thus, for all practical purposes, the dc value is zero, as expected.

Next, the rms value is determined, and the code is RMS(V(1)). Once again, this is a running operation, and the desired value at $t = 1$ s is indicated by cursor C2 as 0.707058. This is very close to the theoretical value of 0.707109.

Although we do not need to employ these functions in this example, note that the code for an integration of V(1) is S(V(1)), and the code for a differentiation of V(1) is D(V(1)).

PSPICE EXAMPLE 10-2

Use PROBE post-processor functions to determine the average and rms values of the function of Example 10–2 [Figure 10–3(a)].

FIGURE 10–31
Code for PSPICE Example 10–1.

```
PSPICE EXAMPLE 10-1
V 1 0 SIN 0 1 1
R 1 0 1
.TRAN 1 1
.PROBE
.END
```

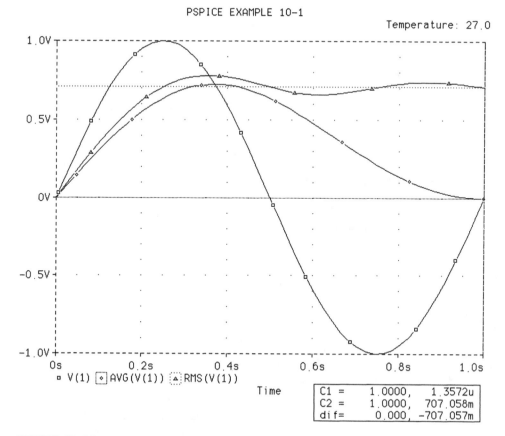

FIGURE 10–32

Waveforms for PSPICE Example 10–1.

Solution

This example is almost like the preceding one, except that the waveform takes much more effort to describe. Referring to Figure 10–33, the PWL function is employed. At each of the transition points, a rise or fall time of 1 μs is assumed. On the time scale involved, this value should be insignificant.

```
PSPICE EXAMPLE 10-2
V 1 0 PWL 0 0 1E-6 60 2E-3 60 2.001E-3 0 4E-3 0 4.001E-3 -30 7E-3 -30 7.001E-3 0
R 1 0 1
.TRAN 8E-3 8E-3
.PROBE
.END
```

FIGURE 10–33

Code for PSPICE Example 10–2.

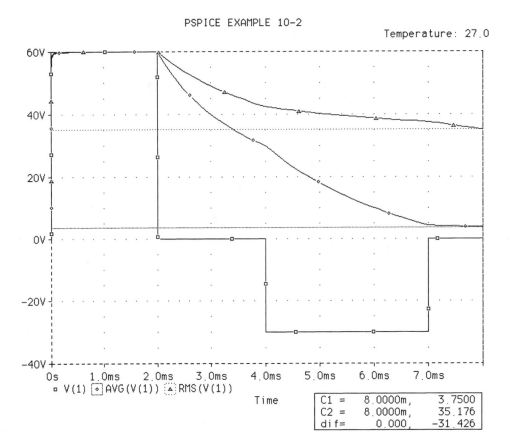

FIGURE 10–34
Waveforms for PSPICE Example 10–2.

The resulting waveform is shown as V(1) on Figure 10–34. The averaging function and the rms function were employed, and the results at the end of one cycle (8 ms) are shown by C1 and C2. The dc value was determined to be 3.7500 V, which is identical with the value calculated in Example 10–2. The rms value was determined to be 35.176 V, which is virtually the same as the value of 35.18 calculated in the text.

DRILL PROBLEMS

10–1. Consider the periodic voltage waveform shown in Figure P10–1 as having a period of 20 ms. Determine **(a)** the dc value, **(b)** the rms value, and **(c)** the form factor. **(d)** If this voltage appears across a 4-Ω resistance, determine the average power.

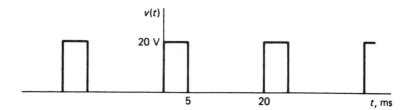

FIGURE P10–1

10–2. Consider the periodic voltage waveform shown in Figure P10–2 having a period of 6 ms. Determine **(a)** the dc value and **(b)** the rms value. **(c)** If this voltage appears across a 50-Ω resistance, determine the average power.

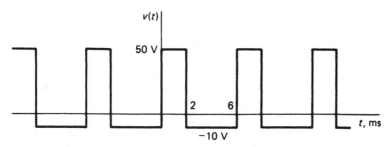

FIGURE P10–2

10–3. Consider the periodic current waveform shown in Figure P10–3 as having a period of 200 ms. (Only one cycle is shown.) Determine **(a)** the dc value and **(b)** the rms value. **(c)** If this current is flowing in a 3-Ω resistance, determine the average power.

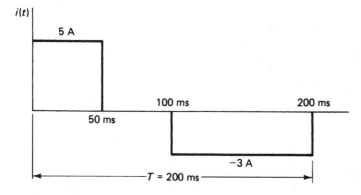

FIGURE P10–3

10–4. Consider the periodic current waveform shown in Figure P10–4 as having a period of 10 ms. (Only one cycle is shown.) Determine **(a)** the dc value and **(b)** the rms value. **(c)** If this current is flowing in a 1000-Ω resistance, determine the average power.

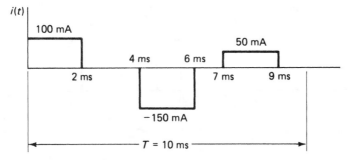

FIGURE P10-4

10-5. A certain dc voltage of 24 V is turned on at $t = 0$. Write an expression for the voltage $v(t)$ in terms of the unit step function.

10-6. A certain dc current of -5 A is turned on at $t = 0$. Write an expression for the current $i(t)$ in terms of the unit step function.

10-7. A certain current $i(t)$ is turned on at $t = 0$, and it decreases linearly at the rate of -3 A/s. Write two separate equation forms for the current: **(a)** by use of the unit step function for switching and **(b)** directly, in terms of the unit ramp function.

10-8. A certain voltage $v(t)$ is turned on at $t = 0$, and it increases linearly at the rate of 7 V/s. Write two separate equation forms for the voltage: **(a)** by use of the unit step function for switching and **(b)** directly, in terms of the unit ramp function.

10-9. Repeat Problem 10-5 if the voltage is delayed and turned on at $t = 5$ s.

10-10. Repeat Problem 10-6 if the current is delayed and turned on at $t = 8$ s.

10-11. Repeat Problem 10-7 if the current is delayed and turned on at $t = 4$ s.

10-12. Repeat Problem 10-8 if the voltage is delayed and turned on at $t = 3$ s.

10-13. Express the voltage waveform of Figure P10-13 as a combination of the basic waveforms of Section 10-4.

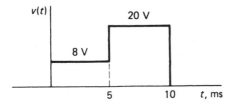

FIGURE P10-13

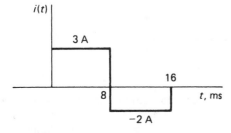

FIGURE P10-14

10–14. Express the current waveform of Figure P10–14 as a combination of the basic waveforms of Section 10–4.

10–15. Express the current waveform of Figure P10–15 as a combination of the basic waveforms of Section 10–4.

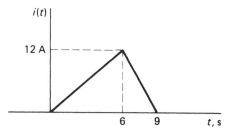

FIGURE P10–15

10–16. Express the voltage waveform of Figure P10–16 as a combination of the basic waveforms of Section 10–4.

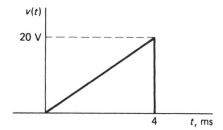

FIGURE P10–16

10–17. In Example 10–9, the *RC* circuit of Example 4–11 was analyzed using superposition. In this problem, a similar process will be used to analyze the circuit of Example 4–9. By expressing the input pulse as the algebraic sum of two step functions and applying superposition, obtain an expression for $v_2(t)$. Show that this result is equivalent to the solution obtained in Example 4–9.

10–18. Assume that when a certain circuit is excited by a unit step function of input voltage, the output voltage is $v_0(t) = 3(1 - e^{-200t})u(t)$. Assume that the voltage of Problem 10–13 (Figure P10–13) is applied as the input voltage. Using superposition, write an expression for the output voltage.

10–19. Determine the capacitor voltage $v_c(t)$ in the circuit of Figure P10–19.

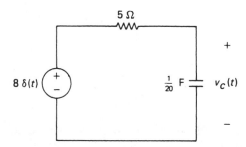

FIGURE P10–19

10–20. Determine the current $i(t)$ in the circuit of Figure P10–20.

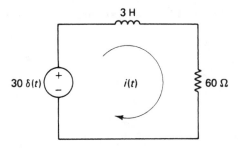

FIGURE P10–20

DERIVATION PROBLEMS

10–21. The waveform of Figure P10–21 is a *half-wave rectified* sine wave. Prove that the dc value V_{dc} and rms value V_{rms} are

$$V_{dc} = \frac{V_p}{\pi} = 0.318V_p$$

and

$$V_{rms} = \frac{V_p}{2} = 0.5V_p$$

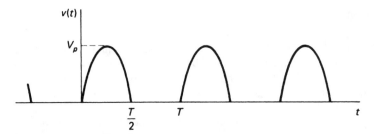

FIGURE P10–21

10–22. The waveform of Figure P10–22 is a *full-wave rectified* sine wave. Prove that the dc value V_{dc} and rms value V_{rms} are

$$V_{dc} = \frac{2V_p}{\pi} = 0.637V_p$$

and

$$V_{rms} = \frac{V_p}{\sqrt{2}} = 0.707V_p$$

(*Hint:* Although the original period of the sine wave was T, a slight reduction in the analysis may be made by recognizing that the actual period of the rectified waveform is $T/2$.)

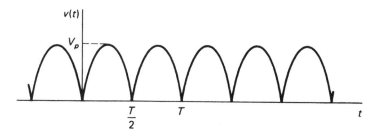

FIGURE P10–22

10–23. The waveform of Figure P10–23 is a *triangular wave*. Prove that the dc value V_{dc} and rms value V_{rms} are

$$V_{dc} = \frac{V_p}{2} = 0.5 V_p$$

$$V_{rms} = \frac{V_p}{\sqrt{3}} = 0.577 V_p$$

(Note that these are the same results as for the sawtooth waveform of Example 10–3.)

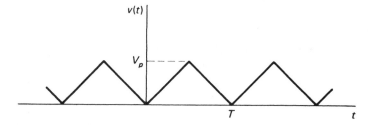

FIGURE P10–23

10–24. The waveform of Figure P10–24 is a *pulse train* with *duty cycle d*, where *d* is defined as $d = \tau/T$. Prove that the dc value V_{dc} and rms value V_{rms} are

$$V_{dc} = d V_p$$
$$V_{rms} = \sqrt{d} V_p$$

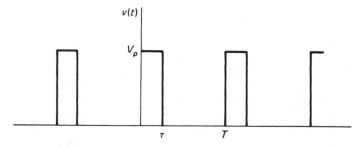

FIGURE P10–24

10–25. Assume that a given voltage $v(t)$ is composed of a sine wave at a frequency ω plus a sine wave at a frequency 2ω as follows:

$$v(t) = V_{p1} \sin \omega t + V_{p2} \sin 2\omega t$$

Prove that the rms value V_{rms} is given by

$$V_{rms} = \sqrt{V_{rms1}^2 + V_{rms2}^2}$$

where $V_{rms1} = V_{p1}/\sqrt{2}$ and $V_{rms2} = V_{p2}/\sqrt{2}$. *Hint:* Return to the basic definition and perform the integration. Make use of the identity

$$\sin A \sin B = \tfrac{1}{2} \cos (A - B) - \tfrac{1}{2} \cos (A + B)$$

10–26. Assume that a given voltage $v(t)$ is composed of a sine function plus a cosine function at a given frequency ω, i.e.,

$$v(t) = V_{ps} \sin \omega t + V_{pc} \cos \omega t$$

Prove that the rms value V_{rms} is given by

$$V_{rms} = \sqrt{V_{rmss}^2 + V_{rmsc}^2}$$

where $V_{rmss} = V_{ps}/\sqrt{2}$ and $V_{rmsc} = V_{ps}/\sqrt{2}$.

10–27. The input–output relationship of an ideal delay line is given by

$$y(t) = x(t - T)u(t - T)$$

where $x(t)$ is the input, $y(t)$ is the output, and T is the delay time. **(a)** Derive the s-domain transfer function $G(s)$ of the ideal transfer function, i.e.,

$$G(s) = \frac{Y(s)}{X(s)}$$

(b) Determine the steady-state transfer function $G(j\omega)$. **(c)** Determine the amplitude response $A(\omega)$ and the phase response $\beta(\omega)$. (*Hint:* The notation $A(\omega)\underline{/\beta(\omega)}$ is a shortcut notation for $A(\omega)e^{j\beta(\omega)}$.) **(d)** What are the practical implications of the results of (c)?

10–28. Prove the following statements: **(a)** To establish an instantaneous step of voltage across a capacitor, an impulse of current would be required. **(b)** To establish an instantaneous step of current in an inductor, an impulse of voltage would be required.

APPLICATION PROBLEMS

10–29. The RC coupling network shown in Figure P10–29(a) can be used to extract the dc component of a periodic signal while coupling the time-varying component to the output. One way to analyze this effect is to recognize that for any periodic signal, the steady-state average capacitor current must be zero. Since this current flows through R, the average output voltage must be zero. The average capacitor voltage must then be the same as the average value of the input voltage. For minimum distortion of the time-varying component, the product RC must be very large compared with the period. Under this condition, the capacitor voltage v_c will be nearly

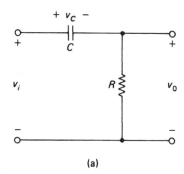

(a)

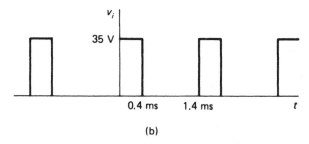

(b)

FIGURE P10-29

a constant value equal to the average input voltage, and the output $v_o = v_i - v_c$ will be the time-varying component of the input.

For the input waveform of Figure P10-29(b), sketch the forms of v_c and v_o, and label pertinent levels. Assume a very large time constant, as discussed.

10-30. When a periodic signal is being observed on an oscilloscope with the vertical channel set for *dc coupling*, the entire signal is displayed at its level relative to the baseline trace. However, when the oscilloscope is set to *ac coupling*, the signal is coupled through an *RC* circuit like that of Problem 10-29. The output then shifts in a direction such that its dc component is reduced to zero. This allows the time-varying component to be amplified without the disturbing presence of the dc level.

Assume that the presentation shown in Figure P10-30 is being observed with dc coupling. **(a)** When the oscilloscope is switched to ac coupling, by how many

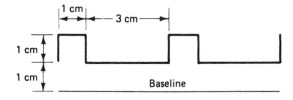

Vertical sensitivity   = 5 V/cm
Horizontal sensitivity = 20 ms/cm

FIGURE P10-30

centimeters and in which direction does the signal shift? **(b)** Sketch the form of the signal about the baseline after the shift.

10–31. Many traditional voltmeters and ammeters measure sinusoidal ac waveforms by the method to be described here. The basic dc meter movement responds to the average or dc value of the waveform being observed. Since the average value of a sinusoid is zero, such a meter will not respond to ac. However, if the ac is first rectified, the result will have a dc component, and the meter will respond to it. The meter can then be calibrated to read a property of the ac signal, for example, the rms value. The rms value of a sine wave will then be correct as read from the calibrated scale. However, the rms value will generally not be correct for waveforms other than a sine wave. (A true rms meter would be required in that case.)

Either half-wave or full-wave rectification could be used. In the remainder of this problem, full-wave rectification will be assumed, and the results of Problem 10–22 can be used as required.

Assume that a given dc voltmeter is to be modified so that ac voltages can be read. Assume that an ideal full-wave rectifier is employed. For this purpose, neglect diode drops. Assume that a new ac scale is to be placed above the dc scale with a scaling factor S defined by

$$V_{rms} = SV_{dc}$$

Thus, if S were 1.5, a dc value $V_{dc} = 10$ V would correspond to a reading $V_{rms} = 15$ V on the ac scale. Determine the scale factor S with full-wave rectification.

10–32. Repeat Problem 10–31 with half-wave rectification. You may employ the results of Problem 10–21.

10–33. Determine the form factor for the full-wave rectified sine wave of Problem 10–22.

10–34. Determine the form factor for the half-wave rectified sine wave of Problem 10–21.

11

FOURIER ANALYSIS

OBJECTIVES

After completing this chapter, the reader should be able to:

- Determine the values of the fundamental frequency and the harmonic frequencies contained in a periodic signal.
- State the three forms of the Fourier series.
- For a given periodic signal, determine the three forms of the Fourier series.
- Convert among the various forms of the Fourier series.
- Construct both one-sided and two-sided frequency spectrum plots from a given Fourier series.
- Inspect a given periodic signal to determine any possible symmetry conditions.
- Indicate the implications of any symmetry conditions on a periodic signal in terms of the resulting spectral content and simplification of the spectrum computation.
- State the definitions of the Fourier transform and inverse transform.
- Discuss the difference between the Fourier series and transform in terms of applicable time functions and the resulting spectral properties.
- Inspect a given nonperiodic function to determine any possible symmetry condition.
- Indicate the implications of any symmetry condition in a nonperiodic function in terms of the resulting spectral content and simplification of the spectrum computation.
- Discuss the various Fourier transform operation pairs and the effect on the spectrum in each case.
- Determine the roll-off rate of the spectrum of a given time function from a knowledge of the relative continuity of the function and its derivatives.

11–1 FOURIER SERIES

The concept of the Fourier series is based on representing a periodic signal as the sum of harmonically related sinusoidal functions. As we are already aware, sinusoidal functions are among the most important waveforms that arise in electrical systems. With the aid of Fourier series, complex periodic waveforms can be represented in terms of sinusoidal functions, whose properties are familiar to us. Thus, the response of a system to a complex waveform can be viewed as the response to a series of sinusoidal functions.

Without realizing it, the reader may already be familiar with some aspects of Fourier theory through interests in audio or stereo systems. Consider an audio amplifier in which the amplitude response is specified as flat from, say, 20 Hz to 20 kHz. This specification is based on a test of the amplitude response using a sinusoidal input as discussed in Chapter 9. A sinusoidal function certainly does not represent a signal of any interest for listening. However, the use of the sinusoid for testing is based on the fact that a complex signal can be represented as a combination of sinusoids. By studying the response of the system to individual frequencies, it is possible to infer properties of the system for complex waveforms such as music.

Periodic Function

The Fourier series is best applied to a periodic function, and the treatment in this text will be limited to that case. Consider the arbitrary periodic signal $x(t)$ shown in Fig. 11–1. The function could represent either a voltage or a current waveform. [The symbol $f(t)$ is not used here since f is the designation for frequency.] According to Fourier theory, this signal may be represented by the sum of a series of sine and/or cosine functions plus a dc term. The resulting series is called a *Fourier series.*

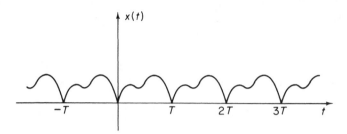

FIGURE 11–1
Example of an arbitrary periodic signal for which a Fourier series representation could be obtained.

Fundamental Frequency

The lowest frequency (other than dc) of the sinusoidal components is a frequency f_1 given by

$$f_1 = \frac{1}{T} \tag{11-1}$$

This frequency is referred to as the *fundamental* component, and it is the same as that of the waveform itself. Thus, a periodic signal with a period of 1 millisecond (ms) will have a fundamental component with a frequency $f_1 = 1$ kHz.

Harmonics

All other frequencies in the signal will be integer multiples of the fundamental. These various components are referred to as *harmonics,* with the *order* of a given harmonic indicated by the ratio of its frequency to the fundamental frequency. Thus, for the waveform suggested in the preceding paragraph, frequencies appearing in the signal above 1 kHz would, in general, be a second harmonic at 2 kHz, a third harmonic at 3 kHz, a fourth harmonic at 4 kHz, and so on.

Bandwidth

Any transmission system through which a given signal passes must have a bandwidth sufficiently large to pass all significant frequencies of the signal. In a purely mathematical sense, many common waveforms theoretically contain an infinite number of harmonics. The preceding two sentences would then lead one to believe that an infinite bandwidth would be required to process such signals, which is impossible. The key to this apparent contradiction is the term *significant* in the first sentence.

It is easy to predict the theoretical frequencies present in a given periodic signal, as we have just seen. However, it is much more difficult to predict the relative magnitudes of the components in order to determine which of them are significant and which are not. From a signal transmission point of view, if we can predict the frequency range over which the magnitudes are significant in size, we can then estimate the bandwidth requirements. In this sense, we simply ignore all harmonics of the signal that are outside the range and assume that their exclusion produces no noticeable degradation in the signal quality.

We will now consider the mathematical structure of the Fourier series. There are three forms of the Fourier series. They are (a) the *sine–cosine* form, (b) the *amplitude-phase* form, and (c) the *complex exponential* form. Each form will be considered individually.

Sine–Cosine Form

The sine–cosine form is the one most commonly presented first in circuits and mathematics texts. This form represents a periodic signal $x(t)$ as a sum of sines and cosines in the form

$$x(t) = A_0 + \sum_{n=1}^{\infty} (A_n \cos n\omega_1 t + B_n \sin n\omega_1 t) \qquad \textbf{(11–2)}$$

where

$$\omega_1 = 2\pi f_1 = \frac{2\pi}{T} \qquad \textbf{(11–3)}$$

is the fundamental *angular* frequency in radians per second (rad/s). The *n*th harmonic *cyclic* frequency (in hertz) is nf_1, and the corresponding angular frequency is $n\omega_1$. As a result of the equalities in Equation (11–3), the argument for either the sine or cosine function in Equation (11–2) can be expressed in either of the following forms:

$$n\omega_1 t = 2\pi n f_1 t = \frac{2\pi n t}{T} \qquad \textbf{(11–4)}$$

The term A_0 represents the dc term and, as we will see shortly, it is simply the average value of the signal over one cycle. (*Note*: Some authors define the dc value as $A_0/2$ in order to make some of the later formulas apply to this case, but the A_0 form is easier to interpret.) Other than dc, there are two components appearing at a given harmonic frequency in the most general case: a cosine term with an amplitude A_n, and a sine term with an amplitude B_n.

The major task involved in a Fourier analysis is in determining the A_n and B_n coefficients. The dc term A_0 is simply the average value and is given by

$$A_0 = \frac{1}{T} \int_0^T x(t)\, dt = \frac{\text{area under curve in one cycle}}{\text{period } T} \qquad \textbf{(11–5)}$$

as discussed in Section 10–1. Formulas for A_n and B_n are derived in applied mathematics books, and are summarized as follows:

$$A_n = \frac{2}{T} \int_0^T x(t) \cos n\omega_1 t\, dt \qquad \textbf{(11–6)}$$

for $n \geq 1$ but *not* for $n = 0$.

$$B_n = \frac{2}{T} \int_0^T x(t) \sin n\omega_1 t\, dt \qquad \textbf{(11–7)}$$

for $n \geq 1$. In all three of the preceding integrals, the limits of integration may be changed for convenience, provided that the interval of integration is over one complete cycle in a positive sense. A common alternative range is from $-T/2$ to $T/2$.

The formulas for A_n and B_n indicate that the quantities are determined by first multiplying the signal by a cosine or sine term of the corresponding frequency at which the coefficient is desired, and then determining the area of the resulting product function over one cycle. In many cases, a general expression for A_n or B_n can be determined from one integration, with n appearing as a parameter. In other cases, particularly for experimental data, a separate integration may be required at each frequency. Computational means have been developed for performing such operations efficiently on a computer, and many scientific computer systems have programs available for determining the Fourier series of experimental data signals. If a signal $x(t)$ is defined in different forms over different parts of a cycle, the evaluation of either of the integrals may require expansion into several integrals over shorter portions of the cycle.

Amplitude-Phase Form

The sine–cosine form of the Fourier series given in Equation (11–2) is usually the easiest form from which to evaluate the coefficients, and it is the form most commonly tabulated in reference books. However, it suffers from the fact that there are, in a sense, two separate components at a given frequency, each of which has a separate amplitude. When we measure the magnitude of a given spectral component with a frequency-selective instrument, which component do we obtain? The fact is that the actual magnitude that would be measured with most instruments would be neither A_n nor B_n, but rather a special combination of the two, as will be seen shortly.

The amplitude-phase form of the Fourier series is based on the concept developed in Section 8–3 in which the sum of two or more sinusoids of a given frequency is equivalent to a single sinusoid at the same frequency. Specifically, the sine–cosine form involves a single sine function plus a single cosine function at each frequency in the series. The approach in Section 8–3 emphasized expressing the sum as a sine function, since that form is usually more convenient in analyzing circuit problems. Traditionally, the cosine function has probably been used more in Fourier analysis, so we will consider both variations in this development.

The amplitude-phase form of the Fourier series can be expressed as either

$$x(t) = C_0 + \sum_{n=1}^{\infty} C_n \cos(n\omega_1 t + \phi_n) \tag{11–8}$$

or

$$x(t) = C_0 + \sum_{n=1}^{\infty} C_n \sin(n\omega_1 t + \theta_n) \tag{11–9}$$

The amplitudes of corresponding components in either of the preceding variations are the same, but the phase angles ϕ_n and θ_n differ, since one representation involves cosine functions and the other involves sine functions. The first term C_0 is the dc value and is the same as given by Equation (11–5); that is, $C_0 = A_0$. It has been redefined here as C_0 in order to maintain a consistent form of notation.

In the form of either Equation (11–8) or Equation (11–9), a given C_n represents the *net* amplitude of a given component at the frequency nf_1. Since sine and cosine phasor forms are always perpendicular to each other, the net amplitude is

$$C_n = \sqrt{A_n^2 + B_n^2} \tag{11-10}$$

The angles ϕ_n or θ_n are best determined from the procedures developed in Sections 8–2 and 8–3. The quantity θ_n represents the angle measured from the positive sine axis, and ϕ_n represents the angle measured from the positive cosine axis. If the exact phasor form established in those sections is used, θ_n will represent the exact phasor angle, whereas ϕ_n would be determined by adding $-90°$ to the exact phasor angle based on the sine reference.

We return now to the question of which component is measured with a frequency-selective instrument. With the majority of common instruments of this type, the reading is proportional to C_n, the net amplitude. (The instrument may actually be calibrated to read the rms value of C_n, which is $C_n/\sqrt{2}$.) There are, however, certain special phase-sensitive instruments that can be used to obtain A_n and B_n separately, so the peculiarities of such instruments should be understood before measurements are made.

There are a number of cases of well-known waveforms in which either A_n or B_n (but obviously not both) is identically zero at all possible frequencies for the signal. Indeed, as we will see later, it is possible in some cases to choose the time origin in a way that will force this result. With such signals, the sine–cosine form is identical with one of the variations of the amplitude-phase form.

Complex Exponential Form

The complex exponential form of the Fourier series is the most difficult form for many persons to perceive, primarily because it represents a step away from the domain of real signals into a domain of complex mathematical representations. However, there are some developments that can be done much more easily with the exponential form than with either of the earlier forms. In particular, the exponential form is a direct link with the concept of the Fourier transform, which is very important in dealing with nonperiodic signals, as we will see later.

The exponential form of the Fourier series is related to the fact that both the sine and cosine functions can be expressed in terms of exponential functions with purely imaginary arguments. The basis for this is Euler's formula, which is written in two separate forms as

$$e^{jn\omega_1 t} = \cos n\omega_1 t + j \sin n\omega_1 t \tag{11-11}$$

$$e^{-jn\omega_1 t} = \cos n\omega_1 t - j \sin n\omega_1 t \tag{11-12}$$

Alternate addition and subtraction of Equations (11–11) and (11–12) result in the following two expressions for cosine and sine functions:

$$\cos n\omega_1 t = \frac{e^{jn\omega_1 t} + e^{-jn\omega_1 t}}{2} \tag{11-13}$$

$$\sin n\omega_1 t = \frac{e^{jn\omega_1 t} - e^{-jn\omega_1 t}}{2j} \tag{11-14}$$

Note that both forms contain an exponential function with a $(jn\omega_1 t)$ argument and an exponential function with a $(-jn\omega_1 t)$ argument. The first term may be thought of as a "positive frequency" term corresponding to a frequency nf_1 (assuming n is positive), and the second term may be considered as a "negative frequency" term corresponding to a frequency $-nf_1$. Both terms are required to completely describe the sine or cosine function.

The exponential form can be developed by expanding the sine and cosine functions according to these exponential definitions and regrouping. This general process is somewhat detailed, and will not be given here. We will concentrate here on the results and the corresponding interpretations.

The general form of the complex exponential form of the Fourier series can be expressed as

$$x(t) = \sum_{n=-\infty}^{\infty} \overline{X}_n e^{jn\omega_1 t} \tag{11–15}$$

where the bar above $\overline{X}_n$ indicates that it is, in general, a complex value. The Fourier coefficient at a given frequency nf_1 is the complex quantity $\overline{X}_n$. An expression for determining $\overline{X}_n$ is

$$\overline{X}_n = \frac{1}{T} \int_0^T x(t) e^{-jn\omega_1 t}\, dt \tag{11–16}$$

Some interpretation of the preceding results is in order. Note in Equation (11–15) that the exponential series is summed over both negative and positive frequencies (or negative and positive values of n), as previously discussed. At a given real frequency kf_1, $(k > 0)$, the spectral representation consists of

$$\overline{X}_k e^{jk\omega_1 t} + \overline{X}_{-k} e^{-jk\omega_1 t}$$

The first term is thought of as the "positive frequency" contribution, whereas the second is the corresponding "negative frequency" contribution. Although either one of the two terms is a complex quantity, they add together in such a manner as to create a real function, and this is why both terms are required to make the mathematical form complete. On the other hand, all the spectral information can be deduced from either the $\overline{X}_k$ term or the $\overline{X}_{-k}$ term, since there is a direct relationship between them. Let $\overline{\overline{X}}$ represent the complex conjugate of $\overline{X}$. Then it can be shown that

$$\overline{X}_{-n} = \overline{\overline{X}}_n \tag{11–17}$$

Thus, the negative frequency coefficient is the complex conjugate of the corresponding positive frequency coefficient.

Although the coefficient $\overline{X}_n$ can be calculated from Equation (11–16) directly, it turns out that $\overline{X}_n$ can also be calculated directly from A_n and B_n of the sine–cosine form. The relationship reads

$$\overline{X}_n = \frac{A_n - jB_n}{2}, \quad \text{for } n \neq 0 \tag{11–18}$$

Even though A_n and B_n are interpreted only for positive n in the sine–cosine form, their functional forms may be extended for both positive and negative n in applying Equation (11–18). Alternatively, Equation (11–18) may be applied for positive n, and Equation (11–17) may be used to determine the corresponding coefficients for negative n. The dc component $\overline{X}_0$ is simply

$$\overline{X}_0 = \frac{1}{T}\int_0^T x(t)\, dt = A_0 = C_0 \tag{11–19}$$

which is the same in all the Fourier forms.

In many situations involving the complex form of the Fourier series, it is desirable to express $\overline{X}_n$ as a magnitude and an angle. Since $\overline{X}_n$ is a complex quantity having a real and an imaginary part, this can be readily achieved. Thus, $\overline{X}_n$ can be expressed in polar form as

$$\overline{X}_n = X_n e^{j\phi n} = X_n\,\underline{/\phi_n} \tag{11–20}$$

The quantity $X_n = |\overline{X}_n|$ represents the magnitude of the complex Fourier coefficient at a frequency nf_1, and ϕ_n is the corresponding angle or phase expressed in the polar form. For example, if $\overline{X}_n = 3 + j4$ at a given frequency, it can also be expressed as $\overline{X}_n = 5\,\underline{/53.13°}$ with $X_n = 5$ and $\phi_n = 53.13°$.

EXAMPLE 11–1

This example has as a primary objective the conversion between different forms of the Fourier series. A certain periodic bandlimited signal has only three frequencies in its Fourier series representation: dc, 1 kHz, and 2 kHz. The signal can be expressed in sine–cosine form as

$$\begin{aligned} x(t) = 18 &+ 40\cos 2000\pi t - 30\sin 2000\pi t \\ &- 24\cos 4000\pi t + 10\sin 4000\pi t \end{aligned} \tag{11–21}$$

Express the signal in **(a)** amplitude-phase form and **(b)** complex exponential form.

Solution

(a) The amplitude-phase form desired for the signal reads

$$x(t) = 18 + C_1\cos(2000\pi t + \phi_1) + C_2\cos(4000\pi t + \phi_2) \tag{11–22}$$

or

$$x(t) = 18 + C_1\sin(2000\pi t + \theta_1) + C_2\sin(4000\pi t + \theta_2) \tag{11–23}$$

where $\overline{C}_0 = 18$ has been noted by obvious inspection.

The amplitude and angles are determined from the phasor forms. Let $\overline{C}_1$ represent the phasor associated with 1000 Hz. Using the sine function as the basis as established in Chapter 2, $\overline{C}_1$ is shown in Figure 11–2(a), and it can be expressed as

$$\overline{C}_1 = -30 + j40 = 50\,\underline{/126.87°} \tag{11–24}$$

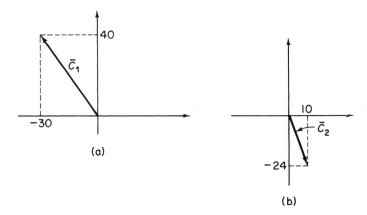

FIGURE 11–2
Phasor diagrams for Example 11–1.

Let $\overline{C}_2$ represent the corresponding phasor associated with 2000 Hz as shown in Figure 11–2(b). We have

$$\overline{C}_2 = 10 - j24 = 26 \,\underline{/-67.38°} \qquad \textbf{(11–25)}$$

The angles associated with Equations (11–24) and (11–25) are θ_1 and θ_2, respectively. The angles $\phi_1 = 126.87° - 90° = 36.87°$, and $\phi_2 = -67.38° - 90° = -157.38°$. The two variations of the amplitude-phase form are

$$x(t) = 18 + 50\cos(2000\pi t + 36.87°) + 26\cos(4000\pi t - 157.38°) \qquad \textbf{(11–26)}$$

and

$$x(t) = 18 + 50\sin(2000\pi t + 126.87°) + 26\sin(4000\pi t - 67.38°) \qquad \textbf{(11–27)}$$

(b) The complex exponential form will be determined from Equation (11–18) in conjunction with the conjugate relationship of Equation (11–17). Thus,

$$\overline{X}_1 = \frac{40 - j(-30)}{2} = 20 + j15 = 25 \,\underline{/36.87°} \qquad \textbf{(11–28)}$$

$$\overline{X}_{-1} = \widetilde{\overline{X}}_1 = 20 - j15 = 25 \,\underline{/-36.87°} \qquad \textbf{(11–29)}$$

$$\overline{X}_2 = \frac{-24 - j10}{2} = -12 - j5 = 13 \,\underline{/-157.38°} \qquad \textbf{(11–30)}$$

$$\overline{X}_{-2} = \widetilde{\overline{X}}_2 = -12 + j5 = 13 \,\underline{/157.38°} \qquad \textbf{(11–31)}$$

The series may then be expressed as

$$\begin{aligned} x(t) = 18 &+ (20 + j15)e^{j2000\pi t} + (20 - j15)e^{-j2000\pi t} \\ &+ (-12 - j5)e^{j4000\pi t} + (-12 + j5)e^{-j4000\pi t} \end{aligned} \qquad \textbf{(11–32)}$$

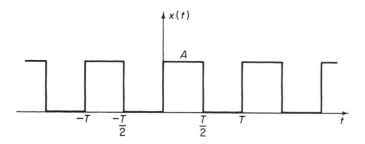

FIGURE 11–3
Waveform for Example 11–2.

in which the coefficients have been expressed in rectangular forms. Alternatively, the polar forms of the coefficients may be used in the expansion, in which the expression becomes

$$x(t) = 18 + 25e^{j(2000\pi t + 36.87°)} + 25e^{-j(2000\pi t + 36.87°)}$$
$$+ 13e^{j(4000\pi t - 157.38°)} + 13e^{-j(4000\pi t - 157.38°)} \tag{11–33}$$

EXAMPLE 11–2
Determine the Fourier series representation for the waveform shown in Fig. 11–3. Express in each of the following forms: **(a)** sine–cosine, **(b)** amplitude-phase, and **(c)** complex exponential.

Solution
(a) The sine–cosine form is usually the easiest form for determining the coefficients directly. The signal $x(t)$ over one complete cycle can be expressed as

$$x(t) = \begin{cases} A & \text{for } 0 < t < T/2 \\ 0 & \text{for } T/2 < t < T \end{cases} \tag{11–34}$$

The dc component A_0 can be determined by inspection, since it is the average value. Thus,

$$A_0 = \frac{\text{area under curve in one cycle}}{T} = \frac{AT/2}{T} = \frac{A}{2} \tag{11–35}$$

The coefficients A_n can be determined from Equation (11–6). Note that, although integration over a complete cycle is required, the function is zero over half of the cycle, so the integral reduces to

$$A_n = \frac{2}{T} \int_0^{T/2} A \cos n\omega_1 t \, dt$$

$$= \frac{2A}{n\omega_1 T} \sin n\omega_1 t \Big]_0^{T/2} = \frac{2A}{n\omega_1 T} \left(\sin \frac{n\omega_1 T}{2} - 0 \right) \tag{11–36}$$

In performing Fourier coefficient evaluations, the product $n\omega_1 T$ appears in virtually every case. When this occurs, the substitution $n\omega_1 T = 2\pi n$ is recommended to the reader as a possible simplification. This equality is readily verified by noting that $n\omega_1 T = n2\pi f_1 T = n2\pi (1/T)T = 2\pi n$. This substitution will be made in various examples that follow without further justification.

The preceding evaluation now reduces to

$$A_n = \frac{2A}{2\pi n} \sin n\pi = 0 \qquad \text{for } n \neq 0 \tag{11-37}$$

since $\sin n\pi = 0$ for n, an integer. Thus, in this particular example, the coefficients of all the cosine terms in the sine–cosine form are zero (except dc).

The B_n coefficients are now determined from the following development:

$$B_n = \frac{2}{T} \int_0^{T/2} A \sin n\omega_1 t \, dt$$

$$= \frac{-2A}{n\omega_1 T} \cos n\omega_1 t \Big]_0^{T/2} \tag{11-38}$$

$$= \frac{-2A}{2\pi n} \left(\cos \frac{n\omega_1 T}{2} - 1 \right) = \frac{A}{n\pi} (1 - \cos n\pi)$$

The quality $\cos n\pi$ satisfies

$$\cos n\pi = \begin{cases} -1 & \text{for } n \text{ odd} \\ +1 & \text{for } n \text{ even} \end{cases} \tag{11-39}$$

When n is even, there is a cancellation inside the parentheses of the last term in Equations (11–38), whereas for n odd, the term $-(-1) = +1$ adds up to the other term. Thus, the expression for B_n reduces to

$$B_n = \begin{cases} \dfrac{2A}{n\pi} & \text{for } n \text{ odd} \\ 0 & \text{for } n \text{ even} \end{cases} \tag{11-40}$$

The resulting sine–cosine form of the Fourier series representation of $x(t)$ can be expressed as

$$x(t) = \frac{A}{2} + \frac{2A}{\pi} \sin \omega_1 t + \frac{2A}{3\pi} \sin 3\omega_1 t$$

$$+ \frac{2A}{5\pi} \sin 5\omega_1 t + \frac{2A}{7\pi} \sin 7\omega_1 t + \cdots \tag{11-41a}$$

$$= \frac{A}{2} + \sum_{\substack{n=1 \\ n \text{ odd}}}^{\infty} \frac{2A}{n\pi} \sin n\omega_1 t \tag{11-41b}$$

(b) Since the A_n coefficients in the sine–cosine form are zero, the sine–cosine series in this case is identical to one form of the amplitude-phase Fourier series. Specifically, the result of (a) is identical to the series form of Equation (11–9) with $\theta_n = 0$.

(c) The complex exponential form will now be developed. Although we could determine the coefficients $\bar{X}_n$ directly from the sine–cosine terms making use of Equation (11–18), it will be more instructive for the reader's sake to start over again using the defining relationship for $\bar{X}_n$ as given by Equation (11–16). First, we note that the dc component is again

$$\bar{X}_0 = A_0 = \frac{A}{2} \tag{11–42}$$

The general coefficient $\bar{X}_n$ is given by

$$\bar{X}_n = \frac{1}{T}\int_0^{T/2} A e^{-jn\omega_1 t}\,dt = \frac{-A}{jn\omega_1 T}e^{-jn\omega_1 t}\Bigg]_0^{T/2}$$

$$= \frac{-A}{j2n\pi}(e^{-jn\omega_1 T/2} - 1) = \frac{A}{j2n\pi}(1 - e^{-jn\pi}) \tag{11–43}$$

$$= \frac{A}{j2n\pi}(1 - \cos n\pi + j\sin n\pi)$$

where Euler's formula was used in the last step. This result can be readily simplified by first noting that $\sin n\pi = 0$ for n, an integer. Furthermore, $1 - \cos n\pi = 2$ for n odd, and $1 - \cos n\pi = 0$ for n even. Thus, $\bar{X}_n$ reduces to

$$\bar{X}_n = \frac{A}{jn\pi} = \frac{-jA}{n\pi} \qquad \text{for } n \text{ odd} \tag{11–44}$$

The reader can readily verify that this same result is quickly obtained by applying Equation (11–18) to the previous results of the sine–cosine form.

The complex Fourier series for $x(t)$ can now be expressed as

$$x(t) = \frac{A}{2} - j\frac{A}{\pi}e^{j\omega_1 t} - j\frac{A}{3\pi}e^{j3\omega_1 t} - \cdots$$

$$+ j\frac{A}{\pi}e^{-j\omega_1 t} + j\frac{A}{3\pi}e^{-j3\omega_1 t} + \cdots \tag{11–45}$$

11–2 FREQUENCY SPECTRUM PLOTS

One of the most useful forms for displaying the Fourier series of a signal is by means of a graphical plot showing the relative strengths of the components as a function of frequency. Such a plot is loosely referred to as the *frequency spectrum* of the given signal. A frequency spectrum plot permits a quick visual determination of the frequencies present in the signal and their relative magnitudes. This is the basis for the spectrum analyzer, which is a widely used instrument in communications systems work providing a cathode-ray tube (CRT) display of the spectrum in much the same form as the graphical technique that will be discussed in this section.

In principle, any of the three forms could be shown graphically. In practice, however, the sine–cosine form is less desirable for this purpose since both A_n and

B_n would have to be plotted separately (unless, of course, one of the two is zero). For the same reason, spectral displays involving the complex exponential form generally focus on the polar form (magnitude and angle) rather than the real and imaginary representation. Consequently, we focus on the amplitude-phase form and the magnitude and angle representation of the complex exponential form in our developments.

Both the amplitude-phase form and the complex exponential form have two quantities to be specified at each frequency (i.e., the amplitude and phase of the spectral components). Although both quantities are necessary for mathematical reconstruction of the signal, in practical spectral displays, the amplitude is almost always the quantity that is emphasized. The reason for this is the simple fact that the relative amplitudes of spectral components are most significant in determining the bandwidth, whereas the phase indicates only the relative time shift of a given component relative to others. In fact, a simple time shift of the signal will readily change the phase terms without affecting the amplitude terms. The amplitude form will be referred to as an *amplitude frequency-spectrum plot*.

One point about the terms *amplitude* and *magnitude* should be noted. By convention, these terms normally indicate a positive, real value. There are instances, however, when it is convenient to allow a given amplitude or magnitude spectrum to assume negative real values, and this concept is used later in the chapter. This is done only as a convenience in simplifying the mathematical form of the function. In the sense of complex numbers, this is permissible provided that the phase is adjusted accordingly. For example, $5 \underline{/-150°}$ can be expressed as $-5 \underline{/30°}$. The first form involves a "positive amplitude," and the second involves a "negative amplitude," but the two results are identical in the sense of complex numbers. Thus, the reader should not be disturbed by negative values appearing in some of the amplitude spectra.

As a result of the discussion thus far, it would seem that a frequency-spectrum plot would most likely consist of a graph of either C_n from the amplitude-phase form, or X_n from the complex exponential form. Recall that the C_n terms are defined only for $n \geq 0$, whereas the X_n terms are defined for both positive and negative n (as well as $n = 0$). For this reason, the plot of C_n as a function of frequency is called a *one-sided spectrum*, whereas the plot of X_n as a function of frequency is called a *two-sided spectrum*. The terms *one-sided* and *two-sided* will be used extensively, since they are easy to remember and quickly alert the reader to the required mathematical form.

Referring back to Equation (11–18) momentarily and evaluating the magnitude of X_n, the result obtained is

$$X_n = |\overline{X}_n| = \sqrt{\frac{A_n^2 + B_n^2}{4}} = \frac{\sqrt{A_n^2 + B_n^2}}{2} \qquad \text{for } n \neq 0 \qquad \textbf{(11–46)}$$

Comparing this result with Equation (11–10), we see that

$$X_n = \frac{C_n}{2} \qquad \text{for } n \neq 0 \qquad \textbf{(11–47)}$$

and, of course, $X_0 = C_0$ as previously noted.

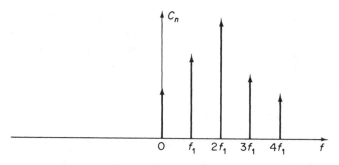

FIGURE 11–4
Typical one-sided amplitude-frequency spectrum.

The result of Equation (11–47) may clear away at least a portion of the mystery surrounding the complex exponential form of the Fourier series. On an amplitude-spectrum basis, the magnitudes of the components in the two-sided spectral form (except for dc) are exactly one-half the values in the one-sided form. An artificial, but rather easy, way to remember this is that, when the terms are displayed on both sides, the one-sided terms are "cut in half" in order to provide the components for the other side. The dc term appears in only one place, so its value does not change.

When a signal is described as a function of time, that is, as $x(t)$, the result is said to be a *time-domain* representation. Conversely, when the spectral information is provided, the result is said to be a *frequency-domain* representation. Either form completely describes the signal, provided that both amplitude and phase are given in the frequency domain.

A typical example of a one-sided frequency spectrum is shown in Figure 11–4. The corresponding two-sided spectrum for the same signal is shown in Figure 11–5. Note that the lengths of all the components except dc in the second case are half the values in the first case, as expected. The dc component, however, is the same in both cases.

An important summary point concerning the frequency spectrum will now be

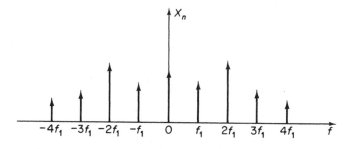

FIGURE 11–5
Two-sided amplitude spectrum corresponding to Figure 11–4.

made: The frequency spectrum of a periodic signal is a *discrete* or *line* spectrum; that is, it contains components only at integer multiples of the repetition frequency of the signal (including dc).

EXAMPLE 11–3

Consider the bandlimited signal of Example 11–1. Plot **(a)** one-sided and **(b)** two-sided frequency spectra for this signal.

Solution

(a) The one-sided spectrum is obtained from the amplitude-phase form of the series, which was developed in Equations (11–26) and (11–27). The amplitudes of the components on a one-sided basis and their frequencies are summarized as follows:

Frequency (Hz)	0	1000	2000
Amplitude	18	50	26

The one-sided plot is shown in Figure 11–6(a).

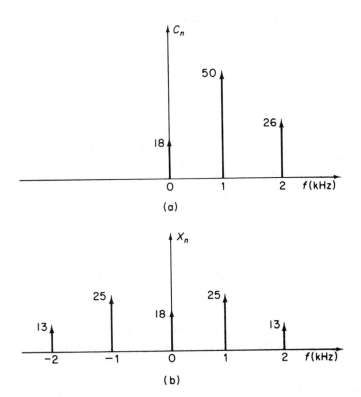

(a)

(b)

FIGURE 11–6
One-sided and two-sided amplitude-frequency spectra of Examples 11–1 and 11–3.

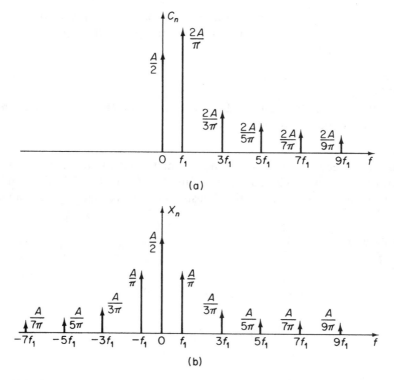

FIGURE 11-7
One-sided and two-sided amplitude-frequency spectra of square wave (see Example 11-4).

(b) Whereas the two-sided plot could be readily determined directly from the one-sided plot, for convenience, the results will be tabulated again. These results could be deduced from Equation (11-47) or from the expansion of Equation (11-33). The results are summarized as follows:

Frequency (Hz)	0	±1000	±2000
Amplitude	18	25	13

The two-sided plot is shown in Figure 11-6(b).

EXAMPLE 11-4
Plot **(a)** the one-sided amplitude-frequency spectrum and **(b)** the two-sided amplitude-frequency spectrum for the square-wave signal considered in Example 11-2.

Solution
(a) The one-sided form is determined by noting the coefficients in Equation (11-41a). The one-sided spectrum out to the ninth harmonic is shown in Figure 11-7(a).

(b) The two-sided form can be determined either directly from the one-sided form or from the results of Equation (11–45). The form of the two-sided spectrum is shown in Figure 11–7(b).

11–3 FOURIER SERIES SYMMETRY CONDITIONS

The various equations developed in the preceding several sections may, in theory, be applied to any signal to determine its spectrum. On the other hand, there are certain properties that may be used to simplify the computation of the spectrum, in many cases. Furthermore, some of these conditions permit many important properties of certain waveforms to be obtained by simple inspection procedures. For some applications, this information might be sufficient without having to compute the spectrum at all. In most cases, the use of these conditions will at least provide some information about the spectrum from a direct inspection.

The types of criteria to be studied in this section are the *symmetry* conditions. All the various symmetry conditions to be considered in this book are summarized in Table 11–1. The proofs of these conditions are given in various texts on applied mathematics, and some are given as exercises at the end of this chapter. Our emphasis is on the practical interpretation and application of these various properties.

The equations at the top of the table are the various forms of the Fourier series that were developed in the preceding sections, along with the relationships for converting from one form to another. The first row of the table provides the general relationships developed in the preceding sections, which can be used in all cases. Subsequent cases apply whenever the waveform possesses one or more symmetry conditions. Let us consider each case individually.

Even Function

A function $x(t)$ is said to be *even* if

$$x(-t) = x(t) \tag{11–48}$$

An example of an even function is shown in Figure 11–8. It can be shown that in this case only cosine terms appear in the spectrum. Furthermore, the integral used in determining A_n has the property that the area under the curve in one half of a cycle is the same as that in the other half. Hence we need only integrate over half a cycle and double the result as indicated in the table. Finally, the integral for $\overline{X}_n$ reduces to an integral involving the product of the time signal and the cosine function rather than the complex exponential. The coefficients $\overline{X}_n$ are all *real* in this case.

Odd Function

A function $x(t)$ is said to be *odd* if

$$x(-t) = -x(t) \tag{11–49}$$

TABLE 11-1
Fourier series symmetry conditions

Sine–cosine form: $x(t) = A_0 + \sum_{n=1}^{\infty}(A_n \cos n\omega_1 t + B_n \sin n\omega_1 t)$, $\qquad \omega_1 = 2\pi f_1 = \dfrac{2\pi}{T}$

Amplitude-phase form: $x(t) = C_0 + \sum_{n=1}^{\infty} C_n \cos(n\omega_1 t + \phi_n) = C_0 + \sum_{n=1}^{\infty} C_n \sin(n\omega_1 t + \theta_n)$, $\qquad C_n = \sqrt{A_n^2 + B_n^2}$

Complex exponential form: $x(t) = \sum_{n=-\infty}^{\infty} \overline{X}_n e^{jn\omega_1 t}$, $\quad \overline{X}_n = \dfrac{A_n - jB_n}{2}$, $\quad$ for $n \neq 0$ $\quad \overline{X}_0 = A_0$

Condition	A_n (except $n = 0$)	B_n	$\overline{X}_n$	Comments
General	$\dfrac{2}{T}\int_0^T x(t)\cos n\omega_1 t\, dt$	$\dfrac{2}{T}\int_0^T x(t)\sin n\omega_1 t\, dt$	$\dfrac{1}{T}\int_0^T x(t)e^{-jn\omega_1 t}\, dt$	
Even function $x(-t) = x(t)$	$\dfrac{4}{T}\int_0^{T/2} x(t)\cos n\omega_1 t\, dt$	0	$\dfrac{2}{T}\int_0^{T/2} x(t)\cos n\omega_1 t\, dt$	One-sided forms have only cosine terms $\overline{X}_n$ terms are real
Odd function $x(-t) = -x(t)$	0	$\dfrac{4}{T}\int_0^{T/2} x(t)\sin n\omega_1 t\, dt$	$\dfrac{-2j}{T}\int_0^{T/2} x(t)\sin n\omega_1 t\, dt$	One-sided forms have only sine terms $\overline{X}_n$ terms are imaginary
Half-wave symmetry $x\left(t + \dfrac{T}{2}\right) = -x(t)$	$\dfrac{4}{T}\int_0^{T/2} x(t)\cos n\omega_1 t\, dt$	$\dfrac{4}{T}\int_0^{T/2} x(t)\sin n\omega_1 t\, dt$	$\dfrac{2}{T}\int_0^{T/2} x(t)e^{-jn\omega_1 t}\, dt$	Odd-numbered harmonics only
Full-wave symmetry $x\left(t + \dfrac{T}{2}\right) = x(t)$	$\dfrac{4}{T}\int_0^{T/2} x(t)\cos n\omega_1 t\, dt$	$\dfrac{4}{T}\int_0^{T/2} x(t)\sin n\omega_1 t\, dt$	$\dfrac{2}{T}\int_0^{T/2} x(t)e^{-jn\omega_1 t}\, dt$	Even-numbered harmonics only

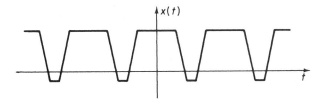

FIGURE 11–8
Example of an even function.

An example of an odd function is shown in Figure 11–9. It can be shown that in this case, only sine terms appear in the spectrum. The integral for B_n need be evaluated over only half a cycle and doubled, and the integral used to determine $\overline{X}_n$ reduces to the product of the time signal and a sine function. Note that all the $\overline{X}_n$ terms are purely imaginary in this case.

This type of symmetry is one of two types that can be "disguised" by the presence of a dc component. Consider the waveform shown in Figure 11–10, which is identical with that of Figure 11–9 except that a dc component has been added. Certainly, logic tells us that the dc component should not affect any other portion of the spectrum except at zero frequency, so there should be only sine terms in the remaining part of the spectrum. However, the basic odd function condition of Equation (11–49) is not satisfied. The dc component can be thought of as a limiting case of a cosine function of zero frequency, and this one "cosine" function obscures the symmetry.

The way around this problem is to inspect each waveform by mentally shifting it up or down to see whether the symmetry condition can be achieved by this process. If so, a new function $x_1(t)$ can be formed as follows:

$$x_1(t) = x(t) - A_0 \qquad\qquad \textbf{(11–50)}$$

where A_0 is the dc component of the signal. The function $x_1(t)$ now satisfies the pertinent symmetry condition and can be integrated according to the form given

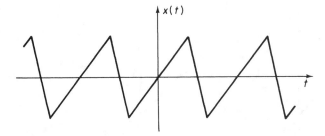

FIGURE 11–9
Example of an odd function.

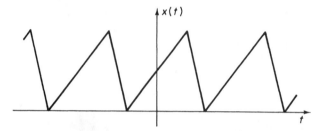

FIGURE 11-10
Function of Figure 11-9 with dc component added.

in the table. Note that if the symmetry condition is employed, the new function $x_1(t)$ should be integrated, rather than the original function.

Half-Wave Symmetry

A function is said to possess half-wave symmetry if

$$x\left(t + \frac{T}{2}\right) = -x(t) \tag{11-51}$$

A typical function having half-wave symmetry is shown in Figure 11-11. It can be shown that a function satisfying this condition will have only *odd-numbered harmonics* (i.e., $n = 1, 3, 5, 7, \ldots$). However, unless one of the previous two conditions is also satisfied, there will be both sine and cosine terms in the expansion. As in the case of even and odd functions, integration need be performed over only half a cycle, and the result is doubled.

One point of confusion regarding terms should be considered. The words *even* and *odd* were used in a different sense entirely for the previous two symmetry conditions, as compared with the present condition and the next one. In the former case, "even" and "odd" referred to definitions regarding the image of a function projected around the vertical axis. In this case and in the next one, "even" and

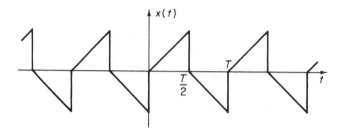

FIGURE 11-11
Example of function with half-wave symmetry.

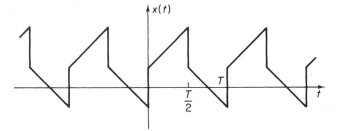

FIGURE 11–12
Function of Figure 11–11 with dc component added.

"odd" refer to the numbers of the harmonics. The two meanings are entirely different. For example, we can have an even function that has only odd-numbered harmonics.

As in the case of an odd function, the presence of a dc component can disguise half-wave symmetry. In this case the dc component is an even-numbered harmonic ($n = 0$), and its presence obscures the symmetry. A function satisfying this property is shown in Figure 11–12. The procedure for handling this case is the same as for the previous condition. The dc component is subtracted, and the symmetry condition is applied to the new function.

Full-Wave Symmetry

A function is said to possess full-wave symmetry if

$$x\left(t + \frac{T}{2}\right) = x(t) \tag{11–52}$$

A typical function having full-wave symmetry is shown in Figure 11–13. It can be shown that a function satisfying this condition will have only *even-numbered harmonics* (i.e., $n = 0, 2, 4, 6, \ldots$).

Actually, the reader may see a flaw in the preceding discussion. From Equation (11–52) and Figure 11–13, the question arises as to whether the period is really T

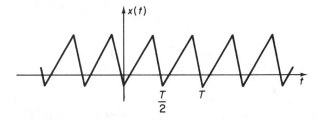

FIGURE 11–13
Example of function with full-wave symmetry.

as assumed, or whether the period is, in fact, $T/2$. To get to the point, the period is really $T/2$, and we could avoid discussing the concept of this symmetry condition altogether by redefining the period as $T/2$. However, this situation frequently arises in conjunction with nonlinear operations on signals where the period is effectively halved. In such cases, it is often desirable to maintain the original base period as T. If we redefined the period as $T/2$, the fundamental frequency would be $2/T$, which is the second harmonic of the original reference fundamental frequency. However, with respect to the original fundamental, there are only even-numbered harmonics.

A given waveform may possess no more than two of the preceding symmetry conditions. The function may be either even or odd (but not both), and the function may possess either half-wave or full-wave symmetry (but not both).

Other properties pertaining to Fourier series will be discussed after the Fourier transform is introduced. The reason we postpone such considerations now is that most of them apply equally well to the Fourier transform.

EXAMPLE 11–5

Consider again the square wave that was analyzed in Examples 11–2 and 11–4. Analyze this waveform to determine any symmetry conditions present, and use such symmetry conditions to simplify the computation of the spectrum.

Solution

The waveform is repeated in Figure 11–14(a) for convenience. As it appears, the reader should verify that neither of the four symmetry conditions is satisfied. However, suppose the dc value $A/2$ is subtracted, and a new signal $x_1(t)$ is formed as shown in Figure 11–14(b); that is,

$$x_1(t) = x(t) - \frac{A}{2} \tag{11–53}$$

In performing this shift, we understand, of course, that the new function $x_1(t)$ will not have a dc component, so the dc component of $x(t)$ has been recognized already and should be tabulated. All other terms in $x_1(t)$ should be the same as those in $x(t)$, since only the dc level has been affected.

Let us now check the symmetry conditions. We quickly establish that $x_1(t)$ is an odd function, since it satisfies Equation (11–49). This indicates that only sine terms will appear in the spectrum for $x_1(t)$, so $A_n = 0$.

Note also that the half-wave symmetry condition of Equation (11–51) is satisfied. This means that only odd-numbered harmonics will appear in the spectrum for $x_1(t)$.

Relating the observation for $x_1(t)$ back to $x(t)$, we can conclude that the Fourier series for $x(t)$ consists of a dc component and sine components at odd-numbered harmonic frequencies (including the fundamental, i.e., $n = 1$). This conclusion

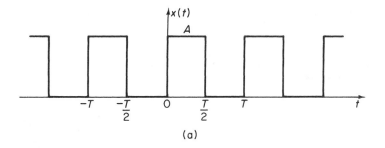

(a)

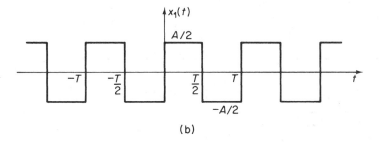

(b)

FIGURE 11–14
Removal of dc component from square wave to allow use of symmetry condition.

obviously agrees with the result of Example 11–2, as the reader may quickly verify. However, this inspection process provides valuable information in advance of a detailed calculation and can save some of the steps involved.

As a final step in this example, let us verify that appreciation of the symmetry condition in the computation of the B_n coefficients for the modified function $x_1(t)$ produces the same results as were obtained in Example 11–2 for $x(t)$. According to Table 11–1, we should integrate only over half a cycle and double the result. However, the amplitude of $x_1(t)$ is half that of $x(t)$, so the integral is

$$B_n = \frac{4}{T}\int_0^{T/2} \frac{A}{2} \sin n\omega_1 t \, dt \tag{11-54a}$$

$$= \frac{2A}{T}\int_0^{T/2} \sin n\omega_1 t \, dt \tag{11-54b}$$

Comparison of Equation (11–54b) with the first expression of Equation (11–38) reveals that the two expressions are exactly the same, so we need not proceed further. The "doubling" in front of the integral cancelled the "halving" of the function amplitude after shifting, so that the B_n coefficients are the same as before.

TABLE 11–2

Some common periodic signals and their Fourier series

Signal $x(t)$	Fourier Series
Square wave	$\dfrac{4A}{\pi}\left(\cos \omega_1 t - \dfrac{1}{3}\cos 3\omega_1 t\right.$ $\left. + \dfrac{1}{5}\cos 5\omega_1 t - \dfrac{1}{7}\cos 7\omega_1 t + \cdots\right)$
Triangular wave	$\dfrac{8A}{\pi^2}\left(\cos \omega_1 t + \dfrac{1}{9}\cos 3\omega_1 t\right.$ $\left. + \dfrac{1}{25}\cos 5\omega_1 t + \cdots\right)$
Sawtooth wave	$\dfrac{2A}{\pi}\left(\sin \omega_1 t - \dfrac{1}{2}\sin 2\omega_1 t\right.$ $\left. + \dfrac{1}{3}\sin 3\omega_1 t - \dfrac{1}{4}\sin 4\omega_1 t + \cdots\right)$
Half-wave rectified cosine	$\dfrac{A}{\pi}\left(1 + \dfrac{\pi}{2}\cos \omega_1 t + \dfrac{2}{3}\cos 2\omega_1 t\right.$ $- \dfrac{2}{15}\cos 4\omega_1 t + \dfrac{2}{35}\cos 6\omega_1 t$ $\left. - \cdots (-1)^{n/2+1}\dfrac{2}{n^2 - 1}\cos n\omega t + \cdots\right)$ $n \text{ even}$
Full-wave rectified cosine	$\dfrac{2A}{\pi}\left(1 + \dfrac{2}{3}\cos 2\omega_1 t - \dfrac{2}{15}\cos 4\omega_1 t\right.$ $+ \dfrac{2}{35}\cos 6\omega_1 t - \cdots (-1)^{n/2+1}\dfrac{2}{n^2 - 1}\cos n\omega_1 t$ $\left. + \cdots\right) n \text{ even}$
	$Ad\left[1 + 2\left(\dfrac{\sin \pi d}{\pi d}\cos \omega_1 t\right.\right.$ $+ \dfrac{\sin 2\pi d}{2\pi d}\cos 2\omega_1 t + \dfrac{\sin 3\pi d}{3\pi d}\cos 3\omega_1 t$ $\left.\left. + \cdots\right)\right]$ $d = \tau/T$

11-4 COMMON PERIODIC WAVEFORMS AND THEIR FOURIER SERIES

To finalize the discussion of Fourier series, some common periodic waveforms and their Fourier series are summarized in Table 11–2. Carefully note the location of the origin in each case, since a shift to the left or to the right of any waveform will change the exact form of the spectrum (but not the net magnitude). It turns out that all the waveforms as shown are either even or odd, so that each series has only sine or cosine terms, but not both. Note also the presence or lack of a dc component in some forms. The derivations of some of these functions will be given as problems at the end of the chapter.

11-5 FOURIER TRANSFORM

The emphasis on spectral analysis thus far has centered on periodic functions, whose spectra consist of discrete components at integer multiples of the fundamental repetition frequency. We consider next the process of spectral analysis for nonperiodic signals, which is achieved with the *Fourier transform*. To satisfy certain mathematical restrictions, assume that all nonperiodic signals of interest have finite energy.

To illustrate qualitatively the concept of the Fourier transform, refer to Figure 11–15. Some arbitrary, periodic, pulse-type signal $x(t)$ and its assumed amplitude spectrum X_n are shown in (a). The time function is periodic, with period T, and the spectrum is therefore discrete. The fundamental component is $f_1 = 1/T$, and spectral components appear at integer multiples of that frequency.

In Figure 11–15(b), the pulse width and shape remain the same, but the period is doubled by inserting a space between successive pulses. An expression for $\bar{X}_n$ would be the same as before, since the integrand has not changed. What has changed, however, is the fundamental frequency f_1. When T is doubled, f_1 is halved, so the spacing between spectral lines is halved as shown.

In Figure 11–15(c), the pulse width and shape again remain the same, but the period is doubled again. The spacing between spectral lines is again halved.

The effect of this trend is that the relative shape of the envelope of the $\bar{X}_n$ coefficients remains the same, but the number of components in a given frequency interval increases as the period increases. (The level of the envelope changes, but the *relative shape* remains constant. Each spectral plot is assumed to be adjusted in amplitude accordingly.) Simultaneously, the increment f_1 between successive frequency components decreases. In the limit as $T \rightarrow \infty$, as shown in Figure 11–15(d), the frequency difference approaches zero.

In this limiting form, the spectral lines all merge together, so it is no longer desirable to display the spectrum as a group of lines. Instead, the points representing the amplitudes of the lines effectively all merge together and form a continuous curve. Thus, in the limit as the period approaches infinity, the spectrum becomes a continuous spectrum. Whereas the spectrum of a *periodic* signal is *discrete*, and components appear only at integer multiples of the repetition rate, the spectrum

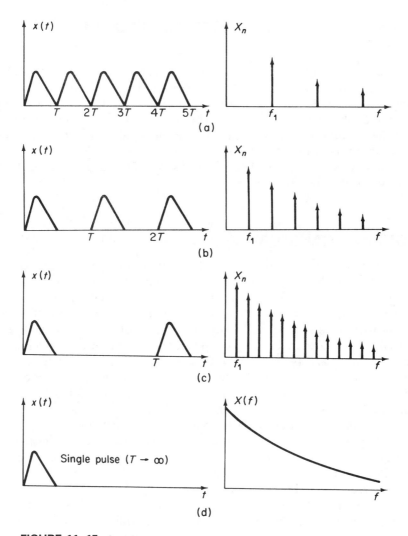

FIGURE 11–15
Concept of the Fourier transform as a limiting case of the Fourier series when the period becomes infinite.

of a *nonperiodic* signal is *continuous* and could, in theory, appear at any frequency. Of course, a bandlimited signal would necessarily have a spectrum only over a finite frequency range, but the point is that there are no restrictions of the type that exist for the periodic signal.

The *Fourier transform* is the commonly used name for the mathematical function that provides the frequency spectrum of a nonperiodic signal. Assume that a nonperiodic time signal $x(t)$ is given. The Fourier transform is designated as $\overline{X}(f)$, and f is the cyclic frequency in hertz. [The overbar on $\overline{X}(f)$ emphasizes that it is a complex quantity.] Thus, $\overline{X}(f)$ is the mathematical function expressed as a function of frequency that indicates the relative spectrum, which could exist at any arbitrary frequency f. $\overline{X}(f)$ for the nonperiodic signal corresponds to the two-sided spectrum $\overline{X}_n$ for the periodic case. Although it would be possible to obtain Fourier transform forms that would relate more directly to the one-sided discrete spectral forms, almost all the available results are more directly related to the two-sided exponential form, and that approach will be followed here.

The process of Fourier transformation of a time function is designated symbolically as

$$\overline{X}(f) = \mathscr{F}[x(t)] \tag{11-55}$$

The inverse operation is designated symbolically as

$$x(t) = \mathscr{F}^{-1}[\overline{X}(f)] \tag{11-56}$$

The actual mathematical processes involved in these operations are as follows:

$$\overline{X}(f) = \int_{-\infty}^{\infty} x(t)e^{-j\omega t}\, dt \tag{11-57}$$

$$x(t) = \int_{-\infty}^{\infty} \overline{X}(f)e^{j\omega t}\, df \tag{11-58}$$

Note that the argument of $\overline{X}(f)$ and the differential of Equation (11–58) are both expressed in terms of the cyclic frequency f (in hertz), but the arguments of the exponentials in Equations (11–57) and (11–58) are expressed in terms of the radian frequency ω, where $\omega = 2\pi f$. These are the most convenient forms for manipulating and expressing the given functions. Most spectral displays are made in terms of f, although the analytical expressions are often easier to deal with in terms of ω. This should present no serious problem as long as the 2π scale factor is understood: $\omega = 2\pi f$.

The Fourier transform $\overline{X}(f)$ is, in general, a complex function and has both a magnitude and an angle. Thus, $\overline{X}(f)$ can be expressed as

$$\overline{X}(f) = X(f)e^{j\phi(f)} = X(f)\,\underline{/\phi(f)} \tag{11-59}$$

where $X(f)$ represents the *amplitude spectrum* and $\phi(f)$ is the *phase spectrum*. A typical amplitude spectrum is shown in Figure 11–16.

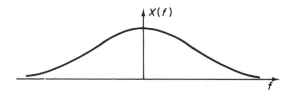

FIGURE 11–16
Typical amplitude spectrum of nonperiodic signal.

TABLE 11–3
Fourier transform symmetry conditions

Condition	$\overline{X}(f)$	Comment
General	$\displaystyle\int_{-\infty}^{\infty} x(t)e^{-j\omega t}\,dt$	
Even function $x(-t) = x(t)$	$\displaystyle 2\int_{0}^{\infty} x(t)\cos\omega t\,dt$	$\overline{X}(f)$ is an even, real function of f
Odd function $x(-t) = -x(t)$	$\displaystyle -2j\int_{0}^{\infty} x(t)\sin\omega t\,dt$	$\overline{X}(f)$ is an odd, imaginary function of f

Again, a point that should be stressed for the *nonperiodic* signal is that its spectrum is *continuous,* and, in general, it consists of components at *all* frequencies in the range over which the spectrum is present.

As in the case of periodic signals, certain symmetry conditions can be applied to aid in the computation of the Fourier transform. However, not all the periodic signal properties are applicable to a nonperiodic signal. Furthermore, because of established convention, the results will be interpreted in slightly different forms. For example, the coefficients A_n and B_n are widely used in dealing with Fourier series, but the corresponding forms in the Fourier transform (which would be proportional to the real and imaginary parts of the transform) are not used nearly as often in signal analysis and will not be discussed here.

The symmetry conditions to be considered are summarized in Table 11–3. These results indicate that for either an even or an odd function, one need integrate only over half the total interval and double the result. Furthermore, the form of the integrand is different in each case. Notice the similarity between the forms of the Fourier transform and Fourier series integrals when the function is either even or odd.

EXAMPLE 11–6
Derive the Fourier transform of the rectangular pulse function shown in Figure 11–17(a).

Solution
The pulse can be defined as

$$x(t) = \begin{cases} A & \text{for } -\tau/2 < t < \tau/2 \\ 0 & \text{elsewhere} \end{cases} \tag{11–60}$$

Since the given signal is even, a symmetry condition from Table 11–3 may be applied. Using the integral form as given in Table 11–3, we have

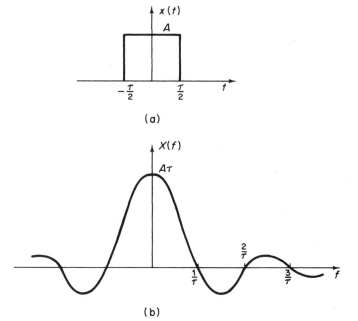

FIGURE 11–17
Rectangular pulse and its Fourier transform as considered in Example 11–6.

$$\overline{X}(f) = 2\int_0^{\tau/2} A \cos \omega t \, dt$$

$$= \frac{2A}{\omega} \sin \omega t \Big]_0^{\tau/2} \tag{11–61}$$

$$= \frac{2A}{\omega} \sin \frac{\omega\tau}{2}$$

By setting $\omega = 2\pi f$ and performing some additional manipulations, we can write

$$\overline{X}(f) = A\tau \frac{\sin \pi f \tau}{\pi f \tau} \tag{11–62}$$

The form of the amplitude spectrum $X(f)$ is shown in Figure 11–17(b). This function is very important in communications signal analysis and transmission. [The phase response is $\Phi(f) = 0$.]

EXAMPLE 11–7
Derive the Fourier transform of the exponential function given by

$$x(t) = \begin{cases} Ae^{-\alpha t} & \text{for } t > 0 \\ 0 & \text{for } t < 0 \end{cases} \tag{11–63}$$

where $\alpha > 0$.

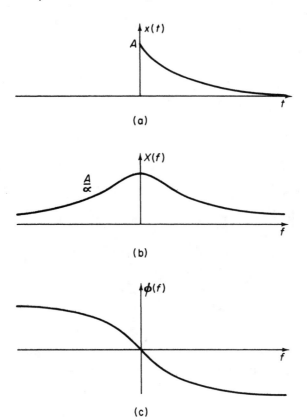

FIGURE 11–18
Exponential function of Example 11–7 and its amplitude and phase spectra.

Solution
The function is shown in Figure 11–18(a). Application of the definition of the Fourier transform as given by Equation (11–57) yields

$$\overline{X}(f) = \int_0^\infty A e^{-\alpha t} e^{-j\omega t} \, dt$$

$$= \left. \frac{A e^{-(\alpha + j\omega)t}}{-(\alpha + j\omega)} \right]_0^\infty = 0 + \frac{A}{\alpha + j\omega} \qquad (11\text{–}64)$$

The result is a complex function, as expected. The amplitude and phase functions are found by determining the magnitude and angle associated with the final result of Equation (11–64). These functions are

$$X(f) = \frac{A}{\sqrt{\alpha^2 + \omega^2}} = \frac{A}{\sqrt{\alpha^2 + (2\pi f)^2}} \qquad (11\text{–}65)$$

and

$$\phi(f) = -\tan^{-1}\frac{\omega}{\alpha} = -\tan^{-1}\frac{2\pi f}{\alpha} \qquad (11\text{--}66)$$

These functions are shown in Figure 11–18(b) and (c) for some arbitrary value of α. As in the case of discrete spectra, the phase spectrum is not as useful as the amplitude spectrum, and will usually not be shown in most developments. However, it was included here to make the development complete.

EXAMPLE 11–8

One property of the impulse function not considered earlier is

$$\int_{-\infty}^{\infty} g(t)\delta(t)\, dt = g(0) \qquad (11\text{--}67)$$

where $g(t)$ is any continuous function. Derive the Fourier transform of the impulse function.

Solution

The impulse function occurring at $t = 0$ is illustrated in Figure 11–19(a). Application of the definition of the Fourier transform yields

$$\mathscr{F}[\delta(t)] = \int_{-\infty}^{\infty} \delta(t)e^{-j\omega t}\, dt \qquad (11\text{--}68)$$

This integral is readily evaluated by the property of the impulse function as given by Equation (11–67).

$$\mathscr{F}[\delta(t)] = 1 \qquad (11\text{--}69)$$

This simple result is shown in Figure 11–19(b).

This result is rather interesting, and should catch the reader's attention. Although the impulse function is more of an ideal mathematical model than a real physical waveform, it is often used to represent noise phenomena (e.g., a function that appears as a sharp pulse with near-zero duration). The spectrum of such a

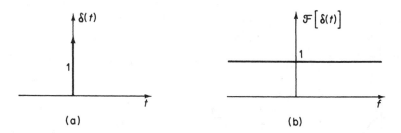

(a)   (b)

FIGURE 11–19
Impulse function and its spectrum as developed in Example 11–8.

signal is extremely wide and, in the theoretical limit, would contain components at all frequencies over an infinite frequency interval.

To illustrate one more point regarding this type of phenomenon, the reader has probably observed interference produced on a radio receiver due to "impulse" sources in the immediate area (e.g., electrical appliances, automobile ignition systems). These "impulses" create broad spectra that can be heard on radio receivers over a wide frequency range, as noted from the nature of the Fourier transform of the function.

11–6 COMMON NONPERIODIC WAVEFORMS AND THEIR FOURIER TRANSFORMS

Some common nonperiodic waveforms and their Fourier transforms are summarized in Table 11–4. Note the location of the origin in each case, since a shift to the left

TABLE 11–4
Some common signals and their Fourier transform

Signal $x(t)$	Spectrum $\overline{X}(f)$
Rectangular pulse	$A\tau \dfrac{\sin \pi f \tau}{\pi f \tau}$
Triangular pulse	$A\tau \left(\dfrac{\sin \pi f \tau}{\pi f \tau} \right)^2$
Sawtooth pulse	$\dfrac{jA}{2\pi f} \left[\dfrac{\sin \pi f \tau}{\pi f \tau} e^{-j\pi f \tau} - 1 \right]$
Cosine pulse	$\dfrac{2A\tau}{\pi} \dfrac{\cos \pi f \tau}{1 - 4f^2 \tau^2}$

or to the right of any waveform will change the exact form of the spectrum (but not the net magnitude). The derivations of some of these functions will be given as problems at the end of the chapter.

11–7 FOURIER TRANSFORM OPERATIONS AND SPECTRAL ROLL-OFF

Time functions are altered in form as they pass through various stages of a signal-processing system or a signal-transmission channel. These operations may be used deliberately to change the form of the signal in some cases, or certain of the operations may arise as a result of natural limitations of a system and could distort the signal. In either event, it is worthwhile to study the effects on the resulting signal from a spectral point of view.

The primary Fourier transform operation pairs of interest for our purposes are summarized in Table 11–5. Certain of these pairs will be derived in problems at the end of this section; others will be left as exercises for analytically inclined readers.

The practical significance of all these operation pairs will now be discussed. The following notational form will be used here and in certain subsequent sections:

$$x(t) \leftrightarrow \overline{X}(f) \tag{11-70}$$

This notation indicates that $x(t)$ and $\overline{X}(f)$ are a corresponding transform pair; that is, $\overline{X}(f) = \mathscr{F}[x(t)]$. The equation numbers of the operation pairs correspond to those of Table 11–5.

Superposition Principle (O–1) The first transform pair in the table is

$$ax_1(t) + bx_2(t) \leftrightarrow a\overline{X}_1(f) + b\overline{X}_2(f) \tag{O-1}$$

TABLE 11–5
Fourier transform operation pairs

$x(t)$	$\overline{X}(f) = \mathscr{F}[x(t)]$	
$ax_1(t) + bx_2(t)$	$a\overline{X}_1(f) + b\overline{X}_2(f)$	(O–1)
$\dfrac{dx(t)}{dt}$	$j2\pi f\overline{X}(f)$	(O–2)
$\displaystyle\int_{-\infty}^{t} x(t)\,dt$	$\dfrac{\overline{X}(f)}{j2\pi f}$	(O–3)
$x(t - \tau)$	$e^{-j2\pi f\tau}\overline{X}(f)$	(O–4)
$e^{-j2\pi f_0 t}x(t)$	$\overline{X}(f - f_0)$	(O–5)
$x(at)$	$\dfrac{1}{a}\overline{X}\left(\dfrac{f}{a}\right)$	(O–6)

This result specifies the basic property that the Fourier transform integral is a linear operation and thus obeys the principle of superposition as far as the level of a signal and the combination of several signals are concerned.

Differentiation (O–2)   The differentiation Fourier transform pair is

$$\frac{dx(t)}{dt} \leftrightarrow j2\pi f \overline{X}(f) \tag{O–2}$$

This theorem, which will be derived in Example 11–9, indicates that each time a signal is differentiated, the spectrum is multiplied by $j2\pi f$. Multiplication by $j2\pi f$ has the effect of decreasing the relative level of the spectrum at low frequencies and increasing the relative level at higher frequencies. Note that a pure dc component is eliminated. A sketch illustrating the general effect on the amplitude spectrum resulting from differentiating a time signal is shown in Figure 11–20.

Integration (O–3)   The integration Fourier transform pair is

$$\int_{-\infty}^{t} x(t)\, dt \leftrightarrow \frac{\overline{X}(f)}{j2\pi f} \tag{O–3}$$

This theorem, which is the reverse of (O–2), indicates that when a signal is integrated, the amplitude spectrum is divided by $2\pi f$. Division by $2\pi f$ has the effect of increasing the relative level of the spectrum at low frequencies and decreasing the relative level at higher frequencies. A sketch illustrating the general effect on the amplitude spectrum resulting from integrating a time signal is shown in Figure 11–21. In a sense, integration is a form of low-pass filtering, in that high-frequency components of the spectrum are attenuated. However, a pure integrator is not often used as a low-pass filter due to the pronounced accentuation effect at very low frequencies. Nevertheless, there are situations in which the integrator is considered as a form of low-pass filter.

Time Delay (O–4)   The time-delay transform operation, which is derived in Example 11–10, is

$$x(t - \tau) \leftrightarrow e^{-j2\pi f \tau}\, \overline{X}(f) \tag{O–4}$$

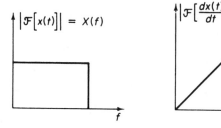

FIGURE 11–20
Effect on the spectrum of differentiating a time signal.

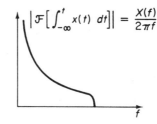

FIGURE 11–21
Effect on the spectrum of integrating a time signal.

The function $x(t - \tau)$ represents the delayed version of a signal $x(t)$ as illustrated in Figure 11–22. This operation could occur as a result of passing a signal through an ideal delay line with delay τ, for example. It can readily be shown that the amplitude spectrum is not changed by the shifting operation, but the phase spectrum is shifted by $-2\pi f\tau$ radians. Certainly, one would expect the amplitude spectrum of a given fixed signal to be independent of the time at which the signal occurs, but the phase shifts of all the components are increased to reflect the result of the time delay.

Modulation (O–5)   This operation bears the same relationship to the frequency domain as the time-delay theorem does to the time domain. This theorem, whose derivation will be left as an exercise (Problem 11–27), is

$$e^{j2\pi f_0 t}x(t) \leftrightarrow \overline{X}(f - f_0) \tag{O–5}$$

This result has profound implications in the study of amplitude modulation in communications theory. If a time signal is multiplied by a complex exponential, the spectrum is translated to the right by the frequency of the exponential, as shown in Figure 11–23. In practical cases, complex exponentials occur in pairs with a term of the form of (O–5) along with its conjugate.

Time Scaling (O–6)   The time-scaling operation transfer pair is

$$x(at) \leftrightarrow \frac{1}{a}\overline{X}\left(\frac{f}{a}\right) \tag{O–6}$$

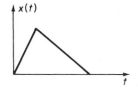

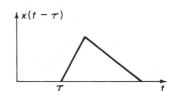

FIGURE 11–22
Time signal and its delayed form.

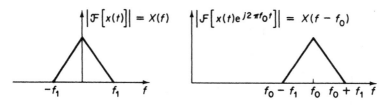

FIGURE 11–23
Effect on the spectrum of the modulation operation.

The derivation of this transform pair will be left as an exercise (Problem 11–28). If $a > 1$, $x(at)$ represents a "faster" version of the original signal, whereas if $a < 1$, $x(at)$ represents a "slower" version. In the former case, the spectrum is broadened, whereas in the latter case, it is narrowed. These concepts are illustrated in Figure 11–24.

We now turn to the problem of determining the *spectral roll-off rate* or, in more formal mathematical terms, the *convergence* of the Fourier spectrum. We have seen that the Fourier transforms and series for many common waveforms appear to be infinitely wide. Obviously, in the real world this result cannot be true.

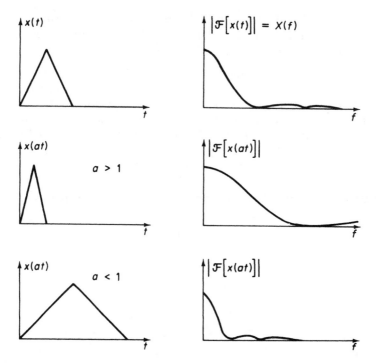

FIGURE 11–24
Effect on the spectrum of the time-scaling operation.

In reality, spectral components above a certain frequency range are simply ignored, since their contributions are negligible, and the practical bandwidth is established at the minimum level for reasonable reproduction of the signal.

An important factor that can be used qualitatively in estimating the relative bandwidths of different signals is the spectral roll-off rate. The roll-off rate is an upper-bound (worst-case) measure of the rate at which the spectral components diminish with increasing frequency. The basic way to specify the roll-off rate is a $1/f^k$ variation for a Fourier transform or a $1/n^k$ variation for a Fourier series, where k is an integer. As k increases, the spectrum diminishes more rapidly. Thus, a signal with a $1/f^3$ roll-off rate would normally have a narrower bandwidth than a signal with a $1/f^2$ rate.

A common, practical way to specify the roll-off rate is in decibels/octave, in somewhat the same manner as for Bode plots. A roll-off rate of $1/f^k$ can be readily shown to correspond to a slope of $-6k$ dB/octave, where an octave corresponds to a doubling of the frequency. Thus, a $1/f^3$ roll-off rate corresponds to -18 dB/octave, and a $1/f^2$ rate corresponds to -12 dB/octave.

The roll-off rate refers to the worst case or upper bound of the magnitude spectrum of the signal. It is an estimate of the worst-case effect, and should not be interpreted as an exact formula for predicting spectral components. For example, the spectrum of a signal may contain components that have a -6 dB/octave roll-off rate, plus components that have a -12 dB/octave roll-off rate. Eventually, the latter components will be so small that they may be ignored, and the -6 dB/octave components will dominate. The tests we will present would predict a -6 dB/octave roll-off rate, and the result has to be interpreted as a worst-case bound.

Before discussing the particular tests, the following qualitative points are worthwhile for dealing with signals in a general sense and should be carefully noted.

1. Time functions that are relatively "smooth" (i.e., no discontinuities or jumps in the signal or its lower-order derivatives) tend to have higher roll-off rates and corresponding narrower bandwidths.
2. Time functions with discontinuities in the signal tend to have lower roll-off rates and corresponding wider bandwidths.

An example of a smooth signal is the sinusoid whose bandwidth is so narrow that it has only one component. Conversely, a square wave has finite discontinuities in each cycle, and its spectrum is very wide.

The roll-off rates for a wide variety of common signals may be estimated from the information provided in Table 11–6. The left column provides a test to apply to the signal, and the right columns provide the appropriate roll-off rates. Observe that one column on the right is applicable to nonperiodic signals analyzed with the Fourier transform, whereas the other is applicable to periodic signals analyzed with the Fourier series. Some of the information in the table will now be discussed.

The first condition is hypothetical, and applies when the signal is assumed to contain one or more ideal impulse functions. This situation could never actually exist in practice, but there are situations in which the assumption of ideal impulse functions is convenient. In such a case, the spectrum will contain a portion having

TABLE 11–6
Spectral roll-off rates of Fourier transforms and Fourier series

Smoothness of Function	Roll-off Rate	
	Fourier Transform	Fourier Series
$x(t)$ has impulses	No spectral roll-off	No spectral roll-off
$x(t)$ has finite discontinuities	$\dfrac{1}{f}$ or -6 dB/octave	$\dfrac{1}{n}$ or -6 dB/octave
$x(t)$ is continuous, $x'(t)$ has finite discontinuities	$\dfrac{1}{f^2}$ or -12 dB/octave	$\dfrac{1}{n^2}$ or -12 dB/octave
$x(t)$ is continuous, $x'(t)$ is continuous, $x''(t)$ has finite discontinuities	$\dfrac{1}{f^3}$ or -18 dB/octave	$\dfrac{1}{n^3}$ or -18 dB/octave

no roll-off at all; that is, the spectrum would theoretically be infinite in bandwidth.

If the signal has finite discontinuities or jumps, the spectrum has a -6 dB/octave roll-off rate. The ideal square wave fits this case, as will be illustrated in Example 11–11.

If the signal is continuous (i.e., no jumps), but the first derivative or slope changes abruptly at one or more points, the spectrum has a -12 dB/octave roll-off rate. The triangular wave is a good example of this type (see Problems 11–13b and 11–14b).

If the signal and its first derivative are both continuous, but the second derivative has a finite discontinuity, the spectrum has a -18 dB/octave roll-off rate. It is nearly impossible to detect this condition visually, but a knowledge of the fact may occur as a result of other information.

The table could be continued, but the range shown covers most of the normal requirements for common waveforms. In general, if a function and its first $k - 1$ derivatives are continuous, but its kth derivative has a finite discontinuity, the spectral roll-off rate will be $-6(k + 1)$/dB/octave.

EXAMPLE 11–9
Derive the Fourier transform of the first derivative of a time signal [i.e., pair (O–2) of Table 11–5].

Solution
The theorem is best derived by considering the inverse transform of $\overline{X}(f)$.

$$x(t) = \int_{-\infty}^{\infty} \overline{X}(f)e^{j\omega t}\,df \tag{11–71}$$

Both sides of this equation are now differentiated with respect to time. This yields

$$\frac{dx(t)}{dt} = \int_{-\infty}^{\infty} j\omega\overline{X}(f)e^{j\omega t}\,df \tag{11–72}$$

By comparing Equation (11–72) with Equation (11–71), we see that the quantity $j\omega \overline{X}(f)$ represents the Fourier transform of the derivative on the left. By induction, this result is readily extended to the case of the nth derivative.

EXAMPLE 11–10
Derive operation pair (O–4) of Table 11–5.

Solution
Application of the definition of the Fourier transform to $x(t - \tau)$ yields

$$\mathscr{F}[x(t - \tau)] = \int_{-\infty}^{\infty} x(t - \tau)e^{-j\omega t} \, dt \tag{11–73}$$

A change in variables will now be made. Let $u = t - \tau$, which results in $du = dt$. Substitution of these values yields

$$\begin{aligned} \mathscr{F}[x(t - \tau)] &= \int_{-\infty}^{\infty} x(u)e^{-j\omega u}e^{-j\omega \tau} \, du \\ &= e^{-j\omega \tau} \int_{-\infty}^{\infty} x(u)e^{-j\omega u} \, du = e^{-j\omega \tau}\overline{X}(f) \\ &= e^{-j2\pi f\tau}\overline{X}(f) \end{aligned} \tag{11–74}$$

EXAMPLE 11–11
Using Table 11–6, determine the spectral roll-off rates for the functions corresponding to the example problems indicated below. Verify by comparing with the results of **(a)** Example 11–7, **(b)** Example 11–8, and **(c)** Example 11–2.

Solution
(a) The function of Example 11–7 has a finite discontinuity at $t = 0$. Hence the spectrum should display a roll-off rate of $1/f$, or -6 dB/octave. From Equation (11–65), the high-frequency asymptotic behavior of $X(f)$ approaches

$$X(f) \approx \frac{A}{2\pi f} \qquad \text{for } f \gg \frac{\alpha}{2\pi} \tag{11–75}$$

which is a $1/f$ form, as predicted.
(b) The function of Example 11–8 is an impulse, which should display no roll-off at all. Indeed, this prediction is readily verified by Equation (11–69).
(c) The function of Example 11–2 has two finite discontinuities in each cycle, a positive jump and a negative jump. The function is periodic, so it should display a roll-off rate of $1/n$. Observe from Equation (11–40) that this property is true, so the spectrum has a -6 dB/octave roll-off. The absence of the even harmonics in the spectrum should not confuse the issue, since the roll-off rate is correct for those components present.

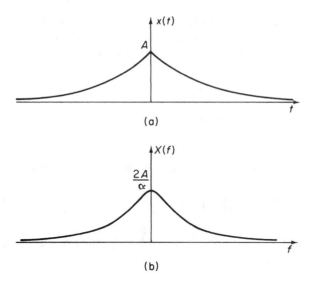

FIGURE 11–25
Exponential function of Example 11–12 and its transform.

EXAMPLE 11–12

Derive the Fourier transform of the two-sided exponential function shown in Figure 11–25(a) having a damping factor α. Sketch the form of the spectrum.

Solution
This function may be expressed as

$$x(t) = Ae^{-\alpha|t|} \tag{11–76}$$

or

$$x(t) = \begin{cases} Ae^{\alpha t} & \text{for } t < 0 \\ Ae^{-\alpha t} & \text{for } t > 0 \end{cases} \tag{11–77}$$

where $\alpha > 0$. Since the function is even, the transform may be evaluated by integrating only over positive time, and this operation is

$$\overline{X}(f) = 2 \int_0^\infty Ae^{-\alpha t} \cos \omega t \, dt \tag{11–78}$$

This integral may be evaluated by parts, or the result may be found in a standard integral table. The result is

$$\overline{X}(f) = \frac{2Ae^{-\alpha t}(-\alpha \cos \omega t + \omega \sin \omega t)}{\alpha^2 + \omega^2} \bigg]_0^\infty$$

$$= \frac{2\alpha A}{\alpha^2 + \omega^2} = \frac{2\alpha A}{\alpha^2 + 4\pi^2 f^2} \tag{11–79}$$

It is interesting to point out that the spectrum is real and is an even function of frequency, as expected. Furthermore, the function itself is continuous, but its first derivative has a finite discontinuity at $t = 0$. This property indicates that the spectrum should have a roll-off rate of $1/f^2$, which is readily observed in Equation (11–79). A sketch of the spectrum is shown in Figure 11–25(b). This two-sided exponential function and its spectrum can be compared with the one-sided exponential function and its spectrum, as considered in Example 11–7. In the present example, the spectrum has a roll-off rate of $1/f^2$ (or -12 dB/octave); in the earlier example, the discontinuity at $t = 0$ resulted in a $1/f$ (or -6 dB/octave) roll-off rate.

11–8 PSPICE EXAMPLES

PSPICE contains several interesting utilities for performing spectral analysis. In this section we investigate some of the capabilities as they relate to the treatment within the chapter.

PSPICE EXAMPLE 11–1

The basic Fourier analysis capability of PSPICE will be demonstrated in this example and the next several by generating a function whose Fourier coefficients are known at the outset. Consider the voltage given by

$$v(t) = 12 \sin 2\pi \times 10t + 5 \sin 2\pi \times 20t + 3 \sin 2\pi \times 30t \qquad \textbf{(11–80)}$$

Write a PSPICE program to generate the waveform over one cycle and use the basic Fourier command to determine the spectrum.

Solution
It must be stressed again that the function is already in a basic Fourier series form, and by inspection, we see that $A_n = 0$ for all n, $B_1 = 12$, $B_2 = 5$, $B_3 = 3$, and all other $B_n = 0$. The fundamental frequency is $f = 10$ Hz, and the other two nonzero components are the second and third harmonics, with frequencies of 20 and 30 Hz, respectively. Our purpose is to show how PSPICE analyzes the spectral content.

In order to perform a Fourier analysis, it is necessary first to perform a transient analysis to generate the waveform for which the spectrum is to be determined. A simple code to achieve this purpose is shown in Figure 11–26. Three separate sinusoidal voltages are defined at nodes 1, 2, and 3 with respect to ground. A VCVS is then defined at node 4 with rrespect to ground, using a VALUE function. The value of the voltage is the sum of the values of the three voltage sources.

To utilize the Fourier command, the pertinent waveform must be generated over at least one cycle of the fundamental. If the circuit contains a transient interval, a sufficient number of cycles should be employed to ensure that steady-state conditions have been reached. In this case, steady-state conditions are reached immediately, so only one cycle will be used in this example. The fundamental frequency

FIGURE 11-26
Code for PSPICE Example 11–1.

```
SUM OF THREE SINE FUNCTIONS
.OPTIONS NOECHO NOPAGE NOBIAS
V1 1 0 SIN 0 12 10
V2 2 0 SIN 0 5 20
V3 3 0 SIN 0 3 30
E 4 0 VALUE={V(1)+V(2)+V(3)}
.TRAN 1E-4 0.1
.PROBE
.FOUR 10 V(4)
.END
```

is $f = 10$ Hz, and the period is $T = 1/f = 1/10 = 0.1$ s. Thus, the transient interval selected is 0.1 s.

It is this author's experience that better results are obtained in Fourier analysis when the step interval is forced to be sufficiently small to ensure at least 100 points per cycle. In this example, 1000 points per cycle were chosen with the transient command

```
.TRAN 1E-4 0.1
```

The Fourier command used in this example is as follows:

```
.FOUR 10 V(4)
```

Following the .FOUR command, the fundamental (10 Hz) is listed. This is followed by one or more node voltages at which the analysis is to be performed. The program will use the last segment of the response, equal to $1/f$. For example, if there had been a transient interval of about 0.4 s, and the transient had been run for 0.5 s, PSPICE would use the interval from $t = 0.4$ s to 0.5 s for the purpose of performing the Fourier analysis. The default Fourier command generates all components up through the ninth harmonic in the amplitude-phase form and provides a tabular output. (It can be modified to provide more components, if desired.) The series obtained can thus be expressed as

$$v(t) = C_0 + \sum_{n=1}^{9} C_n \sin(n\omega_1 t + \theta_n) \tag{11-81}$$

If a cosine series is desired, the angles in Equation (11–8) are readily determined to be

$$\phi_n = \theta_n - 90° \tag{11-82}$$

The waveform obtained with PROBE is shown in Figure 11–27. Obviously, the function does not look anything like a sine wave, and it would be impossible for anyone to determine the number of significant components and their relative magnitudes or phases from a direct inspection.

The Fourier coefficients are determined from the output data file, which is shown in Figure 11–28 for this example. Although we know that the waveform has no dc component, the program produces a minuscule value close to $C_0 = -3.2 \times$

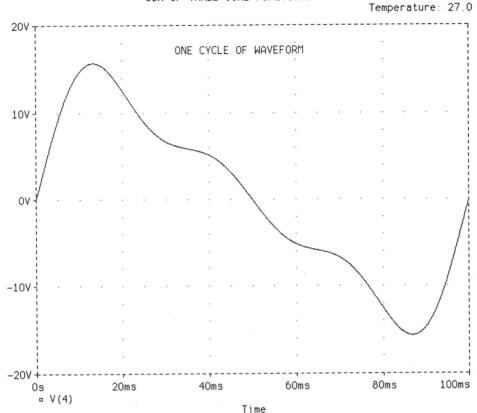

FIGURE 11–27
Waveform for PSPICE Examples 11–1 and 11–2.

10^{-8}, which should be interpreted as zero, for all practical purposes. The fundamental C_1 is listed as "HARMONIC 1," and it has a frequency of 10 Hz and an amplitude of 12, as expected. Theoretically, the phase angle should be zero, but a value of about -7.7×10^{-8} degrees is given. Again, this is a result of numerical roundoff and should be ignored.

The two columns beginning with "NORMALIZED" represent magnitude and phase, adjusted with respect to the fundamental. In this column, the normalized magnitude of the fundamental will always be set at 1, and the normalized phase of the fundamental is set at 0°. The second harmonic, normalized magnitude is 0.4167, which is the ratio of the second harmonic magnitude to the fundamental, namely, C_2/C_1. The normalized phase is the difference between the second harmonic phase angle and the fundamental, that is, $\theta_2 - \theta_1$. For the fundamental, second, and third harmonics, the values are all essentially zero.

```
******************************** Evaluation PSpice ************************
  SUM OF THREE SINE FUNCTIONS


  ****      CIRCUIT DESCRIPTION


*****************************************************************************


.OPTIONS NOECHO NOPAGE NOBIAS


  ****      FOURIER ANALYSIS                  TEMPERATURE =   27.000 DEG C



FOURIER COMPONENTS OF TRANSIENT RESPONSE V(4)



  DC COMPONENT =  -3.215331E-08

  HARMONIC   FREQUENCY     FOURIER      NORMALIZED      PHASE       NORMALIZED
    NO         (HZ)       COMPONENT     COMPONENT       (DEG)      PHASE (DEG)
     1       1.000E+01    1.200E+01     1.000E+00     7.667E-08     0.000E+00
     2       2.000E+01    5.000E+00     4.167E-01     1.234E-06     1.158E-06
     3       3.000E+01    3.000E+00     2.500E-01     5.542E-06     5.465E-06
     4       4.000E+01    3.231E-08     2.693E-09    -6.248E+01    -6.248E+01
     5       5.000E+01    6.130E-09     5.109E-10    -2.881E+01    -2.881E+01
     6       6.000E+01    4.397E-08     3.664E-09    -2.704E+01    -2.704E+01
     7       7.000E+01    2.063E-08     1.719E-09     4.226E+01     4.226E+01
     8       8.000E+01    7.802E-09     6.502E-10    -3.686E+01    -3.686E+01
     9       9.000E+01    1.940E-08     1.617E-09     5.561E+01     5.561E+01


     TOTAL HARMONIC DISTORTION =   4.859043E+01 PERCENT


         JOB CONCLUDED

         TOTAL JOB TIME          13.73
```

FIGURE 11–28
Output data for PSPICE Example 11–1.

The total harmonic distortion (THD) is defined as

$$\text{THD} = \frac{\left(\sqrt{\sum_{n=2}^{9} C_n^2} \right)}{C_1} \times 100\% \qquad \textbf{(11–83)}$$

For the waveform given, the value is about 48.6%. This quantity is significant when one is determining the harmonic content generated by a nonlinear system on a sinusoidal input. However, for a general analysis of a nonsinusoidal waveform, it may or may not be significant.

PSPICE EXAMPLE 11-2

Use the Fourier analysis command of PROBE to plot the spectrum of the function of PSPICE Example 11-1.

Solution

The PROBE program of PSPICE provides many outstanding display features, of which Fourier analysis of the resulting screen time function is one. The program employs the *fast Fourier transform* (FFT) algorithm, which is widely used as a computationally efficient means for determining a spectrum. A full treatment of this operation is not within the scope of our present coverage, but it may be readily employed as part of the PSPICE package.

A few points about the FFT and the manner in which PROBE adjusts the data should be noted.

1. The PSPICE version plots only the magnitude, not the phase.
2. The level of the magnitude is adjusted by PROBE to be compatible with the amplitude-phase form and *not* the exponential form. (In general, FFT algorithms may be adjusted to be compatible with the exponential form, or any other form, for that matter.)
3. In the basic FFT algorithm, the spectrum is a line spectrum with spacing between lines equal to the reciprocal of the record length. *However, PROBE provides a straight-line connection between points.* This latter property can result in some confusion in interpreting the spectrum.

The program in this case is the same as for PSPICE Example 11-1, and the applicable time waveform is Figure 11-27. The horizontal axis command is activated, and Fourier analysis is invoked.

The resulting spectrum is shown in Figure 11-29. At the frequencies 10, 20, and 30 Hz, the amplitudes are about 12, 5, and 3, respectively, as expected. However, although the spectrum *appears* to be continuous, this is a result of the straight-line connections previously mentioned.

PSPICE EXAMPLE 11-3

Repeat PSPICE Example 11-2 using two cycles of the waveform.

Solution

The code is essentially the same as in Figure 11-26, except for two changes. First, since a tabular output is not desired, the .FOUR statement can be eliminated (PROBE provides all the capability for a graphical spectral analysis.) Second, the transient command was modified to read

```
.TRAN 1E-4 0.2
```

The resulting waveform is shown in Figure 11-30.

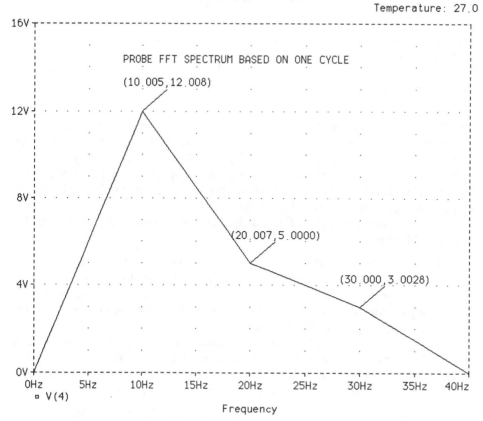

FIGURE 11–29
PROBE spectrum for PSPICE Example 11–2.

Since the record length is now 0.2 seconds, the fundamental frequency is interpreted as 5 Hz, although the value at that frequency and at odd integer multiples of that frequency is zero. (The waveform has *full-wave* symmetry over the time interval shown.) The resulting spectrum is shown in Figure 11–31. The straight-line extrapolation between frequencies is again evident, but the spectrum is now beginning to assume some of the characteristics of a line spectrum.

PSPICE EXAMPLE 11–4
Repeat PSPICE Example 11–2 using 10 cycles of the waveform.

Solution
The transient command was modified in this case to read

```
.TRAN 1E-4 1
```

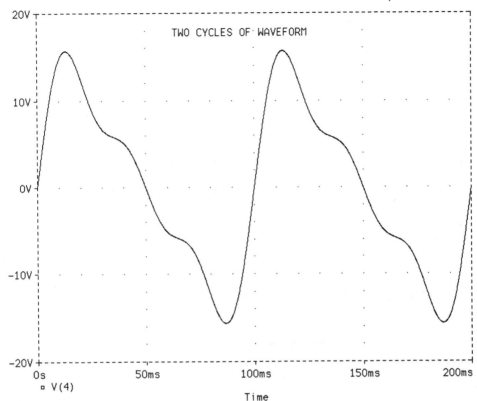

FIGURE 11–30
Waveform for PSPICE Example 11–30.

The waveform is shown in Figure 11–32, and the spectrum is shown in Figure 11–33. The lines used to extrapolate between components reach out only to 1 Hz above and 1 Hz below each of the actual components. The spectral display is, therefore, much closer to the true line spectrum that would result if the time signal continued indefinitely.

From this and previous examples, the following conclusions can be drawn from the PROBE Fourier analysis.

1. If the signal being analyzed is truly periodic, the waveform should be generated and displayed over several cycles, concluding with an integer number of cycles. As the number of cycles increases, the display will more closely resemble the true line spectrum intended.
2. If the function is not periodic within the display window (i.e., either a nonperiodic signal or one cycle of a periodic signal), the spectrum appears

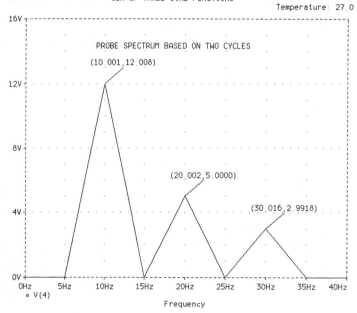

FIGURE 11–31
PROBE spectrum for PSPICE Example 11–3.

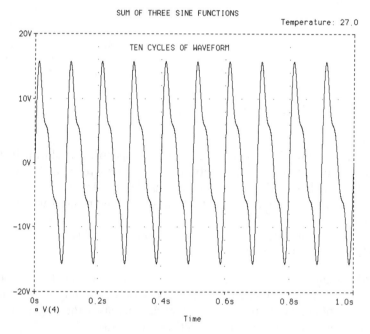

FIGURE 11–32
Waveform for PSPICE Example 11–3.

644

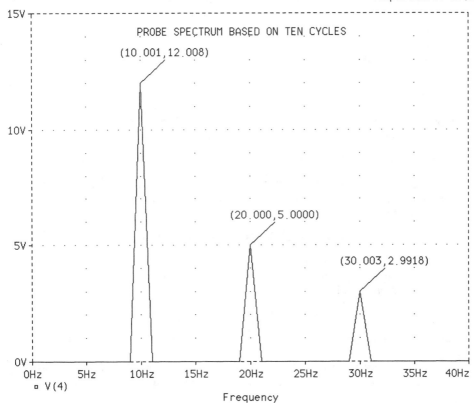

SUM OF THREE SINE FUNCTIONS

Temperature: 27.0

PROBE SPECTRUM BASED ON TEN CYCLES

(10.001,12.008)

(20.000,5.0000)

(30.003,2.9918)

□ V(4)

Frequency

FIGURE 11–33
PROBE spectrum for PSPICE Example 11–4.

as a continuous spectrum. However, since a straight-line extrapolation is used between successive components, the spectral shape may differ somewhat from that of the true Fourier transform.

3. The magnitude scaling is that of the amplitude-phase form of the Fourier series.

GENERAL PROBLEMS

11–1. A certain periodic signal, as viewed on an oscilloscope, has a period of 2 ms. The net positive area in one cycle appears to be greater than the net negative area. List the five lowest frequencies on a one-sided basis that would probably appear in the spectrum.

11–2. A certain periodic signal, as viewed on an oscilloscope, has a period of 5 μs. The net negative area in one cycle appears to be greater than the net positive area. List the five lowest frequencies on a one-sided basis that would probably appear in the spectrum.

11–3. A certain periodic, bandlimited signal has the following frequencies in the Fourier series: dc, 20 Hz, and 40 Hz. The signal can be expressed in sine–cosine form as

$$x(t) = 5 + 6 \cos 40\pi t + 8 \sin 40\pi t - 5 \cos 80\pi t + 12 \sin 80\pi t$$

Express the signal in **(a)** amplitude-phase form and **(b)** complex exponential form.

11–4. A certain periodic, bandlimited signal has the following frequencies in the Fourier series: dc, 500 Hz, and 1 kHz. The signal can be expressed in sine–cosine form as

$$x(t) = 12 + 20 \cos 1000\pi t - 20 \sin 1000\pi t + 12 \cos 2000\pi t - 6 \sin 2000\pi t$$

Express the signal in **(a)** amplitude-phase form and **(b)** complex exponential form.

11–5. A certain periodic, bandlimited signal has the following Fourier series amplitude-phase representation:

$$x(t) = 6 + 10 \sin(100\pi t + 30°) + 8 \sin(200\pi t - 120°)$$

Express the signal in **(a)** sine–cosine form and **(b)** complex exponential form.

11–6. A certain periodic, bandlimited signal has the following Fourier series amplitude-phase representation:

$$x(t) = 20 + 12 \cos(2000\pi t - 60°) + 6 \cos(4000\pi t + 150°)$$

Express the signal in **(a)** sine–cosine form and **(b)** complex exponential form.

11–7. A certain periodic voltage has the form of the square wave shown in Table 11–2. The amplitude is $A = 10$ V, and the period is $T = 1$ ms. Prepare a table listing all frequencies below 10 kHz, and the corresponding amplitude of the components on a one-sided basis.

11–8. A certain periodic voltage has the form of the triangular wave shown in Table 11–2. The amplitude is $A = 20$ V, and the period is $T = 2$ μs. Prepare a table listing all frequencies below 5 MHz, and the corresponding amplitudes on a one-sided basis.

11–9. For the voltage of Problem 11–7, plot the **(a)** one-sided and **(b)** two-sided spectra for the frequency range -10 to 10 kHz.

11–10. For the voltage of Problem 11–8, plot the **(a)** one-sided and **(b)** two-sided spectra for the frequency range -5 to 5 MHz.

11–11. A certain rectangular pulse has the form shown in Table 11–4. The amplitude is $A = 10$ V, and the pulse width is $\tau = 1$ ms.
(a) Write an equation for the Fourier transform.
(b) Sketch the form of the amplitude spectrum between dc and 2 kHz.

11–12. A certain triangular pulse has the form shown in Table 11–4. The amplitude is $A = 20$ V, and the pulse width is $2\tau = 1$ ms.
(a) Write an equation for the Fourier transform.
(b) Sketch the form of the amplitude spectrum between dc and 4 kHz.

11–13. By inspection, determine the spectral roll-off rates for the periodic signals of Table 11–2.
(a) Square wave
(b) Triangular wave
(c) Sawtooth wave

(d) Half-wave rectified wave

(e) Full-wave rectified wave

(f) Pulse train

Express the results both as $1/n^k$ and in decibels per octave. Check conclusions with the tabulated results.

11–14. By inspection, determine the spectral roll-off rates for the nonperiodic signals of Table 11–4.

(a) Rectangular pulse

(b) Triangular pulse

(c) Sawtooth pulse

(d) Half-cycle cosine pulse

Express the results both as $1/f^k$ and in decibels per octave. Check your conclusions with the tabulated results.

DERIVATION PROBLEMS

11–15. Starting with the expression for $\overline{X}_n$ as given by Equation (11–16), apply Euler's formula and establish the formula for Equation (11–18).

11–16. Starting with the exponential form of the Fourier series in Equation (11–15), apply Equations (11–17) and (11–18) and show that the exponential series is equivalent to the sine–cosine series as given by Equation (11–2).

11–17. Derive the even-function symmetry condition for a Fourier series; that is, show that if $x(-t) = x(t)$, integration need be performed over only half a cycle, and the result is doubled. (*Hint:* This theorerm is easier to derive by performing the basic integration from $-T/2$ to $T/2$.)

11–18. Derive the odd-function symmetry condition for a Fourier series; that is, show that if $x(-t) = -x(t)$, integration need be performed over only half a cycle, and the result is doubled (see the hint in Problem 11–17).

11–19. Without using any symmetry conditions, derive the sine–cosine form of the Fourier series of the square wave of Table 11–2. Note that the amplitude-phase form is the same in this case.

11–20. Without using any symmetry conditions, derive the sine–cosine form of the Fourier series of the triangular wave of Table 11–2. Note that the amplitude-phase form is the same in this case.

11–21. Repeat the derivation of Problem 11–19 using a symmetry condition to simplify the analysis.

11–22. Repeat the derivation of Problem 11–20 using a symmetry condition to simplify the analysis.

11–23. Determine the complex exponential form of the series for Problems 11–19 and 11–21.

11–24. Determine the complex exponential form of the series for Problems 11–20 and 11–22.

11–25. Without using any symmetry conditions, derive the Fourier transform of the triangular pulse of Table 11–4.

11–26. Repeat the derivation of Problem 11–25 using a symmetry condition to simplify the analysis.

11–27. Derive the modulation theorem as given by (O–5) of Table 11–5.

11–28. Derive the time-scaling transform pair as given by (O–6) of Table 11–5.

A

DETERMINANTS

Cramer's rule will be given here as a means for solving a set of linear simultaneous equations. Particular emphasis will be placed on the special cases of two and three unknowns. With the modern availability of computer circuit analysis programs such as described in the text, as well as available programs for solving simultaneous equations, few people will be required to solve "by hand" arrays larger than these common cases. There are numerous tricks for simplifying arrays of simultaneous equations to expedite the solution, but these techniques will not be covered here.

Consider the set n simultaneous equations given by

$$
\begin{aligned}
a_{11}x_1 + a_{12}x_2 + \cdots + a_{1n}x_n &= b_1 \\
a_{21}x_1 + a_{22}x_2 + \cdots + a_{2n}x_n &= b_2 \\
a_{n1}x_1 + a_{n2}x_2 + \cdots + a_{nn}x_n &= b_n
\end{aligned}
\tag{A-1}
$$

The quantities $x_1, x_2, \ldots, x_n$ are assumed to be unknown, and all the a and b quantities are assumed to be known. The x quantities will normally be either voltages or currents in most circuit analysis problems.

Cramer's rule provides a solution to the set as follows:

$$
x_1 = \frac{\Delta_1}{\Delta}
$$

$$
x_2 = \frac{\Delta_2}{\Delta}
\tag{A-2}
$$

$$
x_n = \frac{\Delta_n}{\Delta}
$$

where

$$\Delta = \begin{vmatrix} a_{11} & a_{12} & \cdots & a_{1n} \\ a_{21} & a_{22} & \cdots & a_{2n} \\ \cdot & \cdot & & \cdot \\ \cdot & \cdot & & \cdot \\ a_{n1} & a_{n2} & \cdots & a_{nn} \end{vmatrix} \tag{A-3}$$

$$\Delta_1 = \begin{vmatrix} b_1 & a_{12} & \cdots & a_{1n} \\ b_2 & a_{22} & \cdots & a_{2n} \\ \cdot & \cdot & & \cdot \\ \cdot & \cdot & & \cdot \\ b_n & a_{n2} & \cdots & a_{nn} \end{vmatrix} \tag{A-4}$$

$$\Delta_2 = \begin{vmatrix} a_{11} & b_1 & \cdots & a_{1n} \\ a_{21} & b_2 & \cdots & a_{2n} \\ \cdot & \cdot & & \cdot \\ \cdot & \cdot & & \cdot \\ a_{n1} & b_n & \cdots & a_{nn} \end{vmatrix} \tag{A-5}$$

$$\Delta_n = \begin{vmatrix} a_{11} & a_{12} & \cdots & b_1 \\ a_{21} & a_{22} & \cdots & b_2 \\ \cdot & \cdot & & \cdot \\ \cdot & \cdot & & \cdot \\ a_{n1} & a_{n2} & \cdots & b_n \end{vmatrix} \tag{A-6}$$

The preceding quantities are called determinants. Observe that a given Δ_k is obtained from Δ of Equation (A–3) by replacing the kth column of Δ by the coefficients $b_1, b_2, \ldots, b_n$.

A given $n \times n$ determinant can be evaluated by expanding it into $n\ (n-1) \times (n-1)$ determinants, and this process is repeated until the resulting determinants are reduced to first order. Consider, for example, Δ as given by Equation (A–3). It can be evaluated by expanding along the first row.

$$\Delta = a_{11}M_{11} - a_{12}M_{12} + a_{13}M_{13} - a_{14}M_{14} + \cdots + (-1)^{n+1}a_{1n}M_{1n} \tag{A-7}$$

where M_{ij} is an $(n-1) \times (n-1)$ determinant obtained from Δ by crossing out the ith row and jth column. (Alternately, the determinant can be evaluated by expanding along the first column.)

A–1 $n = 2$

A 2×2 determinant has the value

$$\begin{vmatrix} a_{11} & a_{12} \\ a_{21} & a_{22} \end{vmatrix} = a_{11}a_{22} - a_{12}a_{21} \tag{A-8}$$

A-2 $n = 3$

A 3×3 determinant can be expanded as

$$
\begin{vmatrix} a_{11} & a_{12} & a_{13} \\ a_{21} & a_{22} & a_{23} \\ a_{31} & a_{32} & a_{33} \end{vmatrix} = a_{11} \begin{vmatrix} a_{22} & a_{23} \\ a_{32} & a_{33} \end{vmatrix} - a_{12} \begin{vmatrix} a_{21} & a_{23} \\ a_{31} & a_{33} \end{vmatrix} + a_{13} \begin{vmatrix} a_{21} & a_{22} \\ a_{31} & a_{32} \end{vmatrix}
$$

$$
= a_{11}(a_{22}a_{33} - a_{23}a_{32}) - a_{12}(a_{21}a_{33} - a_{23}a_{31}) \tag{A-9}
$$
$$
+ a_{13}(a_{21}a_{32} - a_{22}a_{31})
$$

B*

COMPLEX ALGEBRA

B-1 INTRODUCTION

The use of complex algebra greatly facilitates the solution of many circuit problems involving both the steady-state sinusoidal approach and the general Laplace transform approach. The reader who is acquainted with steady-state ac circuit theory will undoubtedly be familiar with this subject, and this treatment will serve as a review. The reader not familiar with ac circuit theory should find this treatment adequate to deal with the complex number operations required in the text.

B-2 COMPLEX NUMBERS

The concept of a complex number system is derived from a two-dimensional representation as shown in Figure B–1. In this representation we define a complex number $\overline{A}$ as

$$\overline{A} = a_1 + ja_2 \tag{B-1}$$

where $j = \sqrt{-1}$.

The quantity a_1 is called the *real part* of $\overline{A}$, and the quantity a_2 is called the *imaginary part* of $\overline{A}$. Of course, both a_1 and a_2 are themselves real numbers. The real part a_1 is read along the x-axis, and the x-axis is called the *real axis*. The imaginary part a_2 is read along the y-axis, and the y-axis is called the *imaginary axis*. Complex numbers are often called *phasors* when applied to ac circuit theory. The bar above $\overline{A}$ emphasizes that the quantity is complex.

---

* W. D. Stanley, *Transform Circuit Analysis for Engineering and Technology,* 3d ed., 1997, Appendix A. Reprinted by permission of Prentice Hall, Inc., Upper Saddle River, NJ.

FIGURE B-1
Representation of the complex number
system.

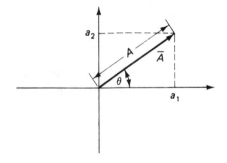

Thus, we may represent a complex number by specifying its real and imaginary parts, two quantities being required to completely specify the number. In the event that the imaginary part is zero, the locus of all complex numbers is the real axis. Thus, the real axis represents the domain of all real numbers in the complex-number system. The form of a complex number in which the real and imaginary parts are specified is called the *rectangular* or *cartesian* form of a complex number.

Referring back to Figure B-1, we observe that if the complex number is interpreted as being a phasor or vector from the origin to the point (a_1, a_2), we can also describe the complex number by specifying its *magnitude A* and its angle (also called the *argument*) with respect to the x-axis. (Positive angles are interpreted in a counter-clockwise rotational direction.) From the geometry, it can be seen that

$$a_1 = A \cos \theta \qquad \textbf{(B-2)}$$

$$a_2 = A \sin \theta \qquad \textbf{(B-3)}$$

Substitution of Equations (B-2) and (B-3) into Equation (B-1) yields

$$\overline{A} = A \cos \theta + jA \sin \theta$$
$$= A(\cos \theta + j \sin \theta) \qquad \textbf{(B-4)}$$

The quantity in parentheses above can be related to a compact expression by means of *Euler's formula*. The proof of this formula is given in most basic calculus books. Euler's formula states that

$$e^{j\theta} = \cos \theta + j \sin \theta \qquad \textbf{(B-5)}$$

Substitution of Equation (B-5) into (B-4) yields

$$\overline{A} = Ae^{j\theta} \qquad \textbf{(B-6)}$$

This form of a complex number is called the *polar form*, in which the magnitude A and the angle θ are directly specified. The beauty of the exponential form is that it obeys all the algebraic laws of exponential functions, and, as we shall see shortly, this form is ideal for multiplying and dividing complex numbers.

Now let us consider a shorthand way of writing the polar form of complex numbers that is universally employed in electrical theory. Referring back to Equation (B–6), we define the following notation:

$$\overline{A} = Ae^{j\theta} \overset{\text{def}}{=} A \underline{/\theta} \qquad \text{(B–7)}$$

In subsequent work, we will use the form defined by Equation (B–7) extensively.

To summarize, a complex number may be expressed in either rectangular or polar form:

$$\overline{A} = a_1 + ja_2 \qquad \text{(B–8)}$$

or

$$\overline{A} = Ae^{j\theta} \overset{\text{def}}{=} A \underline{/\theta} \qquad \text{(B–9)}$$

Conversion from rectangular to polar form is achieved by the relationships

$$A = \sqrt{a_1^2 + a_2^2} \qquad \text{(B–10)}$$

$$\theta = \tan^{-1}\frac{a_2}{a_1} \qquad \text{(B–11)}$$

Conversion from polar to rectangular form is achieved by the relationships

$$a_1 = A \cos \theta \qquad \text{(B–12)}$$

$$a_2 = A \sin \theta \qquad \text{(B–13)}$$

Most scientific calculators have functions for rectangular-to-polar and polar-to-rectangular conversions.

EXAMPLE B–1

Convert each of the following numbers from rectangular to polar form:
(a) $1 + j\sqrt{3}$
(b) $-1 + j\sqrt{3}$
(c) $-1 - j\sqrt{3}$
(d) $1 - j\sqrt{3}$

Solution
The basic angle associated with a base of unity and an altitude $\sqrt{3}$ is $\pi/3$, or 60°. Strictly speaking, the angle should be expressed in radians when placed in the exponential polar form. However, it is a common practice to express the angle in degrees, and we will maintain this practice. Inspection of the signs of the real and imaginary parts of the four numbers results in the answers that follow. Note that (c) and (d) are expressed in both positive and negative angles, since the use of the negative angles results in the smallest possible angle.

(a) $1 + j\sqrt{3} = 2 \underline{/60°}$
(b) $-1 + j\sqrt{3} = 2 \underline{/120°}$
(c) $-1 - j\sqrt{3} = 2 \underline{/240°} = 2 \underline{/-120°}$
(d) $1 - j\sqrt{3} = 2 \underline{/300°} = 2 \underline{/-60°}$

The reader may find it instructive to convert the polar forms obtained back to rectangular forms.

B–3 OPERATIONS WITH COMPLEX NUMBERS

Let us now consider the basic arithmetic operations applied to complex numbers. For reference in subsequent operations, let us define two complex numbers $\overline{A}$ and $\overline{B}$ where

$$\overline{A} = a_1 + ja_2 = A\,\underline{/\theta} \tag{B–14}$$

$$\overline{B} = b_1 + jb_2 = B\,\underline{/\phi} \tag{B–15}$$

as shown in Figure B–2.

Addition

The sum of two complex numbers is given by

$$\overline{A} + \overline{B} = a_1 + b_1 + j(a_2 + b_2) \tag{B–16}$$

According to Equation (B–16), to add complex numbers, we simply add the real parts and the imaginary parts separately. Geometrically, this operation may be represented as shown in Figure B–3, in which one of the phasors is "picked up" and placed on the end of the other phasor. The sum is the resultant phasor between the origin and the terminal point of the second phasor. Thus, complex numbers may be added analytically by Equation (B–16) or by the graphical procedure just discussed, if desired.

Subtraction

The difference between two complex numbers is given by

$$\overline{A} - \overline{B} = a_1 - b_1 + j(a_2 - b_2) \tag{B–17}$$

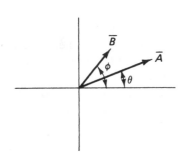

FIGURE B–2
Two complex numbers or phasors.

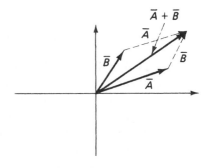

FIGURE B–3
Addition of complex numbers.

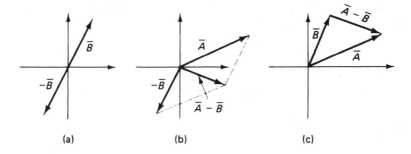

FIGURE B–4
Subtraction of complex numbers.

Since, in an algebraic sense, subtraction is considered as the addition of a negative number, a geometric interpretation is obtained by first forming the quantity $-\overline{B}$ as shown in Figure B–4(a). This is equivalent to rotating the phasor by 180°. Now we add $\overline{A}$ and $-\overline{B}$ as shown in Figure B–4(b). An alternative geometric procedure is to keep the original orientation as shown in Figure B–4(c) and to construct the difference phasor between the terminal points of $\overline{A}$ and $\overline{B}$. This phasor is displaced from the origin, but its magnitude and angle are preserved.

Multiplication

Multiplication of two complex numbers is best achieved by means of the polar forms. We have

$$\overline{A} \times \overline{B} = (A\underline{/\theta})(B\underline{/\phi})$$

$$= AB\underline{/\theta + \phi}$$

(B–18)

Thus, to form the product of two complex numbers, we *multiply the magnitudes* and *add the angles*. The geometric interpretation is shown in Figure B–5.

FIGURE B–5
Multiplication of complex numbers.

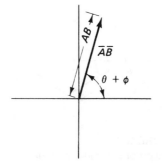

We may also multiply two complex numbers in rectangular form by expanding the products and using the fact that $j^2 = -1$. We thus have

$$\overline{A} \times \overline{B} = (a_1 + ja_2)(b_1 + jb_2) = a_1b_1 - a_2b_2 + j(a_1b_2 + a_2b_1) \quad \text{(B–19)}$$

The reader should not try to remember the result expressed by Equation (B–19), because this can be easily worked out in each case when numbers are given.

Division

Division of two complex numbers is also best achieved by means of the polar forms. We have

$$\frac{\overline{A}}{\overline{B}} = \frac{A\underline{/\theta}}{B\underline{/\phi}} = \frac{A}{B}\underline{/\theta - \phi} \quad \text{(B–20)}$$

Thus, to divide two complex numbers, we *divide the magnitudes* and *subtract the angles*. The geometric interpretation is shown in Figure B–6.

Division may also be accomplished in rectangular form by use of the complex conjugate concept. The complex conjugate of a vector $\overline{A}$ is designated by $\overline{A}^*$. If $\overline{A}$ is given by

$$\overline{A} = a_1 + ja_2 = A\underline{/\theta} \quad \text{(B–21)}$$

then

$$\overline{A}^* = a_1 - ja_2 = A\underline{/-\theta} \quad \text{(B–22)}$$

as shown in Figure B–7.

The product of $\overline{A}$ and $\overline{A}^*$ is given by

$$\overline{A}\,\overline{A}^* = A^2 \quad \text{(B–23)}$$

Thus, the product of a complex number and its conjugate is the square of the magnitude of the number.

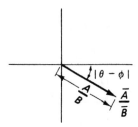

FIGURE B–6
Division of complex numbers.

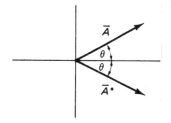

FIGURE B–7
Illustration of the complex conjugate.

In dividing with rectangular forms, an expression of the following form appears:

$$\frac{\overline{A}}{\overline{B}} = \frac{a_1 + ja_2}{b_1 + jb_2} \tag{B-24}$$

If we now multiply both numerator and denominator by the complex conjugate of the denominator, we reduce the denominator to a real number. Hence

$$\frac{\overline{A}}{\overline{B}} = \frac{(a_1 + ja_2)}{(b_1 + jb_2)} \times \frac{(b_1 - jb_2)}{(b_1 - jb_2)} = \frac{a_1b_1 + a_2b_2 + j(a_2b_1 - a_1b_2)}{b_1^2 + b_2^2}$$

$$= \frac{a_1b_1 + a_2b_2 + j(a_2b_1 - a_1b_2)}{B^2} \tag{B-25}$$

EXAMPLE B–2

Given:

$$\overline{A} = 3 + j4 = 5\,\underline{/53.13°}$$

$$\overline{B} = -2 + j2 = 2.828\,\underline{/135°}$$

$$\overline{C} = 2 - j1 = 2.236\,\underline{/-26.57°}$$

Determine **(a)** $\overline{D} = \overline{A} + \overline{B} - \overline{C}$, and **(b)** $\overline{E} = \dfrac{\overline{A} \times \overline{C}}{\overline{B}}$ (two ways).

Solution

(a) Application of the rules for addition and subtraction yields

$$\overline{D} = \overline{A} + \overline{B} - \overline{C} = (3 + j4) + (-2 + j2) - (2 - j1) = -1 + j7 \tag{B-26}$$

(b) Application of the rules for multiplication and division in polar form yields

$$\overline{E} = \frac{\overline{A} \times \overline{C}}{\overline{B}} = \frac{(5\,\underline{/53.13°})(2.236\,\underline{/-26.57°})}{(2.828\,\underline{/135°})}$$

$$= 3.953\,\underline{/-108.44°} \tag{B-27}$$

To perform these operations in rectangular form, we proceed as follows:

$$\overline{E} = \frac{(3 + j4)(2 - j1)}{(-2 + j2)} = \frac{10 + j5}{-2 + j2}$$

$$= \frac{(10 + j5)(-2 - j2)}{(-2 + j2)(-2 - j2)} = \frac{-10 - j30}{8} = -1.25 - j3.75 \tag{B-28}$$

Notice that multiplication and division in polar form resulted in $\overline{E}$ being expressed in polar form, whereas the same operation performed in rectangular form resulted in $\overline{E}$ being expressed in rectangular form. It can be readily shown that the two results are identical.

C

LAPLACE TRANSFORMS

Laplace Transform Pairs Encountered in Circuit Analysis

$f(t)$	$F(s) = \mathcal{L}[f(t)]$	
1 or $u(t)$	$\dfrac{1}{s}$	(T–1)
$e^{-\alpha t}$	$\dfrac{1}{s + \alpha}$	(T–2)
$\sin \omega t$	$\dfrac{\omega}{s^2 + \omega^2}$	(T–3)
$\cos \omega t$	$\dfrac{s}{s^2 + \omega^2}$	(T–4)
$e^{-\alpha t} \sin \omega t$	$\dfrac{\omega}{(s + \alpha)^2 + \omega^2}$	(T–5)*
$e^{-\alpha t} \cos \omega t$	$\dfrac{s + \alpha}{(s + \alpha)^2 + \omega^2}$	(T–6)*
t	$\dfrac{1}{s^2}$	(T–7)
t^n	$\dfrac{n!}{s^{n+1}}$	(T–8)
$e^{-\alpha t} t^n$	$\dfrac{n!}{(s + \alpha)^{n+1}}$	(T–9)
$\delta(t)$	1	(T–10)

* Use when roots are complex.

Laplace Transform Operations Encountered in Circuit Analysis

$f(t)$	$F(s)$	
$f'(t)$	$sF(s) - f(0)$	(O–1)
$\displaystyle\int_0^t f(t)\,dt$	$\dfrac{F(s)}{s}$	(O–2)
$e^{-\alpha t}\,f(t)$	$F(s + \alpha)$	(O–3)
$f(t - T)u(t - T)$	$e^{-sT}\,F(s)$	(O–4)
$f(0)$	$\displaystyle\lim_{s\to\infty} sF(s)$	(O–5)
$\displaystyle\lim_{t\to\infty} f(t)$	$\displaystyle\lim_{s\to 0} sF(s)*$	(O–6)

* Poles of $sF(s)$ must be in left-hand half-plane.

D

ELEMENTS OF PSPICE

D-1 GENERAL DISCUSSION

The original SPICE program was developed at the University of California at Berkeley in the 1970s. SPICE stands for *Simulation Program with Integrated Circuit Emphasis*. This program, along with several variations and improvements, has become an industry and university standard over the years.

PSPICE was introduced by MicroSim Inc.* in the 1980s as a version of SPICE specifically tailored for use on microcomputers. It has received widespread use in colleges and universities as an aid in studying electrical and electronic circuit analysis and design. Its popularity is in part due to the liberal policy of MicroSim in that they supply a student version to faculty members and permit them to reproduce an unlimited number of copies for student distribution.

The student version of PSPICE is limited to 10 transistors, which is much less than the professional version. However, most of the general features of the program are available in the student version. For teaching purposes, the 10-transistor limitation rarely causes a problem, since few student problems will involve circuits with more complexity. All PSPICE examples given in this text can be solved with the student version.

A PSPICE file can be broken into four consecutive parts, as illustrated in Figure D–1. They are (1) the *title;* (2) the *circuit block;* (3) the *control block;* and (4) an *end* statement.

The *title* must be included, but it can be any name that you wish to assign to the circuit. Normally, it will consist of one line only, but it can be extended if necessary.

---

*MicroSim Corporation, 20 Fairbanks, Irvine, CA 92718.

FIGURE D-1
Form of a PSPICE file.

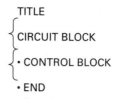

The *circuit block* is usually the longest block (except for very simple circuits), and it contains a complete description of the circuit in a special format that the program will translate to the appropriate mathematical formulation for solution. As a general rule, there will be *one line for each branch* (although more than one line could be used if necessary).

The *control block* provides a series of one or more lines that supply instruction to the program on (1) the type of analysis to be performed and (2) the nature and form of the output results. All lines in this block must begin with a period (.).

The *end* statement must appear as the last line in the file. Like the lines in the control block, it should begin with a period (.). Thus, the correct form is .END.

D-2 EXAMPLE OF PSPICE CIRCUIT CODE

The process of analyzing a circuit is illustrated with a simple dc circuit example. The circuit to be analyzed is shown in Figure D-2(a). The circuit is purely resistive, and it contains one dc voltage source and one dc current source. Although many circuit variables can be determined with PSPICE, the desired variables in this case will be the node voltages and the resistive branch currents.

The circuit must be properly labeled for analysis, and this process is shown in Figure D–2(b). First, all nodes must be numbered. A ground node is selected and it is numbered as 0. When an actual working circuit is analyzed, the ground node should normally be selected to conform to the actual circuit ground.

In the present circuit, the bus along the bottom is chosen as the ground. In this text, PSPICE nodes are enclosed by circles for clarity. Other nodes are labeled as 1, 2, and 3 as shown.

There is no particular order required on the node numbering. They could just as easily have been numbered as 3, 2, 1 from left to right, or as 1, 3, 2, for that matter. However, some logical pattern is recommended as an aid in checking the circuit. This author follows a left-to-right and bottom-to-top pattern when possible.

Note in the circuit form of Figure D–2(b) and the code of Figure D–2(c) that various labels and values have been changed somewhat in form. First, the various subscripts on parameter names are written on the same levels as the first letter in the code. Subscripts may not be entered on the computer as they appear in a textbook, so all entries must be at the same level. For example, V_s must be entered on the computer as VS, R_1 must be entered as R1, and so forth.

Although there are various prefixes available in PSPICE for entering values, to simplify this abbreviated treatment, *we will always enter values in their basic*

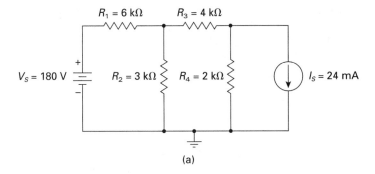

(a)

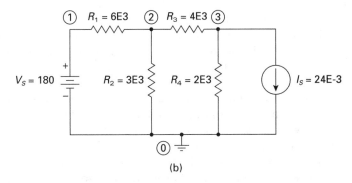

(b)

```
PSPICE SIMPLE DC CIRCUIT EXAMPLE
VS 1 0 DC 180
R1 1 2 6E3
R2 2 0 3E3
R3 2 3 4E3
R4 3 0 2E3
IS  3 0 DC 24E-3
.DC VS 180 180 1
.PRINT DC V(1) V(2) V(3)
.PRINT DC I(R1) I(R2) I(R3) I(R4)
.OPTIONS NOPAGE
.END
```

(c)

FIGURE D–2
Example used to illustrate PSPICE code.

units. When numbers are very small or very large, they will be expressed in scientific or engineering notation form using floating-point values, and the basic units will be understood. For example, a resistance of value 2 MΩ will be expressed as 2×10^6 Ω, and the proper format we will employ for entering it on the computer will be 2E6. As a second example, a capacitor of value 0.5 μF will be expressed as 0.5×10^{-6} F, and the proper format for computer entry will be 0.5E-6.

Now let us turn our attention to the actual circuit description (or *code*), which is shown in Figure D–2(c). The first line is the *title,* which has been given the name PSPICE SIMPLE DC CIRCUIT EXAMPLE.

The next six lines constitute the *circuit block,* and a complete description of the circuit is given. Note that there are six branches, so there is one line for each branch.

The first line in the circuit block reads

```
VS 1 0 DC 180
```

There are five separate items on the line. The first item identifies the type of circuit element in the branch. The symbol *V* as the first letter is the code for a voltage source, and subsequent letters or numbers (up to 7) *without a space* are additional identifying information. (Since there is only one voltage source in this example, we could actually get by with the symbol *V.*) One or more spaces following VS identify the end of the first item, and this space pattern will continue after additional items.

The next two items in the voltage source description are the integers 1 0, and they identify the two nodes to which the element is connected. For a voltage source, the order is important in that the first node listed is the positive node of the voltage source. Thus, if the two integers had been entered in the order 0 1, it would tell the program that the voltage had its positive terminal at ground.

The next item in the line is the designation DC. This tells the computer that the source VS is a dc source. As you might guess, AC would be the designation for an ac phasor source; many other types of sources are also available in PSPICE.

The last item is the quantity 180, and this is the value of the dc voltage source in volts. Incidentally, if the node order had been expressed as 0 1, a value of −180 would, in effect, still create a positive value at node 1. (We have the algebraic negative of a negative!) However, that form is awkward and not recommended unless necessary for some reason.

The second line in the circuit block reads

```
R1 1 2 6E3
```

There are four separate items on this line. Once again, the first item identifies the type of circuit element in the branch. The symbol *R* as the first letter is the code for a resistance, and the 1 corresponds to the subscript.

The next two items in the description are integers 1 2, and they identify the two nodes to which the resistor is connected. The order of the two nodes represents the reference direction of assumed current flow, which in this case is from node 1 to node 2. As always in circuit analysis, if the current is actually flowing from node 2 to node 1, the current value would turn out to be negative.

As a general rule, if the desired output data are to be voltages only, the order of the nodes for *R*, *L*, and *C* is arbitrary.

The last item is the value 6E3, and this represents a resistance of 6×10^3 Ω, or 6 kΩ.

The next three lines describe the remaining three resistive branches, and the code in each case follows the format of R1.

The last line in the circuit block reads

```
IS 3 0 DC 24E-3
```

As in the case of the voltage source, the first letter I of the first item identifies the element as a current source, and the letter S is the subscript. (Since there is only one current source, I would have been sufficient.)

The next two items in the current source description are the integers 3 0, and they identify the two nodes to which the element is connected. As in the case of the voltage source, the order is important for a current source.

In the case of a current source, *the first node listed is the node at which the current enters the source, and the second node listed is the node at which the current leaves the source.* Thus, the direction of flow in this case is from 3 to 0 for the source. If the order of the numbers were reversed, the computer would assume that the current source would be directed upward.

The next item on the line is the designation DC, which, of course, identifies the source as a dc source. Finally, the value 24E-3 is the current source value in amperes.

We now direct our attention to the *control* block. The first line reads

```
.DC VS 180 180 1
```

This line is a bit clumsy looking, and it represents one of the eccentricities of PSPICE when a single-point analysis is desired. Basically, dc analysis was intended for sweeping an input voltage or current over a range of values. When analysis at a single source value is desired, it is necessary to "fool" the program so that the "sweep" turns out to be a single value.

The first two items represent the type of analysis .DC and the name of the source that is to be "swept," which has been designated VS. (Since there are two sources in this circuit, *IS* could also have been selected for this purpose.) The three data values are given in order as 180 180 1. The first value represents the beginning value of the sweep, which is 180 V. The second value represents the ending value of the sweep, which is also 180 V in this case. The third value is the increment between successive sweeps, and for this single-point case, any nonzero value would satisfy the computer. The simplest choice is the value 1.

The next line reads

```
.PRINT DC V(1) V(2) V(3)
```

The command .PRINT tells the program that it should provide an output to a printer, and DC indicates that it is the dc data that are desired, which is the only type of data in this example. The next three items represent desired output variables, which for this line are the voltages at nodes 1, 2, and 3, respectively.

A single print statement can handle up to five different output variables in a block form. In this case, there are seven desired output variables (three node voltages and four branch currents). Although two of the currents could have been put on the same line with the voltages, for convenience in the output data, all currents have been entered in a second .PRINT statement. As long as no more than five output variables are to be grouped together, a single .PRINT statement will suffice.

On the second .PRINT line, note the manner in which the currents are listed. I(R1) represents the current through R_1; I(R2) represents the current through R_2, and so forth.

The next line reads

```
.OPTIONS NOPAGE
```

This statement is optional and could have been eliminated without affecting the analysis. In general, when a program requires more than one page of printout, much of the header information concerning title, date, time, and so forth, is repeated page after page. Unless there is a reason for wanting to have this information on each separate page, the use of this option will generally reduce the volume of paper and provide a more compact package of data. We use it extensively throughout the text when tabulated data are desired.

All the printer output for this example fits on one page, and it is shown in

```
PSPICE SIMPLE DC CIRCUIT EXAMPLE

****       CIRCUIT DESCRIPTION

***********************************************************************

VS 1 0 DC 180
R1 1 2 6E3
R2 2 0 3E3
R3 2 3 4E3
R4 3 0 2E3
IS 3 0 DC 24E-3
.DC VS 180 180 1
.PRINT DC V(1) V(2) V(3)
.PRINT DC I(R1) I(R2) I(R3) I(R4)
.OPTIONS NOPAGE
.END

****       DC TRANSFER CURVES            TEMPERATURE =   27.000 DEG C

VS          V(1)        V(2)        V(3)
1.800E+02   1.800E+02   3.300E+01  -2.100E+01

****       DC TRANSFER CURVES            TEMPERATURE =   27.000 DEG C

VS          I(R1)       I(R2)       I(R3)       I(R4)
1.800E+02   2.450E-02   1.100E-02   1.350E-02  -1.050E-02

        JOB CONCLUDED
        TOTAL JOB TIME              4.23
```

FIGURE D–3
Output data for simple dc circuit example.

Figure D–3. The title of the program (PSPICE SIMPLE DC CIRCUIT EXAMPLE) appears near the top of the page, and the remainder of the code appears under CIRCUIT DESCRIPTION. The remainder of the circuit description is exactly the same as that of the code in Figure D–2.

For each .PRINT statement, there is a collection of data listed under the heading of DC TRANSFER CURVES. As explained earlier, this heading refers to a possible sweep of the input voltage, and it does not quite describe the situation for a single-point analysis. The designation of a temperature refers to the default temperature (27°C) for semiconductor and temperature-dependent parameter variation. This designation should be ignored for the present analysis.

The first line of data provides a listing of the input voltage (because it was the source that was "swept") and all node voltages referred to ground. All values are expressed in floating-point form in their basic units. Note that V(1) = VS = 180 V, which was obvious from the circuit diagram. The other voltages are V(2) = 33 V and V(3) = −21 V. Note that the latter voltage is negative with respect to ground.

The second line of data is keyed to the second .PRINT statement, and the value of VS is repeated. The next four values are the various branch currents in amperes, and they are read as I(R1) = 2.45×10^{-2} A = 24.5 mA; I(R2) = 1.1×10^{-2} A = 11 mA; I(R3) = 1.35×10^{-2} A = 13.5 mA; I(R4) = -1.050×10^{-2} A = −10.5 mA.

The last two lines tell us that the analysis is concluded and give the time required in seconds. In this case the analysis took 4.23 s.

D–3 ELEMENT CODES

A general summary of some of the most common circuit models and their PSPICE codes is given in this section. For the most part, these codes are explained through examples in the various figures provided in this section.

Passive Elements

The forms of the circuit models and codes for resistance (R), capacitance (C), and inductance (L) are provided through examples in Figure D–4. In each case, arbitrary numbers for the nodes, subscripts, and values have been assumed in a way that no single value appears more than once in the example. This author has seen examples in some texts where the number for the node might, for example, be the same as a parameter value, and this might lead to misinterpretation. Basically, the forms for all three passive elements are the same, except for the first letter $(R, C, $ or $L)$.

Independent dc and ac Voltage Sources

The forms and codes for independent dc and phasor ac voltage sources are shown in Figure D–5. Basically, they follow the same format except that one more entry,

$R_3 = 100\ \Omega$

⑧ —————⟋⟍⟍⟍⟋————— ⑨

R3 8 9 100

Resistor R_3 connected between nodes 8 and 9 with a value of 100 Ω.

(a)

$C_7 = 0.39\ \mu\text{F}$

③ ————— ⊣⊢ ————— ⑥

C7 3 6 0.39E-6

Capacitor C_7 connected between nodes 3 and 6 with a value of 0.39 μF.

(b)

$L_4 = 1.5\ \text{mH}$

⑦ —————⟋⟍⟍⟍⟍⟋————— ⑥

L4 7 6 1.5E-3

Inductor L_4 connected between nodes 7 and 6 with a value of 1.5 mH.

(c)

FIGURE D–4

PSPICE codes for three passive elements.

representing the angle of the voltage source, might appear as the last item on the ac source code. If this value is omitted, PSPICE will assume a value of 0°. The first entry in each case must be V. For the two node numbers, the first one represents the positive reference terminal of the source. Following the node connections, either DC or AC should be entered. If this quantity is omitted, PSPICE defaults to DC.

One point about the ac magnitude should be noted. If the desired calculations do not involve power, the magnitude may be interpreted as either the peak or the rms value, as desired, and the resulting output will be interpreted the same way. If power computations are to be performed, however, the magnitude should be the rms value. For example, if the peak value of a phasor voltage is 10 V and power computations are to be performed, the magnitude of this voltage should be entered as $10/\sqrt{2} = 7.071$ V.

Independent dc and ac Current Sources

The forms and code for independent dc and phasor ac current sources are shown in Figure D–6. As in the case of voltage sources, their formats are similar except for the possible additional entry for the angle of an ac source. The first entry in each case is the letter I. For the two-node numbers, the current source is assumed to be flowing from the first node to the second node. Once again, either DC or AC should be entered.

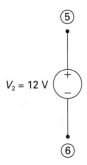

V2 5 6 DC 12
Voltage source V_2 is connected between
nodes 5 and 6 with the positive terminal
at node 5. It is a dc source with value 12 V.

(a)

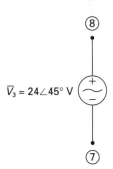

V3 8 7 AC 24 45
Voltage source $\overline{V}_3$ is connected between nodes 8 and 7
with the positive terminal at node 8. It is an ac source
with value 24 V and angle 45°.

(b)

FIGURE D–5
PSPICE codes for independent dc and ac voltage sources.

The point concerning peak or rms values for voltage sources applies equally to current sources. Thus, if power computations are desired, current source values should be entered as rms values.

Dependent Sources

PSPICE permits the four forms for dependent sources to be modeled. These forms are summarized in Figure D–7. Since V and I are used for independent sources, other symbols are used for dependent sources. These symbols are listed as follows:

E   voltage-controlled voltage source (VCVS)
G   voltage-controlled current source (VCIS)
H   current-controlled voltage source (ICVS)
F   current-controlled current source (ICIS)

For voltage-controlled sources (VCVS and VCIS), there are four integer numbers following the source designation. The first two integers specify the nodes to which the source is connected, with the first representing the positive reference for a voltage source and the node at which the current enters the source for a current source. The third and fourth integers represent the nodes that define the controlling voltage, with the third representing the positive reference for the voltage.

For current-controlled sources (ICVS and ICIS), there are only two integer numbers following the source designation, and they specify the connection nodes in the same manner as for the voltage-controlled sources. However, following the node number is the specification of a voltage source through which the controlling current flows. This voltage source may be a dummy source having a value of zero.

FIGURE D–6
PSPICE codes for independent dc and ac current sources.

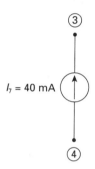

③

$I_7 = 40$ mA

④

I7 4 3 DC 40E-3
Current source I_7 is connected between nodes 4 and 3 with current flow from 4 to 3. It is a dc source with value 40 mA.

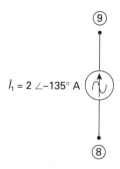

⑨

$\bar{I}_1 = 2 \angle{-135°}$ A

⑧

I1 8 9 AC 2 -135
Current source $\bar{I}_1$ is connected between nodes 8 and 9 with current flow from 8 to 9. It is an ac source with magnitude 2 A and angle $-135°$.

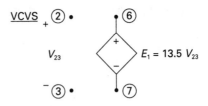

E1 6 7 2 3 13.5

The dependent voltage source E_1 is connected between nodes 6 and 7 with the positive terminal at node 6. It is controlled by the voltage between nodes 2 and 3 (positive reference at 2) and the voltage gain is 13.5.

(a)

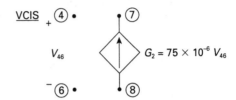

G2 8 7 4 6 75E-6

The dependent current source G_2 is connected between nodes 8 and 7 with current flow from 8 to 7. It is controlled by the voltage between nodes 4 and 6 (positive reference at 4) and the transconductance is 75 μS.

(b)

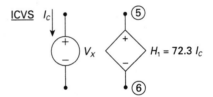

H1 5 6 VX 72.3

The dependent voltage source H_1 is connected between nodes 5 and 6 with the positive terminal at node 5. It is controlled by the current I_C flowing through the source V_X and the transresistance is 72.3 Ω.

(c)

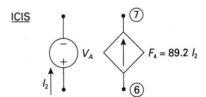

F4 6 7 VA 89.2

The dependent current source F_4 is connected between nodes 6 and 7 with current flow from 6 to 7. It is controlled by the current I_2 flowing through the source V_A, and the current gain is 89.2.

(d)

The last entry is the controlling factor ("gain") for the particular type of source. The following quantities are represented by this value:

VCVS   voltage gain A (no dimensions)
VCIS   transconductance g_m (siemens)
ICVS   transresistance r_m (ohms)
ICIS   current gain β (no dimension)

The form of the source (dc, ac, or otherwise) is not specified for a dependent source, since the form will have the same nature as the controlling variable.

Study the four examples in Figure D–7 carefully so that you will recognize the forms when they appear in the text.

ANSWERS TO SELECTED ODD-NUMBERED PROBLEMS

Chapter 1

1–1. (a) 43.2×10^6 J (b) 12 kWh (c) $1.02

1–3. 15¢

1–5. (a) 1.5 kWh (b) 5.4×10^6 J

1–7. (a) 18 W delivered (b) 24 W absorbed

1–9. 2.681 mA

1–11. (a) 24 W (b) 72 W (c) 96 W

1–13. 6 V, 20 mA, 11 mA, 1636 Ω

1–15. 15 V, 3 Ω

1–17. 6 Ω

1–19. 8 Ω

1–21. 10 V, 30 V, 80 V

1–25.

R	2 Ω	5Ω	6Ω	8Ω	4 Ω	16 Ω
I	10 A	8 A	1.333 A	0.667 A	0.667 A	2 A
V	20 V	40 V	8 V	5.333 V	2.667 V	32 V

1–33. 29 kΩ, 99 kΩ, 299 kΩ, 999 kΩ

1–35. 18 kΩ, 6 kΩ

Chapter 2

2–1. (a) 3 A (b) 12 V, 9 V, 15 V (c) Total power = 150 W

2–3. (a) 2 A (b) 4 V, 8 V, 10 V (c) -16 V (d) Total power = 56 W

2–5. (a) 24 V (b) 1 A, 2 A, 3 A (c) Total power = 240 W

2–7. (a) 20 V (b) 4 A, 10 A (c) 5 A (d) Total power = 340 W

2–9. 10 V

2–11. (a) 28 W (b) 20 W

2–13. (b) 2 A, $-$ 3 A, 5 A

(c)

R	3Ω	2 Ω	4 Ω	5 Ω	6 Ω
I	2 A($\rightarrow$)	5 A($\downarrow$)	2 A($\rightarrow$)	8 A($\uparrow$)	5 A($\rightarrow$)

2–17. 7 V in series with 7 Ω. 1 A in parallel with 7 Ω

2–19. (a) 3 A (b) 15 V, 12 V, 18 V (c) 15 V (d) Total power = 135 W

2–21. (a) 20 V (b) 3.333 A, 6.667 A (c) 5 A (d) Total power = 240 W
2–31. 10.91 Ω
2–33. $R_1 = 4950$ Ω, $R_2 = 50.505$ Ω
2–35. 9 kΩ, 1 kΩ, 1 W
2–37. 8 V, 60 kΩ

Chapter 3
3–1.

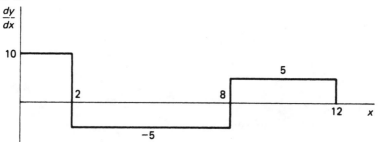

3–3.

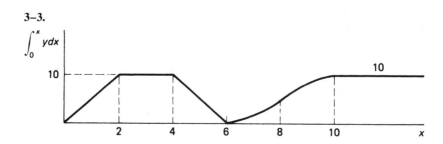

3–5.

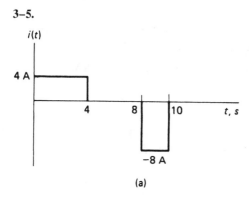

(a)

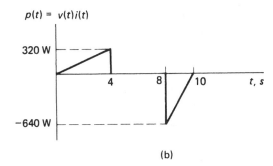

$p(t) = v(t)i(t)$

320 W

4   8   10 t, s

−640 W

(b)

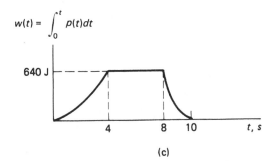

$w(t) = \int_0^t p(t)dt$

640 J

4   8   10 t, s

(c)

3–7.

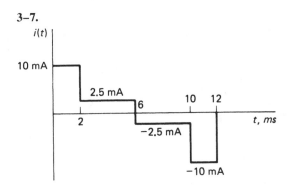

$i(t)$

10 mA

2.5 mA

6   10   12

2

t, ms

−2.5 mA

−10 mA

3–9.

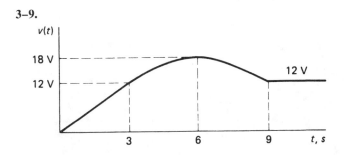

$v(t)$

18 V

12 V

12 V

3   6   9 t, s

3–11. The shape of $v(t)$ has the form of the result of Problem 3–9, but it starts at 5 V, climbs to a peak of 23 V, and drops to a final level of 17 V.

3–13.

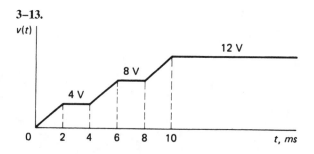

3–15.

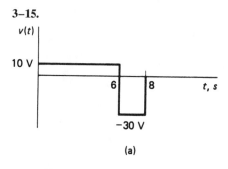

(a)

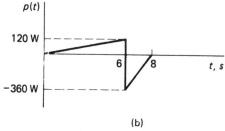

(b)

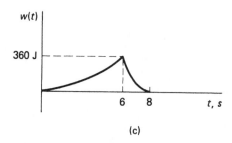

(c)

3–17.

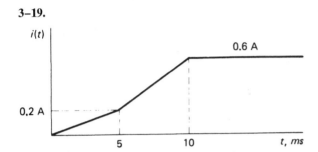

3–19.

3–21. 5 μF

3–23. 32 V, 8 V, 24 V

3–25. (b) $-v_1 + L_1\dfrac{di_1}{dt} + M\dfrac{di_2}{dt} = 0$

$$L_2\dfrac{di_2}{dt} + M\dfrac{di_1}{dt} + Ri_2 = 0$$

3–27. $v_1 = 24 \sin \omega t$, $i_1 = 0.125 \sin \omega t$
$v_2 = 6 \sin \omega t$, $i_2 = 0.5 \sin \omega t$

3–29. (a) 88.9%   (b) 99.5%

3–31. $\omega CV_p \cos \omega t$

3–33. $\omega LI_p \cos \omega t$

3–43. A 0.25-μF capacitor will perform the conversion.

3–45. -1.1 V

3–47. (a) $KV_0t - V_{\text{ref}}$ (b) $\dfrac{2V_{\text{ref}}}{KV_0}$

(d) $v_1(t)$ is a square-wave oscillating between $-V_0$ and $+V_0$, and $v_2(t)$ is a triangular wave oscillating between $-V_{\text{ref}}$ and V_{ref}.

Chapter 4

4–1. $i_{R1} = i_C = 4.8$ A, $v_L = v_{R1} = 24$ V, $v_C = v_{R2} = 0$, $i_{R2} = 0$

4–3. $i_R = i_C = i_L = 3$ A, $v_R = 6$ V, $v_C = 8$ V, $v_L = 10$ V

4–5. $i_{R1} = i_L = i_{R2} = i_{R3} = 2$ A, $v_{R1} = 6$ V, $v_{R2} = 8$ V, $v_{R3} = v_{C2} = 10$ V, $v_{C1} = 18$ V, $i_{C1} = 0$, $i_{C2} = 0$, $v_L = 0$

4–7. (a) $i_L = 2$ A, $v_{C1} = 20$ V, $v_{C2} = v_{R2} = 20$ V, $i_{R2} = 2$ A, $i_{C2} = 0$, $v_L = 0$, $v_{R1} = -5$ V, $i_{R1} = -1$ A, $i_{C1} = -3$ A
(b) $i_{R1} = i_L = i_{R2} = 1$ A, $v_{R2} = v_{C2} = v_{C1} = 10$ V, $v_{R1} = 5$ V, $v_L = 0$, $i_{C1} = 0$, $i_{C2} = 0$

4–9. (b) $i(t) = 10^{-3}e^{-0.1t}$, $v_C(t) = 100(1 - e^{-0.1t})$

4–11. (b) $i(t) = 0.1(1 - e^{-100t})$, $v_L(t) = 100e^{-100t}$

4–13. $i(t) = 1.2e^{-2t}$, $v_C(t) = 20 - 12e^{-2t}$

4–15. $v_C(t) = 5 + 20e^{-2t}$

4–17. $v_C(t) = 8(1 - e^{-0.5t})$

4–19. $i(t) = 4 - 2e^{-2t}$, $v_l(t) = 24e^{-2t}$

4–21. 0.693 ms

4–29. (a) 11 μs (b) 0.55 μs

4–31. 12.78 μs

Chapter 5

5–1. $\dfrac{6}{s}$

5–3. $\dfrac{32}{s^2 + 16}$

5–5. $\dfrac{7(s + 2)}{s^2 + 4s + 13}$

5–7. $\dfrac{8}{s + 2}$

5–9. $2 + \dfrac{3}{s}$

5–11. $6e^{-4t}$

5–13. $5 \sin 4t$

5–15. $8e^{-t} \sin 2t$

5–17. $5e^{-2t} \cos 3t + 4e^{-2t} \sin 3t$

5–19. $3\delta(t) + 2t^2$

5–21. $4e^{-t} + 2e^{-3t}$

5–23. $2 + 3e^{-t} - 4e^{-2t}$

5–25. $3 - 9e^{-t} + 9e^{-2t} - 3e^{-3t}$

5–27. $20e^{-1000t} - 10e^{-2000t}$

5–29. $7.692 + 9.245e^{-2t} \sin(3t - 123.7°)$

5–31. $7.692 - 10e^{-t} + 2.924e^{-2t} \sin(3t + 127.9°)$

5–33. $5 + 5 \sin(2t - 90°) = 5 - 5 \cos 2t$

5–35. $2.076 \sin(2t - 41.63°) + 1.384e^{-2t} \sin(3t + 94.76°)$

5–37. $20e^{-t} + 20t - 20$

5–45. Roots are -1, $-0.5 \pm j0.8660$.

5–47. Roots are -1, $-0.3090 \pm j0.9511$, $-0.8090 \pm j0.5878$.

Chapter 6

6–1. (a) 50 (b) $10^8/s$ (c) $0.2s$

6–3. Voltage source $12/s$ in series with impedance $2 \times 10^6/s$
Current source 6×10^{-6} in parallel with impedance $2 \times 10^6/s$

6–5. Current source $6/s$ in parallel with impedance $0.2s$
Voltage source 1.2 in series with impedance $0.2s$

6–7. $\left(3s + \dfrac{10}{s}\right) I_1(s) - \dfrac{2}{s} I_2(s) = \dfrac{24}{s}$

$-\dfrac{2}{s} I_1(s) + \left(6s + 5 + \dfrac{2}{s}\right) I_2(s) = 12 + \dfrac{4}{s}$

6–9. $\left(2s + \dfrac{1}{4} + \dfrac{8}{15s}\right) V_1(s) - \left(2s + \dfrac{1}{5s}\right) V_2(s) = \dfrac{8}{s} - \dfrac{20s}{s^2 + 4}$

$-\left(2s + \dfrac{1}{5s}\right) V_1(s) + \left(5s + \dfrac{1}{5} + \dfrac{1}{5s}\right) V_2(s) = 36 + \dfrac{6}{s}$

6–11. $\dfrac{2(s^2 + 2)}{s^3 + 8s}$

6–13. Voltage source $\dfrac{40}{s(s + 2)}$ in series with impedance $\dfrac{4}{s + 2}$

6–15. (a) natural response = $5e^{-20t} \cos(40t + 30°)$
forced response = $12 \sin(50t + 60°)$
(b) The natural response is a transient response, and the forced response is a steady-state response.

6–17. $i(t) = (V_s/R)e^{-t/RC}$, $v_C(t) = V_s(1 - e^{-t/RC})$

6–19. (a) $i(t) = -4.8e^{-2.667t} + 6 \sin(2t + 53.13°)$
$v_C(t) = 14.4e^{-2.667t} + 24 \sin(2t - 36.87°)$
(b) $i(0^+) = 0$, $v_C(0^+) = 0$

6–21. (a) $i(t) = -8.1333e^{-2.667t} + 6 \sin(2t + 53.13°)$
$v_C(t) = 24.4e^{-2.667t} + 24 \sin(2t - 36.87°)$
(b) $i(0^+) = -3.333$ A, $v_C(0^+) = 10$ V

6–23. (a) $i(t) = 1.6e^{-1.5t} + 2 \sin(2t - 53.13°)$
$v_L(t) = -9.6e^{-1.5t} + 16 \sin(2t + 36.87°)$
(b) $i(0^+) = 0$, $v_L(0^+) = 0$

6–25. $i(t) = 5e^{-t} \sin 2t$
$v_C(t) = 20 + 22.36e^{-t} \sin(2t - 116.56°)$

6–27. $20e^{-t} \sin 2t$

6–29. $2.5e^{-t} \sin(2t + 126.87°) + 2.236 \sin(5t - 63.43°)$

6–37. 16.67 mH, 7.5 μF

6–39. (b) 555.6 Ω

Chapter 7

7–1. $\dfrac{10^5}{s + 10^5}$

7–3. Same as 7–1

7–5. (a) $20 - 20e^{-10^5 t}$

(b) $-19.802e^{-10^5t} + 19.901 \sin(10^4t + 84.289°)$
$v_2(0^+) = 0$ in both cases

7–7. $\dfrac{1}{s^3 + 2s^2 + 2s + 1}$

7–9. $\dfrac{2s}{s^2 + 2s + 5}$

7–11. $\dfrac{1.5858}{s^2 + 1.4142s + 1}$

7–13. (a) $A_1 + A_2e^{-3t} + A_3e^{-4t} + B_1e^{-2t} \sin(5t + \theta_1)$
(b) $B_2 \sin(4t + \theta_2) + A_4e^{-3t} + A_5e^{-4t} + B_3e^{-2t} \sin(5t + \theta_3)$
The first term in each case is the forced response.

7–15. $\dfrac{A(s^2 + 6s + 13)}{(s + 4)(s^2 + 4s + 20)}$

7–17. Three finite poles: $-4, -3 \pm j5$
Two finite zeros: $0, -2$
One zero at $s = \infty$

7–19. Stable system

7–21. Stable system

7–23. $\dfrac{G_1(s)G_4(s)}{1 + G_1(s)G_2(s)G_3(s)}$

7–25. $\omega_n = 2.236$ rad/s, $\zeta = 0.4472$

7–27. $\omega_n = 2.236$ rad/s, $\zeta = 0.4472$

7–35. 0.4 H, 0.5 F

7–37. 0.4 H, 0.5 F

Chapter 8

8–1. $200 \underline{/30°}$

8–3. (a) $v(t) = 20 \sin(100t + 60°)$, $\overline{V} = 20 \underline{/60°}$
(b) $i(t) = 12 \sin(200t + 90°)$, $\overline{I} = 12 \underline{/90°}$
(c) $v(t) = 20 \sin(150t + 150°)$, $\overline{V} = 20 \underline{/150°}$
(d) $f(t) = 15 \sin(75t - 90°)$, $\overline{F} = 15 \underline{/-90°}$
(e) $v(t) = 170 \sin(377t + 165°)$, $\overline{V} = 170 \underline{/165°}$
(f) $i(t) = 4 \sin(50t - 155°)$, $\overline{I} = 4 \underline{/-155°}$
(g) $f(t) = 25 \sin(200t - 130°)$, $\overline{F} = 25 \underline{/-130°}$
(h) $v(t) = 10 \sin(200t - 70°)$, $\overline{V} = 10 \underline{/-70°}$

8–5. (a) $-20 \sin(100t) - 120°)$
(b) $-12 \sin(200t - 90°)$
(c) $-20 \sin(150t - 30°)$
(d) $-15 \sin(75t + 90°)$
(e) As given
(f) As given
(g) $-25 \sin(200t + 50°)$
(h) $-10 \sin(200t + 110°)$

8–7. $13 \sin(1000t + 97.380°)$

8–9. $11.327 \sin(100t + 61.98°)$

8–11. $1.732V_p \sin(377t - 30°)$

8–13. $207.8\sqrt{2} \cos 377t$

8–15. (a) $R = 120\ \Omega$, $X = 50\ \Omega$
(b) Resistance of 120 Ω in series with inductive impedance of $j50\Omega$
(c) $\overline{Y} = 7.101 \times 10^{-3} - j2.959 \times 10^{-3}$, $G = 7.101 \times 10^{-3}$ S, $B = -2.959 \times 10^{-3}$ S
(d) Conductance of 7.101 mS (140.8 Ω) in parallel with inductive susceptance of -2.959 mS (338 Ω).

8–17. (a) $31.416\ \Omega$, -0.031831 S
(b) $785.40\ \Omega$, -1.2732×10^{-3} S

8–19. (a) $-15,915\ \Omega$, 62.832×10^{-6} S
(b) $-636.62\ \Omega$, 1.5708×10^{-3} S

8–21. $i(t) = 0.04 \sin(5000t + 36.87°)$
$v_R(t) = 16 \sin(5000t + 36.87°)$
$v_L(t) = 52 \sin(5000t + 126.87°)$
$v_C(t) = 64 \sin(5000t - 53.13°)$

8–23. $i_R(t) = 0.02 \sin \omega t (\omega = 125.66 \times 10^3\ \text{rad/s})$
$i_C(t) = 0.025133 \cos \omega t$
$i_L(t) = -0.015915 \cos \omega t$
$i_0(t) = 0.02202 \sin (\omega t + 24.743°)$

8–25. $\overline{I}_1 = 4.8423\ \underline{/35.727°}$, $\overline{I}_2 = 2.4914\ \underline{/4.764°}$

8–27. See answer to 8–25.

8–29. 0.32 W

8–39. 0.002533 μF

8–41. 717 pF

8–43. 693.7 pF, 34.69 μH

Chapter 9

9–1. (a) $i(t) = 6.4e^{-4t} + 6 \sin(3t + 143.13°)$
$v(t) = -19.2e^{-4t} + 24 \sin(3t + 53.13°)$

9–3. $G(j\omega) = \dfrac{j\omega}{10^4 + j\omega}$, $A(\omega) = \dfrac{\omega}{\sqrt{10^8 + \omega^2}}$,
$A_{dB}(\omega) = 20 \log_{10}\left(\dfrac{\omega}{\sqrt{10^8 + \omega^2}}\right)$
$\beta(\omega) = 90° - \tan^{-1}(\omega/10^4)$

9–5. $G(j\omega) = \dfrac{1}{1 - \omega^2 + j1.4142\omega}$, $A(\omega) = \dfrac{1}{\sqrt{1 + \omega^4}}$,
$A_{dB}(\omega) = -10 \log_{10}(1 + \omega^4)$,
$\beta(\omega) = -\tan^{-1}\dfrac{1.4142\omega}{1 - \omega^2}$

9–7. Flat level of 26 dB below $\omega = 10$ rad/s. Changes to a slope of -6 dB/octave at $\omega = 10$ rad/s and reaches a level of 6 dB at $\omega = 100$ rad/s. Changes to a slope of -12 dB/octave at $\omega = 100$ rad/s and continues at that slope.

9–9. Flat level of -12 dB below $\omega = 20$ rad/s. Changes to a slope of -6 dB/octave at $\omega = 20$ rad/s and reaches a level of -26 dB at $\omega = 100$ rad/s. Changes to a slope of -12 dB/octave at $\omega = 100$ rad/s and continues at that slope.

9–11. Flat level of 0° below $\omega = 1$ rad/s. Changes to a slope of $-45°$/decade at $\omega = 1$ rad/s and reaches a level of $-45°$ at $\omega = 10$ rad/s. Changes to a slope of $-90°$/

decade at $\omega = 10$ rad/s and reaches a level of $-135°$ at $\omega = 100$ rad/s. Changes back to a slope of $-45°$/decade at $\omega = 100$ rad/s and reaches a level of $-180°$ at $\omega = 1000$ rad/s. Flat level of $-180°$ above $\omega = 1000$ rad/s.

9–13. 159.15 Hz

9–15. (a) $\dfrac{4 \times 10^6}{s^2 + 2000s + 4 \times 10^6}$ (b) $\zeta = 0.5, f_n = 318.31$ Hz

9–19. (a) $\dfrac{R_2(1 + sR_1C_1)}{R_1 + R_2 + sR_1R_2}$ (b) $f_{bn} = (2\pi R_1 C)^{-1}$,

$f_{bd} = (2\pi R_{eq}C)^{-1}$ where $R_{eq} = R_1\|R_2$

(c) Flat response of $20 \log_{10}(R_2/[R_1 + R_2])$ in dB below f_{bn}. Changes to a slope of $+6$ dB/octave at f_{bn} and reaches a level of 0 dB at f_{bd}. Flat level of 0 dB above f_{bd}.

9–23. 1592 Ω

9–25. 4711 Ω, 1.88 pF

9–27. $R_1 = 7500$ Ω, $R_2 = 833.3$ Ω

9–29. 994.7 Ω

Chapter 10

10–1. (a) 5V (b) 10 V (c) 2 (d) 25 W

10–3. (a) -0.25 A (b) 3.279 A (c) 32.25 W

10–5. $24u(t)$

10–7. (a) $-3tu(t)$ (b) $-3r(t)$

10–9. $24u(t - 5)$

10–11. (a) $-3(t - 4)u(t - 4)$ (b) $-3r(t - 4)$

10–13. $8u(t) + 12u(t - 5 \times 10^{-3}) - 20u(t - 10 \times 10^{-3})$

10–15. $2r(t) - 6r(t - 6) + 4r(t - 9)$

10–19. $32e^{-4t}$

10–27. (a) e^{-sT} (b) $e^{-j\omega T}$ (c) $A(\omega) = 1, \beta(\omega) = -\omega T$

10–29. The capacitor voltage is nearly a dc voltage $v_C = 10$ V, and $v_0 = v_i - 10$. Thus, v_0 has two levels: 25 V and -10 V.

10–31. 1.11

10–33. 1.11

Chapter 11

11–1. dc, 500, 1000, 1500, 2000 Hz

11–3. (a) $5 + 10 \sin(40\pi t + 36.87°) + 13 \sin(80\pi t - 22.62°)$

(b) $5 + (3 - j4)\varepsilon^{j40\pi t} + (3 + j4)\varepsilon^{-j40\pi t} + (-2.5 - j6)\varepsilon^{j80\pi t} + (-2.5 + j6)\varepsilon^{-j80\pi t}$

11–5. (a) $6 + 5 \cos 100\pi t + 8.660 \sin 100\pi t - 6.928 \cos 200\pi t - 4 \sin 200\pi t$

(b) $6 + (2.5 - j4.330)\varepsilon^{j100\pi t} + (2.5 + j4.330)\varepsilon^{-j100\pi t} + (-3.464 + j2)\varepsilon^{j200\pi t} + (-3.464 - j2)\varepsilon^{-j200\pi t}$

11–7.

f, kHz	1	3	5	7	9
Amplitude, V	12.732	-4.244	2.546	-1.819	1.415

11–11. (a) $0.01 \dfrac{\sin 10^{-3}\pi f}{10^{-3}\pi f}$

11–13. (a) $1/n$ (b) $1/n^2$ (c) $1/n$ (d) $1/n^2$ (e) $1/n^2$ (f) $1/n$

INDEX

690